EYNLY

AF606493

SPRINGER SERIES IN SURFACE SCIENCES 46

SPRINGER SERIES IN SURFACE SCIENCES

Series Editors: G. Ertl, H. Lüth and D.L. Mills

This series covers the whole spectrum of surface sciences, including structure and dynamics of clean and adsorbate-covered surfaces, thin films, basic surface effects, analytical methods and also the physics and chemistry of interfaces. Written by leading researchers in the field, the books are intended primarily for researchers in academia and industry and for graduate students.

Please view available titles in *Springer Series in Surface Sciences*
on series homepage http://www.springer.com/series/409

Pietro Ferraro
Adam Wax
Zeev Zalevsky
(Editors)

Coherent Light Microscopy

Imaging and Quantitative Phase Analysis

With 228 Figures

Springer

Volume Editors:

Dr. Pietro Ferraro
CNR, Ist. Nazionale di Ottica Applicata
Via Campi Flegrei 34, 80078 Pozzuoli Napoli, Italy
e-mail: pietro.ferraro@inoa.it

Prof. Adam Wax
Duke University, Dept. Biomedical Engineering
Durham, North Carolina, USA
e-mail: a.wax@duke.edu

Zeev Zalevsky
Bar-Ilan University, Dept. Engineering
Ramat Gan, Israel
e-mail: zalevsz@eng.biu.ac.il

Series Editors:

Professor Dr. Gerhard Ertl
Fritz-Haber-Institute der Max-Planck-Gesellschaft, Faradayweg 4–6,
14195 Berlin, Germany

Professor Dr. Hans Lüth
Institut für Schicht- und Ionentechnik
Forschungszentrum Jülich GmbH,
52425 Jülich, Germany

Professor Douglas L. Mills, Ph.D.
Department of Physics, University of California,
Irvine, CA 92717, USA

Additional material to this book can be downloaded from http://extras.springer.com

Springer Series in Surface Sciences ISSN 0931-5195
ISBN 978-3-642-15812-4 e-ISBN 978-3-642-15813-1
DOI 10.1007/978-3-642-15813-1
Springer Heidelberg Dordrecht London New York

Cover design: Integra Software Services Pvt. Ltd., Pondicherry

Printed on acid-free paper

Springer is part of Springer Science+Business Media (www.springer.com)

Preface

Since Dennis Gabor introduced the hologram in 1947, the coherence imaging approach has been used to generate stunning images and has found application in diverse fields such as optical recording media and security encoding. With the development of modern photonics technology, the field of holography has been greatly advanced with the laborious need for recording and developing photographic films replaced by the use of digital photography. In fact, the new field of digital holography has enabled a rapid expansion in the use of coherence methods for imaging and microscopy, thanks to the development of digital technology and of some key elements such as high-resolution pixelated detectors in all regions of the electromagnetic spectrum, from UV to long IR, high-power compact lasers, spatial light modulators, and especially the increased computational power of modern PC processors. Moreover, the incredibly improved capability for data storage allows the possibility of capturing and managing huge numbers of images. In combination with the development of efficient computational algorithms for image processing and new strategies for conception and design of optical and optoelectronic systems, these advances have enabled novel methods for coherent imaging and microscopy to become useful in biology and microfluidics.

This text seeks to provide an overview of the current state of the field for the application of digital holography for microscopic imaging. One of the aims of this book is to present the best "work in progress" in microscopy based on using coherent light sources to people outside the optics community, to provide readers with the tools for understanding these novel techniques and thus the ability to judge what new capabilities will be important and potentially challenging for their research.

The text has been divided into three sections, covering areas of active research in this field. The first section presents an overview of recent advances in the methods of digital holography. Subjects examined in this section include the basis of image formation in digital holography and the role of coherence, such as the degree of coherence in the illumination. This section also includes discussion of the unique ability to numerically manipulate the digital holograms to produce additional visual representations, such as images comparable to those that would be obtained using traditional phase microscopy imaging methods.

The ability to obtain phase information from the recorded data is a significant strength of the digital holography approach. The second section of this text focuses on novel phase microscopy implementations of digital holography. A clear advantage of digital holography is that quantitative phase information is obtained, addressing a shortcoming of traditional phase microscopy methods which make quantitative analysis difficult. In this section, experimental digital holography methods are discussed which have been developed for specific imaging applications, such as imaging of microlens arrays. Additional topics include the use of novel devices such as spatial light modulators and spectral domain detection, as well as application of phase imaging to biological samples and dynamic phenomena.

The third section of this text discusses current research into improving the performance of digital holography. Topics here include an examination of the nature of image formation as a means to improve phase retrieval and enhancing the numerical aperture of the collected signal to improve spatial resolution. The ability to obtain super-resolved imaging information is a compelling topic also covered in this section. The final chapter shows how coherence imaging can be extended to three-dimensional applications by using speckle pattern analysis. We wish to thank Dr. Francesco Merola (CNR-INO) for helping us in the process: his fruitful cooperation has been truly appreciated.

Pozzuoli (Napoli), Italy — Pietro Ferraro
Durham, North Carolina — Adam P. Wax
Ramat-Gan, Israel — Zeev Zalevsky
August 2010

Contents

Part II Phase Microscopy

Contributors

Avigail Amsel School of Engineering, Bar Ilan University, Ramat Gan 52900, Israel, a.avigail@gmail.com

Donatella Balduzzi Istituto Sperimentale Italiano 'Lazzaro Spallanzani' Località La Quercia, Rivolta d'Adda (CR) 26027, Italy, donatella.balduzzi@istitutospallanzani.it

Yevgeny Beiderman Department of Mathematics, Bar-Ilan University, Ramat-Gan 52900, Israel, b.yevgeny@gmail.com

Annick Brandenburger RUBIO – IRIBHM, IBMM – ULB Aéropole – CP 300, Université Libre de Bruxelles, B-6041 Charleroi – Gosselies, Belgium, abranden@ulb.ac.be

Gary Brooker Johns Hopkins University Microscopy Center, Rockville, MD 20850, USA, gbrooker@jhu.edu

Natacha Callens Microgravity Research Center, Université Libre de Bruxelles, B-1050 Brussels, Belgium, natacha.callens@ulb.ac.be

Luis Camacho Departamento de Óptica, Universitat de Valencia, 46100 Burjassot, Spain, catoluis@alumni.uv.es

Giuseppe Coppola Istituto per la Microelettronica e Microsistemi del CNR, 80131 Napoli, Italy, giuseppe.coppola@cnr.it

Sara Coppola Istituto Nazionale di Ottica del CNR (INO-CNR), 80078 Pozzuoli (NA), Italy, sara.coppola@ino.it

Giuseppe Di Caprio Istituto per la Microelettronica e Microsistemi del CNR, 80131 Napoli, Italy, giuseppe.dicaprio@na.imm.cnr.it

Mehdi DaneshPanah Electrical and Computer Engineering Department, University of Connecticut, Storrs, CT, USA, mehdi@engr.uconn.edu

Huafeng Ding Quantitative Light Imaging Laboratory, Department of Electrical and Computer Engineering, Beckman Institute for Advanced Science and

Technology, University of Illinois at Urbana-Champaign, Urbana, IL 61801, USA, ding@illinois.edu

Frank Dubois Microgravity Research Center, Université Libre de Bruxelles, B-1050 Brussels, Belgium, frdubois@ulb.ac.be

Audrey K. Ellerbee Department of Electrical Engineering, Stanford University, Stanford, CA 94305-4088, USA, akellerbee@gmail.com

Pietro Ferraro Istituto Nazionale di Ottica del CNR (INO-CNR), 80078 Pozzuoli (NA), Italy, pietro.ferraro@ino.it

Carlos Ferreira Departamento de Óptica, Universitat de Valencia, 46100 Burjassot, Spain, carlos.ferreira@uv.es

Andrea Finizio Istituto Nazionale di Ottica del CNR (INO-CNR), 80078 Pozzuoli (NA), Italy, andrea.finizio@ino.it

Dror Fixler School of Engineering, Bar Ilan University, Ramat Gan 52900, Israel, fixeled@mail.biu.ac.il

Andrea Galli Istituto Sperimentale Italiano 'Lazzaro Spallanzani' Località La Quercia, Rivolta d'Adda (CR) 26027, Italy, andrea.galli@istitutospallanzani.it

Javier García Departamento de Óptica, Universitat de Valencia, 46100 Burjassot, Spain, javier.garcia.monreal@uv.es

Pascuala García-Martínez Departamento de Óptica, Universitat de Valencia, 46100 Burjassot, Spain, pascuala.garcia@uv.es

Simonetta Grilli Istituto Nazionale di Ottica del CNR (INO-CNR), 80078 Pozzuoli (NA), Italy, simonetta.grilli@ino.it

Hansford C. Hendargo Department of Biomedical Engineering, Duke University, Durham, NC 27708, USA, hansford.hendargo@duke.edu

Joseph A. Izatt Department of Biomedical Engineering, Duke University, Durham, NC 27708, USA, jizatt@duke.edu

Bahram Javidi Electrical and Computer Engineering Department, University of Connecticut, Storrs, CT, USA, bahram@engr.uconn.edu

Manfred H. Jericho Department of Physics and Atmospheric Science, Dalhousie University, Halifax, NS, Canada, Manfred.Jericho@dal.ca

Shan Shan Kou Division of Bioengineering, Optical Bioimaging Laboratory, National University of Singapore, Singapore 117576, Singapore; Graduate School for Integrative Sciences and Engineering (NGS), National University of Singapore, Singapore 117597, Singapore, shanshan.kou@gmail.com

H. Jürgen Kreuzer Department of Physics and Atmospheric Science, Dalhousie University, Halifax, NS B3H 3J5, Canada, H.J.Kreuzer@DAL.CA; kreuzer@fizz.phys.dal.ca

Shalin B. Mehta Division of Bioengineering, Optical Bioimaging Laboratory, , National University of Singapore, Singapore 117576, Singapore; Graduate School for Integrative Sciences and Engineering (NGS), National University of Singapore, Singapore 117597, Singapore, shalin@nus.edu.sg

Francesco Merola Istituto Nazionale di Ottica del CNR (INO-CNR), 80078 Pozzuoli (NA), Italy, francesco.merola@ino.it

Lisa Miccio Istituto Nazionale di Ottica del CNR (INO-CNR), 80078 Pozzuoli (NA), Italy, lisa.miccio@ino.it

Vicente Micó Departamento de Óptica, Universitat de Valencia, 46100 Burjassot, Spain, vicente.mico@uv.es

Christophe Minetti Microgravity Research Center, Université Libre de Bruxelles, B-1050 Brussels, Belgium, cminetti@ulb.ac.be

Inkyu Moon Electrical and Computer Engineering Department, University of Connecticut, Storrs, CT, USA, inkyu.moon@uconn.edu

Melania Paturzo Istituto Nazionale di Ottica del CNR (INO-CNR), 80078 Pozzuoli (NA), Italy, melania.paturzo@ino.it

Thomas Podgorski Laboratoire de Spectrométrie Physique, Université Joseph Fourier – Grenoble, Grenoble, France, tpodgors@spectro.ujf-grenoble.fr

Gabriel Popescu Quantitative Light Imaging Laboratory, Department of Electrical and Computer Engineering, Beckman Institute for Advanced Science and Technology, University of Illinois at Urbana-Champaign, Urbana, IL 61801, USA, gpopescu@illinois.edu

Roberto Puglisi Istituto Sperimentale Italiano 'Lazzaro Spallanzani' Località La Quercia, Rivolta d'Adda (CR) 26027, Italy, roberto.puglisi@istitutospallanzani.it

Patrick Queeckers Microgravity Research Center, Université Libre de Bruxelles, B-1050 Brussels, Belgium, pqueeck@ulb.ac.be

Shakil Rehman Division of Bioengineering, Optical Bioimaging Laboratory, National University of Singapore, Singapore 117576, Singapore; Singapore Eye Research Institute, Singapore 168751, Singapore, shakil@nus.edu.sg

Matthew T. Rinehart Department of Biomedical Engineering, Fitzpatrick Institute for Photonics, Duke University, Durham, NC 27708, USA, matt.rinehart@duke.edu

Joseph Rosen Department of Electrical and Computer Engineering, Ben-Gurion University of the Negev, Beer-Sheva 84105, Israel, rosen@ee.bgu.ac.il

Natan T. Shaked Department of Biomedical Engineering, Fitzpatrick Institute for Photonics, Duke University, Durham, NC 27708, USA, natan.shaked@duke.edu

Colin J.R. Sheppard Division of Bioengineering, Optical Bioimaging Laboratory, National University of Singapore, Singapore 117576, Singapore; Graduate School for Integrative Sciences and Engineering (NGS), National University of Singapore,

Singapore 117576, Singapore; Department of Biological Sciences, National University of Singapore, Singapore 117576, Singapore, colin@nus.edu.sg

Mina Teicher Department of Mathematics, Bar-Ilan University, Ramat-Gan 52900, Israel, teicher@macs.biu.ac.il

Yaniv Tzadka School of Engineering, Bar Ilan University, Ramat Gan 52900, Israel, yaniv.tzadka@gmail.com

Estela Valero AIDO – Technological Institute of Optics, Color and Imaging, 46980, Paterna, Spain, evalero@aido.es

Veronica Vespini Istituto Nazionale di Ottica del CNR (INO-CNR), 80078 Pozzuoli (NA), Italy, veronica.vespini@ino.it

Adam Wax Department of Biomedical Engineering, Fitzpatrick Institute for Photonics, Duke University, Durham, NC 27708, USA, a.wax@duke.edu

Catherine Yourassowsky Microgravity Research Center, Université Libre de Bruxelles, B-1050 Brussels, Belgium, catherine.yourassowsky@ulb.ac.be

Zeev Zalevsky School of Engineering, Bar Ilan University, Ramat Gan 52900, Israel, zalevsz@eng.biu.ac.il

Part I
Digital Holography

Chapter 1
Point Source Digital In-Line Holographic Microscopy

Manfred H. Jericho and H. Jürgen Kreuzer

Abstract Point source digital in-line holography with numerical reconstruction has been developed into a new microscopy, specifically for microfluidic and biological applications, that routinely achieves both lateral and depth resolution at the submicron level in 3-D imaging. This review will cover the history of this field and give details of the theoretical and experimental background. Numerous examples from microfluidics and biology will demonstrate the capabilities of this new microscopy. The motion of many objects such as living cells in water can be tracked in 3-D at subsecond rates. Microfluidic applications include sedimentation of suspensions, fluid motion around micron-sized objects in channels, motion of spheres, and formation of bubbles. Immersion DIHM will be reviewed which effectively does holography in the UV. Lastly, a submersible version of the microscope will be introduced that allows the in situ study of marine life in real time in the ocean and shows images and films obtained in sea trials.

1.1 Introduction

In the late 1940s, more than a decade after Ernst Ruska had invented the electron microscope, the main obstacle for its practical implementation was the fact that magnetic electron lenses were far from ideal showing substantial aberration effects. Compared to optical glass lenses they had the quality of the bottom of a champagne bottle. It was then that Dennis Gabor [1] came up with the idea that one should get rid of magnetic electron lenses altogether and use the only "perfect" lens available, that is a pinhole or point source of a size less than or about a wavelength. From such a point source a spherical wave emanates within a cone of half angle θ given by the numerical aperture NA of a hole

$$n \sin\theta = \mathrm{NA} = 0.62\frac{\lambda}{r} \tag{1.1}$$

H.J. Kreuzer (✉)
Department of Physics and Atmospheric Science, Dalhousie University, Halifax, NS B3H 3J5 Canada
e-mail: H.J. kreuzer@DAL.CA; kreuzer@Fizz.Phys.dal.ca

P. Ferraro et al. (eds.), *Coherent Light Microscopy*, Springer Series in Surface Sciences 46, DOI 10.1007/978-3-642-15813-1_1,

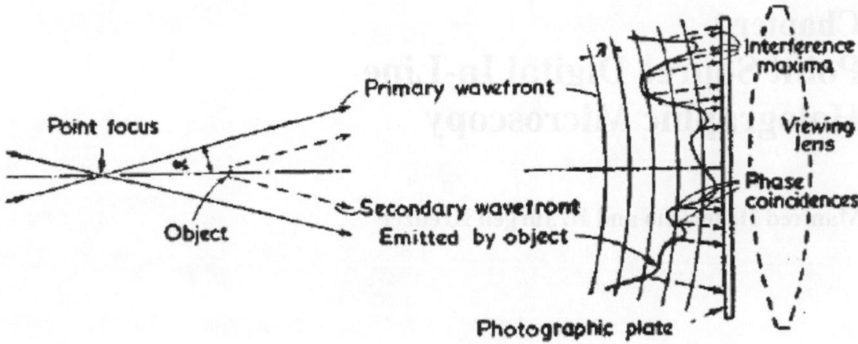

Fig. 1.1 The geometry of holography with a point source in Gabor's Nature paper

where λ is the wavelength of the radiation illuminating the pinhole and r is its radius. Such a point source has the additional advantage that the emanating radiation originates at the same point in space, i.e., it is spatially coherent. Moreover, it is an expanding wave, i.e., it has perfect magnification build in. Gabor's original drawing is reproduced in Fig. 1.1.

Holography in this geometry should be termed more concisely as Point Source In-line Holography to set it apart from Parallel or Collimated Beam In-line Holography advanced after the invention of the laser in the 1960s [2, 3]. Unfortunately, this distinction is often blurred to the extent that in a popular textbook [4] on holography the Collimated Beam geometry is presented as Gabor's original idea. This review is solely concerned with point source in-line holography, and apart from a few historical remarks on electron holography it is exclusively dealing with light as the primary source of radiation. A schematic more suited for the present work is given in Fig. 1.2.

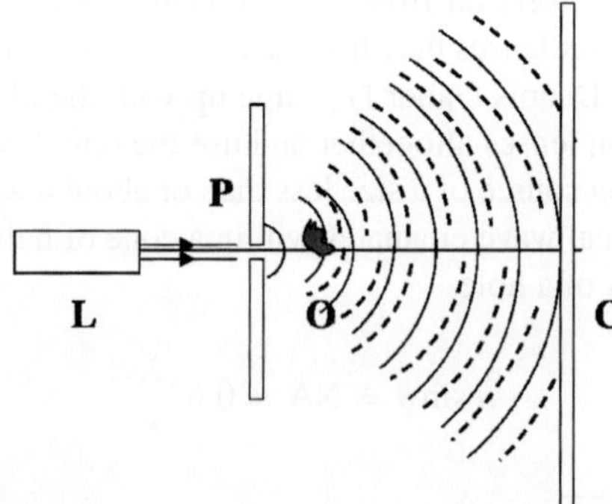

Fig. 1.2 Schematic of DIHM: A laser L is focused onto a pinhole P. The emerging *spherical* wave illuminates the object O and the interference pattern or hologram is recorded on the screen C. The *solid* and *dashed lines* are reference and scattered wave, respectively

The light emanating from the point source is propagating to the screen with some of it being scattered by the object in front of the source. Thus the wave amplitude at the screen is given by

$$A(\mathbf{r}) = A_{\text{ref}}(\mathbf{r}) + A_{\text{scat}}(\mathbf{r}) \tag{1.2}$$

and the intensity recorded on the screen becomes

$$\begin{aligned} I(\mathbf{r}) &= A(\mathbf{r})A^*(\mathbf{r}) \\ &= A_{\text{ref}}(\mathbf{r})A^*_{\text{ref}}(\mathbf{r}) + [A_{\text{ref}}(\mathbf{r})A^*_{\text{scat}}(\mathbf{r}) + A_{\text{scat}}(\mathbf{r})A^*_{\text{ref}}(\mathbf{r})] + A_{\text{scat}}(\mathbf{r})A^*_{\text{scat}}(\mathbf{r}) \end{aligned} \tag{1.3}$$

The first term on the right is the intensity of the unscattered part of the reference wave; the last term is the intensity of the scattered wave; it is the subject of classical diffraction theory in wave optics. The two terms in the square bracket represent the interference between the reference and the scattered waves. This is called holographic diffraction and is the basis of holography.

When are classical or holographic diffraction and interference dominant? Obviously if the object blocks out most of the reference wave the scattered wave can only interfere with itself if sufficiently coherent; a double slit configuration is a case in point. Another example is a simple pinhole camera. In the latter the radius of the pinhole is typically many times the wavelength to allow more light to get into the camera chamber. Sunlight or a lamp illuminates the front of the object, and the reflected light enters the pinhole incoherently. The reference wave (sun or lamp) does not contribute directly to the image inside the camera.

On the other hand, if the object blocks out only a small part of the reference wave holographic diffraction is dominant and leads to a complicated interference pattern on the screen. The fact that the holographic intensity is linear in the scattering amplitude has two important mathematical consequences: (1) the superposition principle holds, i.e., adding two or many holograms does not lead to a loss of information as it would for classical optics where double exposure of a photographic film in a pinhole camera surely reduces or even destroys the information and (2) the scattering amplitude recorded in the hologram can be traced back to the scattering object and extracted including its phase. This is essentially the reason why, by an inverse procedure, one can extract the wave front anywhere at the object thus creating a 3-D image of the object from a 2-D recording on the screen. How is this achieved?

There are two ways of back propagating a wave through a hologram: experimentally and theoretically. Experimentally Gabor's idea was to record or transfer the (electron) hologram onto a semi-transparent film (photographic negative, for instance) and illuminate it from the back with a spherical converging wave. There are three problems with this approach:

(a) it is difficult to produce a large enough spherical wave with a lens without encountering aberration effects; (b) the resulting image as viewed through the hologram is its original size, i.e., no magnification has been achieved; thus this is not a microscope; and (c) when looking at the object image through the hologram one looks straight into the point source reproduced by the converging wave, i.e., no object is visible, it is as invisible as a plane when it flies across the sun.

Gabor's solution to get magnification was simple: take a hologram with electrons of wavelength less than 1 Å, i.e., energies in excess of 100 eV. Then magnify the hologram mechanically by a factor of 5000 and use visible light of the appropriate wavelength to illuminate it. Magnification is now 5000-fold. To avoid the problems with Gabor's geometry of point source holography various schemes of off-line or off-axis holography have been developed [2–5] that rely on parallel light beams as generated by lasers but this is not the topic of this review.

A much easier way to do the back propagation is numerically already contemplated by Gabor [6]. Early implementations of numerical reconstruction were frustrated by the need of simplifications of the reconstruction formula due to the lack of computer power [7–12]. The development of a fast and efficient reconstruction algorithm in the 1990s leads to the emergence of DIHM – Digital In-line Holographic Microscopy as tool for 4-D imaging with submicron length and subsecond timescales of the time-dependent trajectories of microscopic objects in 3-D space [13]. Unlike conventional compound light microscopy, which can give high-resolution information about an object only in the focal plane with a depth of field of less than a micron, digital in-line holographic microscopy (DIHM) offers a rapid and efficient approach to construct high-contrast 3-D images of the whole sample volume from a single hologram.

In-line point source holography with electrons was revived in the late 1980s by Fink et. al. [14–19] when they managed to make stable field emission tips that ended in a single atom thus creating an intense point source for coherent electrons in the energy range from roughly 10 to 300 eV, i.e., for wavelengths from 2 down to 0.5 Å. High-resolution holograms were recorded digitally using a CCD camera and a theory was developed for the numerical reconstruction of these holograms [16, 17]. This must be considered the first success of digital in-line holography with electrons [20–23]. Although the reconstruction algorithm was originally designed for electron holography, its transfer to optical holography is straightforward [24–27].

1.2 DIHM Hardware

DIHM hardware is very simple as illustrated in Fig. 1.2: A low-power laser **L** is focused on a pinhole **P** of typically 1 μm in diameter from which the spherical reference wave emerges. Using a laser diode this light source can be assembled in a standard lens holder which serves as an holographic add-on to any standard microscope. For high-resolution work the object **O** is placed less than 1 mm from the pinhole and the CCD camera **C** is adjusted in position to record the entire emission cone of the pinhole. If only resolution of several microns is required then a larger pinhole of several microns in diameter can be used and the larger sample is placed further from the pinhole. The CCD camera (physical size and pixel number) must be chosen to satisfy the Nyquist–Shannon criteria to avoid aliasing in the reconstruction.

1.3 Hologram Reconstruction

As outlined above, holography is a two-stage process. First a hologram is taken and stored digitally. Second, the role of reconstruction in holography is to obtain the 3-D structure of the object from the 2-D hologram on the screen, or, in technical terms, to reconstruct the wave front at the object. In DIHM this is done numerically based on the theory of wave propagation in optics, i.e., by backward diffraction of the digitally stored pattern on the 2-D hologram via the reference wave. This diffraction process is given in scalar diffraction theory by the Kirchhoff–Fresnel transform:

$$K(\mathbf{r}) = \int_{\text{screen}} \widetilde{I}(\boldsymbol{\xi}) A_{\text{ref}}(\boldsymbol{\xi}) \frac{\exp[-ik|\boldsymbol{\xi} - \mathbf{x}|]}{|\boldsymbol{\xi} - \mathbf{x}|} F_{\text{in}}(\boldsymbol{\xi})\, \mathrm{d}\boldsymbol{\xi} \tag{1.4}$$

Here

$$\widetilde{I}(\boldsymbol{\xi}) = I(\boldsymbol{\xi}) - A_{\text{ref}}(\boldsymbol{\xi}) A^{*}_{\text{ref}}(\boldsymbol{\xi}) \tag{1.5}$$

is the contrast intensity, $\mathbf{x}$ and $\boldsymbol{\xi}$ are vectors from the point source to the object and screen, respectively, $k = 2\pi/\lambda$ is the inverse wave number and χ is the angle between the optical axis from the point source to the center of the screen a distance L away, and the vector $\boldsymbol{\xi}$ on the screen. For a point source the reference wave is spherical, i.e., $A_{\text{ref}}(\boldsymbol{\xi}) = \xi^{-1} \exp[-ik\xi]$. Thus Fresnel's approximation to the inclination factor becomes (for the geometry in which the screen is perpendicular to the optical axis)

$$F_{\text{in}}(\boldsymbol{\xi}) = -\frac{\mathrm{i}}{2\lambda}(1 + \cos\chi) = -\frac{\mathrm{i}}{2\lambda}\left(1 + \frac{L}{\xi}\right) \tag{1.6}$$

Taking the contrast intensity is convenient and advantageous as it removes any unwanted flaws and imperfections in the laser illumination. In addition, by subtracting the laser intensity at the screen in the absence of the object one eliminates this dominant term and also, more importantly, any flaws in the laser or camera. This subtraction can be done as just outlined or, if the removal of the object is not practical, one applies a high-pass filter to the hologram.

$K(\mathbf{r})$ is a complex wave amplitude that can be calculated according to (1.4) anywhere in space, in particular in the volume of the object/sample, thus rendering its 3-D structure if sufficiently transparent. The absolute square $|K(\mathbf{r})|^2$ yields the intensity at the object and its phase gives information about its index of refraction. The numerical evaluation of the diffraction integral is very time consuming even for a "small" hologram of only $10^3 \times 10^3$ pixels. Observing that the Kirchhoff–Fresnel integral is a convolution one is tempted to use fast Fourier transforms for its evaluation. To do this one needs to digitize (1.4) as the hologram itself is of course given in digitized form, $\boldsymbol{\xi} \to (\nu, \mu)a$ where ν and μ enumerate the pixels on the CCD chip and a is the pixel size. Likewise, one needs to digitize the space coordinates $\mathbf{r} \to (nb, mb, z)$, where b is the size of one pixel in the reconstructed image. Unfortunately, to keep the digitized form of (1.4) a convolution one needs to make the identification $b = a$ thus eliminating the opportunity to obtain a magnified image of the object or to use point source DIHM as a viable microscopic technique. To overcome this impasse we explore possibilities to simplify the diffraction integral itself. We observe that in DIHM the distance from the pinhole to the object is typically much smaller than to the screen. Thus we can use an expansion

$$|\boldsymbol{\xi} - \mathbf{x}| \approx \xi \left[1 - \frac{2\boldsymbol{\xi} \cdot \mathbf{x}}{\xi^2} \right]^{1/2} \approx \xi \left[1 - \frac{\boldsymbol{\xi} \cdot \mathbf{x}}{\xi^2} - \frac{1}{2} \left(\frac{\boldsymbol{\xi} \cdot \mathbf{x}}{\xi^2} \right)^2 - \cdots \right] \tag{1.7}$$

Keeping only the linear term yields the Kirchhoff–Helmholtz transform

$$K_{\mathrm{KH}}(\mathbf{r}) = -\frac{i}{2\lambda} \int_{\mathrm{screen}} \left[\widetilde{I}(\boldsymbol{\xi}) \frac{1}{\xi^2} \left(1 + \frac{\xi}{L} \right) \right] \exp\left[-ik\boldsymbol{\xi} \cdot \mathbf{x}/\xi \right] \, \mathrm{d}\boldsymbol{\xi} \tag{1.8}$$

Again, the function $K(\mathbf{r})$ is complex and significantly structured and different from 0 only in the space region occupied by the object. An algorithm has been developed for its evaluation that is outlined in the appendix, faster by many orders of magnitude than the direct evaluation of the double integral. Noteworthy is the fact that in this algorithm the pixel size in the reconstructed image can be chosen arbitrarily to achieve any magnification one wishes.

The complete procedure of DIHM is demonstrated in Fig. 1.3 and described in the figure caption. Noteworthy is the fact that reconstructing the hologram itself (i.e., without subtracting the background) leads to residual interference fringes and artificial features which are totally avoided if reconstruction is done from the contrast hologram. These fringes are, however, not due to the twin image as sometimes claimed in the literature without proof. Although in in-line holography with parallel light the twin image can be a problem, this is not the case in the point source geometry. The argument, substantiated with many experimental data reviewed here and presented elsewhere, is simple: the distance between source and object is typically hundreds or even thousands of wavelengths and the twin image is at the same distance on the other side of the source. The contribution of one on the other is an undetectable uniform background. The situation is vastly different, e.g., in photoelectron holography of surfaces, where source and object are angstroms apart, and the twin image is a problem although ways have been found to minimize its effect even in such unfavorable geometries.

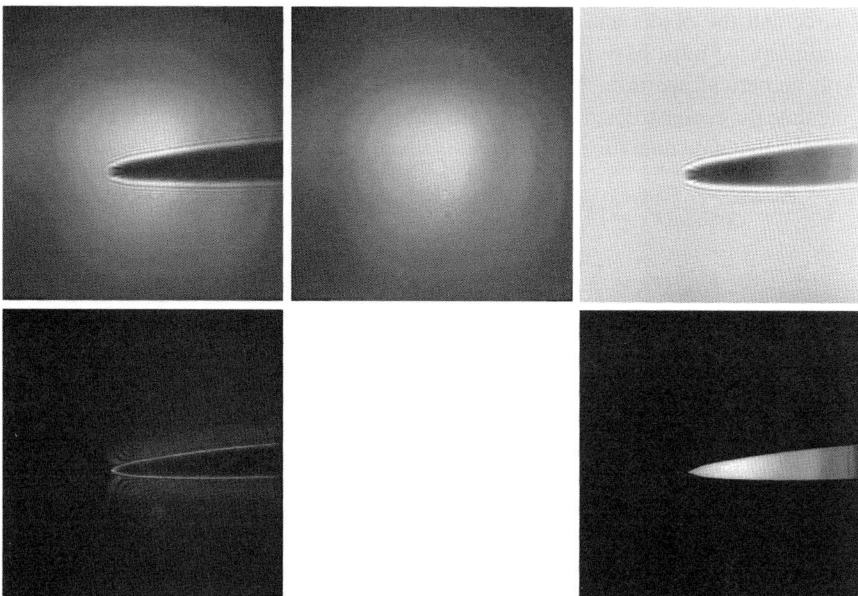

Fig. 1.3 Holography of a needle with blue light ($\lambda = 408$ nm), pinhole diameter 0.5 μm, pinhole–object distance 1.1 mm, pinhole–CCD distance 20 mm, CCD chip size $1.28 \times 1.28\,\text{cm}^2$, NA = 0.3. *Top row*: hologram, background, and contrast hologram. *Bottom row, left*: reconstruction from original hologram and *right*: from contrast hologram. Reconstruction area $685 \times 685\,\mu\text{m}$

1.4 Resolution and Depth of Field

We base our discussion of the lateral and depth resolution of DIHM on the point spread function psf. To obtain the psf we assume a perfect point source emitting a spherical wave $U_{\text{ref}}(\mathbf{r}) = U_0 r^{-1} \exp[ikr]$ with wavenumber $k = 2\pi/\lambda$. In addition we have a point object at a distance $\mathbf{r}_1$ along the optical axis from which scattered spherical waves emerge. The total wave field is thus

$$U(\mathbf{r}) = U_0 \frac{\exp[ikr]}{r} + U_1 \frac{\exp[ik\,|\mathbf{r} - \mathbf{r}_1|]}{|\mathbf{r} - \mathbf{r}_1|} \tag{1.9}$$

The intensity of the contrast image then becomes

$$\begin{aligned} \widetilde{I}(\mathbf{r}) &= I(\mathbf{r}) - \frac{U_0^2}{r^2} \\ &= \frac{U_1^2}{|\mathbf{r} - \mathbf{r}_1|^2} + 2\frac{U_0 U_1}{r\,|\mathbf{r} - \mathbf{r}_1|} \cos[k(r - |\mathbf{r} - \mathbf{r}_1|)] \end{aligned} \tag{1.10}$$

The first term accounts for classical scattering from an isolated object resulting in a smoothly varying background. The second term is due to interference between the source and the object and represents holographic interference. Under holography conditions we must have $U_0 >> U_1$. With the contrast intensity given on a screen with numerical aperture, NA, we can use the Kirchhoff–Helmholtz transform to calculate the reconstructed intensity around the original point source at $\mathbf{r}_1$. This psf can be calculated analytically [19] and is given by

$$\mathrm{psf}_{\mathrm{NA}}(\mathbf{r}) = |K_{\mathrm{NA}}(\mathbf{r})|^2 \tag{1.11}$$

$$K_{\mathrm{NA}}(\mathbf{r}) = \sum_{\mathrm{n}=0}^{\infty} \exp[ni\pi/2] a_n P_n\left(\frac{z - z_0}{|\mathbf{r} - \mathbf{r}_0|}\right) j_n(k\,|\mathbf{r} - \mathbf{r}_0|)$$

$$a_n = (n + 1/2) \int_{\sqrt{1-(\mathrm{NA})^2}}^{1} \mathrm{d}t\, P_n(t) \tag{1.12}$$

The point spread function looks like a prolate spheroid with its long axis along the optical axis. In Fig. 1.4 we show views and cuts through the central point $\mathbf{r}_1$ in a plane perpendicular to the optical axis and along it. In holography it plays the role of the Debye integral in apertured optical systems. Similar arguments [27] result in estimates of the lateral and axial resolution

$$\delta_{\mathrm{lat}} = \frac{\lambda}{2\mathrm{NA}}$$

$$\delta_{\mathrm{ax}} = \frac{\lambda}{2(\mathrm{NA})^2}$$

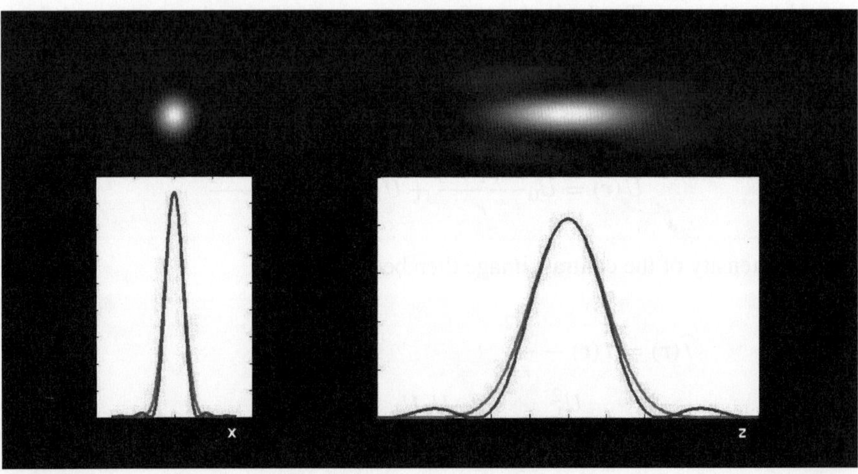

Fig. 1.4 Point spread function psf for NA=0.3. On the *left* (*right*) a view and cut perpendicular to (along) the optical axis

It is quite suggestive that a Gaussian

$$\mathrm{psf}_{\mathrm{NA}}(\mathbf{r}) \approx \frac{1}{2\pi\delta_{\mathrm{lat}}^{2}} \exp\left[-\left(x^{2}+y^{2}\right)/2\delta_{\mathrm{lat}}^{2}\right] \frac{1}{\delta_{\mathrm{ax}}\sqrt{2\pi}} \exp\left[-z^{2}/2\delta_{\mathrm{ax}}^{2}\right] \quad (1.13)$$

represents an acceptable approximation for $\mathrm{psf}_{\mathrm{NA}}$.

For a perfect pinhole the numerical aperture of DIHM depends on the ratio of the detector width to the object – detector distance. The resolution thus increases at shorter wavelength and with larger screen width (larger NA). When the optical path crosses several media with different refractive indexes, hologram formation depends on an effective wavelength that is determined by the length of path through each medium as we will discuss below in connection with immersion holography.

The resolution of a holographic microscope can also be discussed in terms of the number of interference fringes that can be resolved by the recording medium [28, 30]. In the case of digital CCD or CMOS detectors, faithful recording of interference fringes is only possible if the pixel size of the recording chip is much smaller than the smallest fringe spacing that needs to be recorded. Simulations with computer-generated holograms suggested that two scattering centers can be resolved if the number of captured fringes is large (tens of fringes) and that only fringes separated by more than three camera pixels contribute to image reconstruction and resolution. Either too few fringes or fringes that are too close together will lead to reduced resolution. In general, a shorter wavelength will produce more fringes, and usually results in higher resolution. However, the larger number of fringes is spread over a fixed number of pixels on the chip so that at shorter wavelength the average number of pixels per fringe decreases and the resolution may thus be less than expected. A more detailed discussion on fringe number and spacing and its effect on resolution can be found elsewhere [29, 30].

Point source DIHM with a nearly spherical reference wave produces a magnified hologram at the recording screen. This implies that the number of fringes captured, and hence the resolution, depends on the object–screen distance as well. A resolution of 1 μm is easily obtainable with the object close to the point source and the smallest measurable separation of two scattering points increases nearly linearly with source – object distance. The imaged volume is essentially determined by the volume of a pyramid that has the CCD chip as a base and the point source as the apex.

One of the most amazing advantages of holographic microscopy, and in particular of pointsource DIHM, is the fact that a single 2-D hologram produces in reconstruction a 3-D image of the object without loss of resolution. This is in sharp contrast to compound microscopy where the depth of field reduces sharply with improving resolution necessitating refocussing to map out a larger volume. To show this advantage we have embedded 1 μm latex beads in gelatin on a cover slide (to immobilize them), taken one hologram and made five reconstructions at different depths from 300 μm to 3 mm from the cover slide, see Fig. 1.5.

Lastly, we investigate the effect of the pinhole size. In Fig. 1.6 panel A shows the hologram and panels B and C reconstructions at 3 and 2 mm, respectively, with

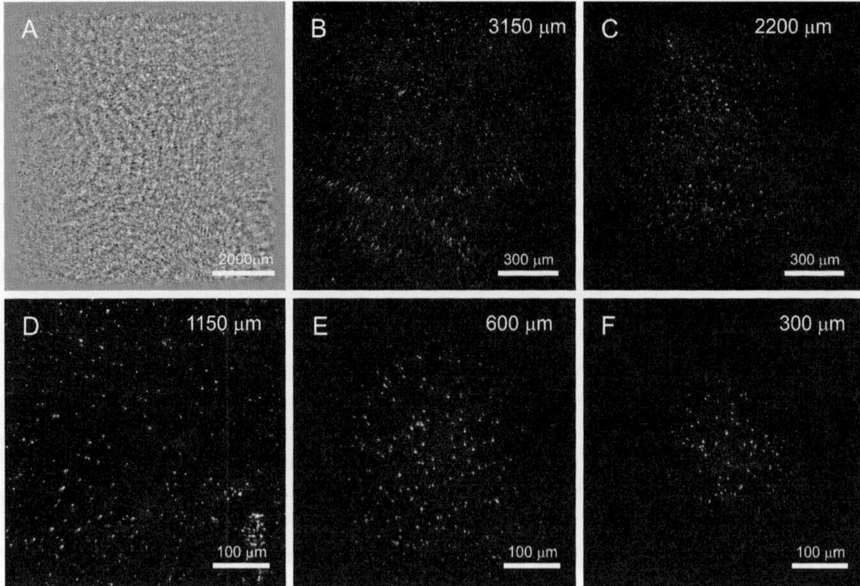

Fig. 1.5 Depth of field: 1 μm latex beads embedded in gelatin on a cover slide (to immobilize them). (**a**) hologram taken with blue light. (**b**–**f**) five reconstructions at different depths from 300 μm to 3 mm from the cover slide

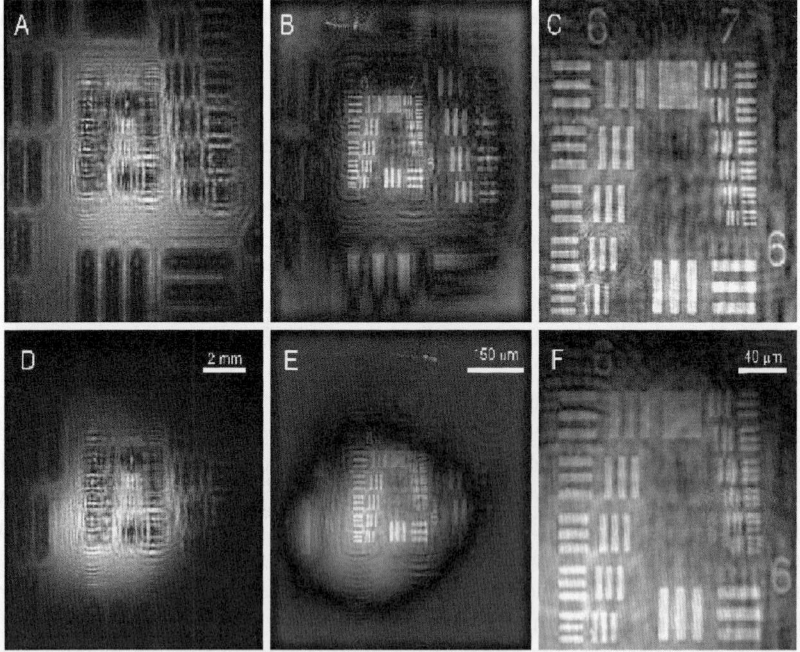

Fig. 1.6 Pinhole size effect: (**a**) Hologram with 0.5 μm pinhole, (**b**) reconstruction, and (**c**) central section of reconstruction only. (**d**–**f**) same for 2 μm pinhole. Laser wavelength 408 nm and numerical aperture NA = 0.208

a field of view of 1.4 mm. The pinhole was 0.5 μm in diameter and the numerical aperture was 0.2. For panels D to F the pinhole was increased to 2 μm thus reducing the numerical aperture so much that the outline of the finite emission cone is seen.

1.5 Deconvolution

As in any optical instrument, resolution and image quality are dictated by the instrument function. In DIHM, limiting factors are the finite extent of the point source and the numerical aperture. For both of these effects it is easy to calculate the point spread function psf given in (1.11) in the previous section. We thus have for the image obtained at finite NA

$$I_{\mathrm{NA}}(\mathbf{r}) = \int d\mathbf{r}' \mathrm{psf}_{\mathrm{NA}}(\mathbf{r} - \mathbf{r}') I(\mathbf{r}') \tag{1.14}$$

where $I(\mathbf{r}')$ is the "perfect" image at NA=1. Taking Fourier transforms we have

$$\mathcal{F}(I_{\mathrm{NA}}) = \mathcal{F}(\mathrm{psf})\mathcal{F}(I) \tag{1.15}$$

and an inverse Fourier transform yields the deconvoluted "perfect" image

$$I = \mathcal{F}^{-1}[\mathcal{F}(I_{\mathrm{NA}})/\mathcal{F}(\mathrm{psf})] \tag{1.16}$$

Note that for the parametrization (1.13) of the psf its Fourier transform is again a product of Gaussians.

An application of the deconvolution procedure is given in Fig. 1.7 for NA=0.3 showing clusters of 1 μm latex beads which after deconvolution are clearly identified.

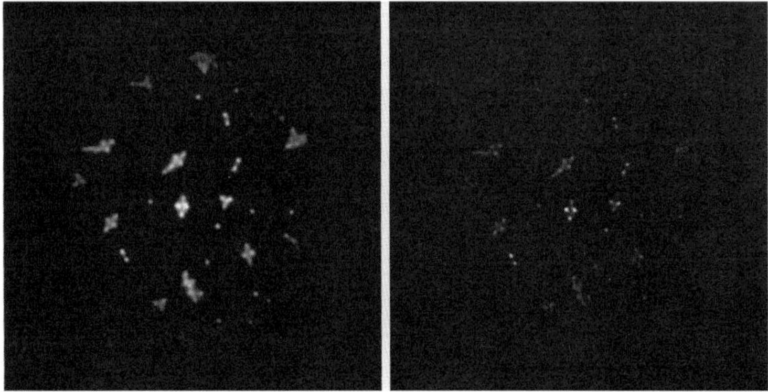

Fig. 1.7 Cluster of 1 μm latex beads. On the *left* taken with numerical aperture NA=0.3. On the *right* after deconvolution

1.6 Immersion Holography

To improve the resolution of DIHM one has two obvious options, as with any type of microscopy:

1. Reduce the wavelength, say into the ultraviolet regime. This produces three challenges: (a) UV lasers are expensive, large and more elaborate, and dangerous; (b) one needs a pinhole with a diameter of a few hundred nanometers; they are expensive, difficult to make, and easily clog up; and (c) one needs a camera sensitive to UV radiation, again an expensive undertaking.
2. Increase the numerical aperture requiring a larger camera chip at more expense.

An elegant solution to this challenge is much akin to the use of oil immersion lenses in compound microscopy. One realizes that the hologram is formed when the scattered wave interferes with the reference wave. This happens in the space region between the object and the detector. If one therefore fills this region with a material of high refractive index n then the hologram is formed at a wavelength λ/n, or, alternatively, with a numerical aperture enlarged by a factor n, see Fig. 1.8. Thus for experiments on biological species in water a blue laser ($\lambda = 408$ nm) is sufficient to do DIHM in the UV range at 308 nm increasing the resolution by 30%. Or using oil ($n = 1.5$) or sapphire glass ($n = 1.8$) we would work at $\lambda/n = 272$ and 227 nm, respectively. If the space between the object and the CCD chip is not completely filled with the high index of refraction material, geometric optics leads to an effective wavelength

$$\lambda_{\text{eff}} = \lambda/[(L - H) + nH]/L] \tag{1.17}$$

An example of immersion holography is shown in Fig. 1.9, again for clusters of latex beads.

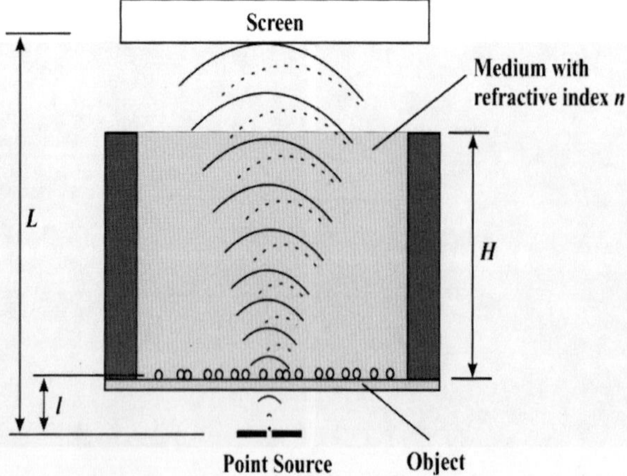

Fig. 1.8 Schematic for immersion holography [32]

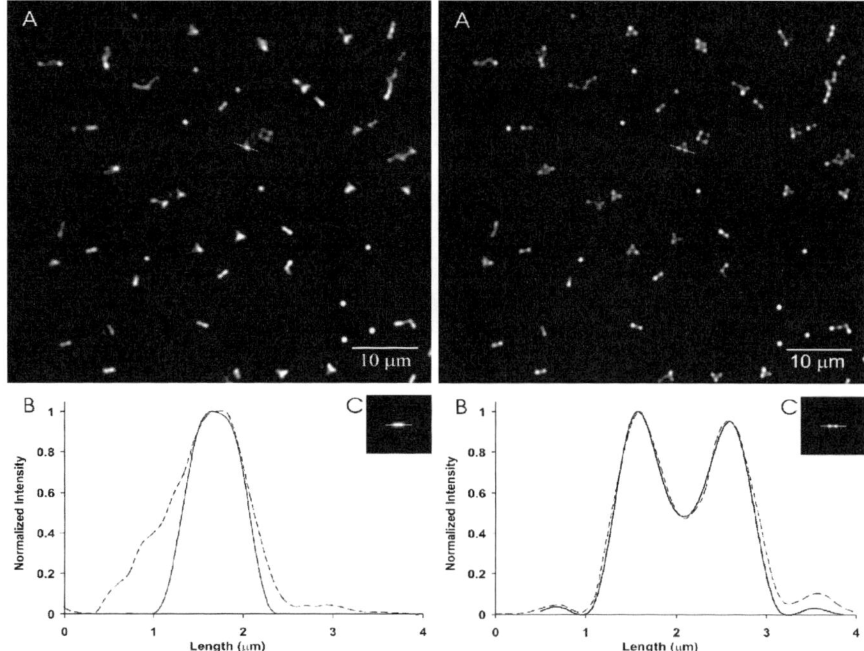

Fig. 1.9 (**a**) Reconstruction of a deposit of 1.09 μm beads on a cover slide showing single beads and small clusters. (**b**) Intensity profile through a dimer. *Left*: with green light and air between cover slide and CCD chip. *Right*: with space filled with oil of $n = 1.5$ [32]

1.7 Phase images

The Kirchhoff–Helmholtz transform yields the wave front at the object with both amplitude and phase information. There are occasionally claims in the literature that point source DIHM is of limited value because it is not capable of producing phase images. These arguments are hard to follow and will not be repeated here. Instead we present an example that shows clearly the capability and superiority of point source DIHM. The task at hand was to map the index of refraction across the core of a fiber optic cable where it is enhanced by about a percent over the value in the glass cladding by doping with rare earth materials. Because the cladding will act as a cylindrical lens one must immerse the fiber into oil with matching refractive index. Panel A of Fig. 1.10 shows the hologram taken with a violet laser ($\lambda = 405$ nm), panel B shows the reconstructed intensity, and panel C the reconstructed phase. One can still see the remnants of the cladding because the oil was not matching the refractive index of the cladding perfectly. Cuts perpendicular to the core showing intensity (panel E) and phase (panel F) confirm that the core is 4.5 μm in diameter as stated by the manufacturer. Assuming that the core is cylindrical we can calculate the optical path length in the core to be $p(r) = 2(R^2 - r^2)^{1/2}$ where R is the radius

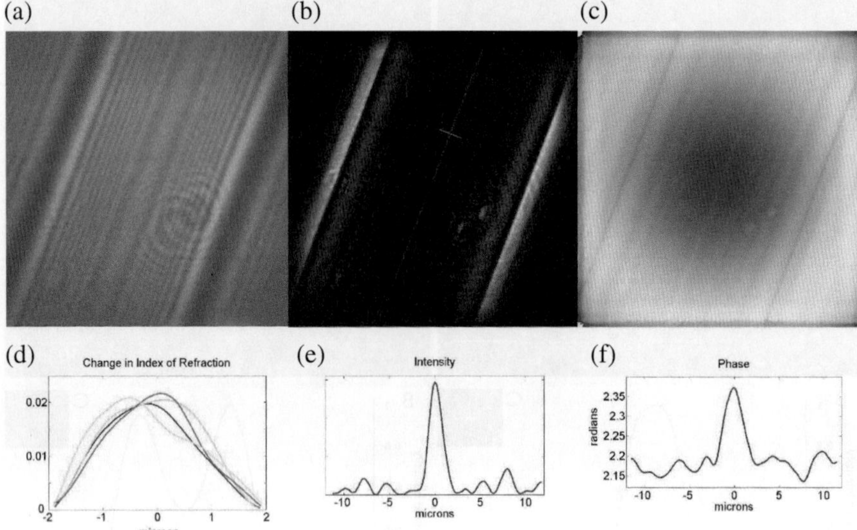

Fig. 1.10 Optical fiber immersed in oil with matching refractive index. *Panel A* shows the hologram taken with a blue laser, *panel B* the reconstructed intensity, and *panel C* the reconstructed phase. Cuts perpendicular to the core showing intensity (*panel E*) and phase (*panel F*) confirm that the core is 4.5 μm in diameter. *Panel D* profiles in the index of refraction at four different positions along the core

of the core and r is the lateral distance from its center. Thus we get for the phase difference with respect to the cladding

$$\Delta\phi(r) = p(r)\Delta n(r) \tag{1.18}$$

where $\Delta n(r)$ is the change of the refractive index of refraction throughout the core which is obtained from $\Delta\phi$ by a simple division. The resulting profiles in the index of refraction at four different positions along the core (Panel D) confirm that the core was tapered with the maximum enhancement of the refraction index about 0.02 with a considerable inhomogeneity along the core and across it.

1.8 4-D Imaging

In many fields of science and technology the microscopist is faced with tracking many objects such as particulates, bubbles, plankton, or bacteria as they move in space, asking for an efficient way of 4-D particle tracking. DIHM is the perfect tool to do this job [31, 32]: 4-D tracking is achieved by (1) recording digitally, e.g., on a CCD chip, a film of N holograms $h_1, h_2, h_3, \ldots$ at times $t_1, t_2, t_3, \ldots$. (2) One constructs digitally a difference hologram $(h_1-h_2)+(h_3-h_4)+(h_5-h_6)+\ldots$, thus retaining only those features in the holograms that correspond to moving objects;

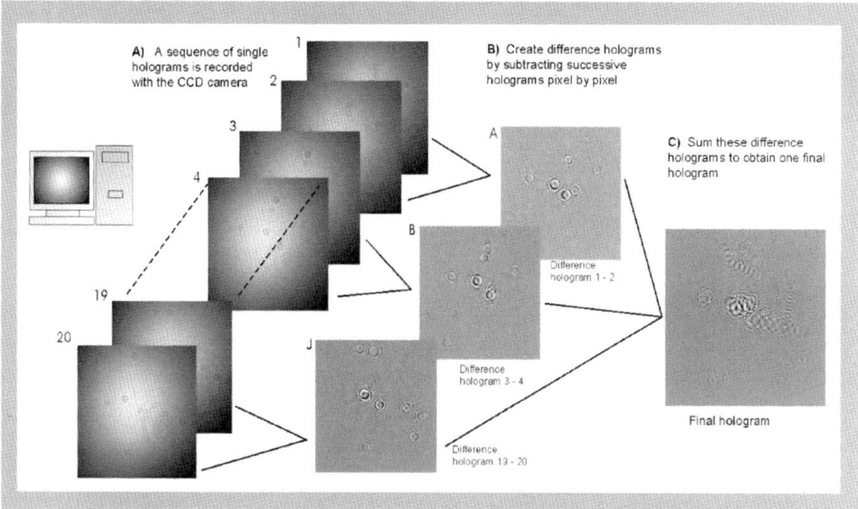

Fig. 1.11 Procedure to obtain the 3-D structures of moving objects from a sequence of holograms by subtracting subsequent holograms, then adding these difference holograms into a final compound hologram from which reconstructed trajectories are obtained with the Helmholtz–Kirchhoff transform

the recording speed must be adjusted accordingly. Because holography is linear in the scattering amplitude the superposition principle holds and adding or subtracting holograms does not lead to a loss of information. By subtracting two holograms and then adding the differences one ensures that (a) the pixels do not saturate and (b) one at least partially reduces the noise in the original holograms. This construct also reduces the size of the data set by a factor of N. Data from different experiments with hundreds of holograms done at different times can be "added" this way for storage. (3) Lastly, one numerically reconstructs a stack of images in a sufficient number of planes throughout the sample volume so that a 3-D rendering displays the tracks of all the moving particles. A schematic of this procedure is shown in Fig. 1.11. One could also reconstruct stacks of images from each frame in the film of holograms and would then be able to produce a 3-D movie of the motion itself.

1.9 Applications in Microfluidics

As an application of 4D tracking of micron-sized objects [33–37] we show in Fig. 1.12 the trajectories of latex beads of 1 μm diameter suspended in water flowing around a 70 μm large bead. The time between successive exposures is 1/15 s. Because we also can measure the distance traveled between exposures we can evaluate the velocity vector field in 3-D.

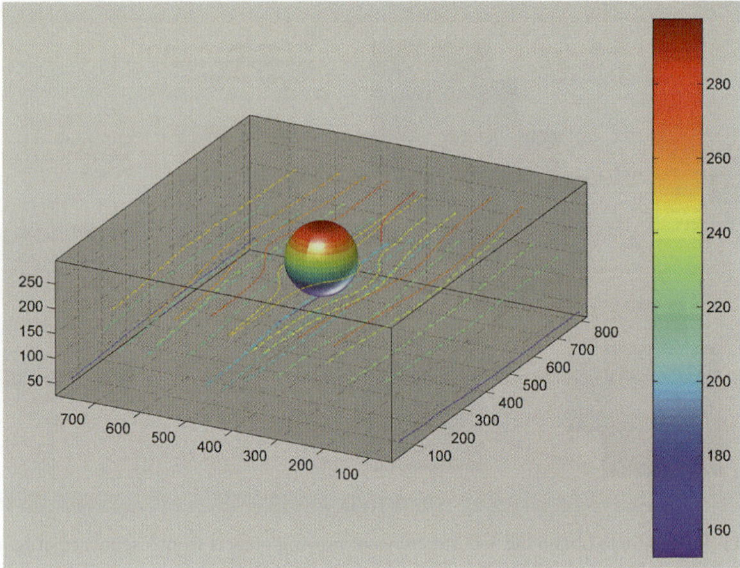

Fig. 1.12 Trajectories of 1 μm latex beads suspended in water flowing around a 70 μm large bead. The time between successive exposures is 1/15 s

1.10 Biology

We show a few examples of imaging biological species with DIHM [38–40] starting in Fig. 1.13 with a high-resolution image of a diatom.

Our next example is an image of *Escherichia coli* bacteria, see Fig. 1.14.

To demonstrate that DIHM is also capable of imaging large objects with micron detail we show in Fig. 1.15 a section through the head of a fruitfly, *Drosophila melanogaster*. Such images reveal the structure of the pigmented compound eye,

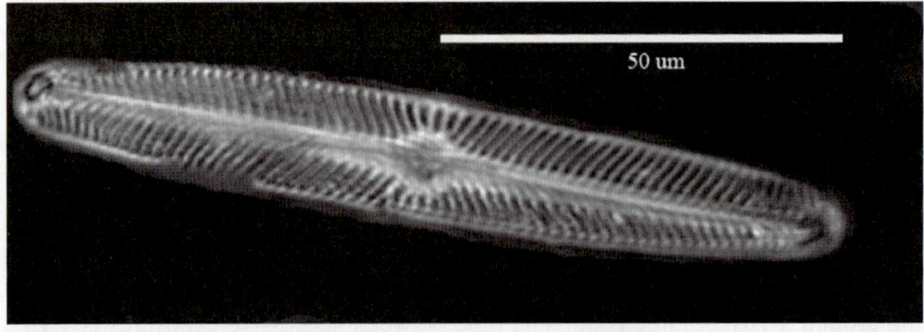

Fig. 1.13 High-resolution reconstruction of a diatom. The hologram was taken with a blue laser at $\lambda = 408$ nm

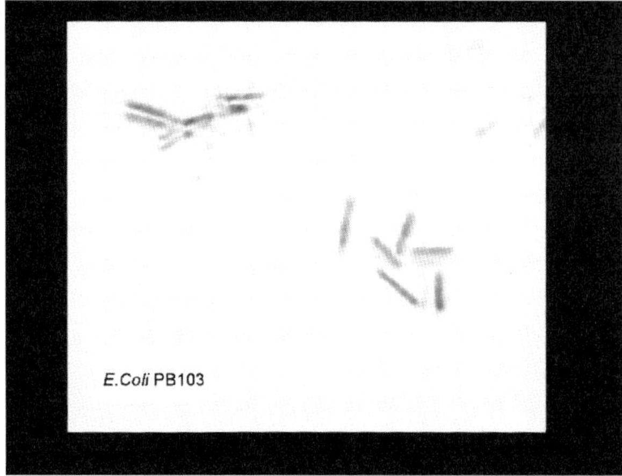

Fig. 1.14 *E.coli* bacteria imaged with a blue laser. Pinhole–object distance 300 μm, pinhole–screen distance 15 mm, pinhole 0.5 μm. Bacteria diameter about 1 μm

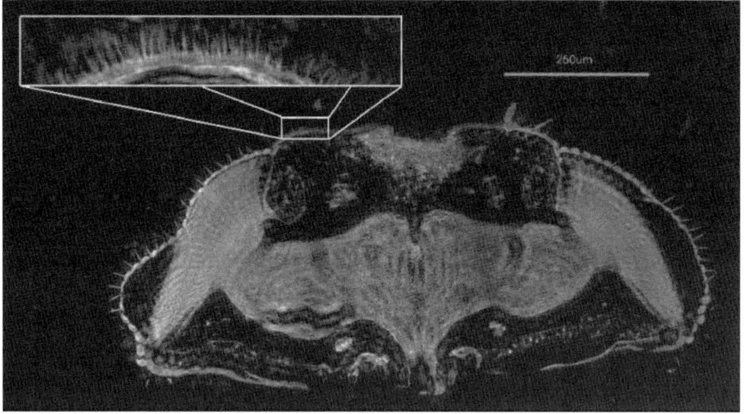

Fig. 1.15 A section through the head of a fruitfly with a blue laser at $\lambda = 408$ nm

and different neuropile regions of the brain within the head cuticle including the optic neuropiles underlying the compound eye.

Our last example in biology is a 3-D rendering of a suspension of *E. coli* bacteria in a stream of growth medium around a big sphere, see Fig. 1.16. Experiments like these allow, for the first time, realistic studies of the colonization of surfaces by bacteria and the mechanism of biofilm formation.

DIHM was also used in a large study of the effect of temperature on motility of three species of the marine dinoflagellate *Alexandrium* which made full use of the 3-D trajectories to extract quantitative information on their swimming dynamics [39, 40].

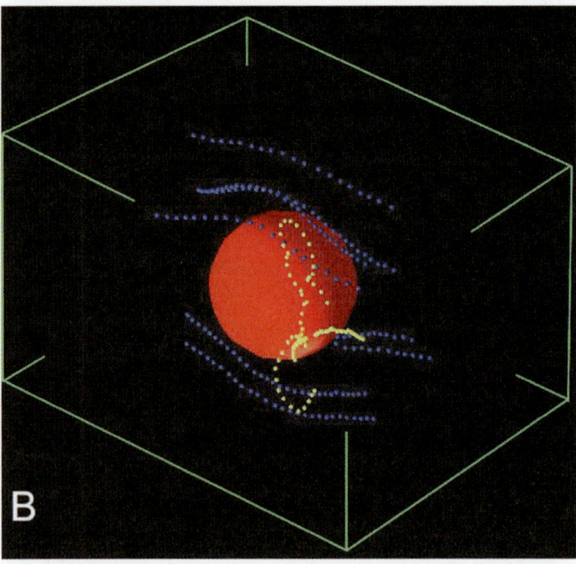

Fig. 1.16 3-D rendering of the tracks of a suspension of *E. coli* bacteria in a stream of growth medium around a big sphere. Holography with blue light ($\lambda = 408$ nm), pinhole diameter 0.5 μm, pinhole–object distance 0.5 mm, pinhole–CCD distance 19 mm, CCD chip size $1.28 \times 1.28\,\text{cm}^2$, NA = 0.3. Reconstruction volume $300 \times 300 \times 300\,\mu\text{m}^3$. Recording the motion at 15 f/s

1.11 Submersible DIHM

To allow observations with DIHM in ocean or lake environments we have constructed an underwater microscope. Its schematic is shown in Fig. 1.17: The microscope consists of two pressure chambers one of which contains the laser and the other the CCD camera (plus power supply). The two chambers are kept at a fixed distance to each other to allow water to freely circulate between them. In the center of the chamber plates facing each other are small windows with the one on the laser chamber having the pinhole. The signal from the CCD camera is transmitted via an underwater USB cable to a buoy or a boat above from where a satellite link can be established for data transmission to a laboratory. Depending on the design of the pressure chamber water depths of several hundred meters are easily accessible and can be extended to thousands of meters with high pressure technology. The prototype was pressure tested to 10 atm and operated at 20 m depth and has given the same performance as far as resolution is concerned as the desktop version of DIHM.

Fig. 1.18 shows a Rotifer (length 200 μm, width 100 μm) swimming at a speed of 2.5 mm/s at a depth of 15 m in the North Atlantic.

As a last example we show in Fig. 1.19 the tracks of algae in permafrost springs on Axel Heiberg Island in the High Arctic. A light weight underwater DIHM was used in this expedition.

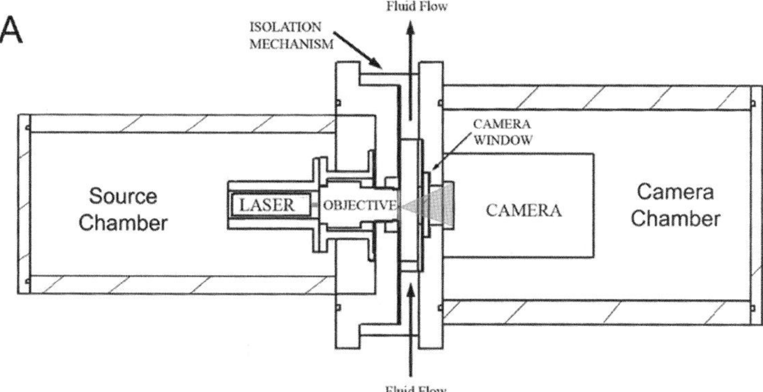

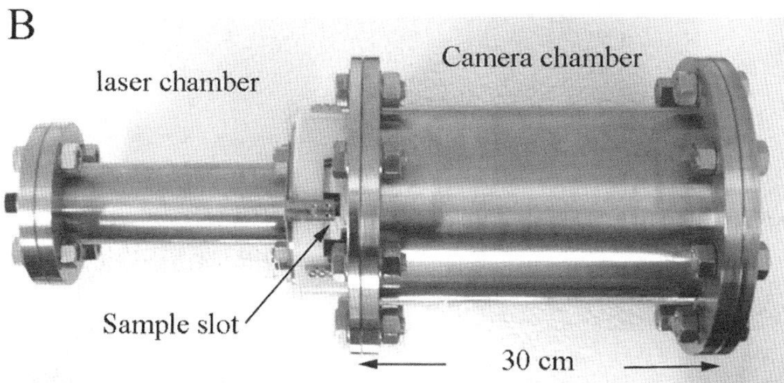

Fig. 1.17 Drawing and instrument of the underwater DIHM microscope, courtesy of Resolution OpticsInc, Halifax NS [30]

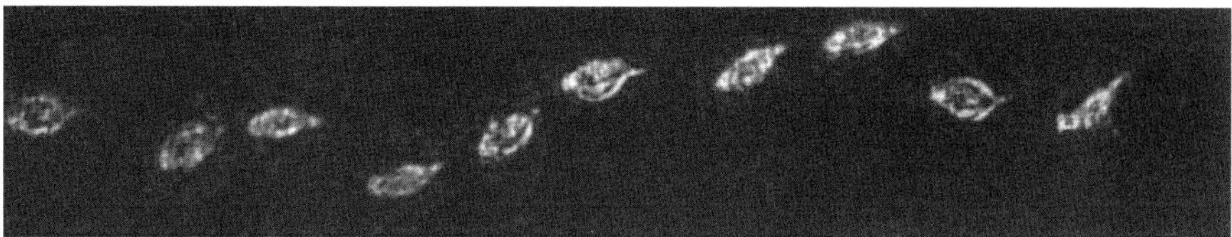

Fig. 1.18 Rotifer (length 200 μm, width 100 μm) swimming at a speed of 2.5 mm/s. Remote observation at a depth of 15 m in the North Atlantic [30]

1.12 Recent progress

In this chapter we want to highlight some recent advances in point source DIHM. We begin with the work by Repetto et al. [41] on lensless digital holographic microscopy with light-emitting diode illumination in which they show that the spatial coherence of light emitting diodes is good enough to obtain resolution of a few microns. Due to the low intensity of LEDs the pinhole has to be rather large (about 5 μm) to get enough illumination. With the advent of more powerful LEDs this opens up a very cheap approach to DIHM [42–44].

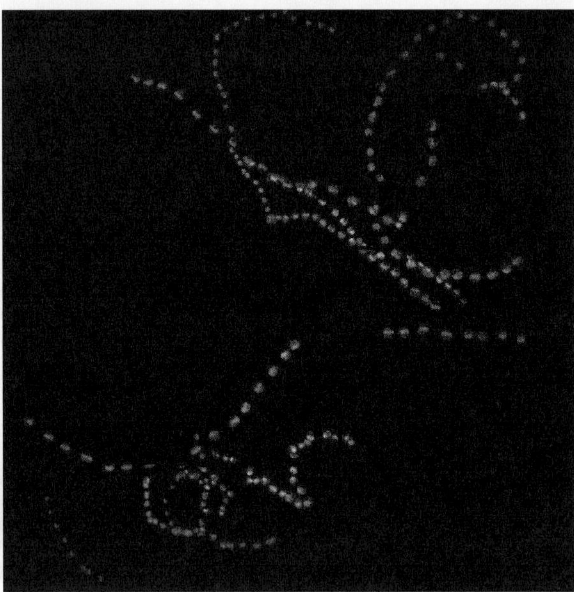

Fig. 1.19 Algae in permafrost springs on Axel Heiberg Island. Holography with blue light (λ = 408 nm), pinhole diameter 0.5 μm, pinhole–object distance 0.8 mm, pinhole–CCD distance 21 mm, CCD chip size $1.28 \times 1.28\,\text{cm}^2$, NA = 0.3. Reconstruction volume $500 \times 500 \times 300\mu\text{m}^3$, framerate 15 f/s [58]

As methods are devised for larger numerical apertures new techniques for rapid reconstruction are also called for. A first step in this direction uses a reorganized hologram with a low number of pixels, the tile superposition technique [45, 46]. The algorithm was applied to imaging of a 2 μm bead cluster and was shown to be superior. It was argued in these papers that this could become the base for high-resolution wide-field imaging by multispot illumination [47–49]. Another new approach advancing a method of multiplane interference and diffraction detection also looks very promising [50–52].

Underwater DIHM has been used for an extensive European study of biofouling phenomena in the marine environment, called AMBIO – Advanced Nanostructured Surfaces for the Control of Biofouling – (http://www.ambio.bham.ac.uk). In the papers by Heydt et al. [53, 54] new ways to cope with a mass of data on moving algae and bacteria have been developed and these observations were combined in general systems approach with surface and polymer chemistry to identify preventative measures to control biofouling [55].

1.13 Summary and Discussion

We want to conclude underlining some remarkable characteristics of DIHM and the underwater DIHM:

1. Simplicity of the microscope: DIHM, as well as underwater DIHM, is microscopy without objective lenses. The hardware required for the desktop version is a laser, a pinhole, and a CCD camera. For the underwater DIHM version we need the same elements contained in submersible hermetic shells.
2. Maximum information: A single hologram contains all the information about the 3-D structure of the object. A set of multiple holograms can be properly added to provide information about 4-D trajectories of samples.
3. Maximum resolution: Optimal resolution, of the order of the wavelength of the laser, can be obtained easily with both versions.
4. Simplicity of sample preparation, particularly for biological samples where no sectioning or staining are required, so that living cells and specimens can be viewed. Indeed, for the underwater DIHM there is no sample preparation at all, and real-time information of living organism can be retrieved remotely.
5. Speed: The kinetics of the sample, such as particle motion or metabolic changes in a biological specimen, can ultimately be followed at the capture rate of the image acquisition system.
6. 4-D tracking: A large number of particles can be tracked simultaneously in 3-D as a function of time.

Regarding the 4-D tracking, feasible in both versions of the DIHM, we emphasize the efficiency in data collection in our procedure. Removal of background effects and construction of summed holograms are easily accomplished so that high-resolution tracking of many particles in 4-D can be obtained from just one difference hologram. Since resolutions on the order of the wavelength of light have been achieved with DIHM, tracking of organisms as small as bacteria is possible, as would the motion of plankton in water or, at lower resolution, the aerial trajectories of flying insects. DIHM can also be used successfully on macroscopic biological specimens, prepared by standard histological procedures, as for a histological section of the head of the fruit fly.

Outside of biology, applications of 4-D DIHM have been demonstrated in microfluidics for particle velocimetry, i.e., tracking of the motion of particles in liquid or gas flows, gas evolution in electrolysis, and in the visualization of structures in convective or turbulent flow. Further applications have dealt with colloidal suspensions, remote sensing and environmental monitoring, investigation of bacterial attachment to surfaces and biofilm formation, and many more. DIHM with its inherent ability to obtain magnified images of objects (unlike conventional off-axis holography) is therefore a powerful new tool for a large range of research fields. Films on 4-D tracking and more examples can be viewed at http://www.physics.dal.ca/~kreuzer and in the supplementary material of this chapter. The results from the expedition to Axel Heiberg island have now been published [58].

This work was supported through grants from the Natural Sciences and Research Council of Canada and the Office of Naval Research, Washington.

Appendix: Kirchoff–Helmholtz Reconstruction Algorithm

Introduction

The role of reconstruction in holography is to obtain the 3-D structure of the object from the 2-D hologram on the screen, or, in technical terms, to reconstruct the wave front at the object. This can be achieved via a Kirchhoff–Helmholtz transform

$$K(\mathbf{r}) = \int_S \mathrm{d}^2\xi \, I(\boldsymbol{\xi}) \exp(ik\boldsymbol{\xi} \cdot \mathbf{r}/\xi) \tag{1.19}$$

where the integration extends over the 2-D surface of the screen with coordinates $\boldsymbol{\xi} = (X, Y, L)$ where L is the distance from the source to the (center of the) screen. $k = 2\pi/\lambda$ is the wave number of the radiation (electromagnetic, acoustic, electrons, or matter) and $I(\boldsymbol{\xi})$ is the contrast image (hologram) on the screen obtained by subtracting the images with and without the object present. The function $K(\mathbf{r})$ is significantly structured and different from 0 only in the space region occupied by the object. By reconstructing the wave front $K(\mathbf{r})$ on a number of planes at various distances from the source in the vicinity of the object, a 3-D image can be built up from a single 2-D hologram. $K(\mathbf{r})$ is a complex function and one usually plots its magnitude to represent the object, although phase information can also be extracted.

The Kirchhoff–Helmholtz transform is sometimes given in polar coordinates

$$\begin{aligned} \widehat{k}_x &= X/\sqrt{X^2+Y^2+L^2} \\ \widehat{k}_y &= Y/\sqrt{X^2+Y^2+L^2} \end{aligned} \tag{1.20}$$

for which (1.1) reads

$$K(\mathbf{r}) = \int_S I(\widehat{k}_x, \widehat{k}_y) \exp(ikz\sqrt{1-\widehat{k}_x^2-\widehat{k}_x^2}) \exp(ik(x\widehat{k}_x - y\widehat{k}_y)) J(\widehat{k}_x, \widehat{k}_y) \mathrm{d}\widehat{k}_x \mathrm{d}\widehat{k}_y \tag{1.21}$$

where the Jacobian is given by

$$J(\widehat{k}_x, \widehat{k}_y) = L^2/(1-\widehat{k}_x^2-\widehat{k}_y^2)^4 \tag{1.22}$$

We should point out that setting this Jacobian equal to 1, which, in some measure, amounts to using an on-axis approximation in half of the integrand, is so crude an approximation that optimal resolution cannot be achieved. Unfortunately it is invoked by several groups but should be avoided at all cost.

The numerical implementation of the full Kirchhoff–Helmholtz transform is very time consuming, even on the fastest computers, because the nonlinear phase factor in the exponential function prohibits the straightforward application of fast Fourier transform (FFT) methods. Our solution of the problem, which is at least a factor of 10^3 faster than the direct evaluation and thus makes numerical reconstruction practical, consists of five steps [56]:

1. A coordinate transformation to cast the Kirchhoff–Helmholtz transform into a standard 2-D Fourier transform.
2. Discretization of the double integral into a double sum.
3. Interpolation of the hologram on an equidistant point grid.
4. Re-writing the Kirchhoff–Helmholtz transform as a convolution.
5. Application of three FFTs to get the result.

This solution was first conceived on March 16, 1992 while at the APS meeting in Indianapolis and was implemented and running within a week.

Coordinate Transformation

To eliminate the nonlinearity in the phase factor we perform a coordinate transformation

$$\begin{aligned} X' &= XL/R \\ Y' &= YL/R \\ R &= (L^2 + X^2 + Y^2)^{1/2} \end{aligned} \tag{1.23}$$

with the inverse transformation

$$\begin{aligned} X &= X'L/R' \\ Y &= Y'L/R' \\ R' &= L^2/R = (L^2 - X'^2 - Y'^2)^{1/2} \end{aligned} \tag{1.24}$$

and the Jacobian

$$\mathrm{d}X\mathrm{d}Y = (L/R')^4 \mathrm{d}X'\mathrm{d}Y' \tag{1.25}$$

This results in

$$K(\mathbf{r}) = \int_{S'} \mathrm{d}X'\mathrm{d}Y' I'(X', \mathbf{Y}') \exp[\mathrm{i}k(xX' + yY')/L] \tag{1.26}$$

with

$$I'(X', Y') = I(X(X', Y'), Y(X', Y'))(L/R')^4 \exp(\mathrm{i}kzR'/L) \tag{1.27}$$

Note that if the screen S is rectangular then S′ becomes barrel shaped with more rounding of the edges for a larger ratio of screen width to L.

Discretization

To evaluate (1.8) with FFT techniques we must rewrite it in discrete form as

$$K_{nm} = \sum_{j,j'=0}^{N-1} I'_{jj'} e^{2\pi i(nj+mj')/N} \tag{1.28}$$

where N, the number of pixels in one direction of the screen, will be arranged to be a multiple of 2. For simplicity we restrict the following presentation to the situation where a quadratic screen is used. We therefore write

$$\begin{aligned} x &= x_n = x_0 + n\delta_x \\ y &= y_n = y_0 + n\delta_y \end{aligned} \tag{1.29}$$

for $n, m = 0, 1, 2, \ldots, N-1$. Likewise we write

$$\begin{aligned} X' &= X'_j = X'_0 + j\Delta'_x \\ Y' &= Y'_j = Y'_0 + j'\Delta'_y \end{aligned} \tag{1.30}$$

for $j, j' = 0, 1, 2, \ldots, N-1$.

Interpolation

Experimental holograms are usually recorded on a flat screen, perpendicular to the optical axis, and digitized on an equidistant grid of pixels

$$\begin{aligned} X &= X_j = X_0 + j\Delta_x \\ Y &= Y_j = Y_0 + j'\Delta_y \end{aligned} \tag{1.31}$$

which, under the transformation (1.5), are transformed into non-equdistant points $(X'_j, Y'_{j'})$, the set of these points is S′.

We therefore construct a rectangular screen S″ that circumscribes S′, and introduce a new set of equidistant points (1.12) where

$$\begin{aligned} X'_0 &= X_0 L/\sqrt{L^2 + X_0^2} \\ Y'_0 &= Y_0 L/\sqrt{L^2 + Y_0^2} \\ \Delta'_x &= \frac{L[(X_0 + (N-1)\Delta_x]}{N\sqrt{L^2 + [(X_0 + (N-1)\Delta_x]^2}} - \frac{LX_0}{N\sqrt{L^2 + X_0^2}} \\ \Delta'_y &= \frac{L[(Y_0 + (N-1)\Delta_y]}{N\sqrt{L^2 + [(Y_0 + (N-1)\Delta_y]^2}} - \frac{LY_0}{N\sqrt{L^2 + Y_0^2}} \end{aligned} \tag{1.32}$$

With this we get the Kirchhoff–Helmholtz transform (1.1) into the discrete form

$$K_{nm} = \Delta'_x \Delta'_y \exp\left\{ik\left[(x_0 + n\delta_x)X'_0 + (y_0 + m\delta_y)Y'_0\right]/L\right\}$$
$$\times \sum_{j,j'=0}^{N-1} I'_{jj'} \exp\left[ik(jx_0\Delta'_x + j'y_0\Delta'_y)/L\right]\exp\left[ik(nj\delta_x\Delta'_x + mj'\delta_y\Delta'_y)/L\right] \tag{1.33}$$

where $I'_{jj'}$ is the image interpolated at the points (1.12). This can be cast into a form suitable for FFT techniques if we choose

$$\delta_x = \frac{\lambda L}{N\Delta'_x}$$
$$\delta_y = \frac{\lambda L}{N\Delta'_y} \tag{1.34}$$

where $\lambda = 2\pi/k$ is the wavelength of the illuminating radiation. We point out that situating the screen symmetrically about the optical axis simplifies all of the above formulae considerably.

Convolution

The advantage of using FFT techniques is obviously their speed, i.e., one gains a factor 10^3 over a direct evaluation of (1.1). The disadvantage, however, is the fact that (1.16) imposes the resolution of the reconstruction. We overcome this limitation by rewriting (1.15) as a convolution and then use the product theorem of Fourier transforms.

We write

$$nj = \left[n^2 + j^2 - (n-j)^2\right]/2 \tag{1.35}$$

and get

$$K_{nm} = \Delta'_x \Delta'_y \exp\left\{ik\left[(x_0 + n\delta_x)X'_0 + (y_0 + m\delta_y)Y'_0\right]/L\right\}$$
$$\times \exp\left\{ik(n^2\delta_x\Delta'_x + m^2\delta_y\Delta'_y)/2L\right\} \tag{1.36}$$
$$\times \sum_{j,j'=0}^{N-1} I'_{jj'} \exp\left[ik(jx_0\Delta'_x + j'y_0\Delta'_y + j^2\delta_x\Delta'_x/2 + j'^2\delta_y\Delta'_y/2)/L\right]$$
$$\times \exp\left[-ik((n-j)^2\delta_x\Delta'_x + (m-j')^2\delta_y\Delta'_y)/2L\right] \tag{1.37}$$

Let us then take the following FFTs:

$$K'_{\nu\nu'} = \sum_{j,j'=0}^{N-1} I'_{jj'} \exp\left[ik(jx_0\Delta'_x + j'y_0\Delta'_y + j^2\delta_x\Delta'_x/2 + j'^2\delta_y\Delta'_y/2)/L\right]$$
$$\times \exp[2\pi \mathrm{i}(\nu j + \nu' j')/N] \tag{1.38}$$

$$R_\nu = \sum_{j=0}^{N-1} \exp\{-ikj^2\delta_x\Delta'_x/2L\} \exp\left[2\pi \mathrm{i}\nu j/N\right] \tag{1.39}$$

An inverse FFT then yields

$$K_{nm} = \Delta'_x\Delta'_y \exp\left\{ik\left[(x_0 + n\delta_x)X'_0 + (y_0 + m\delta_y)Y'_0\right]/L\right\}$$
$$\times \exp\left\{ik(n^2\delta_x\Delta'_x + m^2\delta_y\Delta'_y)/2L\right\} \tag{1.40}$$
$$\times \sum_{\nu,\nu'} K'_{\nu\nu'} R_\nu R_{\nu'} \exp\{-2\pi \mathrm{i}(\nu n + \nu' m)/N\} \tag{1.41}$$

In the numerical implementation care must be exercized in treating end effects by zero padding.

The generalization of the method to nonsymmetric screens and to reconstruction on planes arbitrarily tilted with respect to the optical axis has also been implemented.

For the numerical implementation of the transform, we have developed a very fast algorithm incorporated in a self-contained programme package called DIHM [57] that not only does the numerical reconstruction but also all other procedures connected with data management and visualization.

References

1. D. Gabor, A new microscopic principle. Nature (London)**161**, 777–778 (1948)
2. E.N. Leith, J. Upatnieks, Reconstructed wavefronts and communication theory. J. Opt. Soc. Am. **52**, 1123 (1962); ibid. **53**, 1377 (1963); ibid. **54**, 1295 (1963)
3. J.W. Goodman, R.W. Lawrence, Digital image formation from electronically detected holograms. Appl. Phys. Lett. **11**, 77–79 (1967)
4. P. Hariharan, *Optical Holography* (Cambridge University Press, Cambridge, 1996)
5. T. Kreis, *Holographic Interferometry* (Akademie Verlag, Berlin, 1996)
6. D. Gabor, Microscopy by reconstructed wavefronts. Proc. R. Soc. London, Ser. A **197**, **454** (1949)
7. Y. Aoki, Optical and numerical reconstruction of images from sound-wave holograms. IEEE Trans. Acoust. Speech AU-18, **258** (1970)
8. M.A. Kronrod, L.P. Yaroslavski, N.S. Merzlyakov, Computer synthesis of transparency holograms. Sov. Phys. Tech. Phys-U (USA) **17**, 329 (1972)
9. T.H. Demetrakopoulos, R. Mittra, Digital and optical reconstruction of images from suboptical diffraction patterns. Appl. Opt. **13**, 665 (1974)
10. L. Onural, P.D. Scott, Digital decoding of in-line holograms. Opt. Eng. **26**, 1124 (1987)
11. G. Liu, P.D. Scott, Phase retrieval and twin-image elimination for in-line Fresnel holograms. J. Opt. Soc. Am. A **4**, 159 (1987)

12. L. Onural, M.T. Oezgen, Extraction of three-dimensional object-location information directly from in-line holograms using Wigner analysis. J. Opt. Soc. Am. A **9**, 252 (1992)
13. H.J. Kreuzer, R.P. Pawlitzek, LEEPS, Version 1.2, A software package for the simulation and reconstruction of low energy electron point source images and other holograms (1993–1998)
14. H.-W. Fink, Point source for electrons and ions. IBM J. Res. Dev. **30**, 460(1986)
15. H.-W. Fink, Point source for electrons and ions. Phys. Scripta **38**, 260 (1988)
16. W. Stocker, H.-W. Fink, R. Morin, Low-energy electron and ion projection microscopy. Ultramicroscopy **31**, 379 (1989)
17. H.-W. Fink, W. Stocker, H. Schmid, Holography with low-energy electrons. Phys. Rev. Lett. **65**, 1204(1990)
18. H.-W. Fink, H. Schmid, H.J. Kreuzer, A. Wierzbicki, Atomic resolution in lens-less low-energy electron holography. Phys. Rev. Lett. **67**,15(1991)
19. H.J. Kreuzer, K. Nakamura, A. Wierzbicki, H.-W. Fink, H. Schmid, Theory of the point source electron microscope. Ultramicroscopy **45**, 381 (1992)
20. H.-W. Fink, H. Schmid, H.J. Kreuzer, In: Electron Holography, eds. by A. Tonomura, L.F. Allard, D.C. Pozzi, D.C. Joy, Y.A. Ono, *State of the Art of Low-Energy Electron Holography*, (Elsevier, Amsterdam, 1995), pp. 257–266
21. H.-W. Fink, ,H. Schmid, E. Ermantraut, and T. Schulz, Electron holography of individual DNA molecules, J. Opt. Soc. Am. A 14, 2168 (1997)
22. A. Gölzhäuser, B. Völkel, B. Jäger, M. Zharnikov, H.J. Kreuzer, M. Grunze, Holographic imaging of macromolecules. J. Vac. Sci. Technol. A **16**, 3025 (1998)
23. H. Schmid, H.-W. Fink, H.J. Kreuzer, In-line holography using low-energy electrons and photons; applications for manipulation on a nanometer scale. J. Vac. Sci. Technol. B **13**, 2428 (1995)
24. H.J. Kreuzer, H.-W. Fink, H. Schmid, S. Bonev, Holography of holes, with electrons and photons. J. Microsc. **178**, 191 (1995)
25. H.J. Kreuzer, Low energy electron point source microscopy. Micron **26**, 503 (1995)
26. H.J. Kreuzer, N. Pomerleau, K. Blagrave, M.H. Jericho, Digital in-line holography with numerical reconstruction. Proc. SPIE. **3744**, 65 (1999)
27. M. Born, E. Wolf. *Principles of Optics* (Cambridge University Press, Cambridge, 2006)
28. J. Garcia-Sucerquia, W. Xu, S.K. Jericho, M.H. Jericho, P. Klages, H.J. Kreuzer. Digital in-line holographic microscopy. Appl. Opt. **45**, 836–850 (2006)
29. J. Garcia-Sucerquia, W. Xu, S.K. Jericho, M.H. Jericho, P. Klages, H.J. Kreuzer. Resolution power in digital holography, in ICO20: Optical Information Processing; Y. Sheng, S. Z, Y. Zhang (eds.), Proc. SPIE **6027**, 637–644 (2006)
30. S.K. Jericho, J. Garcia-Sucerquia, Wenbo Xu, M.H. Jericho, H.J. Kreuzer. A submersible digital in-line holographic microscope. Rev. Sci. Instr. **77**, 043706 1–10 (2006)
31. H.J. Kreuzer, M.H. Jericho, Wenbo Xu, Digital in-line holography with numerical reconstruction: three-dimensional particle tracking. Proc. SPIE **4401**, 234, 2001
32. W. Xu, M.H. Jericho, I.A. Meinertzhagen, H.J. Kreuzer, Tracking particles in 4-D with in-line holographic microscopy. Opt. Lett. **28**, 164 (2003)
33. H.J. Kreuzer, M.H. Jericho, I.A. Meinertzhagen, W. Xu, Digital in-line holography with numerical reconstruction: 4D tracking of microstructures and organisms. Proc. SPIE **5005–17**, 299 (2003)
34. J. Garcia-Sucerquia, W. Xu, S.K. Jericho, M.H. Jericho, I. Tamblyn, H.J. Kreuzer, Digital in-line holography: 4-D imaging and tracking of micro-structures and organisms in microfluidics and biology, in ICO20: Biomedical Optics, G. von Bally, Q. Luo (eds.), Proc. SPIE **6026**, 267–275 (2006)
35. J. Garcia-Sucerquia, W. Xu, S.K. Jericho, M.H. Jericho, H.J. Kreuzer, Digital in-line holography applied to microfluidic studies, in Microfluidics, BioMEMS, and Medical Microsystems IV; I. Papautsky, W. Wang (eds.), Proc. SPIE **6112**, 175–184 (2006)
36. J. Garcia-Sucerquia, W. Xu, S.K. Jericho, M.H. Jericho, H.J. Kreuzer, 4-D imaging of fluid flow with digital in-line holographic microscopy. Optik **119**, 419–423 (2008)

37. J. Garcia-Sucerquia, D. Alvarez-Palacio, J. Kreuzer. Digital In-line Holographic Microscopy of Colloidal Systems of Microspheres, in Adaptive Optics: Analysis and Methods/Computational Optical Sensing. Meetings on CD-ROM OSA Technical Digest (CD) (Optical Society of America, 2007), paper DMB4
38. Wenbo Xu, M.H. Jericho, I.A. Meinertzhagen, H.J. Kreuzer, Digital in-line holography for biological applications. Proc. Natl. Acad. Sci. USA **98**, 11301 (2001)
39. N.I. Lewis, A.D. Cemballa, W. Xu, M.H. Jericho, H.J. Kreuzer, Effect of temperature in motility of three species of the marine dinoflagellate Alexandrium, in: Ed. by Bates, S.S. Proceedings of the Eighth Canadian Workshop on Harmful Marine. Algae. Can. Tech. Rep. Fish. Aquat. Sci. 2498: xi + 141, pp. 80–87 (2003)
40. N.I. Lewis, A.D. Cembella, W. Xu, M.H. Jericho, H.J. Kreuzer. Swimming speed of three species of the marine dinoflagellate *Alexandrium* as determined by digital in-line holography. Phycologia **45**, 61–70 (2006)
41. L. Repetto, E. Piano, C. Pontiggia. Lensless digital holographic microscope with light-emitting diode illumination. Opt. Lett. **29** (10), 1132–1134 (2004)
42. J. Garcia-Sucerquia, D. Alvarez-Palacio, H.J. Kreuzer, Partially coherent digital in-line holographic microscopy. OSA/DH/FTS/HISE/NTM/OTA (2009).
43. P. Petruck, R. Riesenberg, M. Kanka, U. Huebner, Partially coherent illumination and application to holographic microscopy, in 4th EOS Topical Meeting on Advanced Imaging Techniques Conference 2009, pp. 71–72, (2009)
44. P. Petruck, R. Riesenberg, R. Kowarschik, Sensitive measurement of partial coherence using a pinhole array, in Proceedings OPTO Sensor+Test', pp. 35–40 (2009)
45. M. Kanka, R. Riesenberg, H.J. Kreuzer, Reconstruction of high-resolution holographic microscopic images. Opt. Lett. **34** (8), 1162–1164 (2009). doi:10.1364/OL.34.001162
46. M. Kanka, A. Wuttig, C. Graulig, R. Riesenberg, Fast exact scalar propagation for an in-line holographic microscopy on the diffraction limit. Opt. Lett. **35**(2), 217–219 (2010)
47. R. Riesenberg, M. Kanka, J. Bergmann, Coherent light microscopy with a multi-spot source, in T. Wilson, (ed.), Proc. SPIE, 66300I, (2007)
48. M. Kanka, R. Riesenberg, Wide field holographic microscopy with pinhole arrays, in Proceedings of OPTO Sensor+Test, pp. 69–72, (2006)
49. R. Riesenberg, A. Wuttig, Pinhole-array and lensless micro-imaging with interferograms, Proceedings of DGaO, 106. Conference, pp. A26, (2005)
50. A. Grjasnow, R. Riesenberg, A. Wuttig, Phase reconstruction by multiple plane detection for holographic microscopy, in T. Wilson, (ed.), Proc. SPIE, 66300J, (2007)
51. A. Grjasnow, R. Riesenberg, A. Wuttig, Lenseless coherent imaging by multi-plane interference detection, in Proceedings of DGaO, 106. Conference, pp. A39, (2005)
52. A. Grjasnow, A. Wuttig, R. Riesenberg, Phase resolving microscopy by multi-plane diffraction detection. J. Microsc. **231**(1), 115–123 (2008)
53. M. Heydt, A. Rosenhahn, M. Grunze, M. Pettitt, M.E. Callow, J.A. Callow, Digital in-line hologrpahy as a three-dimensional tool to study motile marine organisms during thier exploration of surfaces. J. Adhes. **83**, 417–430 (2007)
54. M. Heydt, P. Divos, M. Grunze, A. Rosenhahn, Analysis of holographic microscopy data to quantitatively investigate three-dimensional settlement dynamics of algal zoospores in the vicinity of surfaces. Eur. Phys. J.E. **30**, 141–148 (2009)
55. A. Rosenhahn, S. Schilp1, H.J. Kreuzer, M. Grunze, The role of inert surface chemistry in marine biofouling prevention. Phys. Chem.Chem. Phys. **12**, 4275–4286 (2010). doi: 10.1039/C001968M
56. H.J. Kreuzer, *Holographic microscope and method of hologram reconstruction* US. Patent 6411406 B1, (Canadian patent CA 2376395) 25 June, (2002)
57. DIHM software package, copyright Resolution Optics Inc. Halifax, see also resolutionoptics.com for further information
58. S.K. Jericho, P. Klages, J. Nadeau, E.M. Dumas, M.H. Jericho, H.J. Kreuzer, In-line digital holographic microscopy for terrestrial and exobiological research, Planetary and Space Science **58**, 701–705 (2010)

Chapter 2
Digital Holographic Microscopy Working with a Partially Spatial Coherent Source

Frank Dubois, Catherine Yourassowsky, Natacha Callens, Christophe Minetti, Patrick Queeckers, Thomas Podgorski, and Annick Brandenburger

Abstract We investigate the use of partially spatial coherent illuminations for digital holographic microscopes (DHMs) working in transmission. Depending on the application requirements, the sources are made from a spatially filtered LED or from a decorrelated laser beam. The benefits gained with those sources are indicated. A major advantage is the drastic reduction of the speckle noise making possible high image quality and the proper emulation of phase contrast modes such as differential interference contrast (DIC). For biomedical applications, the DHMs are coupled with fluorescence sources to achieve multimodal diagnostics. Several implementations of biomedical applications where digital holography is a significant improvement are described. With a fast DHM permitting the analysis of dynamical phenomena, several applications in fluid physics and biomedical applications are also provided.

2.1 Introduction – Objectives and Justifications

Optical microscopy is limited by the small depths of focus due to the high numerical apertures of the microscope lenses and the high magnification ratios. The extension of the depth of focus is thus an important goal in optical microscopy. In this way, it has been demonstrated that an annular filtering process significantly increases the depth of focus [1]. A wave front coding method has also been proposed in which a nonlinear phase plate is introduced in the optical system [2]. Another approach is based on a digital holography method where the hologram is recorded with a CCD camera and the reconstruction is performed by computer [3]. The holographic information involves both optical phase and intensity of the recorded optical signal. Therefore, the complex amplitude can be computed to provide an efficient tool to refocus, slice-by-slice, the depth images of a thick sample by implementing the optical beam propagation of complex amplitude with discrete implementations of the Kirchhoff–Fresnel (KF) propagation equation [4, 5]. In addition, the optical

F. Dubois (✉)
Microgravity Research Center, Université Libre de Bruxelles, 50 Av. F. Roosevelt, CP 165/62, B-1050 Brussels, Belgium
e-mail: frdubois@ulb.ac.be

P. Ferraro et al. (eds.), *Coherent Light Microscopy*, Springer Series in Surface Sciences 46, DOI 10.1007/978-3-642-15813-1_2,

phase is the significant information to quantitatively measure the optical thicknesses of the sample which are not available from the measurements with classical optical methods [6]. Digital holographic microscopy (DHM) has been used in several applications of interest such as refractometry [7], observation of biological samples [8–14], living cell culture analysis [15–17], and accurate measurements inside cells such as refractive indexes and even 3D tomography [18–21]. As digital holography determines the complex amplitude signal, it is very flexible to implement powerful processing of the holographic information or of the processed images. As non-exhaustive examples are methods to control the image size as a function of distance and wavelength [22], to correct phase aberration [23–27], to perform 3D pattern recognition [28–31], to process the border artifacts [32, 33], to emulate classical phase-contrast imaging [17, 34], to implement autofocus algorithms [35–41], to perform object segmentation [42], and to perform focusing on tilted planes [43]. The optical scanning holography approach has been introduced with applications, for example, in remote sensing [44]. Holography is very often considered as a 3D imaging technique. However, the 3D information delivered by digital holography is actually not complete [45]. This is the reason why optical tomographic systems based on digital holographic were proposed [40, 46]. It has also to be mentioned that other 3D imaging can be achieved by the integral imaging method [47]. Digital holography provides some 3D information on the sample. Usually, digital holography is implemented with coherent laser beams. However, laser beams are very sensitive to any defect in the optical paths in such a way that the results can be badly corrupted by the coherent artifact noise. For that reason, we developed two DHMs with a partially spatial coherent illumination in Mach-Zehnder configurations in order to reduce this noise [10, 17, 48, 49]. The first DHM is using a LED which is spatially filtered to achieve the spatial coherence. As the spectral bandwidth is about 20 nm [50], the source is also of partially temporal coherence. This feature improves also the noise reduction but imposes to properly align reference and object beams [51]. Therefore, the computation of the optical phase results from the implementation of a phase shifting method. For applications where the phenomena are varying too rapidly to implement the phase shifting, we implemented a DHM where the complex amplitude is computed with the Fourier transform method [52, 53]. In this case, there is an average angle between the reference and the object beam that requests temporal coherence. With respect to that constraint, we realized the partially spatial coherent source from a laser beam focused close to a rotating ground glass. The two configurations are described in the next section. The benefits of using the partial spatial coherence in a DHM are given in Sect 2.3.

2.2 Optical Setups and Digital Holographic Reconstructions

2.2.1 Digital Holographic Microscope Working with a LED

The DHM that benefits from a partially spatial coherent illumination created from a LED is described in Fig. 2.1.

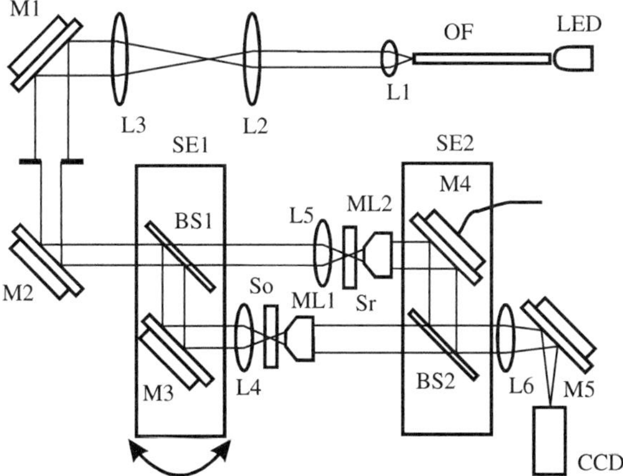

Fig. 2.1 Description of the DHM working with a LED: Light-emitting diode (LED), optical fiber (OF), lenses (L1–L6), beam splitters (BS1–BS2), mirrors (M1–M5), microscope lenses 10× (ML1–ML2), optical flat (Sr), culture dish (So), Hamamatsu Orca camera (CCD). The relative orientation of the two sub-system assemblies (SE1 and SE2) (each hold a beam splitter and a mirror) enables the optical paths in the two arms of the interferometer to be accurately equalized. (M4) mounted on a piezoelectric transducer to implement the four-frame phase stepping technique

A LED beam is injected in a liquid optical fiber in order to homogenize the beam distribution in the front focal plane of a lens L1 which is imaged in the plane of the sample So. On the travel of the beam, there is an aperture between mirrors M1 and M2 in order to increase the spatial coherence. A plane of the sample is imaged on the CCD camera by the couple of lenses ML2–L6.

The sample (So) is placed inside a Mach Zehnder interferometer in order to record the interference patterns between the sample beam and the reference beam.

In addition to the partially spatial coherent nature of the illumination, the LED provides also a low temporal coherence. For samples embedded in scattering media like collagenous gels (Fig. 2.9), this reduces further the noisy coherent contribution by selecting the ballistic photons for the interferometric information. The mirror M4 is placed on a piezoelectric transducer to implement a four image phase stepping method.

The system provides the optical phase and the intensity images on one plane of a sample, with a typical acquisition time of ¼s. The resulting complex optical field is used to refocus the optical intensity and phase fields on parallel planes without any mechanical scanning or loss of time. As the information to perform the refocusing is recorded in the only imaged plane by the CCD, the point spread function of the optical system is constant over the experimental volume, provided that the digital holographic refocusing does not introduce a loss of information. The magnification corresponding to that of a standard optical microscope is about 100×, the lateral resolution computed according to the Rayleigh criteria is 1.3 μm, and the Z resolution, δ, computed with the standard formula $\delta = \lambda/\mathrm{NA}^2$ is 5.4 μm, where λ is the average wavelength of 660 nm and NA the numerical aperture of 0.30. The depth of refocusing by means of digital holographic reconstruction is extended by a factor of about 100 [10].

The resolution of the optical thickness computed on the phase map is about 2 nm. This value is established by assessing the noise that occurs on the phase maps. As the full information on the optical field transmitted by the sample is recorded, it is possible to emulate standard optical microscopy modes such as the differential interference contrast (DIC); this latter is particularly useful for the observation of living cells by providing the scientists with an usual visualization mode. We emphasize that the DIC mode is successfully implemented thanks to the low noise level. We also coupled the digital holographic microscope with a fluorescence excitation source (eGFP) and, as shown in Sect. 2.4.1.2, it is possible for refractive samples to couple the fluorescence and the digital holographic signals to perform some refocusing of fluorescence.

2.2.2 Digital Holographic Microscope Working with a Laser Beam Incident on a Rotating Ground Glass

This section describes the partial spatial coherent microscope working in digital holography and the method to record the complete holographic information on every single video frame as is requested to study fast phenomena. The optical setup is schematized in Fig. 2.2 [40, 49].

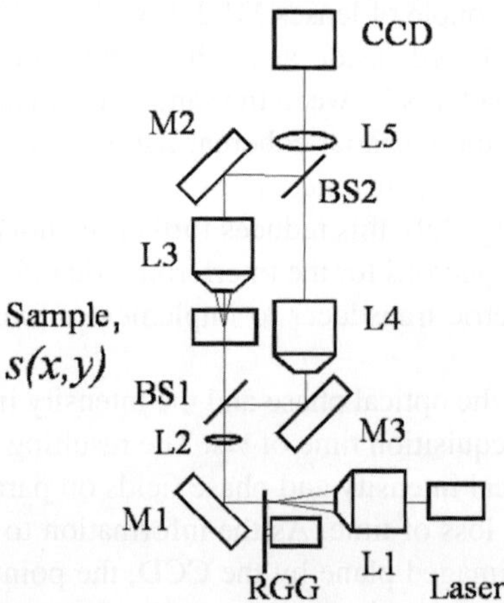

Fig. 2.2 Description of the DHM working with a partially spatial coherent source from a laser beam. L1: focusing lens; RGG: rotating ground glass for spatial coherence reduction; L2: collimating lens; L3, L4: identical microscope lenses (×20); L5: refocusing lens; CCD: charge-coupled device camera with the sensor placed in the back focal plane of L5; M1–3: Mirrors; BS1, BS2: beam splitters; the optional optical compensation in the reference arm is not indicated in the drawing

The coherent source (a mono-mode laser diode, $\lambda = 635\,\text{nm}$) is transformed into a partially spatial coherent source by focusing the beam, by lens L1, close to the plane, with a rotating ground glass (RGG). The partial spatial coherence is adjusted by changing the position of the focused spot with respect to the RGG plane. The lens L2 (focal length $f_2 = 50\,\text{mm}$) collimates the beam that is divided by a beam splitter BS1. The transmitted part, the object beam, illuminates the sample S by transmission. When the RGG is stopped the speckle size in the plane of the sample can be measured to control the spatial coherence. A plane of the sample is imaged by the couple of lenses L3–L5 on the CCD camera sensor. The reference beam, reflected by the beam splitter BS1 and by the mirror M3, is transmitted by the microscope lens L4 (focal length $f_4 = 10\,\text{mm}$), by the beam splitter BS2 and by the lens L5. It interferes with the object beam on the CCD sensor. An optical flat, not indicated in Fig. 2.1, is placed in front of lens L4 to equalize the optical paths of the reference and object beams. The camera is a JAI CV-M4 camera with a CCD array of 1280×1024 pixels. For further processing, a 1024×1024 pixels window is cropped to match the fast Fourier transform computation.

The camera is adjusted with a $100\,\mu\text{s}$ exposure time. The reference beam is slanted with respect to the object beam in such a way that a grating-like thin interference pattern is recorded on the sensor. This is used to implement the Fourier method to compute the complex amplitude of the object beam for every recorded frame [52, 53].

2.2.3 Digital Holographic Reconstruction

With the optical setup described above, we assume that we obtain the complex amplitude $s\,(x, y)$ of the light distribution in the CCD plane. Taking into account that the signals are sampled with a sampling distance Δ in both x and y directions, we have

$$s_d\,(m, n) = s(m\Delta, n\Delta) \tag{2.1}$$

where m and n are integer numbers varying from to 0 to N–1, where N is the total number of sampling points by side.

The first quantity of importance that can be computed is the quantitative phase of $s_d\,(m, n)$. It is simply obtained by computing

$$\varphi\,(m, n) = \arctan_{2\pi}\left\{\frac{\text{Im}\,(s_d\,(m, n))}{\text{Re}\,(s_d\,(m, n))}\right\} \tag{2.2}$$

For the digital holographic reconstruction, knowing the amplitude distribution in a plane P1, the digital holographic reconstruction computes the amplitude distribution in a plane P2 parallel to P1 and separated by a distance d according to

$$s'_d(m', n') = \exp\{jkd\}\left[F^{-1}_{m',n'}\exp\left\{\frac{-jk\lambda^2 d}{2N^2\Delta^2}\left(U^2+V^2\right)\right\}\left[F^{+1}_{U,V}s_d(m,n)\right]\right] \tag{2.3}$$

where m, n, m', n', U, and V are integer numbers varying from 0 to N–1. $F^{\pm 1}$ denotes the direct and inverse discrete Fourier transformations defined by

$$\left[F^{\pm 1}_{m,n}g(m,n)\right] = \frac{1}{N}\sum_{k,l=0}^{N-1}\exp\left\{\mp\frac{2\pi j}{N}(mk+nl)\right\}g(k,l) \tag{2.4}$$

where k, l, m, n are integers in such a way that $k, l, m, n = 0, \ldots, N-1$.

2.3 Benefits of the Partially Spatial Coherence for the DHM in Transmission

2.3.1 Spatial Frequency Filtering

The objective of this section is to investigate the relationship between spatial coherence and digital holographic reconstruction. The theoretical analysis outlines the relationship between coherence and digital holographic reconstruction [48].

In order to have a uniform influence on the field of view, it is very important for the spatial coherence of the illumination to be the same for every point of the field of view (FOV). Therefore, we assume that the spatial coherence of the light distribution illuminating the sample is represented by a stationary function $\Gamma(x_1-x_2, y_1-y_2)$, where (x_1, y_1) are spatial coordinates in the FOV of the microscope in the object channel.

Performing a double Fourier transformation of $\Gamma(x_1-x_2, y_1-y_2)$ on the spatial variables, we obtain

$$F_{x_1,y_1}F^{-1}_{x_2,y_2}\Gamma(x_1-x_2, y_1-y_2) = \gamma(u_2, v_2)\,\delta(u_1-u_2, v_1-v_2) \tag{2.5}$$

where (u_k, v_k) are the spatial frequencies associated with (x_k, y_k).

In (2.5), the presence of the Dirac function δ indicates that the Fourier transformed $\Gamma(x_1-x_2, y_1-y_2)$ can be seen as an incoherent light distribution modulated by the function $\gamma(u_2, v_2)$. Therefore, regardless of the exact way of how the partially spatial coherent illumination is obtained, we can consider that it is built from an apertured spatial coherent source placed in the front focal plane of a lens as depicted in Fig. 2.3. As it corresponds exactly to the way the source is prepared in the microscope working with a laser and rotating ground glass, we will consider that this lens is corresponding to L2, with the focal length f_2 in Fig. 2.2.

We consider a transparency $t(x, y)$ in a plane UP. We assume that the plane FP, the back focal plane of L2, is imaged by the microscope lens on the CCD camera.

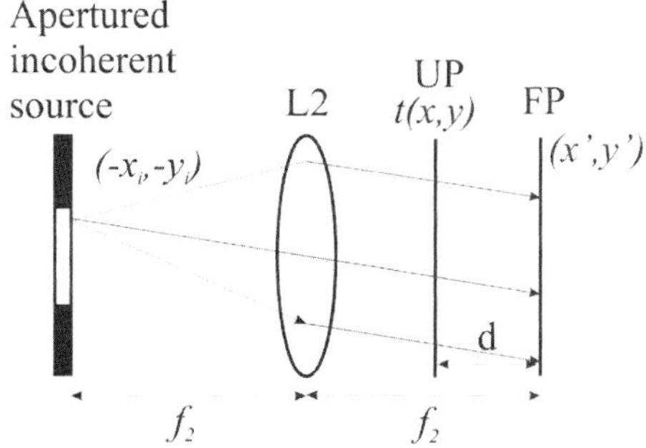

Fig. 2.3 Realization of the partially spatial coherent illumination

We are looking at the diffraction and coherence effects on $t(x, y)$ in the plane FP, separated from UP by a distance d. We take benefit from the fact that the secondary source in the plane of the apertured incoherent source (AIS) is completely incoherent. Therefore, the computation of the partial coherence behavior can be derived as follows.

We consider first an amplitude point source in the AIS plane. Next, we compute the intensity distribution in the output plane due to this point source. Finally, we integrate the result over the source area in the rotating ground glass plane.

Let us consider an amplitude point source $\delta(x + x_i, y + y_i)$ located in the AIS plane. The amplitude distribution that is illuminating the transparency $t(x, y)$ is a tilted plane wave in such a way that the amplitude distribution emerging out of the transparency is expressed by

$$u_{\mathrm{UP}}(x, y) = Ct(x, y)\exp\left(j2\pi\frac{xx_i + yy_i}{\lambda f_2}\right) \tag{2.6}$$

where C is a constant. To determine the amplitude distribution in FP, we apply the free space propagation operator [54] on the amplitude distribution given by (2.6):

$$\begin{aligned} u_{\mathrm{FP}}(x', y') = {} & C'\exp\left\{j2\pi\frac{x_i x' + y_i y'}{\lambda f_2}\right\} \\ & \times \exp\left\{-j\frac{2\pi d}{\lambda f_2^2}\left(x_i^2 + y_i^2\right)\right\} T_f\left(x' - \frac{d}{f_2}x_i, y' - \frac{d}{f_2}y_i\right) \end{aligned} \tag{2.7}$$

with

$$\begin{aligned} & T_f\left(x' - \frac{d}{f_2}x_i, y' - \frac{d}{f_2}y_i\right) \\ & = \iint \mathrm{d}\upsilon_x \mathrm{d}\upsilon_y \exp\left\{j2\pi\left[\upsilon_x\left(x' - \frac{d}{f_2}x_i\right) + \upsilon_y\left(y' - \frac{d}{f_2}y_i\right)\right]\right\} \\ & \quad \times \exp\left\{-j\frac{k\lambda^2 d}{2}\left[\upsilon_x^2 + \upsilon_y^2\right]\right\} T\left(\upsilon_x, \upsilon_y\right) \end{aligned} \tag{2.8}$$

where C' is a constant including terms that are not playing a significant role, $T\left(\upsilon_x, \upsilon_y\right)$ is the continuous spatial Fourier transformation of $t(x, y)$ defined by (2.5), and $\left(\upsilon_x, \upsilon_y\right)$ are the spatial frequencies.

As can be demonstrated on actual examples, we assume that the quadratic phase factor on the right-hand side of (2.7) can be neglected.

$$u_{\mathrm{FP}}(x', y') = C' \exp\left\{ j2\pi \frac{x_i x' + y_i y'}{\lambda f_2} \right\} T_f\left(x' - \frac{d}{f_2} x_i, y' - \frac{d}{f_2} y_i \right) \tag{2.9}$$

Equation (2.8) describes the amplitude distribution obtained in the FP plane for a spatially coherent illumination. The next step consists in evaluating the interference field that is actually recorded. The fringe pattern is recorded by the CCD after the beam recombination. This is achieved by adding to (2.11) the beam originated from the point source $\delta\left(x + x_i, y + y_i\right)$ propagated without transparency in the optical path. As the point source is located in the front focal plane of lens L2, the beam in plane FP is a tilted plane wave that is written as

$$u_{\mathrm{REF}}(x', y') = C'' \exp\left\{ j2\pi \frac{x_i x' + y_i y'}{\lambda f_2} \right\} \tag{2.10}$$

where C'' is a constant.

Equations (2.9) and (2.10) are added and the square modulus of the results is computed to achieve the light intensity

$$i_{\mathrm{FP}}\left(x', y'\right) = \left|u_{\mathrm{REF}}\left(x', y'\right) + u_{\mathrm{FP}}\left(x', y'\right)\right|^2 = \left|C''\right|^2 + \left|C' T_f\right|^2 + A T_f + A^* T_f^* \tag{2.11}$$

where $A = C''^* C'$, and the explicit spatial dependency of T_f was cancelled for the sake of simplicity. The two terms linear with T_f and T_f^*, at the right-hand side of (2.11), are the holographic signals. As we are using a phase stepping technique or the Fourier transform method, the phase and amplitude modulus of T_f are the significant information extracted from (2.11). We consider now the effect of the source extension by integration over the source domain. The signal $v(x', y')$ that will actually be detected can be expressed by

$$v(x', y', d) = A \iint Ip^2\left(x_i, y_i\right) T_f\left(x' - \frac{d}{f_2} x_i, y' - \frac{d}{f_2} y_i \right) \mathrm{d}x_i \mathrm{d}y_i \tag{2.12}$$

Equation (2.13) shows that the signal $v\left(x', y', d\right)$ is the correlation product between T_f and the source function $p^2\left(x_i, y_i\right)$. By performing the change of the integration variables $x_i' = x_i d / f_2$, $y_i' = y_i d / f_2$ and by invoking the convolution theorem, we obtain

$$V\left(\nu_x, \nu_y, d\right) = \left[F^{+1}_{(C)\nu_x,\nu_y} T_f\left(x', y'\right)\right] S\left(\frac{\nu_x d}{f_2}, \frac{\nu_y d}{f_2}\right) \tag{2.13}$$

where V, $\left[F^{+1}_{(C)x',y'} T_f\left(x', y'\right)\right]$, and S are the Fourier transformation of v, T_f, and p^2. Equation (2.13) basically shows that the Fourier transform of the detected signal in partially coherent illumination is the Fourier transform of the signal with a coherent illumination filtered by a scaled Fourier transformation of the source. The scaling factor d/f_2 means that the filtering process increases with d. As $v(x', y', d)$ is the amplitude that is used to perform the digital holographic reconstruction, a loss of resolution of the reconstructed focus image occurs due to the partial spatial coherent nature of the illumination.

Equation (2.13) allows to set accurately the partial coherence state of the source with respect to the requested resolution, refocusing distance, and the location of the optical defects that we want to reject. As an example in biology, it happens often that the selection of the experimental sample cell cannot be made on the only criteria of optical quality but has also to take into account the biocompatibility. Therefore, the experimental cells are often low-cost plastic containers with optical defects. In this situation, the partial coherence of the source can be matched to refocus the sample by digital holography while keeping the container windows at distances where the spatial filtering described above is efficient to reduce the defect influence.

With (2.13), it is expected to have an increasing resolution loss with the distance between the refocused plane and the best focus plane. However, the loss of resolution is controlled by adjusting the spatial partial coherence of the source and, in this way, can be kept smaller than a limit defined by the user. It has also to be emphasized that the reduction in spatial coherence is a way to increase the visibility of the refocused plane by reducing the influence of the out of focus planes. This aspect is particularly useful when the sample highly scatters the light. The adjustment capability of the spatial coherence is then a very useful tool to tune the image visibility and the depth of reconstruction with respect to the sample characteristics.

2.3.2 *Multiple Reflection Removal*

Partial coherent illumination also removes the multiple reflections that can occur with coherent illumination. It is obvious with the LED illumination due to the small temporal coherence. However, it is also true with the microscope working with the RGG and the laser source. In this case, we assume that a reflection introduces an increase of the optical path d. If the distance d introduces a significant decorrelation of the speckle pattern, the contrast of the interference fringe pattern between the reflected beam and the direct beam is reduced. We consider the geometry of Fig. 2.3. As in the RGG plane, the instantaneous amplitude field can be expressed by the product of a random phase function r by an amplitude modulation p. The function r is defined by

$$r\,(x, y) = \exp\{j\phi\,(x, y)\} \tag{2.14}$$

where $\phi\,(x, y)$ is a random function with a constant probability density on the interval $[0, 2\pi]$. It is also assumed that $\phi\,(x, y)$ is very rapidly varying in such a way that

$$\langle r^*\left(x', y'\right) r\,(x, y)\rangle = \delta\left(x' - x, y' - y\right) \tag{2.15}$$

where $\langle\,\rangle$ denotes the ensemble average operation.

Therefore, the instantaneous speckle amplitude field in the object plane is expressed by

$$s\,(x, y) = \exp\{j2kf_2\}\,(P \otimes R)\left(\frac{x}{\lambda f_2}, \frac{y}{\lambda f_2}\right) \tag{2.16}$$

where R and P are the Fourier transformations of, respectively, p and r.

We consider that, in the optical path, we have a double reflection on a window that is introducing an additional optical path d. Let us call this double reflected beam $s'(x', y')$. It can be seen as the $s(x, y)$ beam propagated by a distance d:

$$s'(x', y') = A\exp\left(jkd\right)\left[F^{-1}_{(C)x',y'}\ \exp\left(-\frac{jkd\lambda^2}{2}\left(\nu_x^2 + \nu_y^2\right)\right)\left[F^{+1}_{(C)\nu_x,\nu_y}\,s(x, y)\right]\right] \tag{2.17}$$

where A defines the strength of the double reflection. Inserting (2.16) in (2.17), we obtain

$$s'(x', y') = A\ \exp\left(ikd\right)\left[F^{-1}_{(C)x',y'}\ \exp\left(-\frac{jkd\lambda^2}{2}\left(\nu_x^2 + \nu_y^2\right)\right)\right.$$
$$\left.\times\, r\left(-\nu_x\lambda f_2, -\nu_y\lambda f_2\right) p\left(-\nu_x\lambda f_2, -\nu_y\lambda f_2\right)\right] \tag{2.18}$$

If the quadratic phase factor is slowly varying over the area where the p function in (2.20) is significantly different to zero, $s'\left(x', y'\right)$ is very similar to $s\,(x, y)$ and they mutually interfere to result in a disturbing fringe pattern. On the contrary, when the quadratic phase factor is rapidly varying on the significant area defined by the width of p, $s'\left(x', y'\right)$ and $s\,(x, y)$ are uncorrelated speckle fields and the interference modulation visibility is reduced when the ground glass is moving. Assuming that the laser beam incident on the ground glass has a Gaussian shape, p is expressed by

$$p(x, y) \propto \exp\left\{-\frac{x^2 + y^2}{w^2}\right\} \tag{2.19}$$

where w is the width of the beam. Using (2.19), the width of $p\left(-\nu_x \lambda f_2, -\nu_y \lambda f_2\right)$ in (2.18) is equal to $w^2/\lambda^2 f_2^2$. We assume that we obtain a large speckle decorrelation when the quadratic phase factor in the exponential of (2.18) is larger than π, and when $\left(\nu_x^2 + \nu_y^2\right)$ is equal to the width of $p\left(-\nu_x \lambda f_2, -\nu_y \lambda f_2\right)$. Therefore, a speckle decorrelation between $s'\left(x', y'\right)$ and $s(x, y)$ is achieved when

$$\frac{dw^2}{\lambda f_2^2} >> 1 \tag{2.20}$$

As the width w is adjusted by changing the distance between the lens L1 and the rotating ground glass (RGG), multiple reflection artifacts are removed by a difference path distance d appropriately reduced.

2.3.3 Coherent Noise Removal

A major issue of the full coherent illumination is the coherent speckle noise that arises on the unavoidable defects of the optical system, on the sample container that can be in some cases of poor optical quality, and on the sample itself when it is bulky and scatters the light. The speckle results from the coherent superposition of random contributions that originate from out of focus locations. A way to reduce the strength of the speckle noise is to reduce the spatial coherence of the beam illuminating the sample. The relation between the noise and the spatial and temporal coherence states, although not performed in the frame of digital holography, has already been shown many years ago [55].

Therefore, we will give here an example from a fluid physics experiment. The images of Fig. 2.4 are magnified parts of digital holograms that were recorded with a partially spatial illumination and a fully coherent illumination. The object is identical and consists of 5 μm particles immerged in distilled water. In the partially spatial coherent case, we observed that the diffraction pattern of the unfocused particles is clearly distinguishable. As a result, the digital holographic reconstruction can be applied to refocus the particles in the field of view. On the contrary, with the full coherent illumination, the only visible particles are those that are focused and the background is a speckle field. In that case the digital holographic refocusing is very noisy. The microscope lenses L3 and L4 are 20× microscope lenses with a numerical aperture of 0.3. The magnification, given by the ratio between the focal lengths of L5 and L3, is 15× to provide a 390 μm × 390 μm field of view on the CCD sensor. According to [17], the maximum refocusing distance, in the coherent case, without significant loss of resolution, is approximately ±250 μm around the best focus plane.

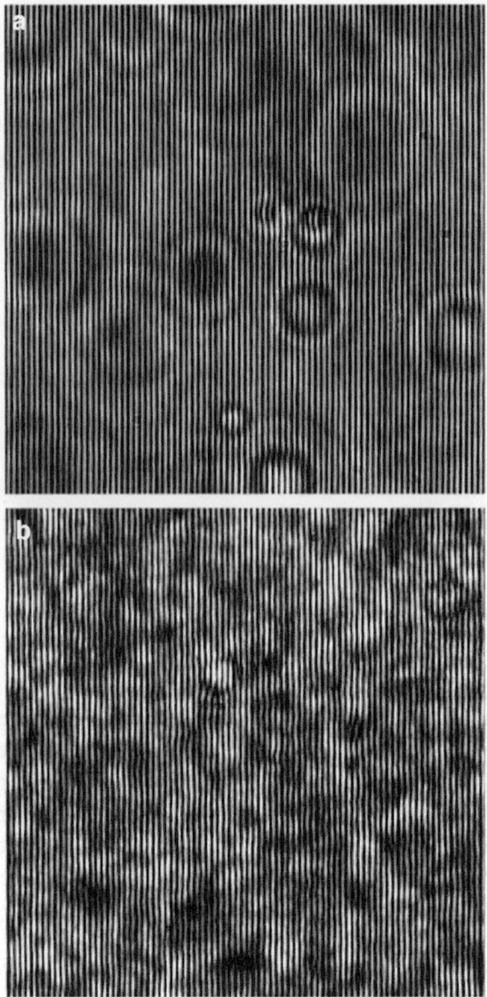

Fig. 2.4 (**a**) Interferometric image of 5 μm particles in distilled water – partially spatial coherent illumination, (**b**) Interferometric image of 5 μm particles in distilled water – spatial coherent illumination

2.4 Experimental Demonstrations and Applications

2.4.1 Biomedical Applications

The biomedical applications that we present below have been performed with the DHM illumination with a LED, as described in Fig. 2.1. Other applications using the DHM configuration working with a partially spatial coherent source from a laser beam are currently under development.

2.4.1.1 Study of Cell Cultures

The DHM that we developed is a powerful tool to study cell cultures and their evolution in time.

Thanks to the partially coherent source, the holographic microscope provides a bright field image quality equivalent to a classical microscope, even with sample container of poor quality such as plastic dishes (Fig. 2.5a).

As the full information on the optical field transmitted by the sample is recorded, it is possible to emulate classical optical microscopy modes such as differential interference contrast (DIC) to improve the visibility of unstained living cells [17] (Fig. 2.5b). The quantitative phase computation is used to analyze dynamically the cell morphology with nanometric accuracies (Fig. 2.5c, d).

2.4.1.2 The Implementation of Fluorescence Imaging in DHM

We developed and patented a DHM and a method to perform refocusing of fluorescent images recorded out of focus [56]. A broadband light source associated with excitation and an emission filter is implemented in the holographic microscope to record the fluorescence image in reflection. A process combines the digital hologram to the fluorescence image in such a way that the latter can be refocused.

The method is illustrated in Fig. 2.6 with fixed green fluorescent protein (GFP)-labeled rat liver cells [14]. Focus and defocus fluorescent images were recorded (Fig. 2.6a, b). The defocus fluorescent image was combined with the digital hologram to refocus it (Fig. 2.6c). We observe that the refocusing process provides an accurate visibility of the individual fluorescent cells.

2.4.1.3 Observation and Morphology Analysis of Cell Fusions and Hybrid Formation

The formation of hybrids between tumor cells and dendritic cells (DC) is an interesting approach for anti-tumor vaccination. Such hybrids combine the expression of specific tumor antigens and the machinery for their optimal presentation, indispensable for the induction of an efficient immune response.

The RUBIO team ('Unité de Recherche en Biothérapie et Oncologie de l'Université Libre de Bruxelles') performs of in vitro fusions between different cell lines some of which are labeled with a fluorophore. In collaboration with this laboratory, we performed observations with the digital holographic microscope of fusion events between Chinese hamster ovary cells (CHO-FMG) and rat glioblastoma cells (9L-eGFP) expressing enhanced green fluorescent protein (eGFP). The digital holographic microscope allows to visualize the fusion events and the formation of hybrids in real time. Accurate analysis of the process is also achieved thanks to the quantitative phase contrast measurement.

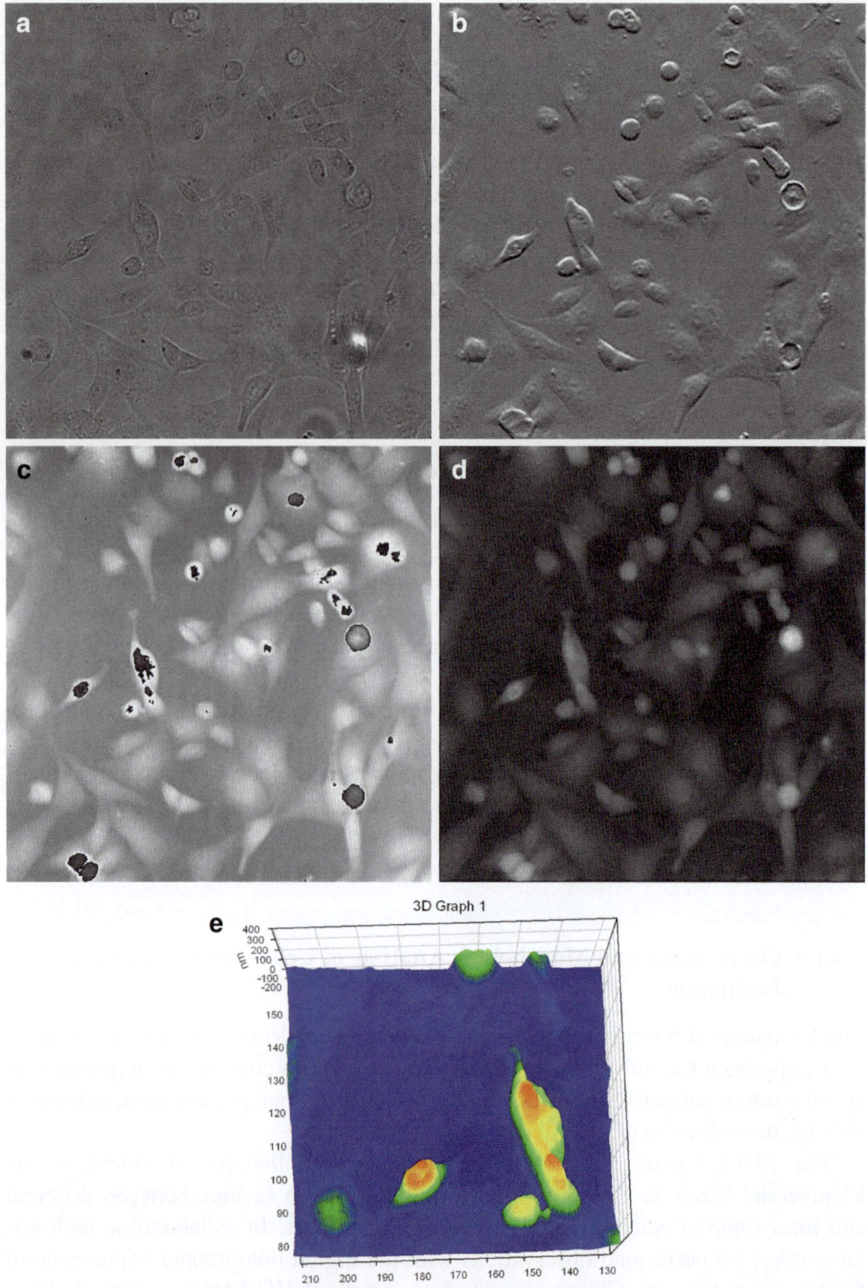

Fig. 2.5 Cell culture in a plastic dish, observed with the digital holographic microscope working with a LED. These quality images illustrate the multimode capabilities of this microscope. It allows to obtain the classical microscopy modes combined with the measurement of the optical thicknesses with a nanometric precision. **(a)** Brightfield image; **(b)** DIC emulation image; **(c)** phase image; **(d)** unwrapped phase image; **(e)** 3D graph of a cropped region from the unwrapped phase image **d**

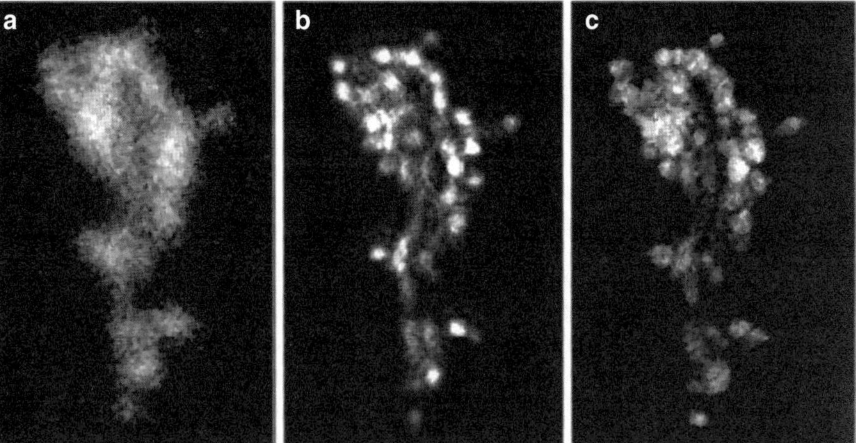

Fig. 2.6 Demonstration of fluorescent refocusing of rat liver cells. (**a**) 80 μm defocus fluorescence image; (**b**) digital holographic refocusing; (**c**) focus fluorescent image

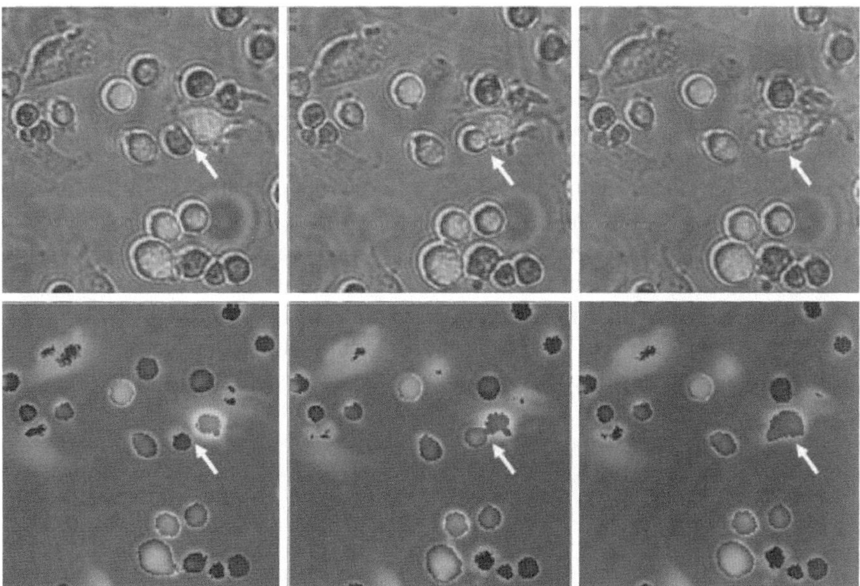

Fig. 2.7 Upper row images combine brightfield (B&W) and fluorescence images (*green*). The lower row images are a combination of the corresponding phase (B&W) and fluorescence images (*green*). The change of color of the left cell outlined by the arrows demonstrates the exchange of biological material between the cells

Different combinations of time-lapse modes show the fusion events between unlabelled CHO-FMG cells and fluorescent 9L-eGFP cells (Fig. 2.7).

Figure 2.8 is a sequence of unwrapped phase images in pseudocolor showing the evolution of the optical thickness of the cells during a fusion process.

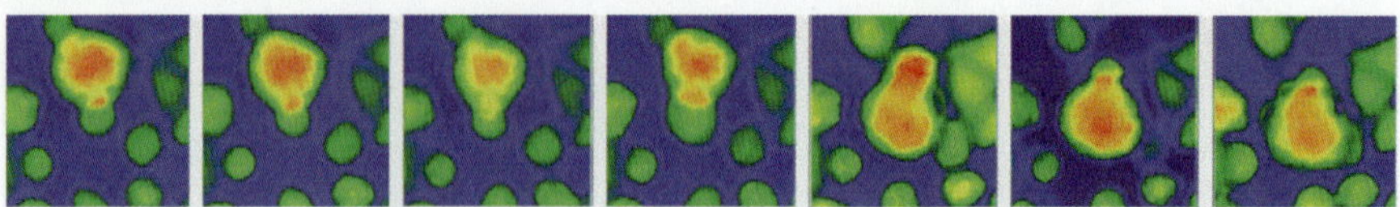

Fig. 2.8 Seven pseudocolour unwrapped phase images extracted from a time lapse

2.4.1.4 Migration of Cancer Cells

The capability to observe unstained living cells in turbid media is shown for the migration of cancer cells embedded in a collagen gel [17]. In collaboration with the Laboratory of Toxicology and the Department of Logical and Numerical Systems of the Université Libre de Bruxelles, we performed time lapse experiments with the digital holographic microscope with the LED illumination on unstained HT-1080 fibrosarcoma cells embedded in a thick 3D collagen matrix (thickness around 1,500 μm). Digital holograms were recorded every 4 minutes for 48 h.

Thanks to the partially coherent nature of the source, one obtains holograms and very accurate images of migrating cancer cells inside the gel and it is possible to follow the behavior of every cell in time. A semi-automated software package allowed to perform a 4D analysis of the cancer cell migration dynamics on the time-lapse sequence based on three visualization modes provided by the holographic microscope, i.e., the bright field, the phase image, and the DIC modes (Fig. 2.9).

A custom software allows to determine a ROI and to displace it in the hologram. Inside this ROI, the operator can refocus with the mouse wheel the different cells embedded in the gel and the recording of the (t, X, Y, Z) locations is carried out with a mouse click on a central point inside the cell body, an usual way to record cell locations in computer-assisted cell tracking experiments. Stacks of 4D images allow to observe the cell displacement during the experiment [17].

With those experiments, we demonstrate that digital holographic microscopy is an efficient non-invasive tool to monitor cells' behaviors in 3D environments closer

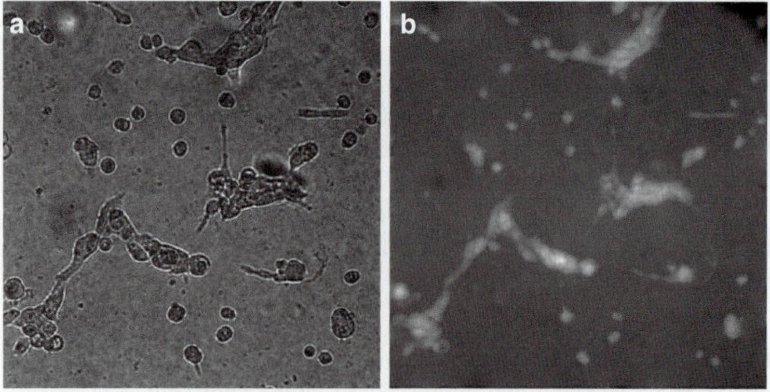

Fig. 2.9 Example of two different visualization modes that can be combined. Bright field image (**a**) and phase image (**b**) of cancer cells inside the collagen gel

to the in vivo conditions. It offers the potentiality to observe action of drugs on 3D marker-free cell cultures in real time.

More generally, thanks to its configuration, this microscope allows to study microscopic elements present in turbid media.

2.4.2 Hydrodynamic Flow of Micron Size Particles

2.4.2.1 Reconstruction of Size-Calibrated Particles in a SPLITT Cell

In this section, we describe the use of the DHM described in Fig. 2.2 to analyze an hydrodynamic flow of micron size particles, passing through a split-flow lateral-transport thin (SPLITT) separation cell [57]. This class of separation performs a continuous and rapid fractionation of macromolecular and particulate materials.

The digital holographic reconstruction in depth of flowing particles provides a measure of the particle location to assess the influence of the hydrodynamic effects [58, 59] on the separation efficiency, as well as particle–particle and particle–wall interactions. DHM permits to obtain the particle diameters, their 3D positions in the channel, and an estimation of their mean velocities. In the case of a mixture of bidisperse particles, DHM can be used to distinguish the two species and their behaviors. A SPLITT cell used to separate particles by their sizes is inserted into the sample holder. The cell is a thin ribbon-like channel of 234 μm thickness, having at its extremities two inlets and two outlets that are separated by splitters. The inlet splitter allows to separately inject the sample through one inlet and the carrier liquid through the other inlet. The outlet splitter divides the flow into two sub-streams in the outlets (Fig. 2.10).

The microscope lenses L3 and L4 are 20× microscope lenses with a numerical aperture of 0.3. The magnification is 20× to provide a 300 μm × 300 μm field of view.

For the experiments, we used different latex bead samples: mono-disperse polymer microspheres of 4 and 7 μm diameter and nominal latex beads of 10 μm.

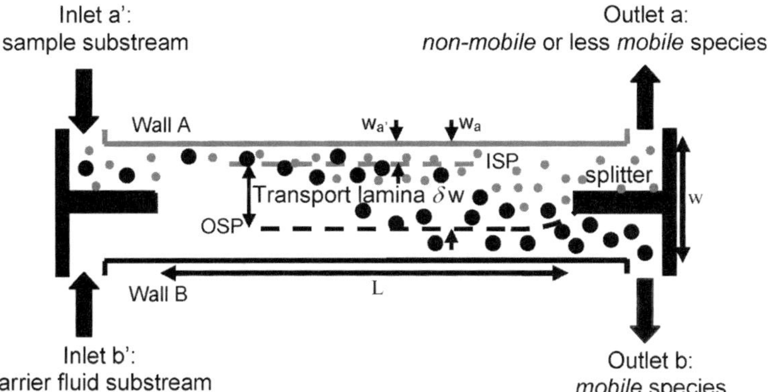

Fig. 2.10 Split-flow lateral-transport thin (SPLITT) separation cell

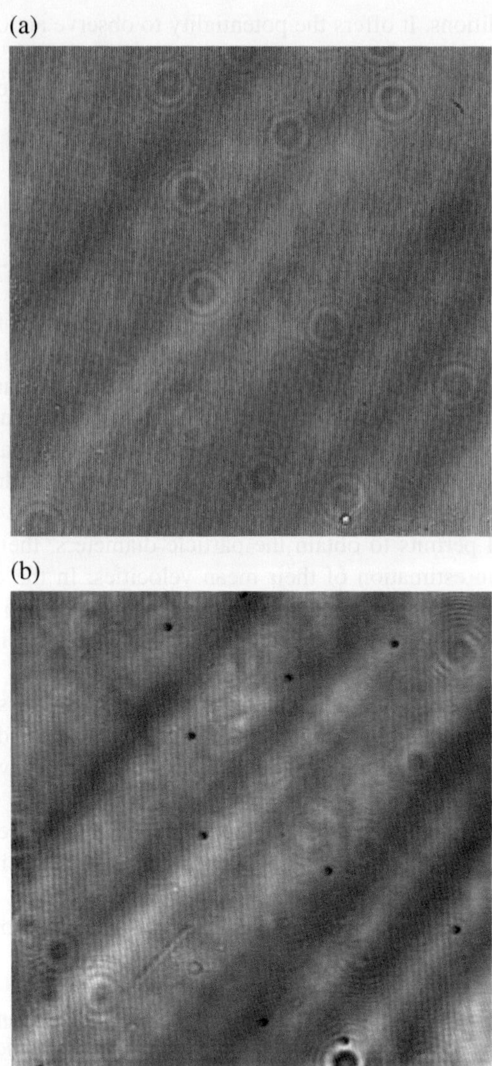

Fig. 2.11 **(a)** Digital hologram of unfocused size-calibrated 4 μm latex beads flowing at a mean velocity of $v = 5.34\,\text{cm/s}$. **(b)** Numerical refocusing of the digital hologram of Fig. 2.11a. The particles have a mean size of $4.2 \pm 0.6\,\mu\text{m}$ and a refocus distance of $23.4 \pm 2.6\,\mu\text{m}$

Double-distilled deionized water is used as carrier fluid. An example of recorded hologram is provided in Fig. 2.11a. The microscope is initially focused on the internal front channel wall. By digital holographic reconstructions, the focus distances and the sizes of the suspended particles are determined (Fig. 2.11b). The measured particle sizes are in agreement with the specifications: $4.2 \pm 0.6\,\mu\text{m}$, $7.2 \pm 0.6\,\mu\text{m}$, and $10.2 \pm 0.6\,\mu\text{m}$. A maximum estimated error of 0.8 μm is obtained. The DHM gives the particle focus distance with an accuracy of $\pm 2.6\,\mu\text{m}$ which means that the particle position over the channel thickness is determined with an accuracy of about 1%.

2.4.2.2 Experimental Measurements of Particles in Flows

When the particle size increases, their transversal migrations, due to hydrodynamic effects, are expected to increase. When a mixture of particles of different sizes is injected, the two species can be differentiated and therefore the respective transverse position can be determined (Fig. 2.12a, b).

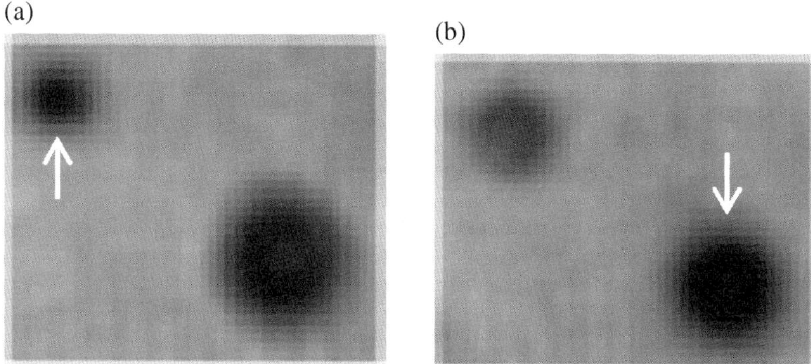

Fig. 2.12 **(a)** The 4 μm latex bead, mean velocity of $v = 3.56\,\mathrm{cm/s}$, has a focus distance of $25 \pm 2.6\,\mu\mathrm{m}$. **(b)** The 7 μm latex bead has a focus distance of $50 \pm 2.6\,\mu\mathrm{m}$

These results underline the capacity of the DHM to study the separation processes of micron size objects in SPLITT channels.

The DHM also gives a direct assessment of particle velocities. In our specific experimental conditions, when the mean flow velocity is greater than 7 cm/s, some particles have too high velocities with respect to the camera exposure time. In that case, particle holograms are blurred by the motion. We tested this operating mode on a suspension of 0.1 % Lichrospher Si60 of 5 μm. Particle traces of different lengths are observed according to the particle transversal position. As the velocity profile is parabolic, a particle flowing near the channel center moves faster than near a wall and therefore leaves a longer trace. By digital holographic reconstruction, traces can be refocused and particle velocities can be deduced by measuring their lengths (Fig. 2.13b). The quantitative agreement between theoretical and experimental results is outlined in Table 2.1.

2.4.3 Dynamics of Phospholipid Vesicles

2.4.3.1 Description

Vesicles are lipid membranes enclosing a fluid and suspended in an aqueous solution (Fig. 2.14).

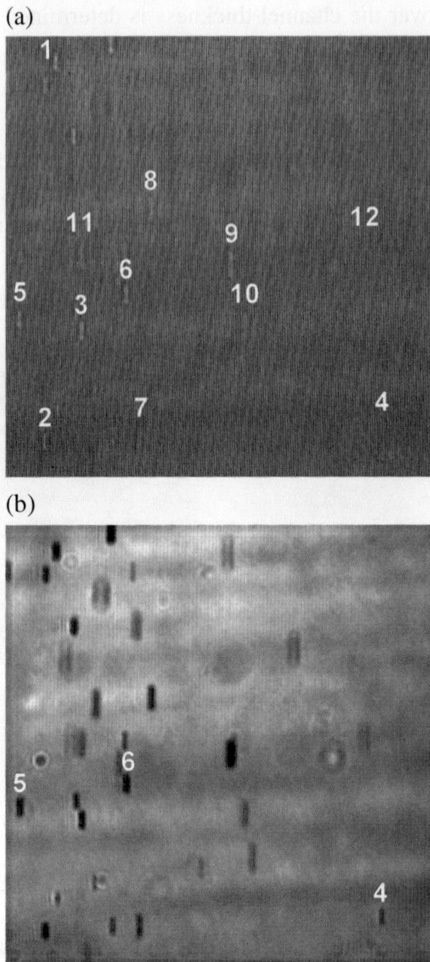

Fig. 2.13 **(a)** Blurred digital hologram of 5 μm Lichrospher particles flowing, at a mean velocity of $v = 12.46\,\mathrm{cm/s}$, in the SPLITT channel. **(b)** Re-constructed image from the digital hologram of Fig. 2.13a. Refocus distance of 40 μm. The sharp trace of some 5 μm particles can be observed

Their size ranges from a few microns to several hundred microns (Fig. 2.15). They can be viewed as a simple model to represent the basic mechanical properties of cells such as red blood cell with which they share several common features and behaviors under flow. Although this is a simplified model (only one constituent in the membrane, no cytoskeleton, homogeneous Newtonian fluid inside), they are an interesting tool for physicists. Parameters such as diameter, volume-to-surface ratio, and viscosity of the filling fluid can be easily varied over significant ranges.

The flow and rheology of complex fluids containing deformable objects such as vesicles have been the subject of several experimental and theoretical studies [58–62]. Nowadays, their behavior under shear flow is not completely understood. When put in a shear flow, vesicles will undergo a lift force and move away from

Table 2.1 Comparison between the experimental and the theoretical values of the 5 μm Lichrospher particle velocities inside the SPLITT channel

Particle number	Transversal position (μm)	v_{exp} (cm/s)	v_{th} (cm/s)	Precision (%)
1	29	7	8.12	13.79
2	35	8	9.51	15.90
3	35	10	9.51	5.12
4	40	9	10.60	15.08
5	40	11	10.60	3.79
6	40	10	10.60	5.65
7	44	13	11.42	13.85
8	44	13	11.42	13.85
9	59	15	14.10	6.37
10	65	14	15.00	6.69
11	70	16	15.68	2.04
12	94	18	17.97	0.14

the wall. Digital holographic microscopy gives the opportunity to investigate the 3D distribution versus time of vesicles flowing inside a shear-flow chamber (of about 200 μm of thickness) and to determine their size and shape.

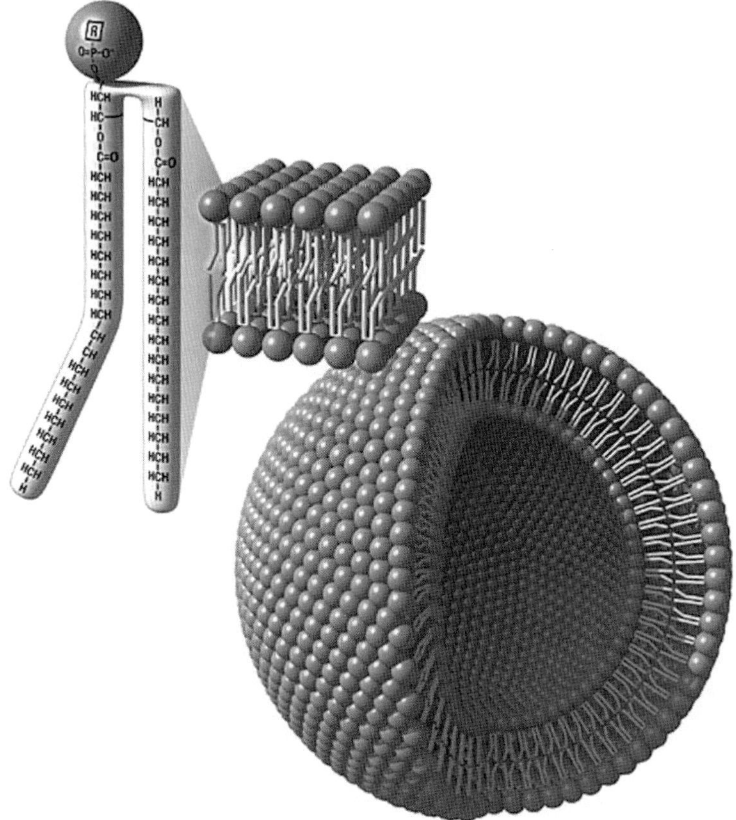

Fig. 2.14 Schematic of a vesicle made of a bilayer of phospholipid molecules
Source: http://astrobiology.nasa.gov/team/images/2006/ciw/1829/1.jpg

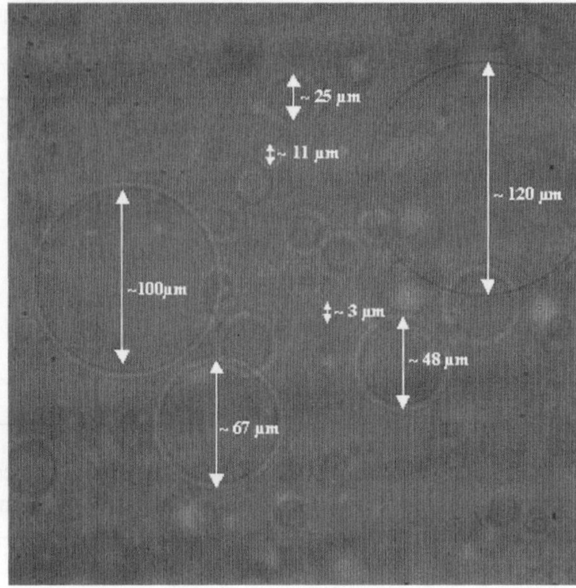

Fig. 2.15 Digital hologram of a suspension of vesicles. Vesicles size ranges from a few microns to several hundred microns. The field of view is 420 × 420 µm

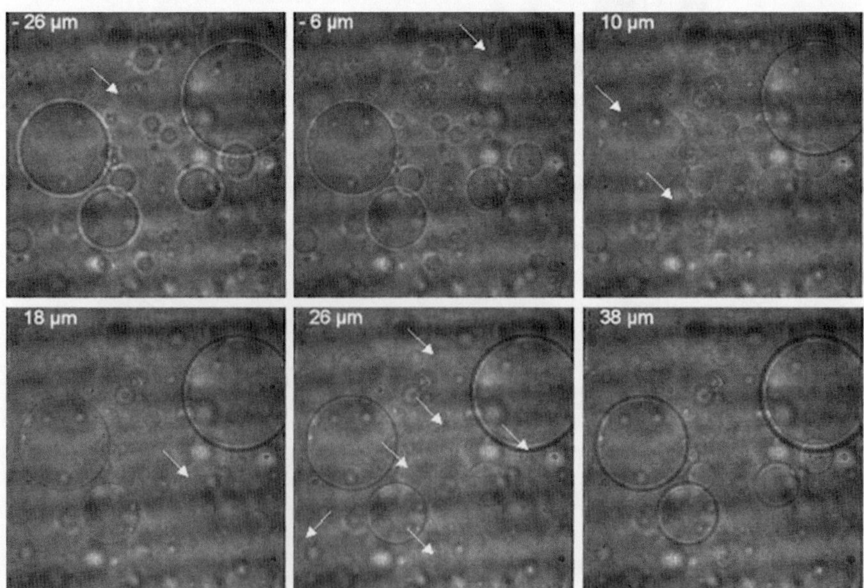

Fig. 2.16 Digital holographic reconstructions of Fig. 2.15 at different refocus distances from 40 µm upstream to 40 µm downstream from the digital holographic microscope focus plan. *Arrows* indicate focused vesicles

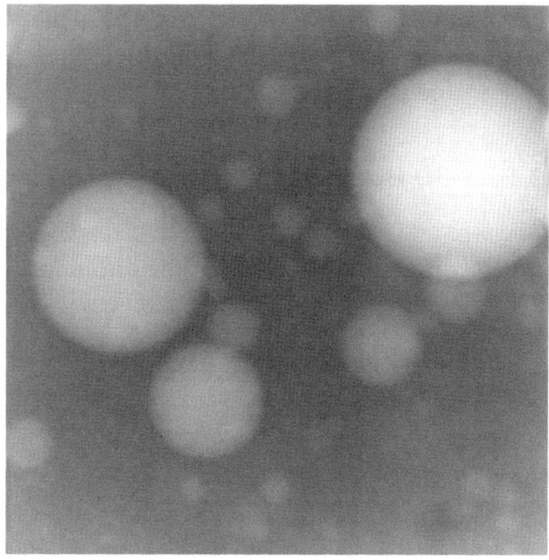

Fig. 2.17 Phase information extracted from Fig. 2.16 after phase compensation

2.4.3.2 Transversal Position

Digital holographic reconstruction determines the transversal position of each vesicle in the field of view (of $420 \times 420\,\mu\text{m}$). Vesicles are phase objects and are completely transparent in intensity when they are focused. When unfocused, they are surrounded by a refraction ring (Fig. 2.16) visible in intensity.

Due to the transparency of vesicles, it is easier to visualize them on the phase images. To determine their size and shape, we use compensated phase images (Fig. 2.17).

When put in an external solution, vesicles deflate by osmosis and their mechanical and hydrodynamic properties change. One of the direct consequences of the deflating is the change of density of the vesicle. Density can be obtained by measuring the sedimentation velocity of one vesicle. Figure 2.18 shows the sedimentation

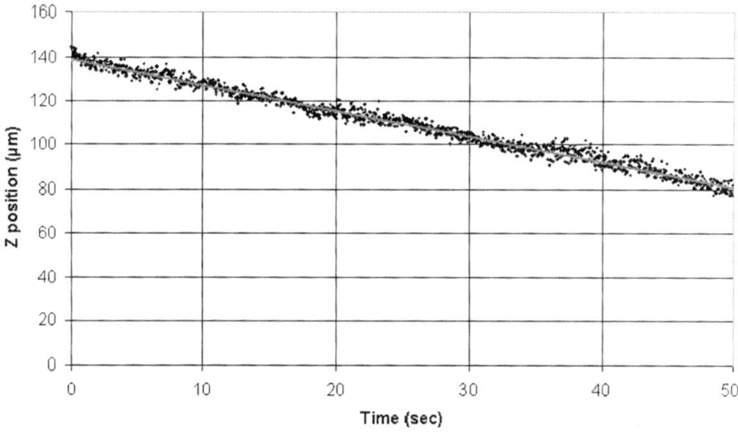

Fig. 2.18 Sedimentation curve obtained by DHM of a spherical vesicle of $15.6\,\mu\text{m}$ of diameter

curve of one spherical vesicle (of 15.6 μm of diameter) obtained by DHM at a frequency of 24 frames/s. The Z position of the vesicle is automatically determined by using a focus metrics [40, 63].

2.4.3.3 3D Distribution

The size and shape distribution of a suspension of vesicles can be obtained by locating the different objects inside the different reconstructed planes and by performing a segmentation on each detected object in its focus plane. The determination of the

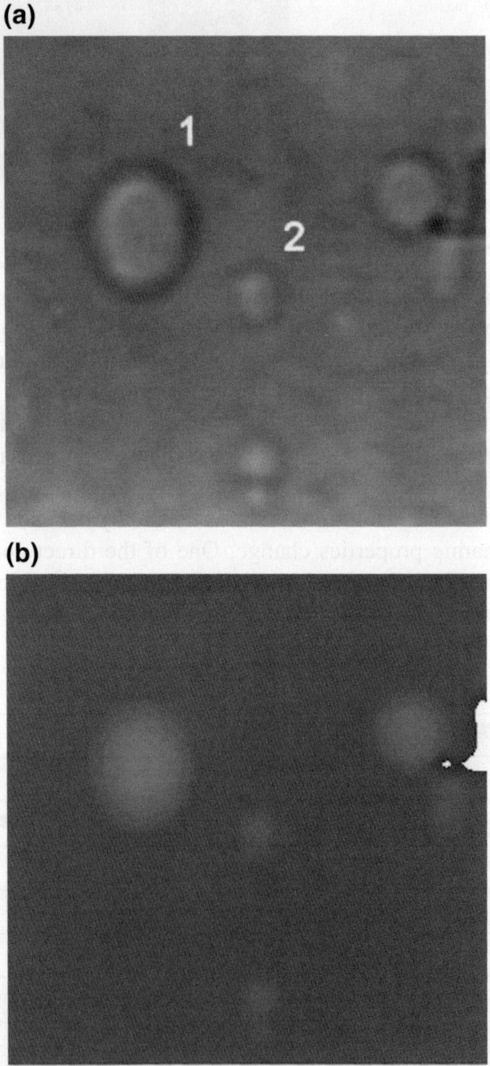

Fig. 2.19 Example of two vesicles with a close (x, y) position and a different z position. Centers of vesicles 1 and 2 are separated by a distance of 24 μm. (**a**) Intensity image and (**b**) compensated phase image

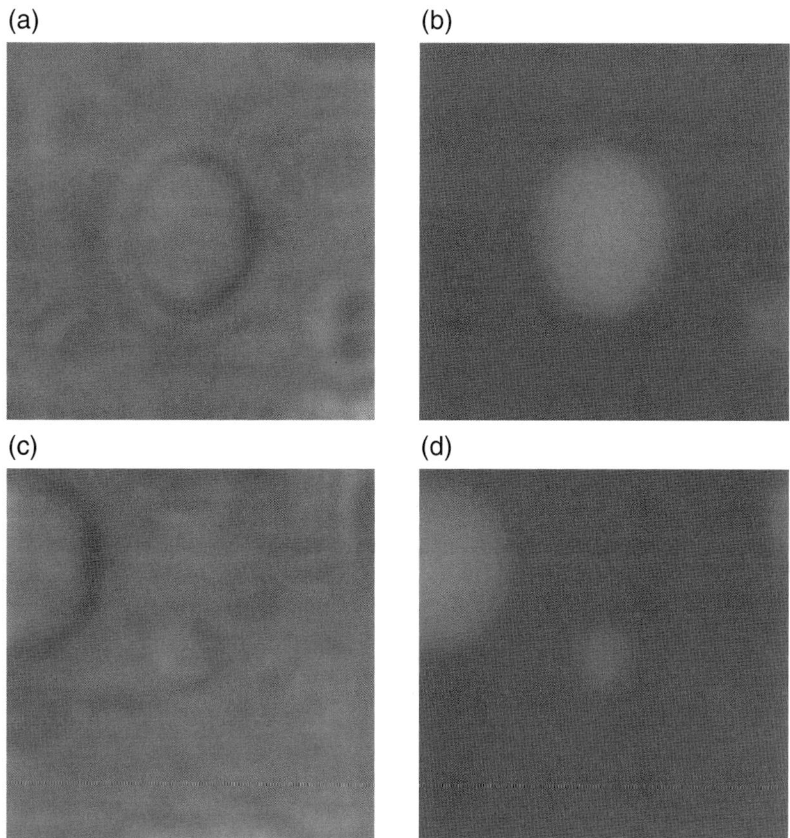

Fig. 2.20 **(a)** Intensity image of vesicle 1. Detected z position : $-87\,\mu$m; big axis : $13.5\,\mu$m; small axis : $11.2\,\mu$m. **(b)** Phase image of vesicle 1. Detected z position : $-87\,\mu$m; big axis : $13.5\,\mu$m; small axis : $11.2\,\mu$m; **(c)** Intensity image of vesicle 2. Detected z position : $-49\,\mu$m; big axis : $4.8\,\mu$m; small axis : $4.2\,\mu$m; **(d)** Phase image of vesicle 2. Detected z position : $-49\,\mu$m; big axis : $4.8\,\mu$m; small axis : $4.2\,\mu$m

transversal position is obtained by using focus plane detection criteria [40]. The feature extraction of each object is performed by segmentation of the compensated phase map (active contour techniques). By fitting the segmented object with an ellipse, we get the size and shape of the vesicle. The following example shows the capacity of the automated software to detect and separate vesicles (of different shapes and sizes) which are close to one another in the intensity image but with different transversal positions (Figs. 2.19 and 2.20).

2.5 Conclusions

The objective of this chapter is to show, through some theoretical aspects and actual implementations of applications, the benefits of digital holography with partial coherent sources compared to classical microscopy and coherent digital holographic microscopy. It is particularly suitable for biomedical applications and for the study of dynamical phenomena where the samples are thick with respect to the classical depth of focus. It becomes possible to perform a full 4D analysis of samples without the need of a mechanical focusing stage. Digital holography microscopy can also be coupled with fluorescence which makes it a very powerful tool for numerous biomedical applications. In addition to the refocusing in depth, digital holography yields the quantitative phase contrast mode that is particularly efficient for the analysis of transparent specimens. The partially coherent nature of the illumination reduces drastically the coherent noise to provide image qualities, even in intensity modes, that are similar to the ones obtained with the best classical microscopes.

References

1. T.-C. Poon, M. Motamedi, Optical/Digital incoherent image processing for extended depth of field. Appl. Opt. **26**, 4612–4615 (1987)
2. E.R. Dowski, Jr., W. Thomas Cathey, Extended depth of field through wave-front coding. Appl. Opt. **34**, 1859–1866 (1995)
3. U. Schnars, W. Jüptner, Direct recording of holograms by a CCD target and numerical reconstruction. Appl. Opt. **33**, 179–181 (1994)
4. I. Yamaguchi, T. Zhang, Phase-shifting digital holography. Opt. Lett. **22**, 1268–1270 (1997)
5. T. Zhang, I. Yamaguchi, Three-dimensional microscopy with phase-shifting digital holography. Opt. Lett. **23**, 1221–1223 (1998)
6. E. Cuche, F. Bevilacqua, C. Depeursinge, Digital holography for quantitative phase contrast imaging. Opt. Lett. **24**, 291–293 (1999)
7. M. Sebesta, M. Gustafsson, Object characterization with refractometric digital Fourier holography. Opt. Lett. **30**, 471–473 (2005)
8. T. Ikeda, G. Popescu, R.R. Dasari, M.S. Feld, Hilbert phase microscopy for investigating fast dynamics in transparent systems. Opt. Lett. **30**, 1165–1167 (2005)
9. G. Popescu, T. Ikeda, C.A. Best, K. Badizadegan, R.R. Dasari, M.S. Feld, Erythrocyte structure and dynamics quantified by Hilbert phase microscopy. J. Biomed. Opt. **10**, 060503 (2005)
10. F. Dubois, L. Joannes, J.-C. Legros, Improved three-dimensional imaging with digital holography microscope using a partial spatial coherent source. Appl. Opt. **38**, 7085–7094 (1999)
11. G. Indebetouw, P. Klysubun, Spatiotemporal digital microholography. J. Opt. Soc. Am. A **18**, 319–325 (2001)
12. I. Yamaguchi, J.-I. Kato, S. Otha, J. Mizuno, image formation in phase-shifting digital holography and applications to microscopy. Appl. Opt. **40**, 6177–6186 (2001)
13. D. Dirksena, H. Drostea, B. Kempera, H. Delerlea, M. Deiwickb, H.H. Scheldb, G. von Bally, Lensless Fourier holography for digital holographic interferometry on biological samples. Opt. Lasers Eng. **36**, 241–249 (2001)
14. F. Dubois, C. Yourassowsky, O. Monnom, in *Microscopie en Holographie Digitale Avec Une Source Partiellement Cohérente*, éd. by M. Faupel, P. Smigielski, R. Grzymala. Imagerie et Photonique pour les sciences du vivant et la médecine. (Fontis Media, Fomartis, 2004), pp. 287–302, Switzerland, Lausanne
15. P. Marquet, B. Rappaz, P.J. Magistretti, E. Cuche Y. Emery, Tristan colomb and christian depeursinge, Digital holographic microscopy: a noninvasive contrast imaging technique allowing quantitative visualization of living cells with subwavelength axial accuracy. Opt. Lett. **30**, 468–470 (2005)

16. D. Carl, B. Kemper, G. Wernicke, G. von Bally, Parameter-optimized digital holographic microscope for high-resolution living-cell analysis. Appl. Opt. **43**, 6536–6544 (2004)
17. F. Dubois, C. Yourassowsky, O. Monnom, J.-C. Legros, O. Debeir, P. Van Ham, R. Kiss, C. Decaestecker, Digital holographic microscopy for the three-dimensional dynamic analysis of *in vitro* cancer cell migration. J. Biomed. Opt. **11**(5), 054032 September/October (2006)
18. N. Lue, G. Popescu, T. Ikeda, R.R. Dasari, K. Badizadegan, M.S. Feld, Live cell refractometry using microfluidic devices. Opt. Lett. **31**, 2759–2761 (2006)
19. F. Charrière, A. Marian, F. Montfort, J. Kühn, T. Colomb, E. Cuche, P. Marquet, C. Depeursinge, Cell refractive index tomography by digital holographic microscopy. Opt. Lett. **31**, 178–180 (2006)
20. F. Charrière, N. Pavillon, T. Colomb, C. Depeursinge, T.J. Heger, E.A.D. Mitchell, P. Marquet, B. Rappaz, Living specimen tomography by digital holographic microscopy: morphometry of testate amoeba. Opt. Express **14**, 7005–7013 (2006)
21. P. Marquet, B. Rappaz, F. Charrière, Y. Emery, C. Depeursinge, P. Magistretti, Analysis of cellular structure and dynamics with digital holographic microscopy. Proc. SPIE **6633**, 66330F (2007)
22. P. Ferraro, S. De Nicola, G. Coppola, A. Finizio, D. Alfieri, G. Pierattini, Controlling image size as a function of distance and wavelength in Fresnel-transform reconstruction of digital holograms. Opt. Lett. **29**, 854–856 (2004)
23. P. Ferraro, S. De Nicola, A. Finizio, G. Coppola, S. Grilli, C. Magro, G. Pierattini, Compensation of the inherent wave front curvature in digital holographic coherent microscopy for quantitative phase-contrast imaging. Appl. Opt. **42**, 1938–1946 (2003)
24. T. Colomb, J. Kühn, F. Charrière, C. Depeursinge, P. Marquet, N. Aspert, Total aberrations compensation in digital holographic microscopy with a reference conjugated hologram. Opt. Express **14**, 4300–4306 (2006)
25. P. Ferraro, D. Alferi, S. De Nicola, L. De Petrocellis, A. Finizio, G. Pierattini, Quantitative phase-contrast microscopy by a lateral shear approach to digital holographic image reconstruction. Opt. Lett. **31**, 1405–1407 (2006)
26. L. Miccio, D. Alfieri, S. Grilli, P. Ferraro, A. Finizio, L. De Petrocellis, S. De Nicola, Direct full compensation of the aberrations in quantitative phase microscopy of thin objects by a single digital hologram. Appl. Phys. Lett. **90**, 041104 (2007)
27. T. Colomb, E. Cuche, F. Charrière, J. Kühn, N. Aspert, F. Montfort, P. Marquet, C. Depeursinge, Automatic procedure for aberration compensation in digital holographic microscopy and applications to specimen shape compensation. Appl. Opt. **45**, 851–863 (2006)
28. T.-C. Poon, T. Kim, Optical image recognition of three-dimensional objects. Appl. Opt. **38**, 370–381 (1999)
29. B. Javidi, E. Tajahuerce, Three-dimensional object recognition by use of digital holography. Opt. Lett. **25**, 610–612 (2000)
30. D. Kim, B. Javidi, Distortion-tolerant 3-D object recognition by using single exposure on-axis digital holography. Opt. Exp. **12**, 5539–5548 (2004)
31. F. Dubois, C. Minetti, O. Monnom, C. Yourassowsky, J.-C. Legros, Pattern recognition with digital holographic microscope working in partially coherent illumination. Appl. Opt. **41**, 4108–4119 (2002)
32. E. Cuche, P. Marquet, C. Despeuringe, Aperture Apodization using cubic spline interpolation: application in digital holography microscopy. Opt. Commun. **182**, 59–69 (2000)
33. F. Dubois, O. Monnom, C. Yourassowsky, J.-C. Legros, Border processing in digital holography by extension of the digital hologram and reduction of the higher spatial frequencies. Appl. Opt. **41**, 2621–2626 (2002)
34. P. Klysubun, G. Indebetouw, *A posteriori* processing of spatiotemporal digital microholograms. J. Opt. Soc. Am. A **18**, 326–331 (2001)
35. L. Yu, L. Cai, Iterative algorithm with a constraint condition for numerical reconstruction of a threedimensional object from its hologram. J. Opt. Soc. Am. A **18**, 1033–1045 (2001)
36. J. Gillespie, R.A. King, The use of self-entropy as a focus measure in digital holography. Pattern Recogn. Lett. **9**, 19–25 (1989)

37. L. Ma, H. Wang, Y. Li, H. Jin, Numerical reconstruction of digital holograms for three-dimensional shape measurement. J. Opt. A: Pure Appl. Opt. **6**, 396–400 (2004)
38. P. Ferraro, G. Coppola, S. De Nicola, A. Finizio, G. Pierattini, Digital holographic microscope with automatic focus tracking by detecting sample displacement in real time. Opt. Lett. **28**, 1257–1259 (2003)
39. M. Liebling, M. Unser, Autofocus for digital Fresnel Holograms by use of a Fresnelet-Sparsity criterion. J. Opt. Soc. Am. **21**, 2424–2430 (2004)
40. F. Dubois, C. Schockaert, N. Callens, C. Yourassowsky, Focus plane detection criteria in digital holography microscopy by amplitude analysis. Opt. Express **14**, 5895–5908 (2006)
41. W. Li, N.C. Loomis, Q. Hu, C.S. Davis, Focus detection from digital in-line holograms based on spectral $l1$ norms. J. Opt. Soc. Am. A **24**, 3054–3062 (2007)
42. C.P. McElhinney, J.B. McDonald, A. Castro, Y. Frauel, B. Javidi, T.J. Naughton, Depth-independent segmentation of macroscopic three-dimensional objects encoded in single perspectives of digital holograms. Opt. Lett. **32**, 1229–1231 (2007)
43. M. Paturzo, P. Ferraro, Creating an extended focus image of a tilted object in Fourier digital holography. Opt. Exp. **17**, 20546–20552 (2009)
44. T. Kim, T-C. Poon, G. Indebetouw, Depth detection and image recovery in remote sensing by optical scanning holography. Opt. Eng. **41**, 1331–1338 (2002)
45. S.S. Kou, C.J.R. Sheppard, Imaging in digital holographicmicroscopy. Opt. Exp. **15**, 13640–13648 (2007)
46. Y. Sung, W. Choi, C. Fang-Yen, K. Badizadegan, R.R. Dasari, M.S. Feld, Optical diffraction tomography for high resolution live cell imaging. Opt. Express **17**, 266–277 (2009)
47. S.-H. Hong, J.-S. Jang, B. Javidi, Three-dimensional volumetric object reconstruction using computational integral imaging. Opt. Exp. **12**, 483–491 (2004)
48. F. Dubois, M.-L.N. Requena, C. Minetti, O. Monnom, E. Istasse, Partial spatial coherence effects in digital holographic microscopy with a laser source. Appl. Opt. **43**, 1131–1139 (2004)
49. F. Dubois, N. Callens, C. Yourassowsky, M. Hoyos, P. Kurowski, O. Monnom, Digital holographic microscopy with reduced spatial coherence for three-dimensional particle flow analysis. Appl. Opt. **45**, 864–871 (2006)
50. D.S. Mehta, K. Saxena, S.K. Dubey C. Shakher, Coherence characteristics of light-emitting diodes. J. Lumin. **130**, 96–102 (2010)
51. B. Kemper1, S. Kosmeier, P. Langehanenberg, S. Przibilla, C. Remmersmann, S. Stürwald, G. von Bally, Application of 3D tracking, LED illumination and multi-wavelength techniques for quantitative cell analysis in digital holographic microscopy. Proc. SPIE **7184**, 1–12 (2009)
52. M. Takeda, H. Ina, S. Kobayashi, Fourier-transform method of fringe-pattern analysis for computer-based topography and interferometry. J. Opt. Soc. Am. **72**, 156–160 (1982)
53. T. Kreis, Digital holographic interference-phase measurement using the Fourier-transform method. J. Opt. Soc. Am. A **3**, 847–855 (1986)
54. J. Shamir, *Optical Systems and Processes* (SPIE Press, Bellingham, WA, 1999)
55. P. Chavel, S. Lowenthal, Noise and coherence in optical image processing. II. Noise fluctuations. J. Opt. Soc. Am. **68**, 721–732 (1978)
56. F. Dubois, C. Yourassowsky, Method and device for obtaining a sample with three-dimensional microscopy. US 7,009,700 B2, Mar.7 (2006)
57. J.C. Giddings, A system based on split-flow lateral-transport thin (SPLITT) for rapid and continuous particle fractionation. Sep. Sci. Technol. **20**, 749–768 (1985)
58. P.M. Vlahovska, T. Podgorski, C. Misbah, Vesicles and red blood cells in flow: From individual dynamics to rheology. C. R. Phys. **10**, 775–789 (2009)
59. N. Callens, C. Minetti, G. Coupier, M.-A. Mader, F. Dubois, C. Misbah, T. Podgorski, Hydrodynamic lift of vesicles under shear flow in microgravity. Europhys. Lett. **83**, 24002 (2008)
60. V. Vitkova, M. Mader, B. Polack, C. Misbah, T. Podgorski, Micro-macro link in rheology of erythrocyte and vesicle suspensions. Biophys. J. **95**(7), 33–35 (2008)
61. G. Danker, T. Biben, T. Podgorski, C. Verdier, C. Misbah, Dynamics and rheology of a dilute suspension of vesicles: higher order theory. Phys. Rev. E **76**, 041905 (2007)

62. M. Mader, V. Vitkova, M. Abkarian, T. Podgorski, A. Viallat, "Dynamics of viscous vesicles in shear flow. Eur. Phys. J. E **19**, 389–397 (2006)
63. C. Minetti, N. Callens, G. Coupier, T. Podgorski, F. Dubois, Fast measurements of concentration profiles inside deformable objects in microlows with reduced spatial coherence digital holography. Appl. Opt. **47**, 5305–5314 (2008)

Chapter 3
Quantitative Phase Contrast in Holographic Microscopy Through the Numerical Manipulation of the Retrieved Wavefronts

Lisa Miccio, Simonetta Grilli, Melania Paturzo, Andrea Finizio, Giuseppe Di Caprio, Giuseppe Coppola, Pietro Ferraro, Roberto Puglisi, Donatella Balduzzi, and Andrea Galli

Abstract In this chapter we show how digital holography (DH) can be efficiently used as a free and noninvasive investigation tool capable of performing quantitative and qualitative mapping of biological samples. A detailed description of the recent methods based on the possibility offered by DH to numerically manage the reconstructed wavefronts is reported. Depending on the sample and on the investigation to be performed it is possible, by a single acquired image, to recover information on the optical path length changes and, choosing the more suitable numerical method, to obtain quantitative phase map distributions. The progress achieved in the reconstruction methods will certainly find useful applications in the field of the biological analysis.

Deep understanding of morphology, behaviour and growth of cells and microorganisms is a key issue in biology and biomedical research fields. Low-amplitude contrast presented by biological samples limits the information that can be retrieved performing optical bright-field microscope measurements. Due to low absorption properties of these specimens light intensity suffers only a little change after passing through them that is not sufficient to distinguish cellular and sub-cellular morphologies. The main effect on light propagating in such objects is in phase, changed with respect to the phase of the light wave propagating in the surrounding medium. Phase contrast (PC) imaging, since its invention [1], has been a powerful optical tool for visualizing transparent and tiny objects. In PC microscopy, interferometric processes are able to transform small phase variation in amplitude modulation so that differences in the beam optical path can be visualized. Many ways exist to obtain PC and, nowadays, optical phase microscopes are commercially available.

L. Miccio (✉)
Istituto Nazionale di Ottica del CNR (INO-CNR), Via Campi Flegrei 34, 80078 Pozzuoli (NA), Italy
e-mail: lisa.miccio@ino.it

P. Ferraro et al. (eds.), *Coherent Light Microscopy*, Springer Series in Surface Sciences 46, DOI 10.1007/978-3-642-15813-1_3,

Differential interference contrast (DIC) microscope is the most popular one because of the high transverse resolution, good depth discrimination and pseudo-3D imaging. Actually, standard PC techniques have the drawback to be qualitative imaging systems because of the nonlinear relationship between the recorded image and the optical path length (OPL).

In recent years digital holography (DH) has been considered as a promising and alternative concept in microscopy thanks to its capability to perform intensity imaging and quantitative phase mapping [2].

Holography is a relatively recent interferometric technique. Dannis Gabor's paper [3] from 1948 is historically accounted as its beginning. From then a large number of improvements have been investigated and still now holography is an upcoming technique [4–6]. The basic principle of holography is the recording of an interference pattern on a photographic plate (classical holography) or by a charge-coupled device (digital holography) and then, from the reconstructed wavefront, it is possible to manage the amplitude as well as the phase of the optical wavefield.

In this chapter the flexibility of DH will be employed to carry out both quantitative phase microscopy (QPM) and qualitative PC imaging. In particular, DIC performance on biological sample will be demonstrated. DH allows to choose, depending on the sample, the phase analysis to be performed and so the aim is to present it as a good candidate for complete specimen analysis in the framework of no invasive microscopy.

3.1 Digital Recording and Numerical Reconstruction in DH

Each optical field consists of amplitude and phase distributions but all detectors or recording materials only register intensity. If two waves of the same frequency interfere, the resulting intensity distribution is temporally stable and depends on the phase difference $\Delta\Psi$. This is used in holography where the phase information is coded by interference into a recordable intensity. To get a temporally stable intensity distribution, $\Delta\Psi$ must be stationary (at least during the recording procedure) which means that wavefields must be mutually coherent. The two interfering waves are commonly named object and reference beams and the light source is a laser whose wavelength depends on the final experimental goal and samples under investigation. Experiments presented in this chapter are performed in 'off-axis' arrangement [7, 8] with the object beam separated from the reference one by an angle θ as shown in Fig. 3.1a. The interference pattern (hologram) between the reference wave and the light reflected or transmitted by the object is recorded. Then the object is removed and the hologram illuminated with the original reference beam (Fig. 3.1b). The complex amplitude of the transmitted wave consists of three terms: the directly transmitted beam (zero diffraction order), a virtual image of the object in its original position and the conjugate image which, in this case, is a real image. In DH the recording material is a charge-coupled device (CCD) camera [9–11], i.e. a 2D rectangular raster of $N \times M$ pixels, with pixel pitches $\Delta\xi$ and $\Delta\eta$ in the two

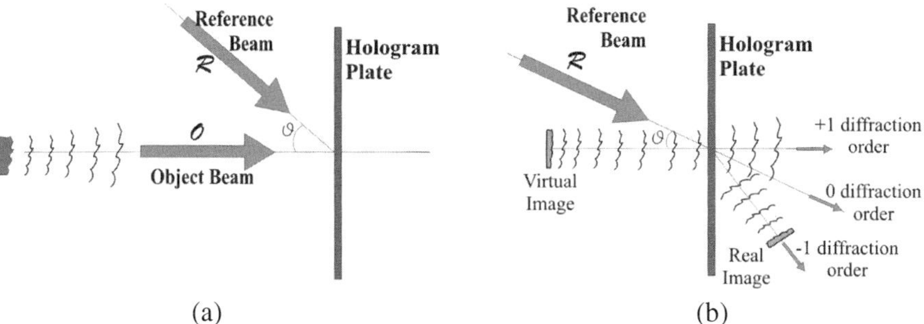

Fig. 3.1 Schematic view of the off-axis configuration for hologram recording (**a**) and image reconstruction (**b**)

directions. Therefore, the wavefront is digitized and stored as an array of zeros and ones in a computer. The stored wavefront can be mathematically described using the following relation:

$$h\left(k\Delta\xi, l\Delta\eta\right) = h\left(\xi, \eta\right)\ \mathrm{rect}\left(\frac{\xi}{N\Delta\xi}, \frac{\eta}{M\Delta\eta}\right)\sum_{k=1}^{N}\sum_{l=1}^{M}\delta\left(\xi - k\Delta\xi, \eta - l\Delta\eta\right) \tag{3.1}$$

where $\delta(\xi, \eta)$ is the 2D Dirac delta function, k and l are integer numbers, $(N \cdot \Delta\xi) \times (M \cdot \Delta\eta)$ is the area of the digitized hologram, rect(ξ, η) is a function defined as a constant amplitude value, if the point of coordinated (ξ, η) is inner to the digitized hologram and is zero elsewhere.

Figure 3.2 shows the relative positions of object, hologram and image planes; z is the optical axis.

The reconstruction process is achieved through numerical back propagation based on the theory of scalar diffraction. During reconstruction, the digitalized hologram, $h(n, m)$, is an amplitude transmittance or aperture that diffracts the

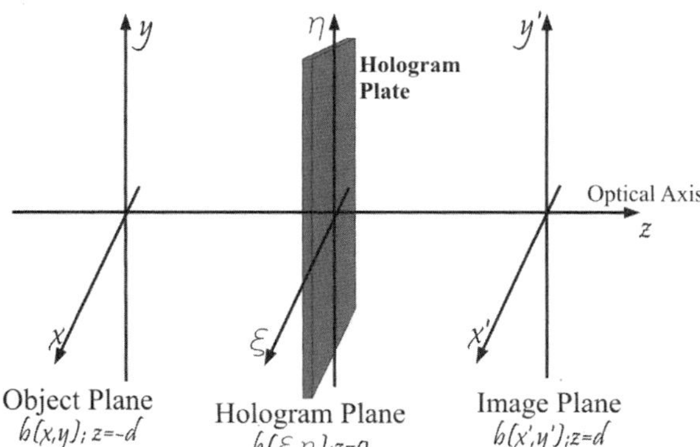

Fig. 3.2 Geometry for digital recording and numerical reconstruction

digitalized reference wave, $r(n, m)$; the propagation of the wavefield $u(n, m) = h(n, m) \cdot r(n, m)$ from the hologram plane to a plane of observation is numerically calculated. Focused image of the original object should be situated at $z = d$ and $z = -d$ provided the reference is a plane wave. A detailed derivation of the mathematical formulae of diffraction theory is given by Goodman [12]. Thus, a digitalized version of the reconstructed wavefront $b'(x', y')$ in the plane of observation can be numerically calculated by a discretization of the Rayleigh–Sommerfeld's diffraction integral in the Fresnel approximation [13, 14], i.e.

$$b'(n,m) = \frac{e^{i\frac{2\pi d'}{\lambda}}}{i\lambda d'} e^{\frac{i\pi\lambda d'}{NM}\left(\frac{n^2}{\Delta\xi^2}+\frac{m^2}{\Delta\eta^2}\right)} \mathrm{DFT}\left\{h(k,l)\, r(k,l)\, e^{i\frac{\pi}{\lambda d'}\left(k^2\Delta\xi^2+l^2\Delta\eta^2\right)}\right\} \quad (3.2)$$

where DFT denotes the discrete Fourier transform, which can be easily calculated by a fast Fourier transform (FFT) algorithm.

Equation (3.2) represents a complex wavefield with intensity and phase distributions I and $\Delta\Psi$ given by

$$I(n,m) = b'(n,m)\, b'*(n,m) \quad (3.3a)$$

$$\Delta\Psi = \arctan\frac{I\left\{b'(n,m)\right\}}{R\left\{b'(n,m)\right\}} \quad (3.3b)$$

$I\left\{b'\right\}$ and $R\left\{b'\right\}$ are imaginary and real parts of b', respectively. The reconstructed image is a matrix with elements (n, m) and steps

$$\Delta x' = \frac{d'\lambda}{N\Delta\xi} \text{ and } \Delta y' = \frac{d'\lambda}{M\Delta\eta} \quad (3.4)$$

$\Delta x'$ and $\Delta y'$ are the reconstruction pixel (RP) sizes and represent the technique resolution. They depend, besides laser wavelength and reconstruction distance, on the physical dimensions of the CCD detector ($N\Delta\xi \times M\Delta\eta$).

Optical setups to record holograms in the experiments described in the following are Mach–Zehnder interferometers working at one or more wavelengths. In Fig. 3.3 schematic drawing of DH setups is reported for two wavelengths.

Laser beam of each wavelength is separated by a beam splitter (BS) in reference and object waves and sample are positioned along the object beam path in transmission geometry in order to allow the light passing through it. In microscope configuration light coming from the specimen is collected by a microscope objective (MO) and made to interfere with the reference beam. The interference pattern (hologram) is recorded by the CCD positioned at distance d from the image plane. Then the hologram is back propagated to calculate intensity and phase of the object beam in the image plane (3.3). In the presence of the MO the RP size depends on its magnification, G, according to the following expressions:

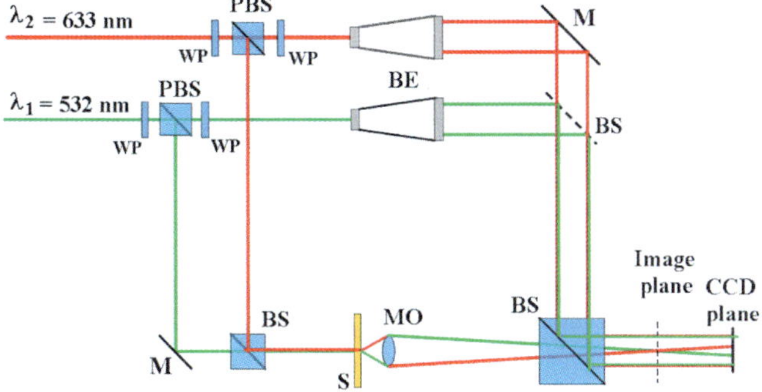

Fig. 3.3 DH off-axis setup with two wavelengths [38]

$$\Delta x' = \frac{\lambda d'}{GN\Delta\xi} \text{ and } \Delta y' = \frac{\lambda d'}{GM\Delta\eta} \tag{3.5}$$

3.2 Phase Contrast Imaging and Quantitative Phase Microscopy by DH

In this section, numerical methods to retrieve information from the calculated phase map will be presented. Depending on the analysis to be carried out, it is possible to choose the numerical code to perform qualitative PCI or QPM.

Metrological measurements adopting an off-axis holographic recording setup were first demonstrated by Cuche et al. [15]. The possibility of managing phase in DH is attractive because aberrations can be removed in a posteriori image analysis. In fact, it has been proved that the wavefront curvature introduced by microscope objective and lenses can be successfully removed and/or compensated [16, 17] as well as spherical aberration [18], astigmatism [19, 20] and anamorphism [21] or even longitudinal image shifting introduced by a cube beam splitter [22]. Complete compensation of aberrations is essential when quantitative phase determination is used in application such as micro-electro-mechanical systems (MEMS) inspection [23–25], and, nevertheless, it is fundamental when phase contrast imaging found applications in the field of biomedicine and life sciences. Holographic metrology is a non-destructive, label-free, full-field and online technique suitable in biology to detect displacement and movements. Information about the integral cellular refractive index, the cell depth and, for adherent grown cells, the cell profile can be determined from DH phase contrast images of in vitro cells [26, 27].

Different techniques for removing unwanted phase factors are described in the following sections. Experimental results on biological sample are presented and comparison among the diverse techniques is discussed.

3.2.1 Double Exposure Technique

The first method consists of the recording of a couple of interference patterns: object and reference holograms [28]. The first exposure is made of the object under investigation, whereas the second one is made of a flat reference surface in proximity of the object. This second hologram contains information about all the aberrations introduced by the optical components including the defocusing due to the microscope objective. By manipulating the two holograms numerically, it is possible to compensate for these aberrations. The phase distribution in the image plane is given by

$$\Psi(n, m) = \arg\left[\frac{b'_{\mathrm{O}}(n, m)}{b'_{\mathrm{R}}(n, m)}\right] \tag{3.6}$$

where $b'_{\mathrm{O}}(n, m)$ and $b'_{\mathrm{R}}(n, m)$ are the complex wavefields calculated at distance d' from the object hologram and the reference hologram, respectively. The requirement for a flat surface is well satisfied when MEMS structures are inspected, because the micro-machined parts are typically fabricated on a flat silicon substrate. In that case the area around the structure offers a good plane surface that can be used as reference. This technique allows to compensate all the undesired phase factors introduced by the optical elements in the setup; so double exposure appears the best way to recover quantitative phase information on sample object. Anyway it has some drawbacks such as it is not applicable in case of samples lacking of flat surrounding surface, i.e. biological sample where high cell density does not permit the selection of a plane area. Moreover, double acquisition and double computation increase the calculation time. Methods discussed further are developed to overcome these limitations.

3.2.2 Self-Referencing Method

In this section, taking advantage of the possibility offered by DH to manage numerically the reconstructed wavefronts, a numerical approach to get quantitative phase maps is presented. The method is based on the creation of a numerical replica of the original complex wavefield. The replica is digitally shifted and its phase is subtracted to the original wavefront in order to obtain the QP map. This approach, that requires the acquisition of a single hologram, is particularly suitable for modern microfluidic bio-chip. In fact, in these structures, microfluidic channels are surrounded by a very wide (compared to the channel area) flat area. This portion can be used as a reference phase map since in this region the sample is flat, without the object and then with a constant phase. The wavefront folding allows to superimpose the portion corresponding to the flat area to that one corresponding to the micro-channel and thus to the biological specimen.

The reconstructed phase map distribution in the image plane can be modelled by the following equation:

$$\Delta\Psi(n,m) = \Psi_0(n,m) + \frac{i\pi}{\lambda R}\left[(n-N_0)^2 + (m-M_0)^2\right]$$
$$\text{for}\begin{Bmatrix} n = N_1,\ldots,N_N \\ m = M_1,\ldots,M_M \end{Bmatrix} \qquad (3.7)$$

where $\Psi_0(n,m)$ is the digitalized phase distribution of the object under observation, N and M are the dimensions of the pixel matrix and the parabolic or defocus term that takes into account the curvature introduced by the microscope objective. The parabolic term is centred on the pixel with coordinates (N_0, M_0).

A region of this phase map, where the contribution of the observed object is null, is selected. This portion can be used as reference phase map since it corresponds to a region in which the sample is flat, without the object and, therefore, with a constant phase.

In order to generate a reference phase map, a numerical replica of this wavefront is laterally shifted. In particular, the obtained self-reference map can be described by

$$\Psi_{\text{ref}}(n,m) = \frac{i\pi}{\lambda R}\cdot\left[(n-N_0+\Delta N)^2 + (m-M_0+\Delta M)^2\right]$$
$$\text{for}\begin{Bmatrix} n = N_1,\ldots,N_N \\ m = M_1,\ldots,M_M \end{Bmatrix} \qquad (3.8)$$

where ΔN (ΔM) is the amount of shift along the x (y) direction. The integer numbers N_N–N_1 and M_M–M_1 allow to define the width of the region of interest relative to the object. The difference between Ψ and Ψ_{ref} is given by the sum of the phase map Ψ_0 with a plane that can be removed by a linear 2D fitting; in fact

$$\Psi(n,m) - \Psi_{\text{ref}}(n,m) =$$
$$\Psi_0(n,m) - \frac{i2\pi}{\lambda R}\cdot\left[\Delta N\cdot\left(n-\left(N_0-\frac{\Delta N}{2}\right)\right) + \Delta M\cdot\left(m-\left(M_0-\frac{\Delta M}{2}\right)\right)\right]$$
$$\text{for}\begin{Bmatrix} n = N_1,\ldots,N_N \\ m = M_1,\ldots,M_M \end{Bmatrix} \qquad (3.9)$$

The method has been applied to retrieve the quantitative phase map of a bovine spermatozoa. The hologram has been acquired by the setup illustrated in Fig. 3.3 employing a single wavelength ($\lambda = 532$ nm) laser source. Moreover, a 50× microscope objective (NA 0.65) has been used to improve the lateral resolution. Bovine sperm cells, prepared by the institute 'Lazzaro Spallanzani,' were injected into a 20 μm wide and 10 μm height polydimethylsiloxane (PDMS)-based micro-channel.

The in-focus intensity map of the object obtained by the reconstruction of the original wavefront is reported in Fig. 3.4a

The reconstructed wavefront has been shifted on the right of about 100 pixels, the corresponding intensity map is reported in Fig. 3.4b. The phases of the two wavefronts are subtracted from each other to obtain the phase map shown in Fig. 3.4c. The regions I and III correspond to the phase map of the quadratic term introduced

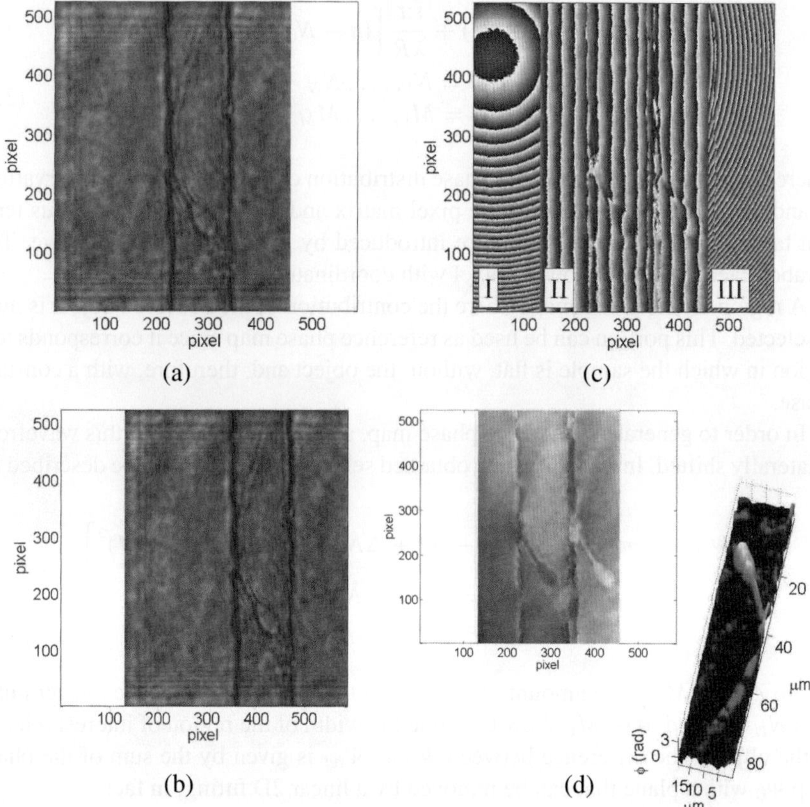

Fig. 3.4 The in-focus intensity map of the reconstructed object as it is (**a**) and shifted on the right of about 100 pixels (**b**); the phase map distribution obtained subtracting the original wavefront and its shifted replica (**c**); the 2D and 3D unwrapped phase map after the elimination of the superimpose plane (**d**) [39]

in (3.7), while region II is a visualization of (3.9). In fact, it is possible to note the wrapped phase distribution both of the spermatozoa and of the superimposed plane. Removing this plane by a fitting procedure, the unwrapped phase map of the object is achieved (see Fig. 3.4d).

A further example to show the flexibility of the proposed method is the reconstruction of the quantitative phase map of an in vitro mouse preadipocyte 3T3-F442A cell line. The aim of this measurement is to monitor the characteristic cell rounding and the lipid droplet accumulation during the cell differentiation. Figure 3.5a shows the in-focus image of the reconstructed intensity map divided into four regions. By means of successively pixel-shifting operations, the wavefront can be folded in both directions as illustrated in Fig. 3.5b. The difference between the phase maps corresponding to the original wavefront, in Fig. 3.5a, and the folded one, in Fig. 3.5b, is shown in Fig. 3.5c. The core of the cell under investigation is present

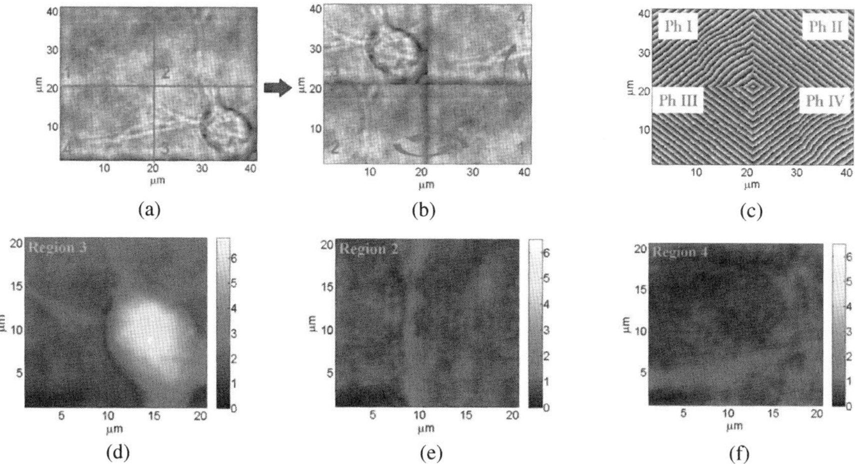

Fig. 3.5 A mouse preadipocyte 3T3-F442A cell line. The in-focus image divided into four regions (**a**) and its replica opportunely shifted (**b**). Difference between the wrapped phase maps relative to (**a**) and (**b**) concerning region 3 (**d**), region 2 (**e**) and region 4 (**f**) [39]

in the zones Ph-I and Ph-IV of Fig. 3.5c. According to (3.9), the superimposed plane is then removed, thus obtaining the quantitative phase map shown in Fig. 3.5d. The same approach has been applied to retrieve the sample phase distribution corresponding to regions 2 and 4 where some filaments of the cell are clearly visible in the amplitude reconstruction (see Fig. 3.5a).

Actually, the wavefront has been folded two more times with the aim to subtract the phase relative to region 1 from the phase relative to region 2 and from the phase relative to region 4, respectively. These operations allow to retrieve the quantitative phase distributions displayed in Fig. 3.5e, f.

3.2.3 2D Fitting Method

In this section a method to estimate quantitatively third-order aberrations and to remove them is presented [29]. It is based on a single hologram acquisition and QPM of thin objects is performed that is appropriate for investigation in biology and biomedical applications. The reconstructed diffracted field $b'(n, m)$ in the reconstruction plane (n, m) at distance d from the hologram is given by (3.2) while the intensity $I(n, m)$ and the phase distribution $\Delta\Psi(n, m)$ of the reconstructed image are determined from (3.3a) and (3.3b). $\Delta\Psi(n, m)$ is reconstructed in the presence of defocus term and higher order aberration contributions that are going to be considered. Indeed, $\Delta\Psi(n, m)$ can be written in the form

$$\Psi\,(n, m) = \Psi_{\mathrm{O}}\,(n, m) + \Delta\Psi_a\,(n, m) \tag{3.10}$$

where $\Psi_{\mathrm{O}}(n, m)$ is the object phase distribution to be recovered and

$$\Psi_a(n,m) = \Psi_D(n,m) + W(n,m) = \mathrm{e}^{\left[\frac{i\pi}{\lambda R}(n^2+m^2)\right]} + W(n,m) \tag{3.11}$$

$\Psi_D(n,m)$ is the quadratic contribution of the defocus aberration and $W(n,m)$ includes higher order aberration terms. If a very thin object is considered, localized in a small area of the reconstructed plane, it could be assumed that the object phase $\Psi_O(n,m)$ is a small perturbation of the aberration contribution to the overall reconstructed phase map, that is, $\Psi_0(n,m) << \Psi_a(n,m)$. Therefore a nonlinear fitting procedure of $\Delta\Psi(n,m)$ allows to obtain a good approximation, $\Delta\Psi_{\mathrm{appr},a}(n,m)$, of the aberration contribution. The object phase displacement can be recovered by subtracting out $\Delta\Psi_{\mathrm{appr},a}(n,m)$ from the reconstructed phase map, i.e. $\Psi_O(n,m) = \Delta\Psi(n,m) - \Delta\Psi_{\mathrm{appro},a}(n,m)$. This approach allows to remove aberrations to obtain the correct QPM phase map by a single hologram.

The setup adopted (Fig. 3.3) uses a single wavelength ($\lambda = 532$ nm) laser source and the microscope objective is a 20× objective having a focal length $f = 9.0$ mm and NA = 0.40. The CCD detector is made of 1280 × 1024 square pixel of 6.7 μm size. The specimen is the mouse preadipocyte 3T3-F442A cell line described in the previous section. Reconstruction of the phase map was performed at the distance $d = 100$ mm using (3.2) and (3.3) where the digital reference wave is set as $r(k,l) = 1$. The QPM wrapped and unwrapped phase maps are shown in Fig. 3.6a, b, respectively. As can be seen in the figures, the reconstructed phase is mainly affected by the contribution of the MO, i.e. the defocus term. In addition two tilts are also present, due to the different phases of the digital reference wave $r(k,l)$ with respect to the real reference wave. It is clearly visible in Fig. 3.6b that the wavefront curvature due to the defocus of MO covers the phase retardation of the specimen since it is very thin compared with the phase retardation introduced by the imaging optics.

A 2D nonlinear fitting procedure of the phase map, $\Delta\Psi(n,m)$, is based on the Levenberg–Marquardt method; the surface employed to fit the phase map is a linear combination of the Zernike polynomials [30]. This procedure provides the linear

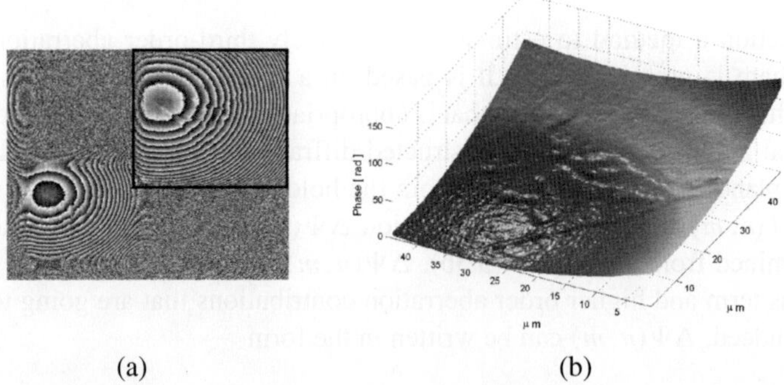

Fig. 3.6 QPM wrapped phase and unwrapped phase after Fresnel reconstruction. (**a**) Wrapped phase and (**b**) unwrapped phase map of the rectangle pointed out in (**a**). In (**b**) the cell shape, embedded in the wavefront curvature, is clearly visible [29]

Table 3.1 Zernike polynomials employed to fit the wavefront curvature and computed coefficients

Zernike polynomials		$a_i^{(O)}$
x	x tilt	0.3700
y	y tilt	−0.2143
$2xy$	Astigmatism at ±45°	−1.7994E-006
$-1+2y^2+2x^2$	Focus	5.6770E-004
y^2-x^2	Astigmatism at 0°	−9.6360E-006
$1-6y^2-6x^2+6y^4+12x^2y^2+6x^4$	Spherical Aberration	8.9254E-009

combination coefficients $a_i^{(O)}$ to calculate

$$\Delta\Psi_{\mathrm{appr},a}(n,m)=\sum a_i^{(O)}Z_i(n,m) \tag{3.12}$$

Table 3.1 shows the Zernike polynomials used in the fit and the computed coefficients. The fitted surface is then subtracted to the phase map to obtain the object phase retardation $\Psi_O(n,m)$.

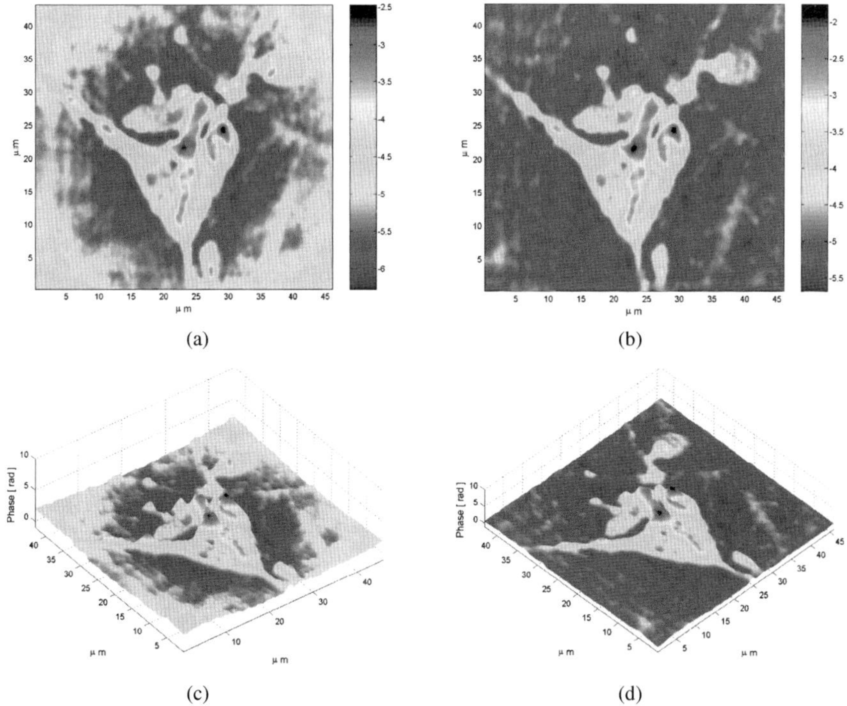

Fig. 3.7 QPM phase map of the tiny cell after the subtraction of the fitted surface which is a linear combination of the Zernike polynomials. (**a**) Only the defocus term and tilts have been considered and (**b**) third-order spherical aberration and astigmatism also have been taken into account. (**c**) and (**d**) Pseudo-3D plot of (**a**) and (**b**), respectively [29]

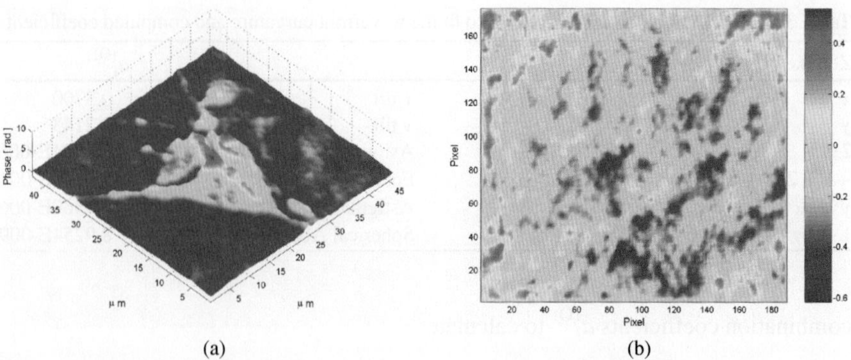

Fig. 3.8 (**a**) Reconstructed phase map of the cell by means of the double exposure technique and (**b**) subtraction between the maps obtained with the two different methods [29]

Figure 3.7 shows the results after subtraction. The surface is fitted by including only the focus term and tilt term in the Zernike polynomials, but it is clear from Fig. 3.7a, c that a residual aberration is still present. Then to clean the phase map of all optical aberrations even the astigmatism and third-order spherical aberration terms are added to fit the phase surface and the final result is displayed in Fig. 3.7b, d. In these conditions it has been assumed that the object is thin, since the phase contribution due to the cell is about 4 rad, while the whole wavefront is about 150 rad, i.e. the cell phase retardation is of the order of 3% of the phase modulation $\Delta\Psi(n, m)$ of the evaluated wavefront. Finally, the correctness of the 2D fitting procedure is evaluated by comparing this method with the double exposure one which makes use of the reference and object hologram. Figure 3.8a is the phase map obtained with double exposure method. The difference between the phase map computed with two different techniques is a plane of about 1 rad (Fig. 3.8b) inside the background noise of the solution in which the cell is living.

The described procedure applied to remove the optical aberration in QPM for thin objects is very simple but very effective for many reasons. It does not require either double recording holograms or iterative procedures to find expansion coefficients and, moreover, it avoids the problem to select a flat area of the hologram to perform the fit of the surface containing the lens aberrations. However, the described technique is particularly useful for inspection and investigation of thin objects. The methods reported in the following sections allow to overcome this limitation.

3.2.4 Numerical Lateral Shearing Method

Numerical lateral shearing (LS) method applied to DH allows to remove the parabolic phase factor introduced by the microscope objective present in the recording setup. These parabolic or defocus terms mainly affect the reconstructed phase distribution of a sample object in respect to the other optical aberrations. Combining LS method with DH it is possible to perform QPM by using a single recorded hologram [31, 32]. The reconstructed wavefront and its replica, obtained digitally by numerical shifting in the image plane, can be subtracted from each other to produce an interferometric shereogram from which the phase map of the object can be retrieved. The process is analogous to lateral shearing interferometry (LSI) usually adopted for retrieving wavefront aberrations in optical testing [33]. The procedure can be applied to transparent objects as well as opaque samples. The reconstructed phase map in the image plane is expressed by (3.7), that is, as the summation of the phase contribution due to the sample object and a quadratic or defocus term introduced by the microscope objective used to image the sample, i.e.

$$\Delta\Psi\left(n,m\right)=\Psi_{\mathrm{O}}\left(n,m\right)+\frac{i\pi}{\lambda R}\left(n^{2}+m^{2}\right) \tag{3.13}$$

To evaluate $\Psi_{\mathrm{O}}\left(n,m\right)$ a numerical lateral shearing in the image plane is introduced for both x and y directions, namely

$$\begin{aligned}\Delta\Psi_{x}&=\Psi\left(n,m\right)-\Psi\left(n-s_{x},m\right)\\ \Delta\Psi_{y}&=\Psi\left(n,m\right)-\Psi\left(n,m-s_{y}\right)\end{aligned} \tag{3.14}$$

The two shearogram maps, $\Delta\Psi_{x}$ and $\Delta\Psi_{y}$, are related to the first-order derivative of the wavefront if the amount of the shear, s_{x} and s_{y} is small. Taking into account (3.13) such derivatives can be written as

$$\frac{\delta\Psi\left(x',y'\right)}{\delta x}\approx\frac{\Delta\Psi_{x}}{s_{x}}=\frac{\Psi_{0}\left(n,m\right)-\Psi_{0}\left(n-s_{x},y'\right)}{s_{x}}+\frac{i2\pi s_{x}}{\lambda R}n \tag{3.15a}$$

$$\frac{\delta\Psi\left(x',y'\right)}{\delta y}\approx\frac{\Delta\Psi_{y}}{s_{y}}=\frac{\Psi_{0}\left(n,m\right)-\Psi_{0}\left(n,m-s_{y}\right)}{s_{y}}+\frac{i2\pi s_{y}}{\lambda R}m \tag{3.15b}$$

Subtraction of the linear terms in the previous equations, representing the defocus aberration, gives the digital shearograms:

$$\begin{aligned}\Delta\Psi_{0,x}&=\Psi_{0}\left(n,m\right)-\Psi_{0}\left(n-s_{x},m\right)\\ \Delta\Psi_{0,y}&=\Psi_{0}\left(n,m\right)-\Psi_{0}\left(n,m-s_{y}\right)\end{aligned} \tag{3.16}$$

From the knowledge of the differences $\Delta\Psi_{0,x}$ and $\Delta\Psi_{0,y}$ along the x and y directions, object phase distribution $\Psi_{0}(n+\Delta n,m+\Delta m)$ at the mesh point $(n+\Delta m,n+\Delta m)$ can be determined by standard numerical integration procedures of its finite-difference approximation:

$$\Psi_{0}(n+\Delta n,m+\Delta m)\approx\Psi_{0}(n,m)+\Delta\Psi_{0}(n,m)\Delta n+\Delta\Psi_{0}(n,m)\Delta m \tag{3.17}$$

LSI combined with DH has been used to monitor lipid droplet accumulation during differentiation in a preadipocyte 3T3-F442A mouse cell line. This investigation is aimed at studying a possible role of endocannabinoid signalling in the control of adipocyte differentiation and function. We expect to identify lipid droplets growth and accumulation detecting modification in the OPL of the laser beam passing thorough each cell. Traditional methods in biology are based on staining the cell culture with the dye Oil Red-O but such method is invasive and can give false responses. LSI and DH allow to calculate sample OPL without any reference digital hologram or cumbersome digital adjusting procedure that is necessary because hundreds of holograms have to be recorded during the observation stage. In fact, for the considered case, a cell line differentiates into adipocytes that were once confluent, which takes approximately 10 days, and media changes have to be made every 48 h.

Figure 3.9 illustrates the results of numerical steps to get quantitative phase contrast map for an in vitro mouse cell. Figure 3.9a, b shows the shearograms obtained from a single digital hologram at the image plane. The shearograms were obtained subtracting digitally two complex wavefields sheared of $s_x = 1$ pixel and $s_y = 1$ pixel along the x and y directions, respectively. Figure 3.9c displays phase map retrieved from the shearograms of Fig. 3.9a, b after the integration step. Finally, Fig. 3.9d shows a pseudo-3D representation of the calculated phase map.

The procedure described is helpful in biology to check the behaviour of time-dependent processes because it does not require the specimen killing and furnishes

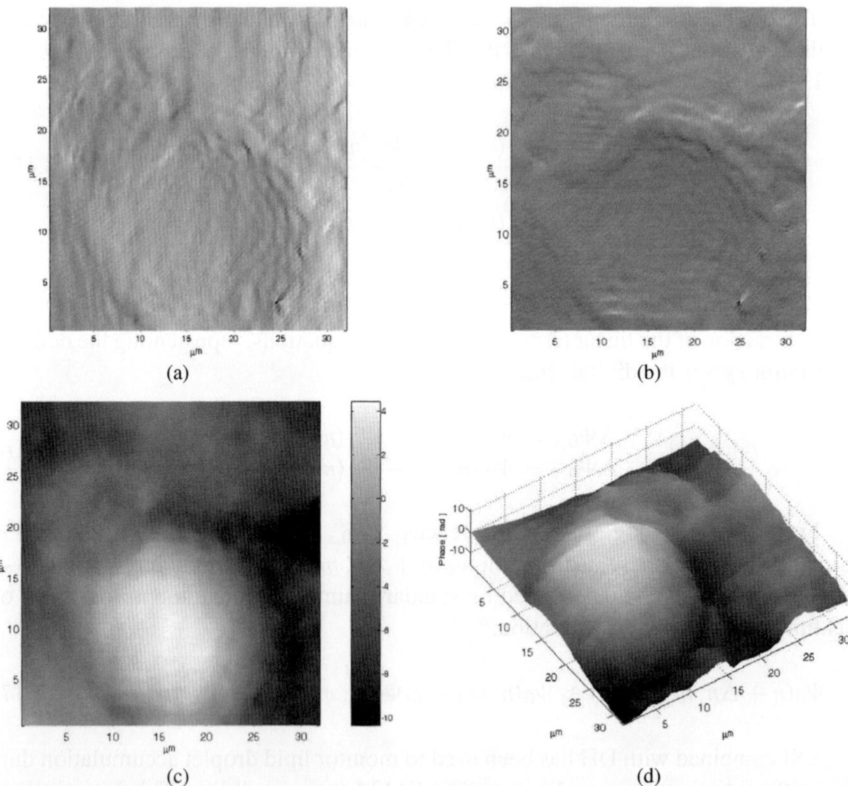

Fig. 3.9 Shearogram along x (**a**) and y (**b**) directions; (**c**) quantitative phase map of a mouse cell with lipid accumulation and its 3D plot (**d**) [31]

a quantitative analysis of its refractive index and thickness variation. However, it allows to compensate only the defocus term due to the microscope objective. If compared with 2D fitting technique LSI could be applied to a wide range of sample because the thin object assumption is not required.

3.2.5 Differential Holographic Interference Contrast

In this section the flexibility of DH will be employed to perform DIC imaging of biological sample. Indeed a novel concept of DIC in microscopy can be envisaged by using DH; we named here DHIC (differential holographic image contrast) the numerical implementation of DIC images starting from DH recording. From a single hologram, it is possible to measure, at the same time, the gradient in phase in all directions in the transverse plane. Moreover, the possibility to image a radial shear will be demonstrated.

Both the preadipocyte 3T3-F442A mouse cell line and bovine spermatozoa described in Sect. 3.2.2 are investigated in order to prove DIC feasibility by DH. In particular, the bovine sperm cells were prepared by the institute 'Lazzaro Spallanzani' after fixation in suspension of the seminal material with 0.2% glutaraldehyde solution in phosphate-buffered saline (PBS) without calcium and magnesium (1:3 v/v). A drop with volume 6 μL has been deposed on a glass slide and then covered with a cover slip (20 mm × 20 mm). The cover slip has been linked to the glass slide by means of a strip of varnish.

Setup for image acquisition is an optical coherent microscope based on DH (see Fig. 3.3) employing a single laser source ($\lambda = 632.8$ nm). After leaving the specimen plane the diffracted light is collected by the MO (20× magnification, 0.4 numerical aperture). Reconstruction of the phase map distribution was performed at distance $d = 170$ mm. The resolution in the image plane, calculated by a test target, is $\Delta x = 0.23\,\mu$m for the mouse cell sample and $\Delta x = 0.1818\,\mu$m for the bovine spermatozoa one.

The reconstruction algorithm is divided into two stages. The first one is the usual back propagation of the recorded holograms based on the diffraction integral in the Fresnel approximation to calculate the complex wavefield (3.2, 3.3). In Fig. 3.10, intensity and OPD $= \Delta\Psi\lambda/2\pi$ for a bovine spermatozoa are displayed. In this first picture the phase map is obtained using the standard procedure of the double exposure, that is, object and reference holograms curvature are subtracted to compensate

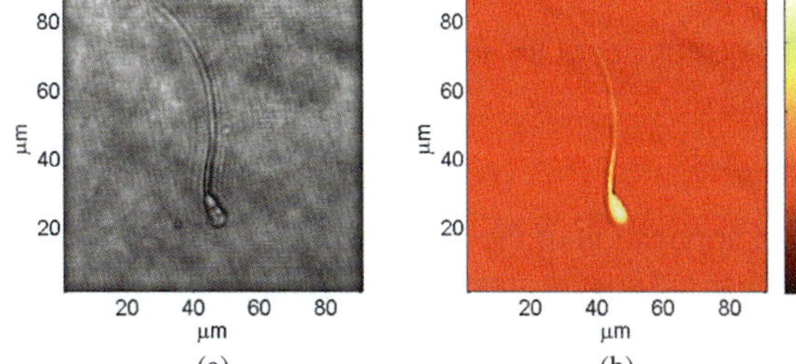

Fig. 3.10 Quantitative intensity (**a**) and phase (**b**) distribution for a bovine spermatozoa [40]

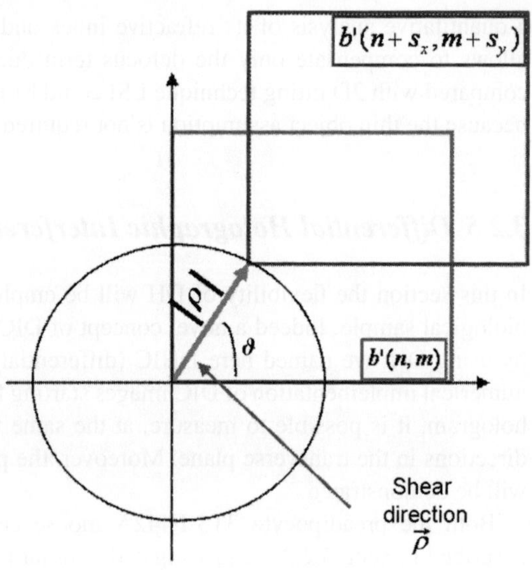

Fig. 3.11 Schematic representation of the digital shearing along a chosen direction [40]

the optical aberration in the setup (Sect. 3.2.1). Double exposure method and LSI combined with DH are well suited and commonly used processes to retrieve quantitative information for this specimen [34].

The next step proposed concerns the possibility to generate DIC images from the object hologram (i.e. from one recording). The complex wavefield $b'(n, m)$ is processed to obtain phase difference contrast images of the sample in several directions. For each direction a replica of $b'(n, m)$ is calculated digitally by numerical shearing in the image plane (see Fig. 3.11). The sheared wavefront is subtracted from the original one to obtain the difference phase image of the sample [31].

If the defocus term is considered as the main contribution to the phase retardation and the higher order aberrations are neglected the calculated phase difference is

$$\Delta\Psi = \Psi_0\,(n, m) - \Psi_0\left(n + s_x, m + s_y\right) - \frac{ik}{2R}\left(2ns_x + s_x^2 + 2ms_y + s_y^2\right) \quad (3.18)$$

The shear quantities depend on the direction, $\vec{\rho}$, as follows:

$$\begin{aligned} s_x &= \rho\cos\theta \\ s_y &= \rho\sin\theta \end{aligned} \quad (3.19)$$

The equivalent DIC images are obtained by the following:

$$\mathrm{DIC}\,(n, m) = 1 - \cos\left(\Delta\Psi\right) \quad (3.20)$$

An application of the routine is performed on the bovine sperm. The modulus of the shear is kept constant and equals 2 pixels, that is, about 364 nm. The proposed routine is very easy and effective, phase gradient in all directions is calculable just

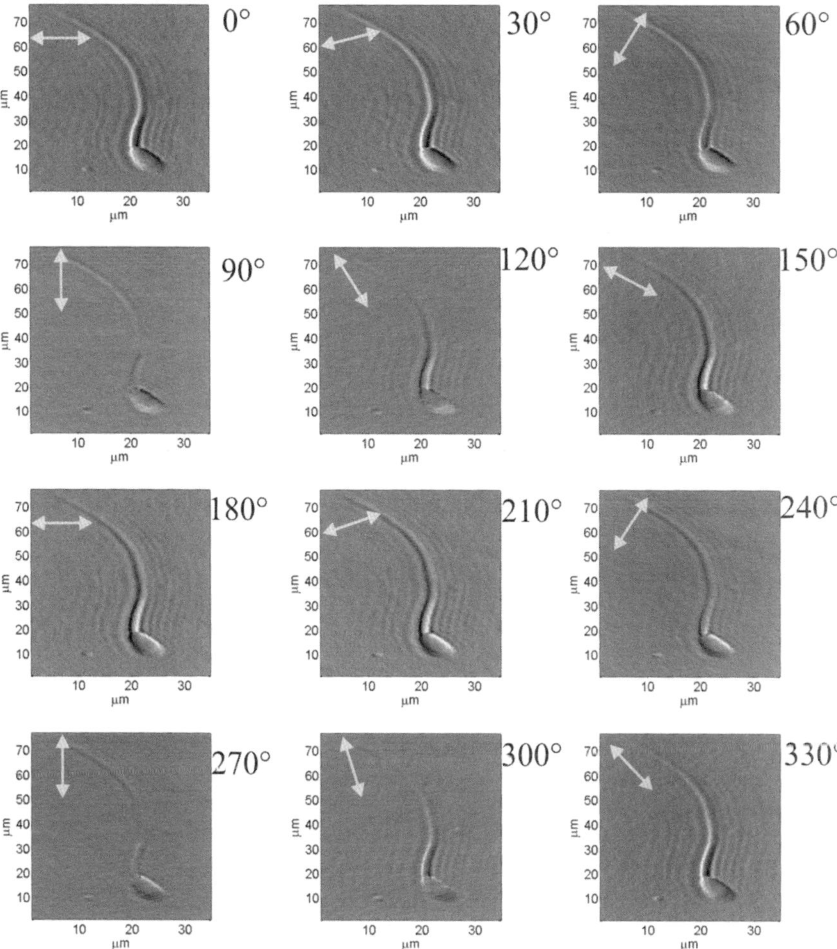

Fig. 3.12 DHIC images in different shear directions of a sperm cell [40]

modifying one parameter: the shear angle ϑ. In Fig. 3.12 the results obtained on the sample are displayed, a shear angle step of 30°C is accomplished. From the picture it is clear that, depending on ϑ, different specimen areas are put in evidence.

A more complete analysis can be carried out on samples having a radial symmetry as in the case of the in vitro mouse cell shown in the following. Besides the direction-dependent shift a radial shear is accomplished that is able to enhance the cell boundaries as a whole and not only an angular portion. The recorded hologram and the corresponding OPL, obtained by digital LSI, are displayed in Fig. 3.13.

DHIC images are retrieved as described before and displayed in Fig. 3.14. Numerical shifting of the complex wavefield is responsible for the presence of a plane whose inclination depends on the shear direction (3.18, 3.19). Converse

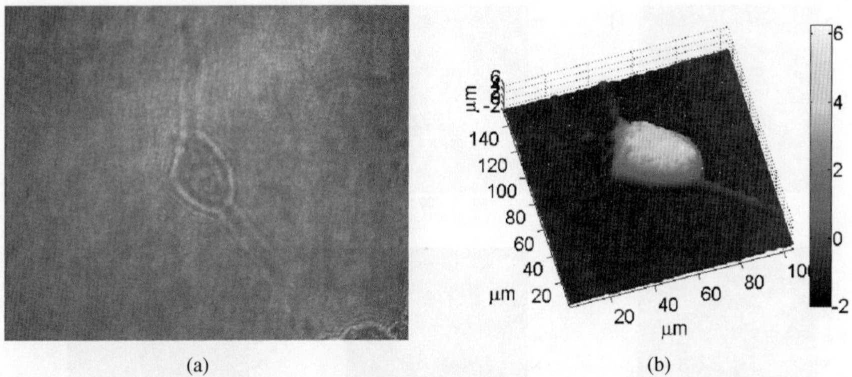

Fig. 3.13 Recorded hologram of a mouse cell (**a**) and its corresponding OPL (**b**) [40]

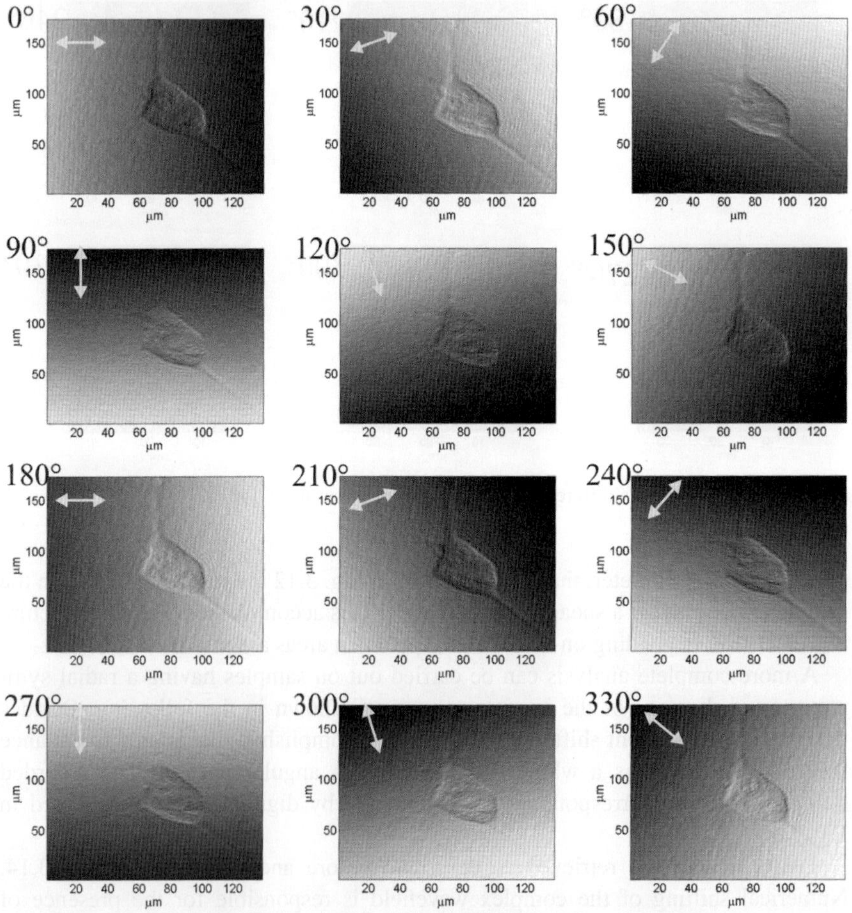

Fig. 3.14 The corresponding DHIC images in the presence of a tilted plane

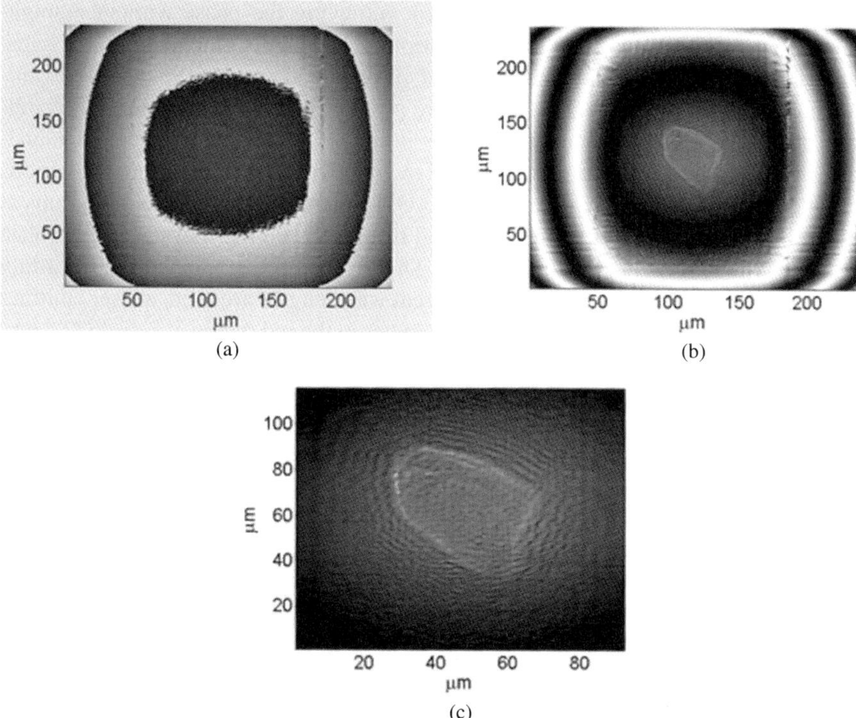

Fig. 3.15 (**a**) Wavefront subtraction of complex wavefield in two different planes; (**b**) the corresponding radial DIC image for a mouse cell; (**c**) magnification of (**b**)

to the spermatozoa images displayed in Fig. 3.12, the DHIC images affected by these planes are shown in order to better visualize the direction-dependent phase gradients.

In order to better highlight the cell boundaries as a whole a radial shear is applied on the cell. Complex wavefield is calculated in two different planes: the image plane at the reconstruction distance d and a slightly out of focus plane positioned at distance $d + \delta d$. The phase difference is calculated as subtraction of the wavefront curvature in these two planes. $\Delta\Psi$ is shown in Fig. 3.15a where the presence of fringes due to the unavoidable difference in curvature is clear. In Fig. 3.15b the corresponding DIC image is presented.

This last evaluation is purely qualitative but it could be useful in checking of cell damage or breaking otherwise not visible.

3.2.6 Synthetic Wavelength Holography

Interferometry often requires an extended range of phase measurement without 2π ambiguity in the phase map due to the long OPL in the tested sample. To overcome this limitation a multi-wavelength configuration is typically used, the aim being to generate a longer synthetic wavelength for retrieving the phase without ambiguities in an extended range [35, 36]. Multiple wavelengths employed in the same optical apparatus lead to the occurrence of chromatic aberrations [37]. DH flexibility, consisting in the numerical re-focusing process, offers the opportunity to compensate aberrations and remove the errors in the QPM without mechanical adjustment. A procedure is implemented to find the relative focal shift among the various wavelengths. Phase map with an extended OPL range using only two holograms at two different wavelengths is presented and results on in vitro mouse cell fibroblast are shown [38]. Figure 3.3 illustrates the DH off-axis setup adopted for recording MWDH holograms. Two lasers with different wavelengths are used, a laser emitting in the green region at $\lambda_1 = 532$ nm and the other in the red region at $\lambda_2 = 632.8$ nm. The optical configuration is arranged to allow the two lasers to propagate almost along the same paths either for reference or for object beams. In order to obtain a good superposition of the reconstructed phase contrast images at different wavelengths, one has to take into account the different wavefront curvatures for diverse wavelengths and the presence of chromatic aberrations due to the optical elements in the setup. Chromatic aberration involves a difference in the magnification and consequently a longitudinal shift of the image plane position on the optical axis as shown in Fig. 3.16.

Let D be the distance between the CCD and the imaging lens and d_{G} the image plane distance for the green wavelength λ_{G} then the reconstruction distance measured backward from the CCD plane is $r_{\mathrm{G}} = |D - d_{\mathrm{G}}|$. After reconstruction of the complex field at distance r_G the total phase is calculated:

$$\phi_{\mathrm{G}}(n, m) = \varphi_0\left(-\frac{n\Delta x_{\mathrm{G}}}{M_{\mathrm{G}}}, -\frac{m\Delta x_{\mathrm{G}}}{M_{\mathrm{G}}}\right) + \frac{\pi}{\lambda_{\mathrm{G}}}\left(1 + \frac{1}{M_{\mathrm{G}}}\right)\frac{n^2 + m^2}{|D - d_{\mathrm{G}}|}\Delta x_{\mathrm{G}}^2 \quad (3.21)$$

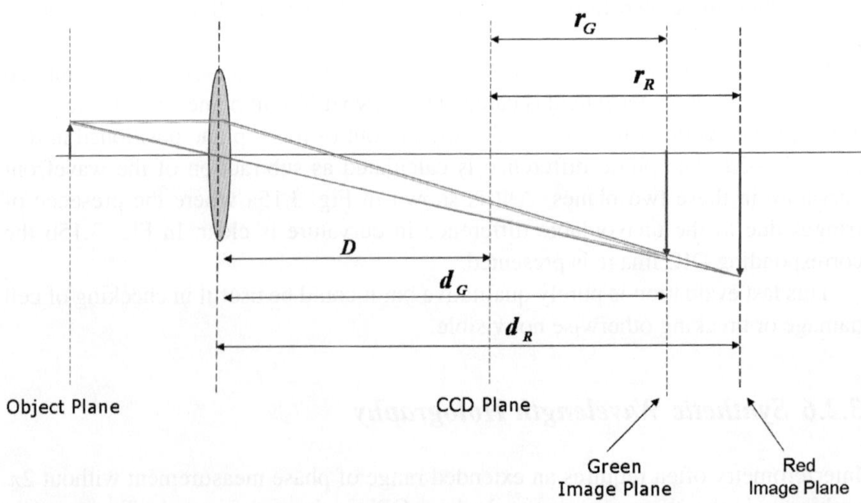

Fig. 3.16 Image plane shift due to chromatic aberration [38]

where n and m are the coordinates at the image plane, φ_0 is the phase retardation introduced by the object and $M_G = d_G/p$ is the magnification at wavelength λ_G which is related to the corresponding focal length and the distance p of the object to the lens plane (thin lens approximation). The phase $\varphi_G(n, m)$ at the image plane is the sum of the object phase, scaled according to M_G and the quadratic term related to the curvature of the wavefront introduced by the magnifying lens. The reconstruction pixel at wavelength λ_G can be expressed as

$$\Delta x_G = \Delta y = \frac{\lambda_G (D - d_G)}{N_G \Delta\xi} \tag{3.22}$$

$\Delta\xi$ is the CCD pixel dimension and N_G the number of pixels employed in the reconstructions. For the red wavelength λ_R we have correspondingly

$$\varphi_R(n, m) = \varphi_0\left(-\frac{n\Delta x_R}{M_R}, -\frac{m\Delta x_R}{M_R}\right) + \frac{\pi}{\lambda_R}\left(1 + \frac{1}{M_R}\right)\frac{n^2 + m^2}{|D - d_R|}\Delta x_R^2 \tag{3.23}$$

where $d_R = d_G + \Delta d_G$ is the position of the image plane at red wavelength which differs from that of the green wavelength by a quantity Δd_G and the corresponding reconstruction pixel is

$$\Delta x_R = \Delta y_R = \frac{\lambda_R (D - d_R)}{N_R \Delta\xi} \tag{3.24}$$

Both phases, (3.21) and (3.23), are expressed as the summation of two terms. The first one takes into account the phase retardation owing to the sample presence, while the second one regards the optical wave propagation from the imaging lens to the image plane. To get the phase map for the equivalent wavelength we subtract the foregoing phase maps:

$$\Delta\varphi = \varphi_R - \varphi_G = \left(\varphi_{O,R} - \varphi_{O,G}\right) + \Delta\varphi_r \tag{3.25}$$

The residual phase, $\Delta\varphi_r$, is a parabolic term that never cancels out and invalidates the difference phase map by the presence of circular fringes. However, $\Delta\varphi_r$ can be minimized to get a difference phase map without circular fringes. We reconstruct the two phase maps in their own image plane and, before subtracting them, we make a padding operation taking into account different magnification distances and wavelengths; in particular, we set

$$\Delta x_G/M_G = \Delta x_R/M_R \tag{3.26}$$

Equation (3.26) assumes that the reconstruction pixel size is the same for both wavelengths and that the residual phase, $\Delta\varphi_r$, has a minimum. The method described can be used as a rapid, even if approximate, graphical technique to find the reconstruction distance for one or more wavelengths (i.e. green and red, in this case) provided that the reconstruction distance for another wavelength (blue, for example) is known. This graphical technique is applied to a biological sample, as illustrated in Fig. 3.17. Physical dimension of the cell is about 25 μm in the $x - y$ plane. In Fig. 3.11a the QPM map of the in vitro cell as it appears at $\lambda_{1,2} = \lambda_1\lambda_2/|\lambda_1 - \lambda_2|$, where $\lambda_1 = \lambda_R = 632.8$ nm and $\lambda_2 = \lambda_G = 473$ nm, is shown; the two maps are reconstructed at the same distance of 110 mm. It is clear from the superimposed fringes that a longitudinal shift exists between the two reconstructed images. These fringes are due to the residual phase factor that is the result of the chromatic aberration and different wavefront curvatures.

If the correct reconstruction distance is known for one wavelength the focus shift can be easily tracked and found automatically by scanning the reconstruction distance. By applying the proposed procedure, i.e. scanning the focus of the blue reconstruction across the red, the fringes due to the chromatic aberration can be effectively nullified in the phase difference obtained by subtracting the phase map of the red hologram reconstructed at the fixed distance of 110 mm and the blue

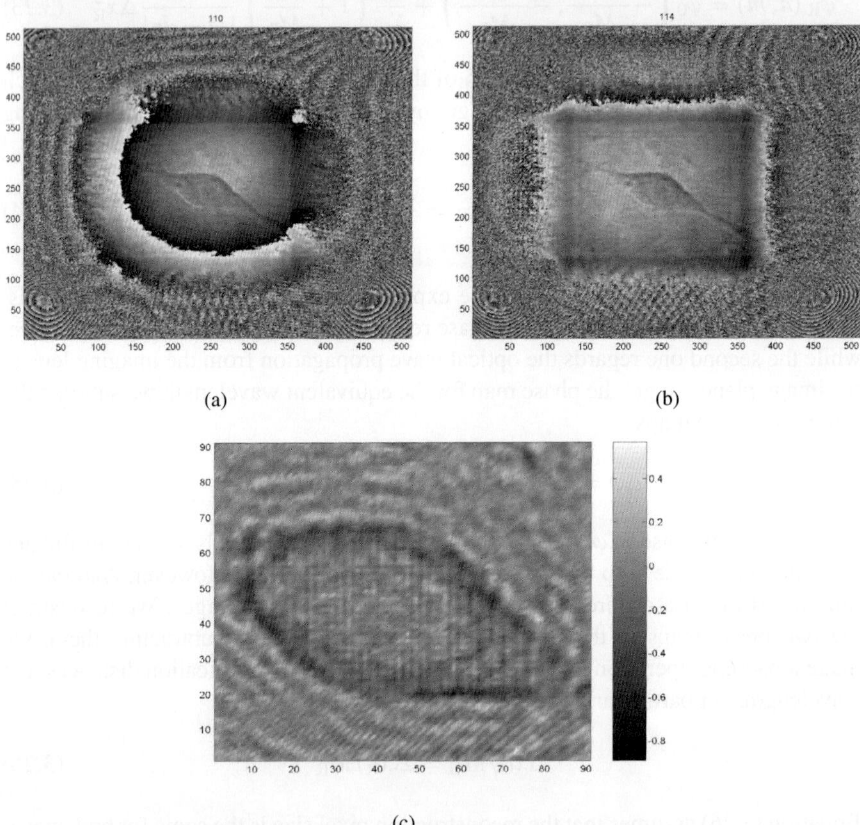

Fig. 3.17 Difference phase map for the in vitro mouse cell. The reconstruction distance for red wavelength is fixed at $d = 110$ mm while the reconstruction distance for blue wavelength is (**a**) $d = 110$ mm and (**b**) $d = 114$ mm; (**c**) difference between two phase maps evaluated in slightly distant planes for the same wavelength [38]

hologram reconstructed at various distances from 90 to 120 mm. The cell is cleared of aberration fringes at the distance of 114 mm for blue image plane, i.e. 4 mm further than the red image plane (Fig. 3.17b). The phase retardation introduced by the cell is about 5.2 rad in the image plane; we have estimated that a slight difference ($\sim$4 mm) in the reconstruction distance for the red wavelength implies a difference in the phase map of about 0.5 rad across the edges of the cell where the defocus produces blurring. Such difference is reported in Fig. 3.17c.

In conclusion, using two or more wavelengths has some drawbacks in the reconstruction process, such as the presence of circular fringes on the phase maps owing to the combined effect of different wavefront curvatures and chromatic aberration into the optical setup. A method has been demonstrated that minimizes this combined effect and retrieves phase maps without circular fringes. The method can be easily applied thanks to the DH feature of numerical re-focusing. The procedure has been employed to find automatically the right reconstruction distances for different wavelengths and makes effective the application of MWDH in cases where an extended measurement range is desired while maintaining interferometric resolution.

3.3 Conclusion

This chapter has described how the DH can be efficiently used as a marker-free and not invasive investigation tool capable of performing quantitative and qualitative mapping of biological samples. A detailed description and discussion of the recent methods based on the possibility offered by DH to manage numerically the reconstructed wavefronts have been reported. Depending on the sample and on the investigation to be performed choosing the more suitable numerical method, it is possible, from one image recording, to recover information on the OPL changes and thus to obtain quantitative phase map distributions. It is believed that the progress achieved in the reconstruction methods will find useful applications in the field of the biological analysis, and we hope they can provide inspiration for further investigations for conceptual developments of new methods and systems useful in this field.

References

1. F. Zernike, Phase contrast, a new method for the microscopic observation of transparent object. Physica **9**, 686–693 (1942)
2. P. Marquet, B. Rappaz, P.J. Magistretti, E. Cuche, Y. Emery, T. Colomb, C. Depeursinge, Digital holographic microscopy: a noninvasive contrast imaging technique allowing quantitative visualization of living cells with subwavelength axial accuracy. Opt. Lett. **30**(5), 468–470 (2005)
3. D. Gabor, A new microscope principle. Nature **161**, 777–778 (1948)
4. P. Ferraro, S. De Nicola, G. Coppola, Digital holography: recent advancements and prospective improvements for applications in microscopy. in *Optical Imaging Sensors and Systems for Homeland Security Applications*, vol. 2, Series ed. by B. Javidi. Advanced Sciences and Technologies for Security Applications. (Springer, Heidelberg, 2005) pp. 47–84

5. P. Ferraro, S. De Nicola, G. Coppola, in *Digital Holography and Three Dimensional Display: Principles and Applications,* ed. by T.C. Poon. Controlling image reconstruction process in digital holography. (Springer, Heidelberg, 2006), pp. 173–212
6. W. Osten, P. Ferraro, in *Optical Inspection of Microsystems*, vol. 109, ed. by W. Osten. Digital holography and its application in MEMS/MOEMS inspection. Optical Science Engineering Series. CRC Press Taylor & Francis Group, Boca Raton, 2006), pp. 351–425
7. E. Leith, J. Upatnieks, Reconstructed wavefronts and communication theory. J. Opt. Soc. Am. **52**, 1123 (1962)
8. E.N. Leith, J. Upatnieks, Wavefront reconstruction with diffused illumination and three dimensional objects. J. Opt. Soc. Am. **54**, 1295–1301 (1964)
9. J.W. Goodman, R.W. Lawrence, Digital image formation from electronically detected holograms. Appl. Phys. Lett. **11**(3), 77–79 (1967)
10. M.A. Kronrod, N.S. Merzlyakov, L.P. Yaroslavsky, Reconstruction of holograms with a computer. Sov. Phys.-Technol. Phys. **17**, 333–334 (1972)
11. U. Schnars, W. Jüptner, Direct recording of holograms by a CCD target and numerical reconstruction. Appl. Opt. **33**(2), 179–181 (1994)
12. J.W. Goodman, *Introduction to Fourier Optics*, 2nd edn. (McGraw-Hill, New York, 1996)
13. T. Kreis, W. Jüptner, Principles of digital holography. Proc. Fringe '97, Akademie Verlag Series in Optical Metrology **3**, 353 (1997)
14. T.M. Kreis, M. Adams, W. Jüptner, Proc. SPIE **3098**, 224 (1997)
15. E. Cuche, F. Bevilacqua, C. Depeursinge, Digital holography for quantitative phase-contrast imaging. Opt. Lett. **24**, 291–293 (1999)
16. E. Cuche, P. Marquet, C. Depeursinge, Simultaneous amplitude-contrast and quantitative phase-contrast microscopy by numerical reconstruction of Fresnel off-axis holograms. Appl. Opt. **38**, 6994–7001 (1999)
17. G. Pedrini, S. Schedin, H.J. Tiziani, Aberration compensation in digital holographic reconstruction of microscopic objects. J. Modern Opt. **48**, 1035–1041 (2001)
18. A. Stadelmaier, J.H. Massig, Compensation of lens aberrations in digital holography. Opt. Lett. **25**, 1630–1633 (2000)
19. S. De Nicola, P. Ferraro, A. Finizio, G. Pierattini, Wavefront reconstruction of Fresnel off-axis holograms with compensation of aberrations by means of phase shifting digital holography. Opt. Lasers Eng. **37**, 331–340 (2002)
20. S. Grilli, P. Ferraro, S. De Nicola, A. Finizio, G. Pierattini, R. Meucci, Whole optical wavefields reconstruction by digital holography. Opt. Express **9**, 294–302 (2001)
21. S. De Nicola, P. Ferraro, A. Finizio, G. Pierattini, Correct image reconstruction in the presence of severe anamorphism by means of digital holography. Opt. Lett. **26**, 974–977 (2001)
22. S. De Nicola, P. Ferraro, A. Finizio, S. Grilli, G. Pierattini, Experimental demonstration of the longitudinal image shift in digital holography. Opt. Eng. **42**, 1625–1630 (2003)
23. X. Lei, P. Xiaoyuan, M. Jianmin, A.K. Asundi, Studies of digital microscopic holography with applications to microstructure testing. Appl. Opt. **40**, 5046–5052 (2001)
24. S. Seebacker, W. Osten, T. Baumbach, W. Juptner, The determination of materials parameters of microcomponents using digital holography. Opt. Laser Eng. **36,** 103–126 (2001)
25. G. Coppola, P. Ferraro, M. Iodice, S. De Nicola, A. Finizio, S. Grilli, A digital holographic microscope for complete characterization of microelectromechanical systems. Meas. Sci. Technol. **15**(3), 529–539 (2004)
26. D. Carl, B. Kemper, G. Wernicke, G. Von Bally, Parameter-optimized digital holographic microscope for high-resolution living-cell analysis. Appl. Opt. **43**, 6536–6544 (2004)
27. C. Mann, L. Yu, C.M. Lo, M. Kim, High resolution quantitative phase contrast microscopy by digital holography. Opt. Express **13**, 8693–8698 (2005)
28. P. Ferraro, S. De Nicola, A. Finizio, G. Coppola, S. Grilli, C. Magro, G. Pierattini, Compensation of the inherent wavefront curvature in digital holographic coherent microscopy for quantitative phase contrast imaging. Appl. Opt. **42**, 1938–1946 (2003)

29. L. Miccio, D. Alfieri, S. Grilli, P. Ferraro, A. Finizio, L. De Petrocellis, S.D. Nicola, Direct full compensation of the aberrations in quantitative phase microscopy of thin objects by a single digital hologram. Appl. Phys. Lett. **90**, 041104 (2007)
30. D. Malacara, S.L. De Vore, in *Optical Shop Testing*, 2nd edn., Chapter 13, ed. by D. Malacara. (Wiley Interscience, New York, NY, 1992), pp. 455–495.
31. P. Ferraro, D. Alfieri, S. De Nicola, L. De Petrocellis, A. Finizio, G. Pierattini, Quantitative phase-contrast microscopy by a lateral shear approach to digital holographic image reconstruction. Opt. Lett. **31**, 1405–1407 (2006)
32. P. Ferraro, C. Del Core, L. Miccio, S. Grilli, S. De Nicola, A. Finizio, G. Coppola, Phase map retrieval in digital holography: avoiding the undersampling effect by a lateral shear approach. Opt. Lett. **32**, 2233–2235 (2007)
33. S. De Nicola, P. Ferraro, Fringe projection based on Moiré method for measuring aberration of axially symmetric optics. Opt. Commun. **185**, 285–293 (2000)
34. G. Di Caprio, M.A. Gioffrè, N. Saffioti, S. Grilli, P. Ferraro, R. Puglisi, D. Balduzzi, A. Galli, G. Coppola, Quantitative label-free animal sperm imaging by means of digital holographic microscopy. IEEE J. Sel. Top. Quantum Electron. **16**, 833–840 (2010)
35. C. Polhemus, Two-wavelength interferometry. Appl. Opt. **12**, 2071–2074 (1973)
36. C. Wagner, W. Osten, S. Seebacher, Direct shape measurement by digital wavefront reconstruction and multiwavelength contouring. Opt. Eng. **39**, 79–85 (2000)
37. M.S. Millán, J. Otón, E. Pérez-Cabré, Dynamic compensation of chromatic aberration in a programmable diffractive lens. Opt. Express **14**, 9103–9012 (2006)
38. P. Ferraro, L. Miccio, S. Grilli, M. Paturzo, S. De Nicola, A. Finizio, R. Osellame, P. Laporta, Quantitative Phase Microscopy of microstructures with extended measurement range and correction of chromatic aberrations by multiwavelength digital holography. Opt. Express **15**, 14591–14600 (2007)
39. G. Coppola, G. Di Caprio, M. Gioffré, R. Puglisi, D. Balduzzi, A. Galli, L. Miccio, M. Paturzo, S. Grilli, A. Finizio, P. Ferraro, Digital self-referencing quantitative phase microscopy by wavefront folding in holographic image reconstruction. Opt. Lett. **35**, 3390–3392 (2010)
40. L. Miccio, A. Finizio, R. Puglisi, D. Balduzzi, A. Galli, P. Ferraro, Dynamic DIC by digital holography microscopy for enhancing phase-contrast visualization, submitted to Opt. Express

Chapter 4
Incoherent Digital Holographic Microscopy with Coherent and Incoherent Light

Joseph Rosen and Gary Brooker

Abstract Holography is an attractive imaging technique as it offers the ability to view a complete three-dimensional volume from one image. However, holography is not widely applied to the regime of fluorescence microscopy, because fluorescent light is incoherent and creating holograms requires a coherent interferometer system. We review two methods of generating digital Fresnel holograms of three-dimensional microscopic specimens illuminated by incoherent light. In the first method, a scanning hologram is generated by a unique scanning system in which Fresnel zone plates (FZP) are created by a coherently illuminated interferometer. In each scanning period, the system produces an on-axis Fresnel hologram. The twin image problem is solved by a linear combination of at least three holograms taken with three FZPs with different phase values. The second hologram reviewed here is the Fresnel incoherent correlation hologram. In this motionless holographic technique, light is reflected from the 3-D specimen, propagates through a spatial light modulator (SLM), and is recorded by a digital camera. Three holograms are recorded sequentially, each for a different phase factor of the SLM function. The three holograms are superposed in the computer, such that the result is a complex-valued Fresnel hologram that does not contain a twin image. When these two types of hologram are reconstructed in the computer, the 3-D properties of the specimen are revealed.

4.1 Introduction

Holographic imaging offers a reliable and fast method to capture the complete 3-D information of the scene from a single perspective. Commonly, there are two phases in the process of holography, namely, the hologram generation and its reconstruction. Usually in the first phase, light from an object is recorded on a certain

J. Rosen (✉)
Department of Electrical and Computer Engineering, Ben-Gurion University of the Negev, P.O. Box 653, Beer-Sheva 84105, Israel
e-mail: rosen@ee.bgu.ac.il

P. Ferraro et al. (eds.), *Coherent Light Microscopy*, Springer Series in Surface Sciences 46, DOI 10.1007/978-3-642-15813-1_4,

holographic medium, whereas in the second phase, an image is reconstructed from the hologram in front of the viewer's eyes. The interaction between holography and the digital world of computers can take place in both phases: in the object acquisition and in the image reconstruction. Basically, there are two methods to generate a hologram, optically or by computers, and there are two methods to reconstruct an image from a hologram, optically or in a computer. The term computer-generated hologram is usually used to indicate the hybrid method in which a hologram is synthesized from computer-generated objects but the reconstruction of the hologram is carried out optically. The term digital hologram (sometimes called electronic hologram) specifies the other hybrid option, in which the hologram is generated optically from real-world objects, by some kind of beam interferences, then digitally processed and reconstructed in the computer. Therefore, in this review, the term incoherent digital hologram means that incoherent light beams reflected or emitted from real existing objects are recorded by a detecting device and digitally processed to yield a hologram. This hologram is reconstructed in the computer, whereas 3-D images appear on the computer's screen. The coherent optical recording is not applicable for the incoherent optics because interference between reference and object incoherent beams cannot occur. Therefore, different holographic acquisition methods should be employed for generating an incoherent digital hologram.

The oldest methods of recording incoherent holograms have made use of the property that every incoherent object is composed of many source points, each of which is self-spatial coherent and therefore can create an interference pattern with light coming from the point's mirrored image. Under this general principle, there are various types of holograms [1–8], including Fourier [2, 6] and Fresnel holograms [3, 4, 8]. The process of beam interfering demands high levels of light intensity, extreme stability of the optical setup, and relatively narrow bandwidth light source. These limitations have prevented holograms from becoming widely used for many practical applications. More recently, three groups of researchers have proposed computing holograms of 3-D incoherently illuminated objects from a set of images taken from different points of view [9–12]. This method, although it shows promising prospects, is relatively slow since it is based on capturing tens of scene images from different view angles. Another method is called scanning holography [13–17], in which a pattern of Fresnel zone plates (FZPs) scans the object such that at each and every scanning position, the light intensity is integrated by a point detector. The overall process yields a Fresnel hologram obtained as a correlation between the object and FZP patterns.

This review concentrates on two techniques of incoherent digital holography that we have been involved recently with its development [17–21]. These two methods are different from each other and they are based on different physical principles. There are only two common aspects that exist in these two methods. First, the system's input signal is always an incoherent light reflected or emitted from a certain 3-D scene. Second, the final product from the two methods is a digital Fresnel hologram.

4.2 General Properties of Fresnel Holograms

The type of hologram discussed in this review is the digital Fresnel hologram, which means that a hologram of a single point has the form of the well-known Fresnel zone plate (FZP). The axial location of the object point is encoded by the Fresnel number of the FZP, which is the technical term for the quantity of the rings density in the FZP.

To understand the operation principle of any general Fresnel hologram, let us look at the difference between regular and Fresnel holographic imaging systems. In classical imaging, image formation of objects at different distances from the lens results in a sharp image at the image plane for objects at only one position from the lens, as shown in Fig. 4.1a. The other objects at different distances from the lens are out of focus. Fresnel holographic system, on the other hand, as depicted in Fig. 4.1b, projects a set of rings known as the FZP onto the plane of the image for each and every point at every plane of the object being viewed. The depth of the points is encoded by the density of the rings such that points that are closer to the system project denser rings than distant points. Because of this encoding method, the 3-D information in the volume being imaged is recorded into the recording medium. Therefore, each plane in the image space reconstructed from a Fresnel hologram is in focus at a different axial distance. The encoding is accomplished by the presence

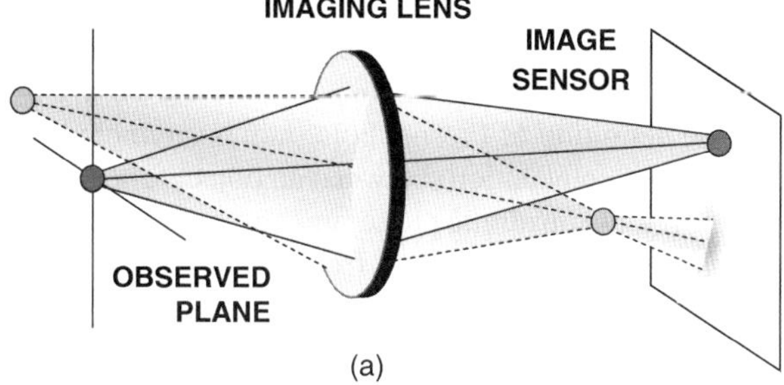

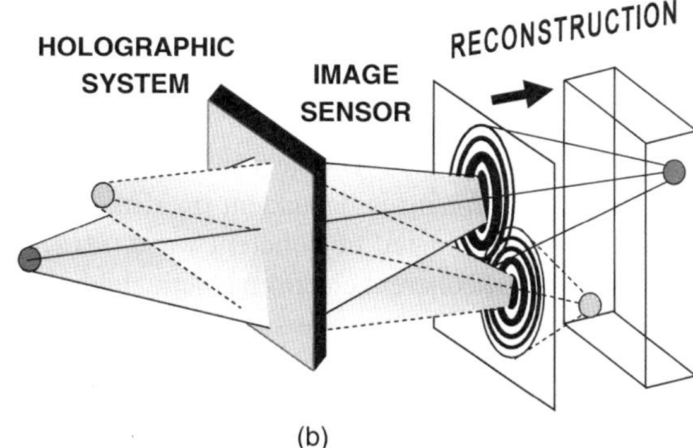

Fig. 4.1 Comparison of the Fresnel holography principle and conventional imaging. (**a**) Conventional imaging system. (**b**) Fresnel holography system

of one of the holographic systems in the image path. Each holographic system, coherent or incoherent, has a different method to project the FZP on the image plane. At this point it should be noted that this graphical description of projecting FZPs by every object's point actually expresses the mathematical 2-D correlation (or convolution) between the object function and the FZP. In other words, the methods of creating Fresnel holograms are different from each other by the way they spatially correlate the FZP with the 3-D scene. Another issue to note is that the correlation should be done with an FZP that is somehow ‘sensitive’ to the axial locations of the object points. Otherwise, these locations are not encoded into the hologram. The systems described in this review satisfy the condition that the FZP is dependent on the axial distance of each and every object point. This means that indeed points, which are far from the system, project FZP with fewer cycles per radial length than nearby points, and by this condition the holograms can actually image the 3-D scene properly.

The FZP is a sum of at least three main functions, a constant bias, a quadratic phase function, and its complex conjugate. The object function is actually correlated with all these three functions. However, the useful information, with which the holographic imaging is realized, is the correlation with just one of the two quadratic phase functions. The correlation with the other quadratic phase function induces the well-known twin image. This means that the detected signal in the holographic system contains three superposed correlation functions, whereas only one of them is the required correlation between the object and the quadratic phase function. Therefore, the digital processing of the detected signal should contain the ability to eliminate the two unnecessary terms.

The definition of Fresnel hologram is any hologram that contains at least, a correlation (or convolution) between an object function and a quadratic phase function. Moreover, the quadratic phase function must be parameterized according to the axial distance of the object points from the detection plane. In other words, the number of cycles per radial distance of each quadratic phase function in the correlation is dependent on the z distance of each object point. Formally, a hologram is called Fresnel hologram if its distribution function contains the following term:

$$H(u,v) = \iiint g(x,y,z) \exp\left[\mathrm{i}\frac{2\pi\beta}{z}\left[(u-x)^2 + (v-y)^2\right]\right] \mathrm{d}x\mathrm{d}y\mathrm{d}z \qquad (4.1)$$

where $g(x, y, z)$ is the 3-D object function and β is a constant. Indeed, in (4.1) the phase of the exponent is dependent on z, the axial location of the object. In case the object is illuminated by coherent wave, $H(u, v)$ given by (4.1) is the complex amplitude of the coherent electromagnetic field directly obtained, under the paraxial approximation [22], by a free space propagation from the object to the detection plane. However, we deal here with incoherent illumination, for which alternative methods to the free propagation should be applied. In fact, in this review we describe two such methods to get the desired correlation with the quadratic phase function given in (4.1), and these methods indeed operate under incoherent illumination.

The more mature technique among the two, and the one that is extensively discussed in the literature, is the scanning holography, pioneered by Poon [13–16]. There are already a textbook [15] and at least one review article [16] on scanning holography. Therefore, in the next section we only summarize briefly the fundamental principles of the classical scanning holography. However, we enlighten a more recent, and less known technique of the scanning holography, called homodyne scanning holography, which we have been involved with its development recently [17].

The second proposed incoherent digital hologram is dubbed Fresnel incoherent correlation hologram (FINCH) [18–21]. The FINCH is actually based on a single-channel on-axis incoherent interferometer. Like any Fresnel holography, in the FINCH the object is correlated with an FZP, but the correlation is carried out without any movement and without multiplexing the image of the scene. Section 4.4 reviews the latest developments of the FINCH in the field of holography and microscopy.

4.3 Scanning Holography

Scanning holography [13–17] has demonstrated the ability to produce a Fresnel hologram of the incoherent light emission distributed in a 3-D structure. As mentioned above, the definition of Fresnel hologram is any hologram that contains at least a correlation (or convolution) between an object function and a quadratic phase function. In scanning holography, the required correlation is performed by a mechanical movement. More specifically, a certain pattern, which is the abovementioned FZP, is projected on the observed object, whereas the FZP moves at a constant velocity relative to the object (or the object moves relative to the FZP). During the movement, the product between the FZP and the object is summed by a lens onto a point detector in discrete times. In other words, the pattern of the FZP scans the object and at each and every scanning position the light intensity is integrated by the detector. The resulting electric signal is a sampled version of the 2-D correlation between the object and the FZP. The dependence of the FZP in the axial position of object points is achieved by interfering two mutually coherent, monochromatic, laser spherical waves on the object surface. The number of cycles per radial distance in each of the spherical waves is dependent on their axial interference location. Since these beams interfere on the object, the axial distance of each object point is stored in the hologram due to the fact that each point is correlated with an FZP the cycle density of which is dependent on the point's 3-D location. Classic scanning holograms [13–16] have been recorded by a heterodyne interferometer in which the holographic information has been encoded on a high carrier frequency. Such method suffers from several drawbacks. On the one hand, trying to keep the scanning time as short as possible requires using carrier frequencies, which may be higher than the bandwidth limit of some, or all of the electronic devices in the system. On the other hand, working with a carrier frequency that is lower than the system limitation extends the scanning time far beyond the minimal time needed to capture the holographic information according to the sampling theorem. Long scanning times limit the system from recording dynamical scenes. In the scanning holography described in [17], the required correlation is performed by scanning the object with a set of frozen-in-time FZP patterns. In this modified system, the hologram is recorded without temporal carrier frequency, using a homodyne interferometer. This offers an improved method of 3-D imaging that can be applied to incoherent imaging in general and to fluorescence microscopy in particular. As mentioned above, the FZP is created by interference of two coherent spherical waves. As shown in Fig. 4.2, the interference pattern is projected on the specimen, scans it in 2-D, and the reflected light from the specimen is integrated on a detector. Due to the line-by-line scanning by the FZP along the specimen, the one-dimensional detected signal is composed of the entire lines of the correlation matrix between the object function and the FZP. In the computer, the detected signal is reorganized in the shape of a 2-D matrix, the values of which actually represent the Fresnel hologram of the specimen. The specimen we consider is 3-D, and its 3-D structure is stored in the hologram by the effect that during the correlation the number of cycles per radial distance of the FZP contributed from a distant object point is slightly smaller than the number of cycles per radial distance of the FZP contributed from a closer object point.

As mentioned above, the FZP is the intensity pattern of the interference between two spherical waves given by

$$F(x,y,z) = Ap(x,y)\left\{2+\exp\left[\frac{i\pi}{\lambda(z_o+z)}\left(x^2+y^2\right)+i\theta\right]+\exp\left[\frac{-i\pi}{\lambda(z_o+z)}\left(x^2+y^2\right)-i\theta\right]\right\} \quad (4.2)$$

where $p(x, y)$ is a disk function with diameter D that indicates the limiting aperture on the projected FZP, A is a constant, θ is the phase difference between the two spherical waves, and λ is the wavelength of the coherent light source. The constant z_o indicates that at a plane $z = 0$, there is effectively interference between two spherical waves, one emerging from a point at $z = -z_o$ and the other converging

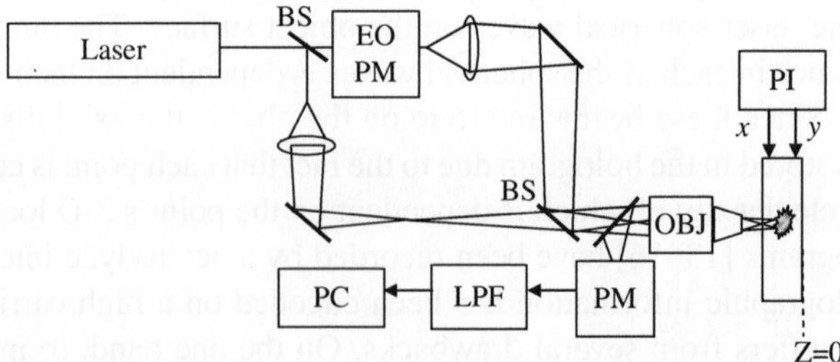

Fig. 4.2 Optical setup of the homodyne scanning holography system: EOPM, electro-optic phase modulator introducing a phase difference between the two beams; BS, beam splitter; PI, piezo X–Y stage; OBJ, objective; PM, photomultiplier tube detector; LPF, lowpass filter; PC, personal computer

to a point at $z = z_o$. This does not necessarily imply that these particular spherical waves are exclusively needed to create the FZP. For a 3-D object $S(x, y, z)$, the correlation with the FZP of (4.2) is

$$\begin{aligned} H(x, y) =& S(x, y, z) * F(x, y, z)\mathrm{d}z = \int S(x, y, z) * p(x, y)\mathrm{d}z \\ &+ \iiint S(x', y', z')p(x' - x, y' - y) \\ &\times \exp\left\{\frac{\mathrm{i}\pi[(x' - x)^2 + (y' - y)^2]}{\lambda(z_o + z')} + \mathrm{i}\theta\right\} \mathrm{d}x'\mathrm{d}y'\mathrm{d}z' \\ &+ \iiint S(x', y', z')p(x' - x, y' - y) \\ &\times \exp\left\{\frac{-\mathrm{i}\pi[(x' - x)^2 + (y' - y)^2]}{\lambda(z_o + z')} - \mathrm{i}\theta\right\} \mathrm{d}x'\mathrm{d}y'\mathrm{d}z' \end{aligned} \quad (4.3)$$

where the asterisk denotes a 2-D correlation. This correlation result is similar to a conventional Fresnel on-axis digital hologram, and therefore, it suffers from the same problems. Specifically, $H(x, y)$ of (4.3) contains three terms, which represent the information on three images, namely, the 0th diffraction order, the virtual image, and the real image. Trying to reconstruct the image of the object directly from a hologram of the form of (4.3) would fail because of the disruption originated from two images out of the three. This difficulty is solved here with the same solution applied in a conventional on-axis digital holography. Explicitly, at least three holograms of the same specimen are recoded, where for each one of them an FZP with a different phase value θ is introduced. A linear combination of the three holograms cancels the two undesired terms and the remaining is a complex-valued on-axis Fresnel hologram, which contains only the information of the single desired image, either the virtual or the real one, according to our choice. A possible linear combination of the three holograms to extract a single correlation between the object and one of the quadratic phase functions of (4.3) is

$$\begin{aligned} H_F(x, y) =& H_1(x, y)\left[\exp(\pm \mathrm{i}\theta_3) - \exp(\pm \mathrm{i}\theta_2)\right] \\ &+ H_2(x, y)\left[\exp(\pm \mathrm{i}\theta_1) - \exp(\pm \mathrm{i}\theta_3)\right] \\ &+ H_3(x, y)\left[\exp(\pm \mathrm{i}\theta_2) - \exp(\pm \mathrm{i}\theta_1)\right] \end{aligned} \quad (4.4)$$

where $H_i(x, y)$ is the ith recorded hologram of the form of (4.3) and θ_i is the phase value of the ith FZP used during the recording process. The choice between the signs in the exponents of (4.4) determines which image, virtual or real, is kept in the final hologram. If, for instance, the virtual image is kept, $H_F(x, y)$ is the final complex-valued hologram of the form

$$H_F(x, y) = \int S(x, y, z') * p(x, y) \exp\left[\frac{-\mathrm{i}\pi}{\lambda(z_o + z')}\left(x^2 + y^2\right)\right]\mathrm{d}z' \quad (4.5)$$

The function $H_F(x, y)$ is the final hologram that contains the information of only one image – the 3-D virtual image of the specimen in this case. Such image $S'(x, y, z)$ can be reconstructed from $H_F(x, y)$ by calculating in the computer the inverse operation to (4.5) as follows:

$$S'(x, y, z) = H_F(x, y) * \exp\left[\frac{i\pi}{\lambda z}\left(x^2 + y^2\right)\right] \quad (4.6)$$

The resolution properties of this imaging technique are determined by the properties of the FZP. More specifically, the FZP diameter D and the constant z_o characterize the system resolution in a similar way to the effect of an imaging lens [22]. Suppose the image is a single infinitesimal point at $z = 0$, then $H_F(x, y)$ gets the shape of a quadratic phase function limited by a finite aperture. The reconstructed point image has a transverse diameter of $1.22\lambda z_o/D$, which defines the transverse resolution, and an axial length of $8\lambda z_o^2/D^2$, which defines the axial resolution. Note also that the width of the FZP's last ring along its perimeter is about $\lambda z_o/D$, and therefore the size of the specimen's smallest distinguishable detail is approximately equal to the width of this ring.

As an example of the homodyne scanning hologram, let us describe the experiment from [17]. The setup shown in Fig. 4.2 was built on a standard widefield fluorescence microscope. The specimen was a slide with two pollen grains positioned at different distances from the microscope objective. The microscope objective was an infinity-corrected 20×, NA = 0.75. The slide was illuminated by the FZP created by the interferometer. A laser beam ($\lambda = 532$ nm) was split into two beams with beam expanders consisting each of a microscope objective and a 12-cm focal-length achromat as a collimating lens. One of the beams passed through an electro-optic phase modulator driven by three (or more) constant voltage values, which induce three (or more) phase difference values between the interfering beams. Note that unlike previous studies [13–16], there is no frequency difference between the two interfering waves since this time we record a hologram with a homodyne interferometer. The two waves were combined by the beam splitter to create an interference pattern in the space of the specimen. The pattern was then reduced in size and projected through the objective onto the specimen. The sample was scanned in a 2-D raster with an X–Y piezo stage. The data were collected by an acquisition system, and data manipulation was performed by programs written in MATLAB.

The three recorded holograms of the specimen taken with phase difference values of $\theta_{1,2,3} = 0, \pi/2$, and π are shown in Fig. 4.3a–c, respectively. In this figures, the dominant term is the low-frequency term [the first in (4.2)], and therefore, without mixing the three holograms in the linear combination that eliminates the low frequency along with the twin image term, there is no possibility to recover the desired image with a reasonable quality. These three holograms are substituted into (4.4) and yield a complex-valued hologram shown in Fig. 4.4. This time the grating lines are clearly revealed in the phase pattern. The computer reconstruction of two pollen grains along the z-axis is shown in Fig. 4.5. As can be seen in this figure, different parts of the pollen grains are in focus at different transverse planes.

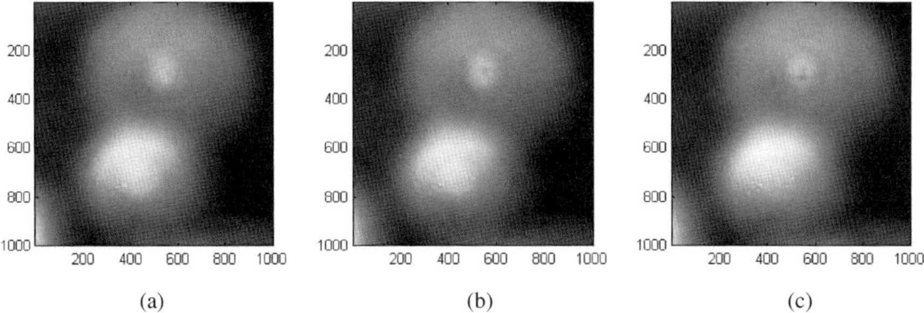

Fig. 4.3 Three recorded holograms with phase difference between the two interferometers arms of (**a**) 0, (**b**) $\pi/2$, and (**c**) π, all obtained from the homodyne scanning holography system

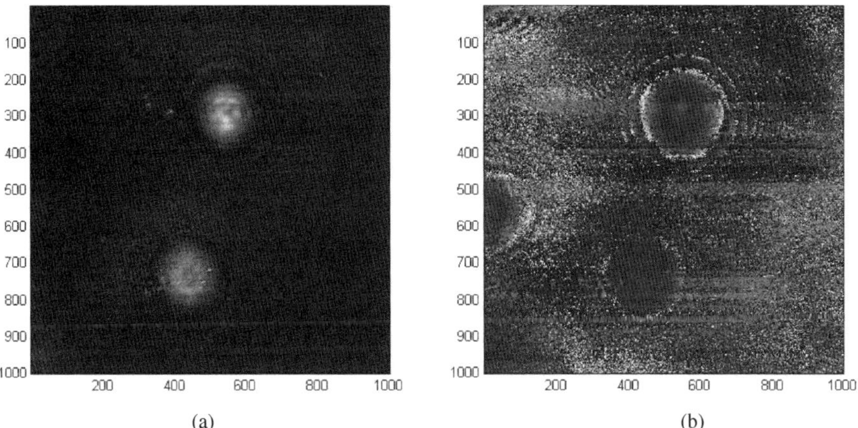

Fig. 4.4 (**a**) The magnitude and (**b**) the phase of the final homodyne scanning hologram

4.4 Fresnel Incoherent Correlation Holography

In this section we describe the FINCH – a different method of recording digital Fresnel holograms under incoherent illumination. Various aspects of the FINCH have been described in [18–21], including FINCH of reflected white light [18], FINCH of fluorescence objects [19], FINCHSCOPE [20] – a holographic fluorescence microscope, and finally SAFE [21] – a process of recording incoherent holograms in a synthetic aperture mode. We briefly review these works in this section.

Generally, in the FINCH system, the reflected incoherent light from a 3-D object propagates through a spatial light modulator (SLM) and is recorded by a digital camera. To solve the twin image problem, three holograms are recorded sequentially, each with a different phase factor of the SLM pattern. The three holograms are superposed in the computer such that the result is a complex-valued Fresnel hologram. The 3-D properties of the object are revealed by reconstructing this hologram in the computer.

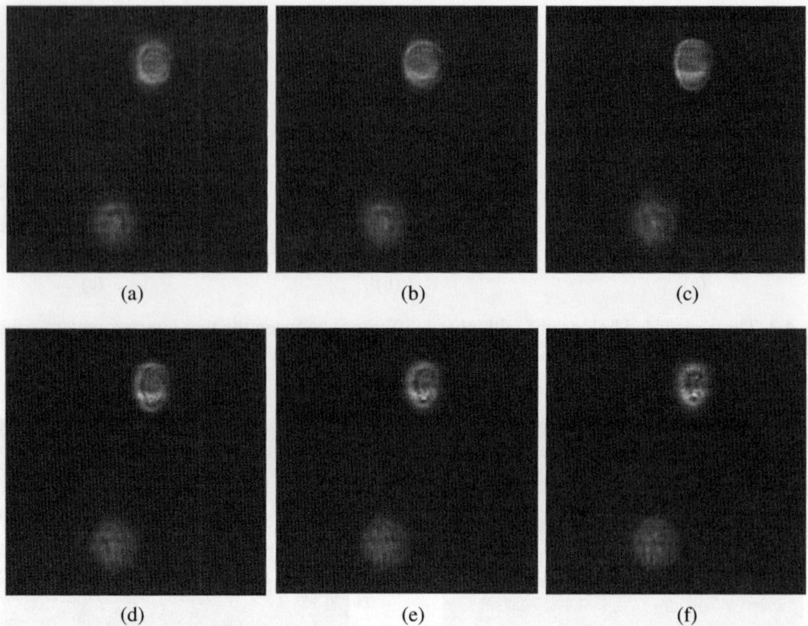

Fig. 4.5 (a–f) Various digital reconstructed images along the light propagation axis obtained by digital reconstruction from the hologram of Fig. 4.4

One of the FINCH systems [18] is shown in Fig. 4.6. A white-light source illuminates a 3-D object, and the reflected light from the object is captured by a CCD camera after passing through a lens L and an SLM. In general, such a system can be analyzed as an incoherent correlator, where the SLM function is considered as a part of the system's transfer function. However, we find it easier to regard the system as an incoherent interferometer, where the grating displayed on the SLM is considered as a beam splitter. As is common in such cases, we analyze the system by following its response to an input object of a single infinitesimal point. Knowing the system's point spread function (PSF) enables one to realize the system operation for any general object. Analysis of a beam originated from a narrowband infinitesimal point source is done using Fresnel diffraction theory [22], since such a source is spatially coherent by definition.

A Fresnel hologram of a point object is obtained when the two interfering beams are, for instance, plane and spherical beams. Such a goal is achieved if the SLM's reflection function $R(x, y)$ is of the form

$$R(x, y) = \frac{1}{2} + \frac{1}{2}\exp\left[-\frac{i\pi}{\lambda a}\left(x^2 + y^2\right) + i\theta\right] = \frac{1}{2} + \frac{1}{2}Q\left(-\frac{1}{a}\right)\exp(i\theta) \quad (4.7)$$

For the sake of shortening, the quadratic phase function is designated by the function Q, such that $Q(b) = \exp[(i\pi b/\lambda)(x^2 + y^2)]$. When a plane wave hits the

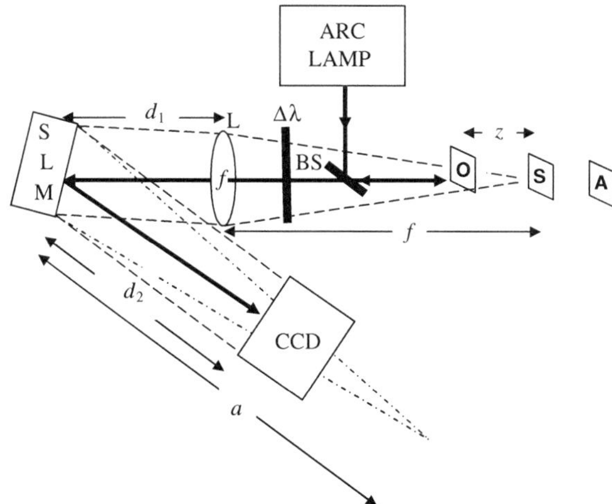

Fig. 4.6 Schematic of FINCH recorder. BS, beam splitters; SLM, spatial light modulator; CCD, charge-coupled device; L is a spherical lens with $f = 25$ cm focal length. $\Delta\lambda$ indicates a chromatic filter with a bandwidth of $\Delta\lambda = 60$ nm

SLM, the first constant term 1/2 in (4.7) represents the reflected plane wave, and the quadratic phase term is responsible for the reflected spherical wave in the paraxial approximation. The angle θ plays an important role later in the computation process to eliminate the twin image and the bias term.

A point source located at the point (x_s, y_s, z_s) a distance z_s from a spherical positive lens, with f focal length, induces on the lens plane a diverging spherical wave of the form of $\mathrm{Q}[1/z_s]$. To simplify the notation we assume that the point sources are located on the front focal plane of the lens, i.e., $z_s = f$. Right after the lens, which has a transmission function of $Q(-1/f)$, the complex amplitude is of a tilted plane wave of the form $C_1(\bar{r}_s)L(-\bar{r}_s/f)$ where the function L stands for a the linear phase function, such that $L(\bar{s}) = \exp[\mathrm{i}2\pi\lambda^{-1}(s_x x + s_y y)]$, $\bar{r}_s = (x_s, y_s)$, and $C_1(\bar{r}_s)$ is a complex constant dependent on the source point's location. After propagating additional distance of d_1 onto the SLM plane, the complex amplitude is the same plane wave besides the constant term. Right after the SLM, with the reflection function given in (4.7), the complex amplitude is equal to $C_2(\bar{r}_s)L(-\bar{r}_s/f)[1 + \exp(\mathrm{i}\theta)Q(-1/a)]$. Finally, in the CCD plane at a distance d_2 from the SLM, the intensity of the recorded hologram is

$$
\begin{aligned}
I_P(x, y) &= A\left|\mathrm{Q}\left(\frac{1}{f}\right)\mathrm{Q}\left(\frac{-1}{f}\right)\mathrm{L}\left(\frac{-\bar{r}_s}{f}\right) * \mathrm{Q}\left(\frac{1}{d_1}\right)\right. \\
&\quad \times \left.\left[1 + \exp(\mathrm{i}\theta)\,\mathrm{Q}\left(\frac{-1}{a}\right)\right] * \mathrm{Q}\left(\frac{1}{d_2}\right)\right|^2 \\
&= A\left|\mathrm{L}\left(\frac{-\bar{r}_s}{f}\right)\left[1 + \exp(\mathrm{i}\theta)\,\mathrm{Q}\left(\frac{-1}{a}\right)\right] * \mathrm{Q}\left(\frac{1}{d_2}\right)\right|^2 \qquad (4.8)
\end{aligned}
$$

where A is a constant. The result of $I_P(x, y)$, after calculating the square magnitude in (4.8), is the PSF for any source point located at any point (x_s, y_s) on the front focal plane of the lens, as follows:

$$I_P(x,y) = A_o\left(2 + \exp\left\{\frac{i\pi}{\lambda(a-d_2)}\left[\left(x - \frac{d_2 x_s}{f}\right)^2 + \left(y - \frac{d_2 y_s}{f}\right)^2\right] + i\theta\right\}\right.$$
$$\left. + \exp\left\{\frac{-i\pi}{\lambda(a-d_2)}\left[\left(x - \frac{d_2 x_s}{f}\right)^2 + \left(y - \frac{d_2 y_s}{f}\right)^2\right] - i\theta\right\}\right) \quad (4.9)$$

The reconstruction distance of the point image from an equivalent optical hologram is $z_r = a - d_2$, although in the present case the hologram is of course digital, and the reconstruction is done by the computer. Note that z_r is obtained specifically in the case that one of the phase masks on the SLM is constant. This choice is used in all the FINCH experiments because practically the fill factor of the SLM is less than 100%, and therefore the constant phase modulation inherently exists in the SLM. Consequently, choosing two diffractive lenses could cause unwanted three, instead of two, waves mixing on the hologram plane, one wave due to the constant phase and the other two from the two different diffractive lenses.

Equation (4.9) is the expression of the transparency function of a hologram created by an object point and recorded by a FINCH system. This hologram has several unique properties. The transverse magnification M_T is expressed as $M_T = \partial x_r/\partial x_s = d_2/f$ for an object located on the front focal plane, and $M_T = d_2/z_s$ for any other plane.

For a general 3-D object $g(x_s, y_s, z_s)$ illuminated by a narrowband incoherent illumination, the intensity of the recorded hologram is an integral of the entire PSF given in (4.9), over all the object intensity $g(x_s, y_s, z_s)$, as follows:

$$H(x,y) \cong A_o\left(C + \iiint g(x_s,y_s,z_s)\exp\left\{\frac{i\pi}{\lambda\gamma(z_s)}\left[\left(x - \frac{d_2 x_s}{z_s}\right)^2\right.\right.\right.$$
$$\left.\left. + \left(y - \frac{d_2 y_s}{z_s}\right)^2\right] + i\theta\right\} dx_s dy_s dz_s$$
$$+ \iiint g(x_s,y_s,z_s)\exp\left\{\frac{-i\pi}{\lambda\gamma(z_s)}\left[\left(x - \frac{d_2 x_s}{z_s}\right)^2\right.\right.$$
$$\left.\left.\left. + \left(y - \frac{d_2 y_s}{z_s}\right)^2\right] - i\theta\right\} dx_s dy_s dz_s\right) \quad (4.10)$$

where $\gamma(z_s)$, the reconstruction distance of each object point, is a complicated expression calculated from (4.8), but this time without the special assumption of $z_s = f$.

Besides a constant term C, (4.10) contains two terms of correlation between an object and a quadratic phase, z_s-dependent, function, which means that the recorded hologram is indeed a Fresnel hologram. In order to remain with a single correlation term out of the three terms given in (4.10), we again follow the usual procedure of on-axis digital holography [17–21]. Three holograms of the same object are recorded each of which with a different phase constant θ. The final hologram H_F is a superposition according to (4.4).

A 3-D image $g'(x, y, z)$ can be reconstructed from $H_F(x, y)$ by calculating the Fresnel propagation formula, as follows:

$$g'(x, y, z) = H_F(x, y) * \exp\left[\frac{i\pi}{\lambda z}\left(x^2 + y^2\right)\right] \tag{4.11}$$

The system shown in Fig. 4.6 was used to record the three holograms [18]. The SLM (Holoeye HEO 1080P) is phase-only, and as so, the desired function given by (4.7) cannot be directly displayed on this SLM. To overcome this obstacle, the phase function $Q(-1/a)$ is displayed randomly on only half of the SLM pixels. These pixels were represented in the second term of (4.7), whereas the rest of the pixels representing the first constant term in (4.7) were modulated with a constant phase. The randomness in distributing the two phase functions has been required because organized nonrandom structure produces unnecessary diffraction orders, and therefore results in lower interference efficiency. The pixels were divided equally, half to each diffractive element, to create two wave fronts with equal energy. By this method the SLM function becomes a good approximation to $R(x, y)$ of (4.7).

The SLM has 1920 × 1080 pixels in a display of 16.6 × 10.2 mm, where only the central 1024 × 1024 pixels were used for implementing the phase mask. The phase distribution of the three reflection masks displayed on the SLM, with phase constants of 0°, 120°, and 240°, are shown in Fig. 4.7a–c, respectively. The other specifications of the system of Fig. 4.6 are $f = 250\,\text{mm}$, $a = 430\,\text{mm}$, $d_1 = 132\,\text{mm}$, and $d_2 = 260\,\text{mm}$.

Three white-on-black letters each of the size 2×2 mm were located at the vicinity of rear focal point of the lens. 'O' was at $z = -24\,\text{mm}$, 'S' was at $z = -48\,\text{mm}$, and 'A' was at $z = -72\,\text{mm}$. These letters were illuminated by a mercury arc lamp. A filter that passed a Poisson-like power spectrum from 574 to 725 nm light with a peak wavelength of 599 nm and a bandwidth (full width at half maximum) of 60 nm was positioned between the beam splitter and the lens L. The three holograms, each for a different phase constant of the SLM, were recorded by a CCD camera and processed by a PC. The final hologram $H_F(x, y)$ was calculated according to (4.4) and its magnitude and phase distributions are depicted in Fig. 4.7e, f, respectively.

The hologram $H_F(x, y)$ was reconstructed in the computer by calculating the Fresnel propagation toward various z propagation distances according to (4.11). Three different reconstruction planes are shown in Fig. 4.7g–i. In each plane, a different letter is in focus as is indeed expected from a holographic reconstruction of an object with a volume.

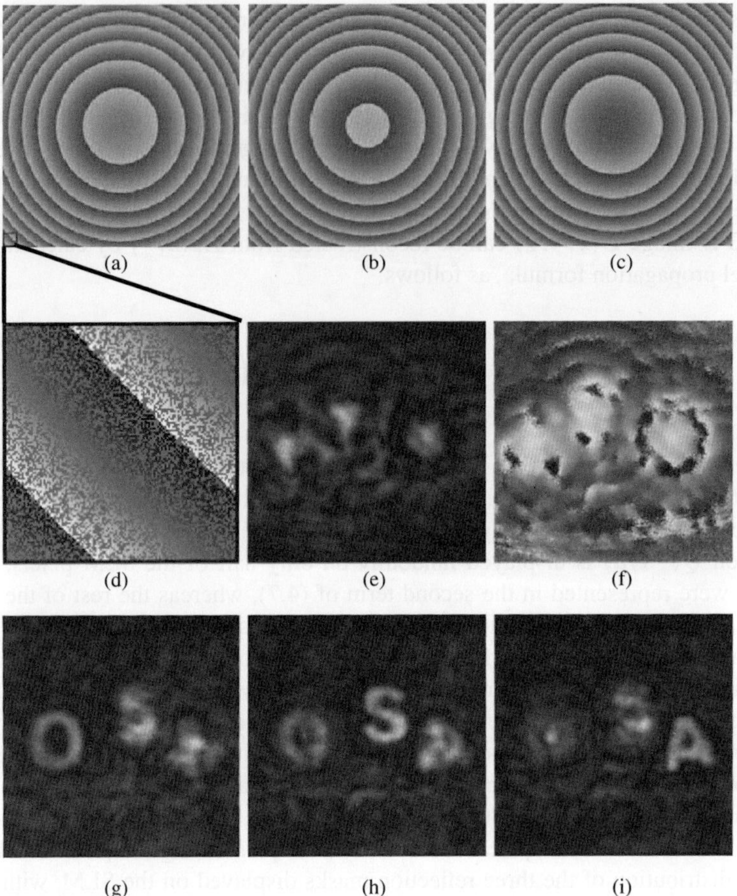

Fig. 4.7 (**a**) Phase distribution of the reflection masks displayed on the SLM, with $\theta = 0°$, (**b**) $\theta = 120°$, and (**c**) $\theta = 240°$. (**d**) Enlarged portion of (**a**) indicating that half (randomly chosen) of the SLM's pixels modulate light with a constant phase. (**e**) Magnitude and (**f**) phase of the final on-axis digital hologram. (**g**) Reconstruction of the hologram of the three letters at the best focus distance of 'O.' (**h**) Same reconstruction at the best focus distance of 'S' and (**i**) of 'A'

In [19], the FINCH has been capable of recording multicolor digital holograms from objects emitting fluorescent light. The fluorescent light, specific to the emission wavelength of various fluorescent dyes after excitation of 3-D objects, was recorded on a digital monochrome camera after reflection from the SLM. For each wavelength of fluorescent emission, the camera sequentially records three holograms reflected from the SLM, each with a different phase factor of the SLM's function. The three holograms are again superposed in the computer to create a complex-valued Fresnel hologram of each fluorescent emission without the twin image problem. The holograms for each fluorescent color are further combined in a computer to produce a multicolored fluorescence hologram and 3-D color image.

An experiment showing the recording of a color fluorescence hologram was carried out [19] on the system shown in Fig. 4.8. The phase constants of $\theta_{1,2,3} = 0°$, 120°, and 240° were introduced into the three quadratic phase functions. The other specifications of the system are $f_1 = 250\,\text{mm}$, $f_2 = 150\,\text{mm}$, $f_3 = 35\,\text{mm}$, $d_1 = 135\,\text{mm}$, and $d_2 = 206\,\text{mm}$. The magnitude and phase of the final complex hologram, superposed from the first three holograms, are shown in Fig. 4.9a, b, respectively. The reconstruction from the final hologram was calculated using the Fresnel propagation formula of (4.11). The results are shown at the plane of the front face of the front die (Fig. 4.9c), and at the plane of the front face of the rear die (Fig. 4.9d). Note that in each plane a different die face is in focus as is indeed expected from a holographic reconstruction of an object with a volume. The second set of three holograms was recorded via a red filter in the emission filter slider F_2, which passed 614–640 nm fluorescent light wavelengths with a peak wavelength of 626 nm and a bandwidth of 11 nm (FWHM). The magnitude and phase of the final complex hologram, superposed from the 'red' set, are shown in Fig. 4.9e, f, respectively. The reconstruction results from this final hologram are shown in Fig. 4.9g, h at the same planes as shown in Fig. 4.9c, d, respectively. Finally, an additional set of three holograms was recorded with a green filter in emission filter slider F_2, which passed 500–532 nm fluorescent light wavelengths with a peak wavelength of 516 nm and a bandwidth of 16 nm (FWHM). The magnitude and phase of the final complex hologram, superposed from the 'green' set, are shown in Fig. 4.9i, j, respectively. The reconstruction results from this final hologram are shown in Fig. 4.9k, l at the same planes as shown in Fig. 4.9c, d, respectively. Compositions of Fig. 4.9c, g, k and Fig. 4.9d, h, l are depicted in Fig. 4.9m,n, respectively. Note that all the colors in Fig. 4.9 are pseudo-colors. These last results yield a complete

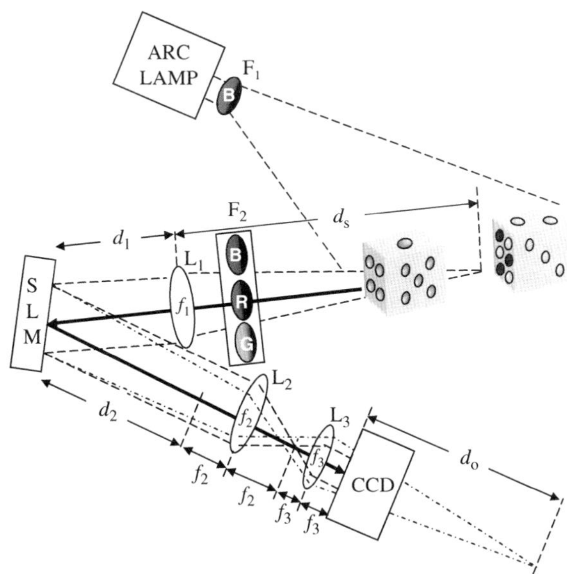

Fig. 4.8 Schematics of the FINCH color recorder. SLM, spatial light modulator; CCD, charge-coupled device; L_1, L_2, L_3 are spherical lenses and F_1, F_2 are chromatic filters

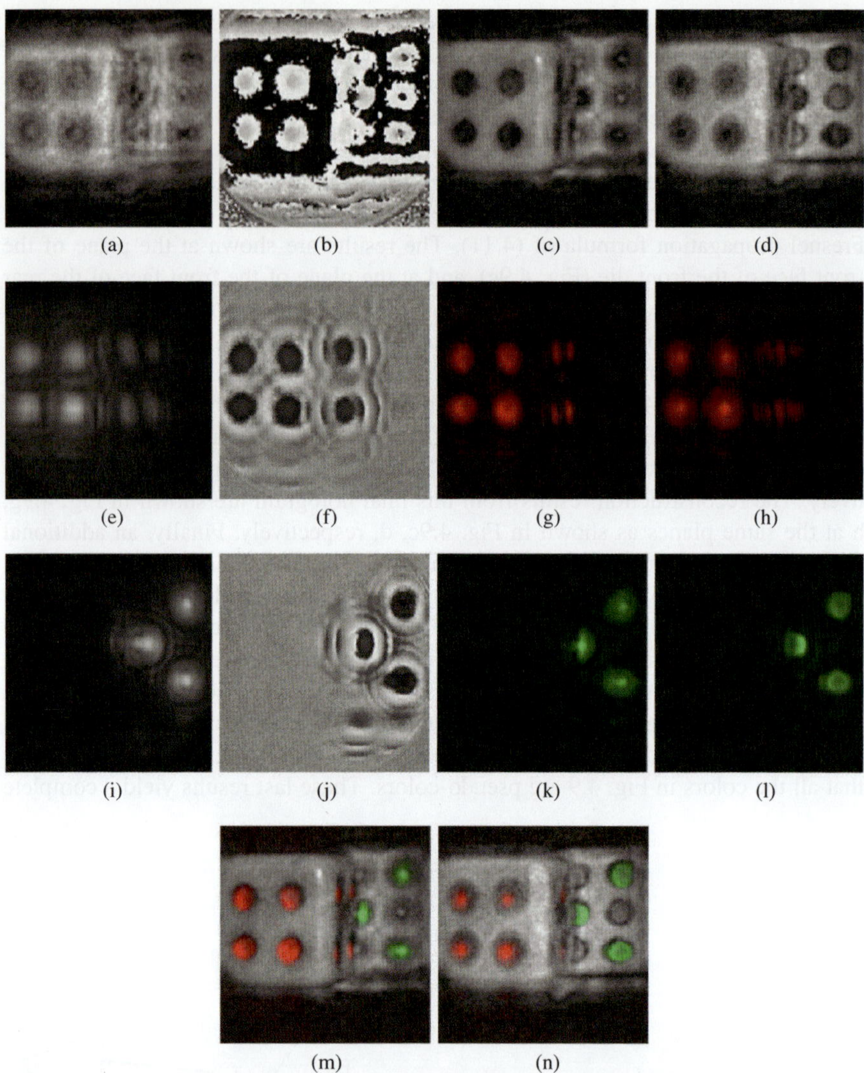

Fig. 4.9 (**a**) Magnitude and (**b**) phase of the complex Fresnel hologram of the dice. Digital reconstruction of the nonfluorescence hologram: (**c**) at the face of the red dots on the die, and (**d**) at the face of the green dots on the die. (**e**) Magnitude and (**f**) phase of the complex Fresnel hologram of the red dots. Digital reconstruction of the red fluorescence hologram: (**g**) at the face of the red dots on the die, and (**h**) at the face of the green dots on the die. (**i**) Magnitude and (**j**) phase of the complex Fresnel hologram of the green dots. Digital reconstruction of the green fluorescence hologram: (**k**) at the face of the red dots on the die and (**l**) at the face of the green dots on the die. Compositions of (**c**), (**g**), (**k**) and (**d**), (**h**), (**l**) are depicted in (**m**), (**n**), respectively

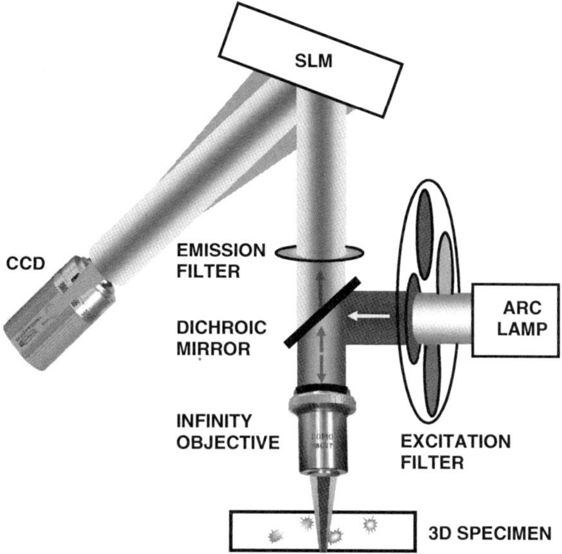

Fig. 4.10 FINCHSCOPE schematic in upright fluorescence microscope. The upright microscope was modified with a reflective SLM positioned at a tilt angle of 11° to reflect emission light from the objective onto the camera

color 3-D holographic image of the object including the red and green fluorescence. While the optical arrangement in this demonstration has not been optimized for maximum resolution, it is important to recognize that even with this simple optical arrangement, the resolution is good enough to image the fluorescent emissions with good fidelity and to obtain good reflected light images of the dice. Furthermore, in the reflected light images in Fig. 4.9c, m, the system has been able to detect a specular reflection of the illumination from the edge of the front dice.

The next system to be reviewed here is the first demonstration of a motionless microscopy system (FINCHSCOPE) based upon the FINCH, and its use in recording high-resolution 3-D fluorescent images of biological specimens [20]. By using high-numerical-aperture lenses, a spatial light modulator, a charge-coupled device camera, and some simple filters, FINCHSCOPE enables the acquisition of 3-D microscopic images without the need for scanning.

A schematic diagram of the FINCHSCOPE for an upright microscope equipped with an arc lamp source is shown in Fig. 4.10. The beam of light that emerges from an infinity-corrected microscope objective transforms each point of the object being viewed into a plane wave, thus satisfying the first requirement of FINCH [18]. An SLM and a digital camera replace the tube lens, reflective mirror, and other transfer optics normally present in microscopes. Because no tube lens is required, infinity-corrected objectives from any manufacturer can be used. A filter wheel was used to select excitation wavelengths from a mercury arc lamp, and the dichroic mirror holder and the emission filter in the microscope were used to direct light to and from the specimen through infinity-corrected objectives.

The ability of the FINCHSCOPE to resolve multicolor fluorescent samples was evaluated by first imaging polychromatic fluorescent beads. A fluorescence bead slide with the beads separated on two separate planes was constructed. FocalCheck polychromatic beads (6 μm) were used to coat one side of a glass microscope slide and a glass coverslip. These two surfaces were juxtaposed and held together at a distance from one another of ~50 μm with optical cement. The beads were sequentially excited at 488, 555, and 640 nm center wavelengths (10–30 nm bandwidths) with emissions recorded at 515–535 nm, 585–615 nm, and 660–720 nm, respectively. Figure 4.11a–d shows reconstructed image planes from 6 μm beads excited at 640 nm and imaged on the FINCHSCOPE with a Zeiss PlanApo ×20, 0.75 NA objective. Figure 4.11a shows the magnitude of the complex hologram, which contains all the information about the location and intensity of each bead at every plane in the field. The Fresnel reconstruction from this hologram was selected to yield 49 planes of the image, 2 μm apart. Two beads are shown in Fig. 4.11b, with only the lower bead exactly in focus. The next image (Fig. 4.11c) is 2 μm into the field in

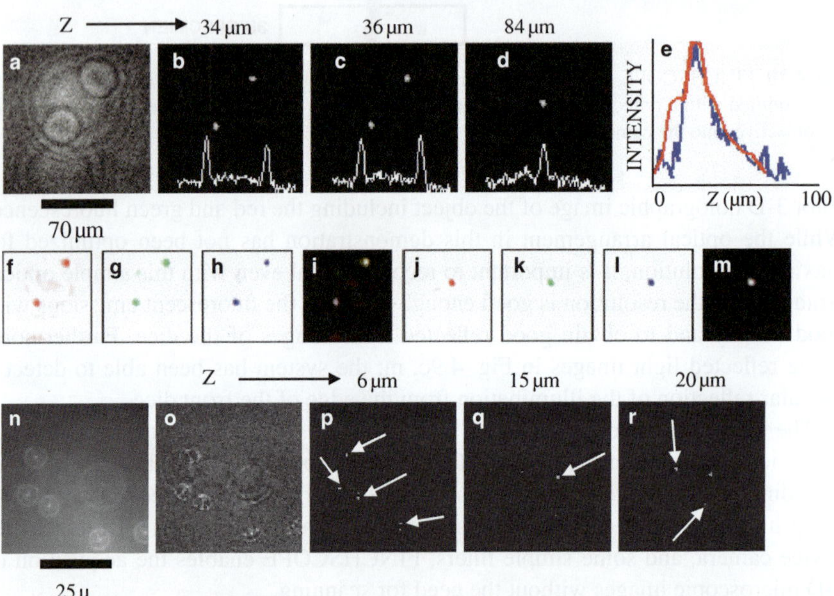

Fig. 4.11 FINCHSCOPE holography of polychromatic beads. (**a**) Magnitude of the complex hologram 6 μm beads. Images reconstructed from the hologram at *z* distances of (**b**) 34 μm, (**c**) 36 μm, and (**d**) 84 μm. Line intensity profiles between the beads are shown at the bottom of panels b–d. (**e**) Line intensity profiles along the *Z* axis for the lower bead from reconstructed sections of a single hologram (*blue line*) and from a widefield stack of the same bead (28 sections, *red line*). (**f–h**) Beads (6 μm) excited at 640, 555, and 488 nm with holograms reconstructed at planes b and (**j–l**) d. (**i, m**) are the combined RGB images for planes b and d, respectively. (**n–r**) Beads (0.5 μm) imaged with a 1.4-NA oil immersion objective: (**n**) holographic camera image; (**o**) magnitude of the complex hologram; (**p–r**) reconstructed image planes 6, 15, and 20 μm. Scale bars indicate image size

the Z-direction, and the upper bead is now in focus, with the lower bead slightly out of focus. The focal difference is confirmed by the line profile drawn between the beads, showing an inversion of intensity for these two beads between the planes. There is another bead between these two beads, but it does not appear in Fig. 4.11b or c (or in the intensity profile), because it is 48 μm from the upper bead; it instead appears in Fig. 4.11d (and in the line profile), which is 24 sections away from the section in Fig. 4.11c. Notice that the beads in Fig. 4.11b, c are no longer visible in Fig. 4.11d. In the complex hologram in Fig. 4.11a, the small circles encode the close beads and the larger circles encode the distant central bead. Figure 4.11e shows that the Z-resolution of the lower bead in Fig. 4.11b, reconstructed from sections created from a single hologram (blue line), is at least comparable to data from a widefield stack of 28 sections (obtained by moving the microscope objective in the Z-direction) of the same field (red line). The colocalization of the fluorescence emission was confirmed at all excitation wavelengths and at extreme Z limits as shown in Fig. 4.11f–m for the 6 μm beads at the planes shown in Fig. 4.11b, f–i and 4.11d, j–m. In Fig. 4.11n–r, 0.5 μm beads (TetraSpeck, Invitrogen) imaged with a Zeiss PlanApo ×63 1.4 NA oil-immersion objective are shown. Figure 4.11n presents one of the holograms captured by the camera and Fig. 4.11o shows the magnitude of the complex hologram. Figure 4.11p–r shows different planes (6, 15, and 20 μm, respectively) in the bead specimen after reconstruction from the complex hologram of image slices in 0.5 μm steps. Arrows show the different beads visualized in different Z image planes. The computer reconstruction along the Z-axis of a group of fluorescently labeled pollen grains (Carolina Biological slide no. 30-4264) is shown in Fig. 4.12b–e. As is expected from a holographic reconstruction of a 3-D object with volume, any number of planes can be reconstructed. In this example, a different pollen grain was in focus in each transverse plane reconstructed from the complex hologram whose magnitude is shown in Fig. 4.12a. In Fig. 4.12b–e, the values of Z are 8, 13, 20, and 24 μm, respectively. A similar experiment was performed with the autofluorescent *Convallaria* rhizome and the results are shown in Fig. 4.12g–j at planes 6, 8, 11, and 12 μm.

The most recent development in FINCH is a new lensless incoherent holographic system operating in a synthetic aperture mode [21]. Synthetic aperture is a well-known superresolution technique, which extends the resolution capabilities of an imaging system beyond the theoretical Rayleigh limit dictated by the system's actual aperture. Using this technique, several patterns acquired by an aperture-limited system, from various locations, are tiled together to one large pattern, which could be captured only by a virtual system equipped with a much wider synthetic aperture.

The use of optical holography for synthetic aperture is usually restricted to coherent imaging [23–25]. Therefore, the use of this technique is limited only to those applications in which the observed targets can be illuminated by a laser. Synthetic aperture carried out by a combination of several off-axis incoherent holograms in scanning holographic microscopy has been demonstrated by Indebetouw et al. [26]. However, this method is limited to microscopy only, and although it is a technique

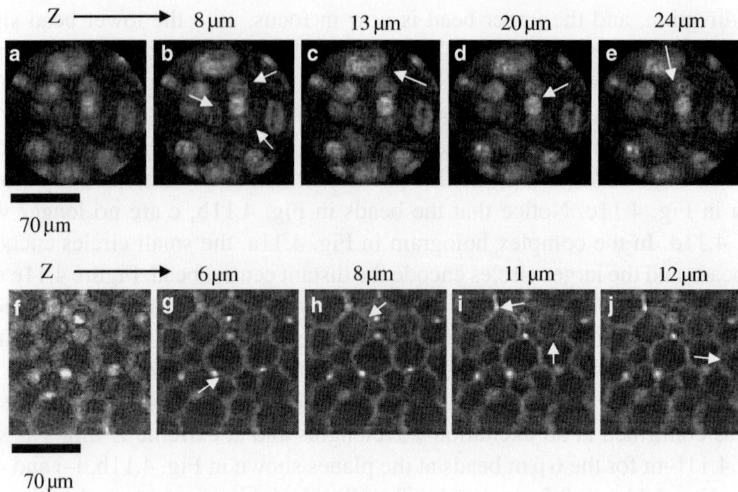

Fig. 4.12 FINCHSCOPE fluorescence sections of pollen grains and *Convallaria* rhizome. The *arrows* point to the structures in the images that are in focus at various image planes. (**b–e**) Sections reconstructed from a hologram of mixed pollen grains. (**g–j**) Sections reconstructed from a hologram of *Convallaria* rhizome. (**a**, **f**) Magnitude of the complex holograms from which the respective image planes were reconstructed. Scale bars indicate image sizes

of recording incoherent holograms, a specimen should also be illuminated by an interference pattern between two laser beams.

Our new scheme of holographic imaging of incoherently illuminated objects is dubbed a synthetic aperture with Fresnel elements (SAFE). This holographic lensless system contains a band-pass filter, a polarizer, an SLM, and a digital camera. SAFE has an extended synthetic aperture in order to improve the transverse and axial resolutions beyond the classic limitations. The term synthetic aperture, in the present context, means time (or space) multiplexing of several Fresnel holographic elements captured from various viewpoints by a system with a limited real aperture. The synthetic aperture is implemented by shifting the BPF–polarizer–SLM–camera set, located across the field of view, between several viewpoints. At each viewpoint a different mask is displayed on the SLM, and a single element of the Fresnel hologram is recorded (see Fig. 4.13). The various elements, each of which is recorded by the real aperture system during the capturing time, are tiled together so that the final mosaic hologram is effectively considered as captured from a single synthetic aperture, which is much wider than the actual aperture.

An example of such system with the synthetic aperture, which is three times wider than the actual aperture, can be seen in Fig. 4.13. For simplicity of the demonstration, the synthetic aperture was implemented only along the horizontal axis. In principle, this concept can be generalized for both axes and for any ratio of synthetic to actual apertures. Imaging with the synthetic aperture is necessary for cases where the angular spectrum of the light emitted from the observed object is wider than the numerical aperture of a given imaging system. In SAFE shown in Fig. 4.13,

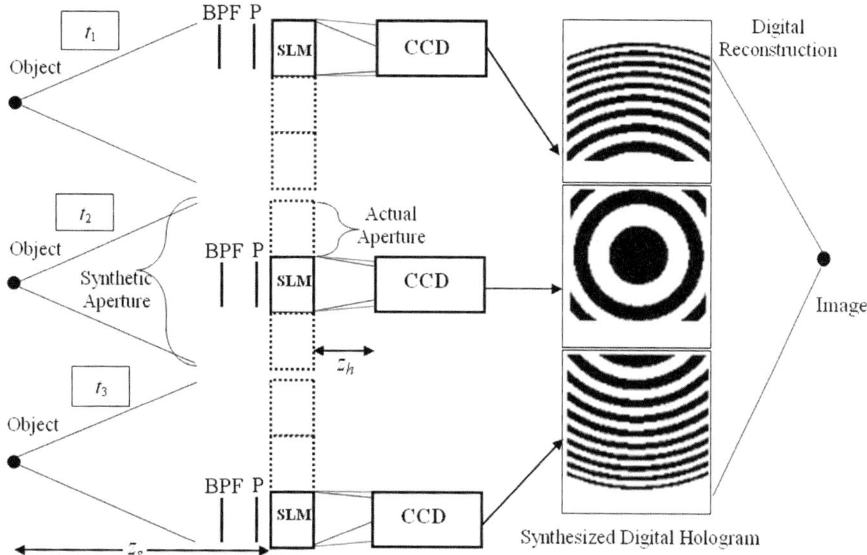

Fig. 4.13 Scheme of SAFE operating as synthetic aperture radar to achieve superresolution. P indicates polarizer

the SLM and the digital camera move in front of the object. The complete Fresnel hologram of the object, located at some distance from the SLM, is a mosaic of three holographic elements, each of which is recorded from a different position by the system with the real aperture of the size $A_x \times A_y$. The complete hologram tiled from the three holographic Fresnel elements has the synthetic aperture of the size $3 \cdot A_x \times A_y$, which is three times larger than the real aperture at the horizontal axis.

The method to eliminate the twin image and the bias term is the same as has been used before; three elemental holograms of the same object and for each point of view are recorded, and each of the holograms has a different phase constant of the SLM's phase mask. The final holographic element is a specific superposition of the three recorded elements. The digital reconstruction of the final complex-valued mosaic hologram is conventionally computed by Fresnel back propagation.

SAFE has been tested in the lab by the system shown in Fig. 4.13. The object in this experiment is a binary grating with cycle length of four lines per mm. The distance from the object to the SLM has been 52 cm, and the distance between the phase-only SLM (Holoeye, PLUTO) and the digital camera (E-VISION, EVC6600SAM-GE5) has been 38.5 cm. A 100 W Halogen ARC lamp has been used for objects illumination, and a BPF (with an 80 nm bandwidth surrounding 550 nm central wavelength) has been placed just in front of the SLM. The results of the experiments are summarized in Fig. 4.14. In the first experiment we have recorded a hologram only by the actual aperture without shifting the system, in the setup shown in Fig. 4.13 at the time t_2. Figure 4.14a shows one of the three masks displayed on the SLM in this experiment. Each of the three masks has one of the three different phase factors: 0°, 120°, or 240°. As mentioned above, these

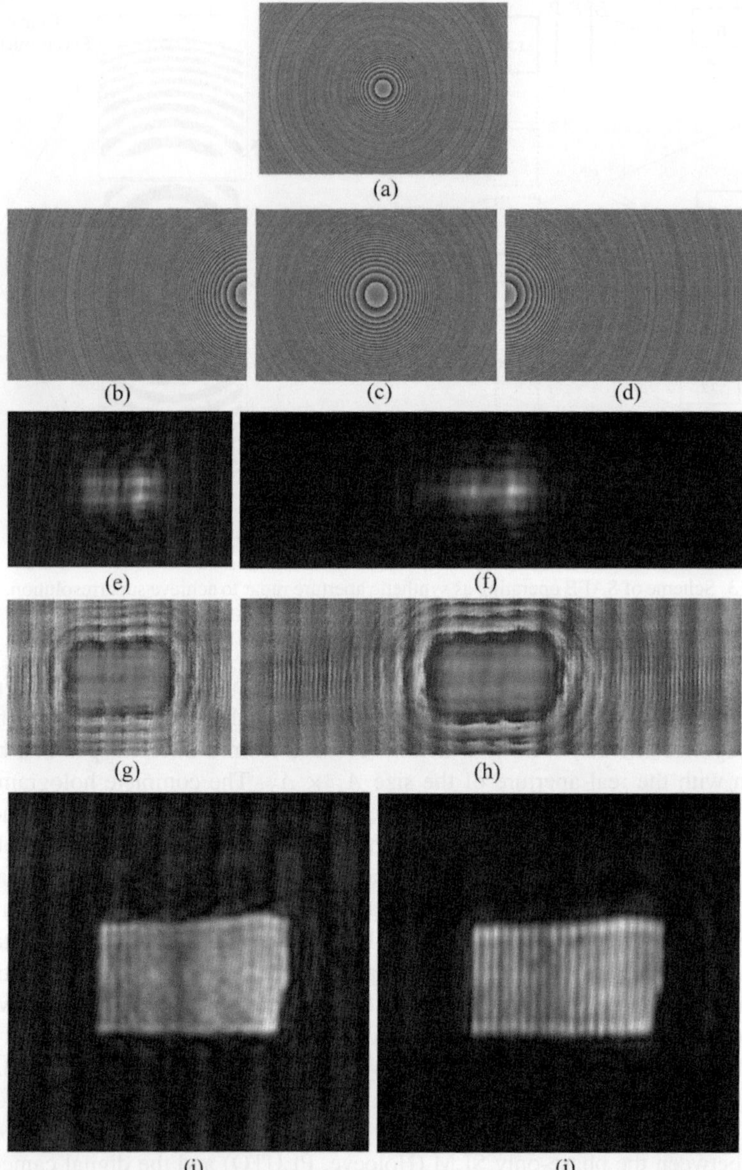

Fig. 4.14 Results of SAFE for the first object with the real and synthetic apertures. (**a**) is the phase distribution of the reflection masks displayed on the SLM at $\theta = 120°$ with the real aperture; (**b–d**) are the same as (**a**) using the synthetic aperture; (**e**) is the magnitude of the final on-axis digital hologram with the real aperture and (**f**) is the same as (**e**) with the synthetic aperture; (**g**) is the phase of the final hologram with the actual aperture and (**h**) is the phase with the synthetic aperture; (**i**) is the reconstruction of the hologram of the binary grating at the best focus distance for the real aperture and (**j**) is for the synthetic aperture

three phase masks with different phase constants are required in order to eliminate the bias term and the twin image from the holographic reconstruction. As stated earlier, another problem with the SLM is that its fill factor is 87%, which means that part of the light is reflected from the SLM without any modulation. In order to avoid the interference of three waves projected on the camera, we have chosen one of the phase elements to be constant. The other phase element has been chosen to be a negative diffractive lens with the shortest focal distance that can be achieved with the SLM having the pixel size of 8 μm. The shortest focal distance guarantees maximum resolution power for a given aperture size. In the case of the actual aperture (1500 $\times$ 1000 pixels) and the synthetic aperture (3000 $\times$ 1000 pixels), the focal distances have been -34 cm and -68 cm, respectively. The $\mathrm{NA}_{\mathrm{in}}$ is 0.0115 and 0.0231 for the real and synthetic apertures, respectively. The NA_{out} is 0.0035 and 0.0044 for the real and synthetic apertures, respectively. Note that the sum of two pure phase functions, i.e., the quadratic phase function $Q[-1/a]$ and the constant phase function in (4.7), is no longer a pure phase function but a complex function with nonconstant magnitude. Since the SLM is a phase-only modulator, we use the previous method of recording general complex function on a phase-only SLM. Each phase function is distributed randomly among half of the SLM pixels.

The three recorded holograms are superposed according to the same superposition (4.4). Figure 4.14e, g is the magnitude and the phase of the superposed holograms for the object. It can be seen that the resolution along the horizontal direction of the reconstructed image, computed by Fresnel back propagation, is damaged in the sense that the image is lacking the original high-frequency gratings along the horizontal direction because the aperture is too narrow to capture the entire gratings spectral content. This damaged reconstructed image is shown in Fig. 4.14i.

In the SAFE experiment, nine different phase masks have been displayed on the SLM, three for each location of the SLM–camera set: left, central, and right. Each of the masks has an actual aperture of 1500 $\times$ 1000 pixels. Each of the three masks at every location has one of the three different phase factors: 0°, 120°, or 240°. In order to avoid edge effects on the recorded holograms, there is an overlap of 750 pixels among the three actual apertures combining the synthetic aperture. For each location of the system, the three recorded holograms have been superposed as mentioned above. Figure 4.14b–d represents three masks out of nine, each of which has been displayed at a different time and at a different location of the setup along the horizontal axis. The superposed complex-valued holographic element from each system's viewpoint is stored in the computer. Upon completing the system movement along the entire synthetic aperture, all three holographic elements are tiled to a single mosaic hologram. Figure 4.14f, h represents the magnitude and the phase of the complete mosaic hologram. The reconstruction result of the mosaic hologram, computed by Fresnel back propagation, is depicted in Fig. 4.14j. The binary grating on the observed objects is seen well in the reconstructed images, indicating that the synthetic aperture is wide enough to acquire most of the horizontal spectral information of the objects.

4.5 Discussion and Conclusions

We have reviewed two different methods of generating incoherent digital Fresnel holograms. The homodyne scanning holography setup has some advantages over the previous designs of scanning holography. The main advantage is that the overall scanning time can be shorter than the case of heterodyne scanning holography. Also, the proposed system is more immune from noise than previous scanning holography systems because in the homodyne scanning holography, only frozen-in-time FZP patterns scan the object. However, the main limitation of scanning holography – the need for a 2-D scan – does exist in the homodyne version as well. In view of this limitation, the scanning holography, in all of its versions, is considered as the slowest method in capturing the scene among all the methods of digital holography.

The second reviewed hologram, the FINCH, is actually recorded by an on-axis, single-channel, incoherent interferometer. This method inherently does not scan the object, neither in the space nor in the time. Therefore, the FINCH can generate the holograms rapidly without sacrificing the system resolution. This system offers the feature of observing a complete volume from a hologram, potentially enabling objects moving quickly in three dimensions to be tracked. The FINCH technique shows great promise in rapidly recording 3-D information in any scene, independently of the illumination. In addition, we have described a rapid, nonscanning holographic fluorescence microscope that produces in-focus images at each plane in the specimen from holograms captured on a digital camera. This motionless 3-D microscopy technique does not require complicated alignment or a laser. The fluorescence emission can be of relatively wide bandwidth because the optical path difference between the beams is minimal in this single-path device. Although at present each reconstructed section is not completely confocal, 3-D reconstructions free of blur could be created by deconvolution of the holographic sections as is typically carried out in widefield microscopy. Time resolution is currently reduced because three holograms need to be captured sequentially. However, in the future, it will be possible to simultaneously capture all three holograms or to overcome the holographic twin image problem and capture only one hologram, as any of the three holograms contains all the essential 3-D information. In the present studies, the image sections were obtained by a process of first capturing three holograms, computing the image z sections from the complex hologram and then, in some cases, further enhancing them by deconvolution. This process could be simplified in the future for real-time display of the holographic image, either with a holographic display system or by algorithms that create the enhanced sections and the 3-D representation directly from the single hologram. There is no need for sectioning or scanning or any mechanical movement. Therefore, this system would be expected ultimately to be faster, simpler, and more versatile than existing 3-D microscopy techniques, which rely on pinhole imaging or deconvolution of stacks of widefield images. At present, the FINCHSCOPE is already considerably faster than conventional 3-D sectioning. For example, the total image capture time for the three FINCHSCOPE images of the pollen grains in Fig. 4.12 was just over 1 s, compared with the 30–45 s needed to create a stack of 48 widefield or spinning disk confocal images. We have also demonstrated fluorescence holography using the high-NA objectives widely used in biological imaging. FINCHSCOPE is able to spatially resolve small beads, biological specimens, and different fluorescence emission colors in x, y, and z planes with perfect registration. The system provides a simple, flexible, cost-effective, and powerful microscopic platform for 3-D imaging. Our demonstration of this advance in microscopy, based on a new, but simple holographic principle, should open up opportunities in many life science and engineering fields, so that living or fixed specimens may be readily observed in three dimensions and possibly at higher resolution than with currently existing techniques.

References

1. A.W. Lohmann, Wavefront reconstruction for incoherent objects. J. Opt. Soc. Am. **55**, 1555–1556 (1965)
2. G.W. Stroke, R.C. Restrick, III, Holography with spatially noncoherent light. Appl. Phys. Lett. **7**, 229–231 (1965)
3. G. Cochran, New method of making Fresnel transforms with incoherent light. J. Opt. Soc. Am. **56**, 1513–1517 (1966)
4. P. Peters, Incoherent holograms with a mercury light source. J. Appl. Phys. Lett. **8**, 209–210 (1966)
5. H.R. Worthington, Jr., Production of holograms with incoherent illumination. J. Opt. Soc. Am. **56**, 1397–1398 (1966)
6. J.B. Breckinridge, Two-dimensional white light coherence interferometer. Appl. Opt. **13**, 2760–2762 (1974)
7. A.S. Marathay, Noncoherent-object hologram: Its reconstruction and optical processing. J. Opt. Soc. Am. A **4**, 1861–1868 (1987)
8. L.M. Mugnier, G.Y. Sirat, D. Charlot, Conoscopic holography: Two-dimensional numerical reconstructions. Opt. Lett. **18**, 66–68 (1993)
9. Y. Li, D. Abookasis, J. Rosen, Computer-generated holograms of three-dimensional realistic objects recorded without wave interference. Appl. Opt. **40**, 2864–2870 (2001)
10. Y. Sando, M. Itoh, T. Yatagai, Holographic three-dimensional display synthesized from three-dimensional Fourier spectra of real-existing objects. Opt. Lett. **28**, 2518–2520 (2003)
11. N.T Shaked, J. Rosen, Multiple-viewpoint projection holograms synthesized by spatially incoherent correlation with broadband functions. J. Opt. Soc. Am. A **25**, 2129–2138 (2008)
12. J.-H. Park, M.-S. Kim, G. Baasantseren, N. Kim, Fresnel and Fourier hologram generation using orthographic projection images. Opt. Exp. **17**, 6320–6334 (2009)
13. T.-C. Poon, A. Korpel, Optical transfer function of an acousto-optic heterodyning image processor. Opt. Lett. **4**, 317–319 (1979)
14. B.W. Schilling, T.-C. Poon, G. Indebetouw, B. Storrie, K. Shinoda, Y. Suzuki, M.H. Wu, Three-dimensional holographic fluorescence microscopy. Opt. Lett. **22**, 1506–1508 (1997)
15. T.-C. Poon, *Optical Scanning Holography with MATLAB* (Springer, New York, NY, 2007)
16. T.-C. Poon, Recent progress in optical scanning holography. J. Hologr. Speckle **1**, 6–25 (2004)
17. J. Rosen, G. Indebetouw, G. Brooker, Homodyne scanning holography. Opt. Exp. **14**, 4280–4285 (2006)
18. J. Rosen, G. Brooker, Digital spatially incoherent Fresnel holography. Opt. Lett. **32**, 912–914 (2007)
19. J. Rosen, G. Brooker, Fluorescence incoherent color holography. Opt. Exp. **15**, 2244–2250 (2007)
20. J. Rosen, G. Brooker, Non-scanning motionless fluorescence three-dimensional holographic microscopy. Nat. Photon. **2**, 190–195 (2008)

21. B. Katz, J. Rosen, Super-resolution in incoherent optical imaging using synthetic aperture with Fresnel elements. Opt. Exp. **18**, 962–972 (2010)
22. J.W. Goodman, *Introduction to Fourier Optics,* 2nd edn. (McGraw-Hill, New York, NY, 1996)
23. S.M. Beck, J.R. Buck, W.F. Buell, R.P. Dickinson, D.A. Kozlowski, N.J. Marechal, T.J. Wright, Synthetic-aperture imaging laser radar: Laboratory demonstration and signal processing. Appl. Opt. **44**, 7621–7629 (2005)
24. V. Mico, Z. Zalevsky, P. García-Martínez, J. García, Synthetic aperture superresolution with multiple off-axis holograms. J. Opt. Soc. Am. A **23**, 3162–3170 (2006)
25. L. Martínez-León, B. Javidi, Synthetic aperture single-exposure on-axis digital holography. Opt. Exp. **16**, 161–169 (2008)
26. G. Indebetouw, Y. Tada, J. Rosen, G. Brooker, Scanning holographic microscopy with resolution exceeding the Rayleigh limit of the objective by superposition of off-axis holograms. Appl. Opt. **46**, 993–1000 (2007)

Part II
Phase Microscopy

Chapter 5
Quantitative Phase Microscopy for Accurate Characterization of Microlens Arrays

Simonetta Grilli, Lisa Miccio, Francesco Merola, Andrea Finizio, Melania Paturzo, Sara Coppola, Veronica Vespini, and Pietro Ferraro

Abstract Microlens arrays are of fundamental importance in a wide variety of applications in optics and photonics. This chapter deals with an accurate digital holography-based characterization of both liquid and polymeric microlenses fabricated by an innovative pyro-electrowetting process. The actuation of liquid and polymeric films is obtained through the use of pyroelectric charges generated into polar dielectric lithium niobate crystals.

5.1 Introduction

In this chapter we will show how digital holography (DH) in the microscope configuration can be the key tool for testing and characterizing micro-optofluidic device and systems through quantitative phase analysis. Among the emerging optofluidics systems, liquid microlens arrays are investigated. The characterization of such optofluidic elements needs the development of accurate technique having the capability to measure the focal length while the lenses are tuned. Moreover, the knowledge of the focal length is not enough while it is highly desirable to have a wide field able to measure and characterize the whole wavefront generated by such liquid microlenses. DH offers the possibility to reconstruct in amplitude and phase the transmitted wavefront furnishing accurate and full information on the lens properties. Furthermore, thanks to the DH feature of numerical focusing, it is possible to retrieve the shape of the wavefront at any location along the optical axis. Consequently, DH has all the attributes to be a prominent method for testing microlenses. Several examples of applications of DH will be given in all the sections of the chapter showing static and dynamic measurement performed on tuneable liquid lenses.

Liquid lenses are becoming important optical devices for a wide variety of applications ranging from mobile-phone cameras [1] to biology. Among others, their foremost advantage is the possibility of changing their shape in order to obtain

F. Merola (✉)
Istituto Nazionale di Ottica del CNR (INO-CNR), Via Campi Flegrei 34, 80078 Pozzuoli (NA), Italy
e-mail: francesco.merola@ino.it

P. Ferraro et al. (eds.), *Coherent Light Microscopy*, Springer Series in Surface Sciences 46, DOI 10.1007/978-3-642-15813-1_5,

different radius of curvature and therefore a variable focus length [2–9]. Nowadays, only a couple of techniques have been implemented for changing the shape of the liquid mass. The first one makes use of a container with a flexible and elastic membrane that changes its shape under hydrostatic and/or pneumatic forces [10–13]. A different approach is based on the use of the electrowetting (EW) effect [1, 14–18], where the shape of the liquid sample is changed by the action of external electric forces, thus adapting the lens focus. Important advancements have been achieved in the development of liquid single lenses with a variable focal length. However, even though many configurations have been demonstrated for single liquid lenses, very few cases are reported for liquid microlens arrays with a variable focus. Microlens arrays can find important applications in medical stereo-endoscopy, imaging, telecommunication and optical data storage. Recently, two significant results have been reported on the design and fabrication of tunable liquid microlens arrays. The first one was developed by using the dielectrophoretic effect on two liquid layers made of water and oil [19]. The focal power of such microlenses was varied by the EW process and the focal lengths were reported in the range from about 1.5 to 2.2 mm. The second noteworthy result [20] consists of an hybrid optical configuration made of solid microlenses combined with liquid filled lenses, thus obtaining a tunable doublet with minimal optical aberrations. The optical characterization was performed as a function of the applied pressure to the microfluidic apparatus.

This chapter presents the results concerning the fabrication of microlenses, also tunable, by a completely new approach based on the use of charges generated pyroelectrically and thus through a technique that we call here pyro-EW. Moreover, a true characterization of these structures is performed by a quantitative phase microscopy (QPM) approach which can provide important information about the optical behaviour of such microlenses. Sections 5.2 and 5.3 describe the fabrication and characterization of spherical liquid microlenses obtained by a pyro-electrowetting-based technique, while Sect. 5.4 deals with the characterization of hemicylindrical microlenses. Section 5.5 presents the results of QPM characterization applied to polymer-based microlenses.

5.2 Arrays of Liquid Microlenses

The most popular and well-established approach to obtain liquid lenses is based on the electrowetting (EW) effect [3]. Such systems usually consist of two immiscible liquids manipulated into special cases made of hydrophobic coatings and electrodes. An external voltage changes the equilibrium at the liquid–liquid and solid–liquid interfaces causing a reshaping and rearrangement of the liquid meniscus. The change of curvature, in modulus and sign, has direct effect on the refraction of the transmitted light, thus allowing to switch between a converging and a diverging lens with flexible focal lengths [1, 2, 9, 21–24]. The basic concept of EW is that a sessile liquid drop free standing on a flat electrode surface can be manipulated by a second needle-like electrode immersed into the drop [25].

Liquid microlenses have been obtained here by an EW effect driven pyroelectrically that we call pyro-EW (PEW). The pyroelectric effect is achieved onto ferroelectric substrates of lithium niobate (LN) crystals. As described above, the operation of conventional EW-based microfluidic devices demands more or less complex electrode geometries to actuate a liquid lens, thus requiring special technological steps and materials for the fabrication. The method developed here simplifies significantly the possibility of functionalizing a specific and appropriate material to get a microfluidic lens array onto a single chip.

LN is a very well-known ferroelectric material widely used as a key element in optical modulators for fibre optic telecommunications [26] and in non-linear optic devices [27]. The spontaneous polarization of LN crystals can be reversed by the electric field poling process [28, 29], thus enabling the fabrication of periodically poled LN (PPLN) crystals. An external voltage exceeding the coercive field of the material (around 21 kV/mm) is necessary to reverse the ferroelectric domains and the inversion selectivity is usually ensured by an appropriate resist pattern generated by photolithography [29]. Figure 5.1 shows the optical microscope image of two PPLN samples fabricated by the electric field poling and used for the lens effect experiments investigated here.

The two samples consist of a square array of bulk reversed domains with a period around 200 μm along both the x and y directions. They differ only for the geometry of the resist openings (see [29] for details) and both of them were used for the same lens effect experiments, in order to test the reliability of the technique. The samples were not etched, contrary to the conventional procedures used to visualize the domain pattern [31, 32]. In fact, the reversed domains are visible thanks to the contrast enhancement provided by the appropriate aperture of the condenser diaphragm, thus improving the diffraction image of the domain walls related to the electro-optic effect induced by the so-called internal field [33] arising after the electric field poling. The PPLN sample was mounted onto a digitally controlled hotplate to ensure a reliable control of the substrate temperature during the experiments. The liquid used

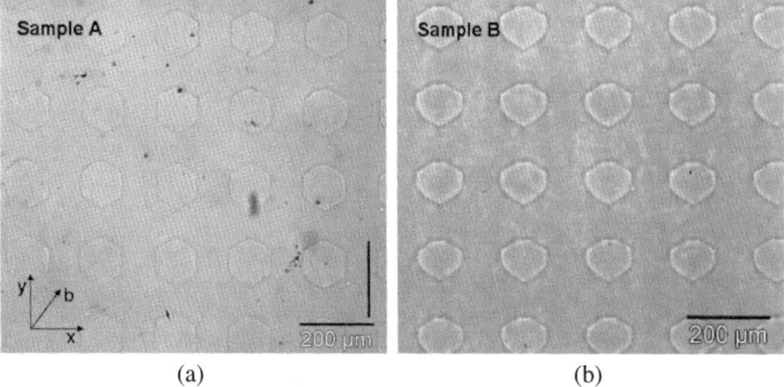

Fig. 5.1 Optical microscope images of two PPLN samples with a square array of reversed domains obtained by using two different lithographic masks consisting of a square array of (**a**) open circles and of (**b**) open hexagons [30]. The period of the structures is around 200 μm in both cases

in this work was a carboxylic acid (pentanoic acid – $C_5H_{10}O_2$) in the form of an oily substance. The values of the dielectric constant and of the refractive index are 2.66 and 1.407–1.410, respectively. The PPLN sample was coated with a thin film of this oil (oil thickness is about 200 μm) and subject first to a heating process up to 100°C at a rate of around 20°C/min and then let cooling down to room temperature. Figure 5.2a–d shows a few frames of the optical microscope movies captured during the cooling process of the oil-coated sample *A*.

Heating process was performed by increasing the temperature from 40°C up to 100°C. The cooling was achieved by letting the temperature to decrease from 100°C down to 40°C, thus with an approximate rate of 12°C/min. The whole process took about 5 min and the liquid lens array kept a stable topography for about 30 min at 40°C, after its formation. The evolution of the oil film topography is clearly visible in the frames in Fig. 5.2. The lens effect is more pronounced in case of the cooling process. This is reasonably due to the different nature of the surface charges that generate the electric potential modulation on the substrate, as will be discussed in the following. The liquid microlenses were formed in correspondence of the hexagonal domains and thus with a lateral dimension of about 100 μm. Practical limitations to the fabrication of smaller lenses are not expected because reversed domains with lateral dimensions down to tens of micron can be reliably obtained in LN substrates. Anyway, the performance of the smaller lenses would be dramatically affected by the consequent enhancement of the diffraction effects due to the fact that the lens aperture can become comparable with wavelength.

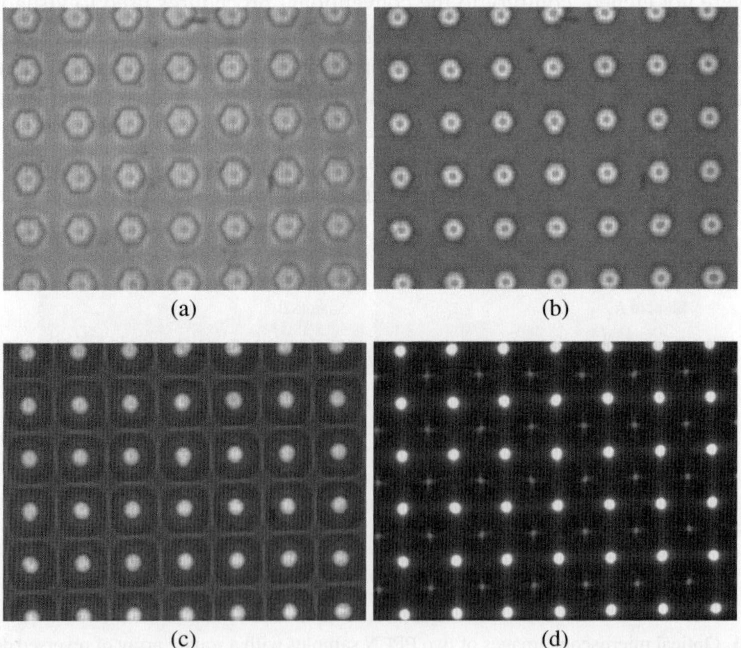

Fig. 5.2 (a–d) Optical microscope frames of the oil-coated sample *A* during cooling process (from [30])

It is important to note that the response of the liquid lens array was quite fast. In fact, the lens array was formed about 1 s after the temperature started to decrease. Moreover, the lens formation followed faithfully the domain grating, thus exhibiting perfect homogeneity over the whole patterned region. In principle, liquid microlenses arrays would be possible with areas as large as desirable depending basically on the area obtainable by the lithographic process used for the domain inversion process. Figure 5.2 shows clearly the degree of uniformity obtainable by the technique, with a field of view including 42 lens elements corresponding to an area of about $(1.4 \times 1.2)\ \text{mm}^2$. The limited field of view corresponding to the chosen magnification prevented to report larger view images.

5.2.1 Interpretation of the Effect

It is well known that LN is a rhombohedral crystal belonging to the point group 3 m at room temperature [34]. The lack of inversion symmetry induces different effects including the pyroelectricity. This is the manifestation of the spontaneous polarization change ΔP_s following a temperature variation ΔT, according to $\Delta \boldsymbol{P}_i = \boldsymbol{p}_i \Delta \boldsymbol{T}$, where ΔP_i is the coefficient of the polarization vector and p_i is the pyroelectric coefficient. At equilibrium, all P_s in the crystal are fully screened by the external screening charge and no electric field exists [35]. The change of the polarization, occurring with temperature variation, perturbs such equilibrium, causing a lack or excess of surface screening charge. Consequently, an electrostatic state appears and generates a high electric field at the crystal surface [36, 37]. Figure 5.3a, b shows the schematic view of the PPLN sample cross section with the charge distribution occurring at the equilibrium state and in case of heating/cooling treatment, respectively. Arrows indicate the orientation of the ferroelectric domains.

According to the pyroelectric effect [35] the heating process makes the polarization magnitude to decrease, thus leaving surface screening charges uncompensated (see Fig. 5.3b, top). These generate a net electric charge distribution depending on the inverted domain structure, with positive and negative sign onto the reversed and un-reversed domain regions, respectively. The screening charges in excess, continuously produced during the heating process, are no more attracted by the polarization charge and consequently are free to diffuse into the oil film. Conversely, the cooling process makes the polarization magnitude to increase, thus generating

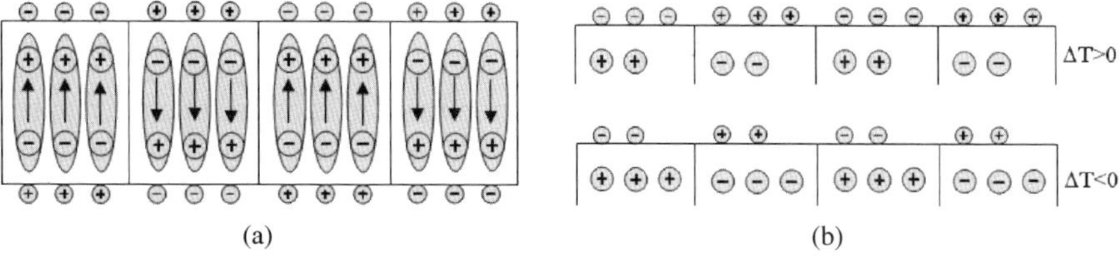

Fig. 5.3 Schematic view of the PPLN sample cross section with the charge distribution exhibited (**a**) at the equilibrium state (**b**) in case of heating (*top*) and (*bottom*) cooling process (from [30])

an electric charge onto the crystal surface (see Fig. 5.3b, bottom) and giving place to the so-called double charge layer at the solid–oil interface [37, 38]. In fact, the uncompensated polarization charge tends to interact with the dipole molecules in the oil, thus redistributing them in proximity of the solid surface. Therefore, the more pronounced lens effect observed in case of cooling (see Fig. 5.2b) is due to the electrostatic action of the uncompensated polarization charge which is more intense compared to that of the screening charges during heating. The lens-like array topography exhibited by the oil film can be considered as the result of the equilibrium condition between the surface tensions and the electric forces related to the charge redistribution on the substrate. In the general case of a sessile drop, the surface tensions at the solid–liquid γ_{sl}, solid–gas γ_{sg} and liquid–gas γ_{lg} interfaces are described by the one-dimensional Young equation:

$$\gamma_{\mathrm{sl}} + \gamma_{\mathrm{lg}} \cos \vartheta = \gamma_{\mathrm{sg}} \tag{5.1}$$

where ϑ corresponds to the contact angle of the drop. The charges at the solid–liquid interface reduce the surface tension according to the Lippman equation [39]:

$$\gamma_{\mathrm{sl}}(V) = \gamma_{\mathrm{sl0}} - \frac{1}{2} c V^2 \tag{5.2}$$

where γ_{sl0} corresponds to zero charge condition and c is the capacitance per unit area assuming that the charge layer can be modelled as a symmetric Helmholtz capacitor [38]. It is important to note that in the present work the surface was not a metal and the liquid was not an electrolyte, as basically assumed by the double charge model [38, 40]. However, a similar model can be as well invoked even in case of dielectric surfaces [41]. Therefore, in the case investigated here, the presence of the net electric charge underneath the crystal surface (see bottom drawing in Fig. 5.3b), generated pyroelectrically, lowers the surface tension due to the repulsion between like charges that make the work for expanding the surface area [37]. The air–liquid interface exhibits a waviness profile to minimize the energy of the whole system. Simulations of the electric potential distribution, generated pyroelectrically, were performed by a finite element-based calculation and Fig. 5.4a shows the result. The plot refers to a section along a diagonal direction (direction b in Fig. 5.1).

The simulation clearly shows that the electric potential is modulated according to the domain structure, thus exhibiting minimum value in correspondence of the hexagon centres. The surface tension profile was then calculated by using (5.2) and the corresponding behaviour, in accordance with the experimental results, is shown in Fig. 5.4b. In fact, the solid–liquid interface tension appears to be modulated according to the electric potential. It is important to consider that the number of charges between two consecutive hexagons is higher along the b direction compared to the horizontal or vertical direction. Therefore, the work done by the charges along the b direction produces a stronger hydrostatic pressure towards hexagon centres, thus leading to the formation of liquid microlenses in correspondence of the hexagons.

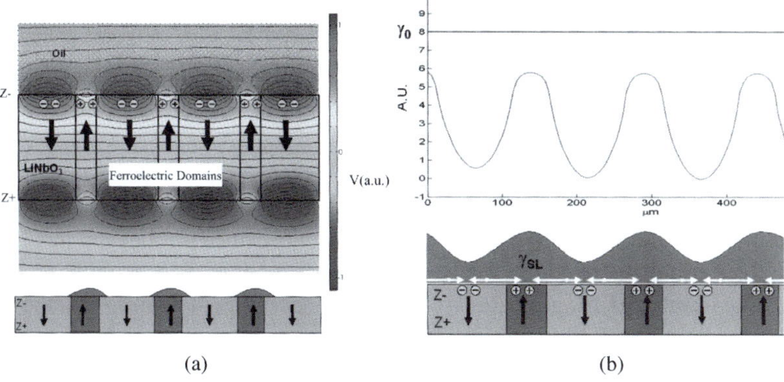

(a) (b)

Fig. 5.4 (**a**) Schematic view of the sample cross section with the simulated electric potential distribution generated pyroelectrically; (**b**) (*top*) surface tension profile and (*bottom*) the schematic view of the corresponding oil film topography. The *black arrows* indicate the orientation of the spontaneous polarization (from [30])

5.2.2 *Characterization by Quantitative Phase Microscopy*

The array of microlenses was observed and investigated during the cooling process by a QPM technique based on DH [42]. The schematic view of the optical setup used for the acquisition of the images is shown in Fig. 5.5a, while Fig. 5.5b shows a detailed view of the sample transmitting the object laser beam.

The wavefront modifications induced by the microlens array onto a collimated laser beam (plane wavefront) were analysed. The phase map of the transmitted wavefront at the exit pupil of the microlens array can be obtained by the numerical reconstruction of digital holograms, which consists in reconstructing the complex wavefront transmitted by the microlens array by back-propagating the diffraction field.

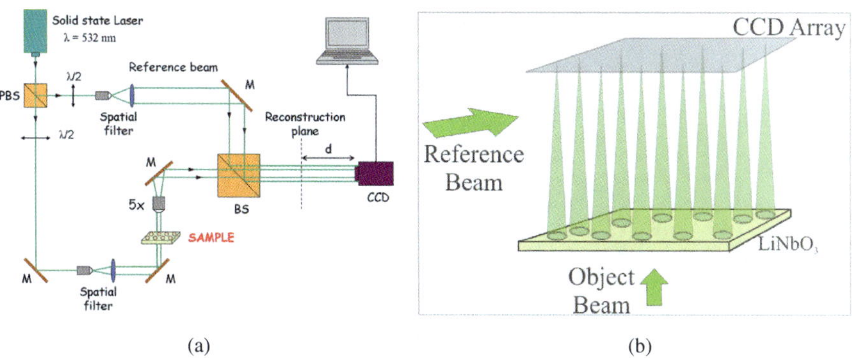

(a) (b)

Fig. 5.5 (**a**) Schematic view of the QPM configuration. PBS, polarizing beam splitter; MO, microscope objective; PH, pin-hole; M, mirror; BS, beam splitter and (**b**) detailed view of the sample during the measurements (from [30, 43])

Amplitude and phase maps of the object wavefront can be retrieved from the complex wavefront. Numerical methods and the principle of operation to get wavefront reconstruction are reported in [29] and [42]. Several holograms were recorded at a rate of 1 image per second. The movie frames in Fig. 5.6 show the wrapped phase maps modulus 2π corresponding to 3×4 lens elements on the incoming collimated beam during the system cool down. The phase curvature indicates the existence of the lens effect. In fact, the curvature of the oil–air interface changes while the sample is cooling, as can be clearly noticed into the images in Fig. 5.6. This effect could be exploited for having an array of microlenses with a variable focus. The number of fringes decreases during the cooling, indicating that the liquid layer is returning back to its initial condition corresponding to a completely erased waviness and thus to an infinite focal length.

Figure 5.7 shows a portion of the mod 2π unwrapped phase map corresponding to an image of Fig. 5.6 and allows to estimate the wavefront curvature in correspondence of 2×2 microlenses of the array.

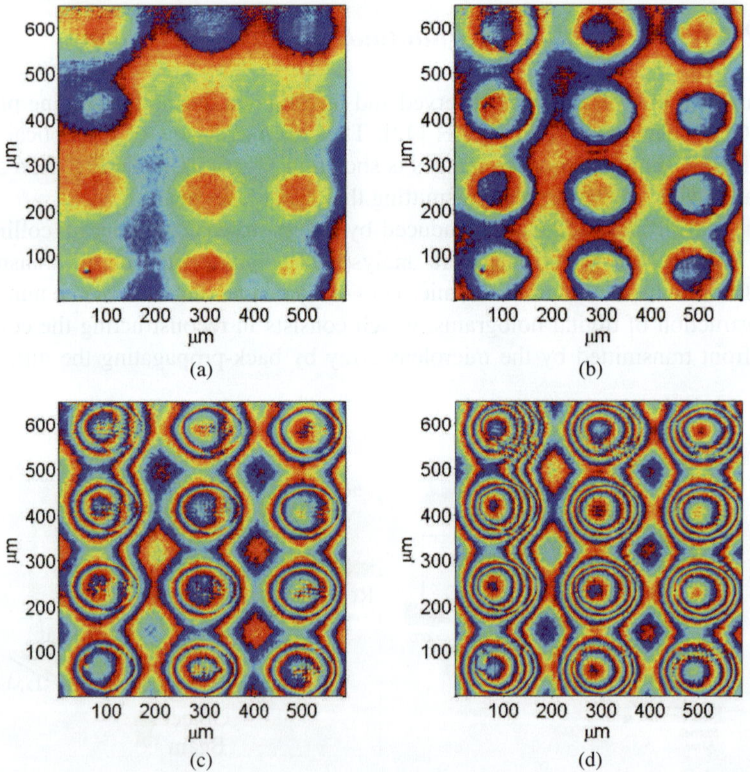

Fig. 5.6 Movie frames of the evolving two-dimensional distribution of the wrapped phase map, modulo 2π, corresponding to 3×4 lens elements on the incoming collimated beam, during cooling in case of the sample B. The lens effect appears in the first frames (**a**, **b**) and is more evident in the last frames (**c**, **d**). The phase map was reconstructed at a distance of 156 mm (from [30])

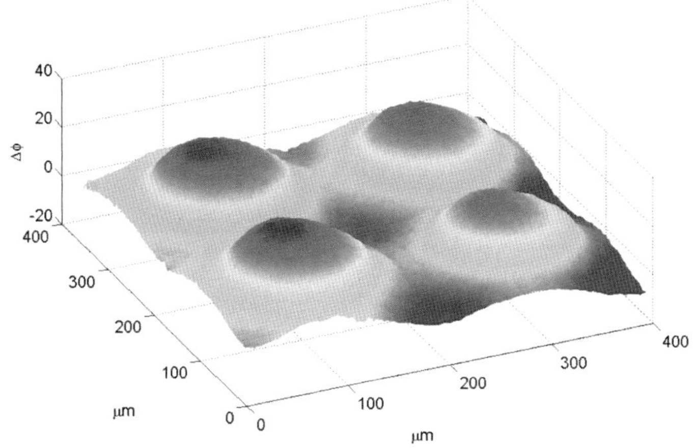

Fig. 5.7 Unwrapped phase map corresponding to a portion of the image in Fig. 5.6 for a fixed temperature during cooling (from [30])

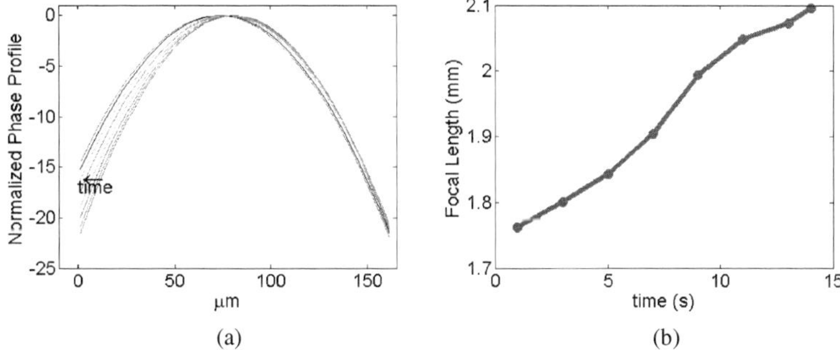

Fig. 5.8 (**a**) Phase profiles of the transmitted wavefront calculated for different frames during cooling; (**b**) focal length values calculated as a function of time during the cooling process (from [30])

Fitted parabolic profiles, calculated during cooling, indicate the presence of variable defocus, namely a variable focal length as shown in Fig. 5.8a, where the phase profiles of the transmitted wavefront, corresponding to different time frames (1, 3, 5, 7, 9, 11, 13, 14 s) during cooling, are reported. The slight tilt of the sample, respect to the microscope objective into the interferometric setup, is revealed by the asymmetry of the curves in Fig. 5.8a. The focal length f of the liquid lenses can be retrieved by fitting the unwrapped phase map $\Phi(x, y)$ to a second-order polynomial according to

$$\Phi(x, y) = \frac{2\pi}{\lambda} \frac{\left(x^2 + y^2\right)}{2f} \tag{5.3}$$

Figure 5.8b shows the variation of the focal length (from 1.75 mm up to 2.1 mm) corresponding to the time frames of Fig. 5.8a during the cooling process. This

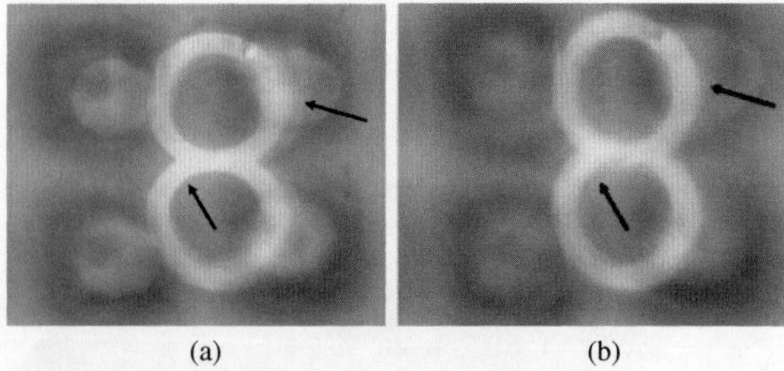

(a) (b)

Fig. 5.9 Optical microscope images of the target "8" observed through the microlens array at two different focal planes imaging the target (**a**) through the region outside the lenses and (**b**) through the up-right microlens, where the focusing capability of the microlenses is clearly visible (from [30])

effect could be used to have an array of microlenses with variable focus. Moreover, an imaging experiment was also performed in order to show the possibility of using this kind of variable focus microlens array for integrated microscope applications.

Figure 5.9 shows the imaging capability of the microlens array on a portion of USAF photo-target. The LN substrate, with the microlens array, was positioned over the target and observed under the optical microscope. The images in Fig. 5.9a, b were acquired at two different focal planes corresponding to focusing the target through the regions outside and inside the aperture of the microlenses, respectively.

Furthermore, different oils with variable densities, such as paraffin oil, primed oil and sweet almond oil, were also used for analogous experiments and the lens effect appeared to work even though exhibiting slight different behaviours depending essentially on the oil density properties. It is important to note that the lens array effect is presented here in case of a relatively simple geometry as the square array, in order to demonstrate the reliability and feasibility of the pyroelectrically driven liquid lens effect. Anyway, more complex geometries, such as hexagonal or circular arrays, could be used for further investigations, by simply providing the appropriate photolithographic mask for the pattern generation.

5.3 Arrays of Tunable Liquid Microlenses

Nowadays important advancements have been achieved in the development of single liquid lenses with a variable focal length. Such components will find wide practical applications even in consumer electronics, such as for the cameras of the cellular phones [1]. Recently, special liquid-based optics, such as axicon lenses, have been developed by using polymeric materials as liquid containers [44]. Cylindrical lenses with hydrodynamic tunability [45] and polymer lenses tuned thermally [46] have been also investigated and tested. However, even though many configurations have been demonstrated for single liquid lenses, very few cases are reported for liquid microlens arrays with a variable focus. Reference [19] reports a dielectrophoretic-based method for inducing lens effect on two liquid layers made of water and oil, where the microlenses have a diameter around 100–140 μm and a focal length ranging between 1.5 and 2.2 mm. The aberrations of such microlenses were calculated numerically by an optical code program, on the basis of the well-known fabrication specifications and material parameters. Another noteworthy result, reported in [20], consists of an hybrid optical configuration made of solid microlenses combined with liquid filled lenses, thus obtaining a tunable doublet with minimal optical aberrations. The optical characterization was performed as a function of the applied pressure to the microfluidic apparatus. In fact, under fluid pressure, the elastomer (PDMS) membrane, containing the liquid, experienced a deformation. An interferometric profilometer was used for measuring the height of the solid lenses, whereas an optical microscope with monochromatic light was adopted to evaluate the f number as a function of the hydrostatic pressure [20]. It is important to note that all of the aforesaid optical systems require the minimization of the aberrations which is achieved by a proper design and a subsequent accurate optical characterization. For example, the fabrication of the liquid microlens array by the two approaches mentioned above requires a relatively complicated process making use of different materials. Other kinds of microlens arrays are based instead on the use of liquid crystals [47, 48], where the birefringence is used to tune the lens through the change of the refractive index. However, the discussion here is limited to tunable liquid microlens arrays that avoid the use of liquid crystals.

This section presents the results concerning the possibility of using the PEW effect described in the previous section for generating an array of tunable liquid microlenses which benefit simultaneously of an electrode-less configuration and of a simplified fabrication process, compared to the above-mentioned techniques. In fact, the more sophisticated setup used here provides a more reliable control of the pyroelectric effect, thus enabling a true characterization of the tunable microlenses. In particular, two different lens regimes that are called here *separated lenses regime* (*SLR*) and *wave-like lenses regime* (*WLR*) have been obtained. In the first case the liquid microlenses are separated from each other due to the breakup of the liquid layer, while in the second case the oil film appears uniformly distributed over the sample with a curved profile in correspondence of specific regions. In addition, the possibility of focus tuning with different optical behaviours was investigated for each kind of microlenses.

5.3.1 Description of the Two Liquid Microlens Regimes: Separated Lenses Regime (SLR) and Wave-Like Lenses Regime (WLR)

The same PPLN sample shown in Fig. 5.1a and the same carboxylic acid (pentanoic acid – $C_5H_{10}O_2$) as the oily substance spread onto PPLN substrates were used in these experiments. Figure 5.10a, b shows the schematic views of the sample cross section under the *WLR* and the *SLR*, respectively. The black arrows indicate the orientation of the spontaneous polarization into the original (light grey) and reversed (dark grey) ferroelectric domains. The case reported in this figure refers to

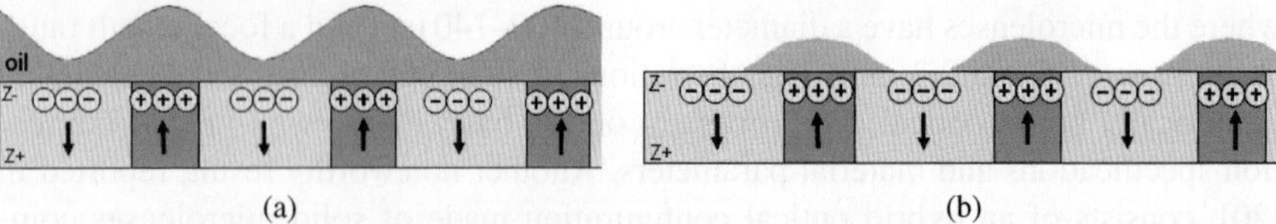

Fig. 5.10 Schematic views of the sample cross section corresponding to the *wave-like lenses regime* (*left*) and to the *separated lenses regime* (*right*) of the microlens array. In both cases the temperature of the substrate is decreasing. The black arrows indicate the orientation of the spontaneous polarization (from [43])

the cooling process, so that uncompensated negative charge is developed in correspondence of the negative side of the spontaneous polarization, and vice versa.

The *WLR* consists of an oil layer with a sinusoidal-like profile according to the electric field generated onto the crystal surface by the pyroelectric effect. Such profile results from the equilibrium condition achieved by the surface tensions at the solid–liquid and liquid–air interfaces in presence of the above-mentioned electric potential, according to the numerical simulations presented in the previous section. Typically the thickness of the liquid layer that generates the *WLR* ranges between 100 and 300 μm. Differently, the *SLR* is obtained by using a thinner layer of liquid onto the crystal surface, obtained by spin coating. Having a thinner oil film the work performed by the charges is able, in this case, to break up the film oil into the regions surrounding the hexagonal regions, thus forming isolated liquid microlenses. The layer thickness in the latter case was estimated to range from 10 to 25 μm. In the case of *WLR* the lens effect is due to a temporary waviness obtained at the liquid–air interface. Higher is the amplitude of the waviness stronger is the lens effect. The focal length in this case of each liquid lens goes from infinity to the minimum value obtained by the shorter curvature at the air–liquid interface. However, the *WLR* is clearly a temporary configuration that disappears when the temperature returns back to its initial value. The tunable effect exploits this temporary effect, and the obtainable focal range depends from the temperature change as well as from the liquid properties. Conversely, the *SLR* exhibits a more stable behaviour. Nevertheless, due to the breakup of the liquid layer, the shape of the microlenses is different and the tunability varies correspondingly.

5.3.2 Characterization of WLR Tunable Lenses by Quantitative Phase Microscopy

Both regimes were investigated by QPM through a DH microscope setup as in Sect. 5.2.2, in order to get a complete characterization of the microlens array for the two regimes in terms of focusing behaviour and optical aberrations. The sample is positioned on a thermo-controlled plate and put into the object arm of the interferometer, while the light coming from the array is collected by a microscope objective and made to interfere with the reference beam. The resulting interference patterns are captured by the CCD camera that is positioned in correspondence of an out of focus plane. An appropriate numerical manipulation of such interferograms allows one to retrieve information about the complex wavefront, and thus about the phase and the intensity of the optical wavefield transmitted by the sample. Two sets of holograms were acquired during the temperature variation, corresponding to the heating and the cooling process. Such holograms were then elaborated numerically in order to reconstruct the phase distribution of the transmitted wavefront, according to standard DH procedures that can be found in [42]. The reconstructed phase images corresponding to different instants of the heating or the cooling process were collected into movies showing the evolution of the wavefront curvature. These results allow one to retrieve information about the focal length variation *versus* the substrate temperature. The aberration characteristics of the liquid microlenses were obtained by a numerical fitting of the phase maps with appropriate polynomials describing different aberration terms. The temperature of the thermo-controlled plate under the sample was varied with a rate of 10°C/min that is lower than that estimated in the previous section (i.e. 20°C). Moreover, in this work the temperature was controlled with an accuracy of 0.1°C and the interferometric characterization was performed by registering the temperature value for each digital hologram captured during the thermal treatment. Figure 5.11a–c and 5.11d–f presents movie frames showing the temporal evolution of the mod 2π spatial phase distribution corresponding to the heating and the cooling process, respectively. The temperature values range from 30 to 90°C in both cases. The images clearly show that the wrapped phase maps exhibit a different behaviour. In case of rising temperature (see Fig. 5.11a–c) the number of fringes increases only during the first part of the process, up to a temperature value of about 65°C, and then stabilizes. Conversely, during the cooling process (see Fig. 5.11d–f), the number of fringes increases during

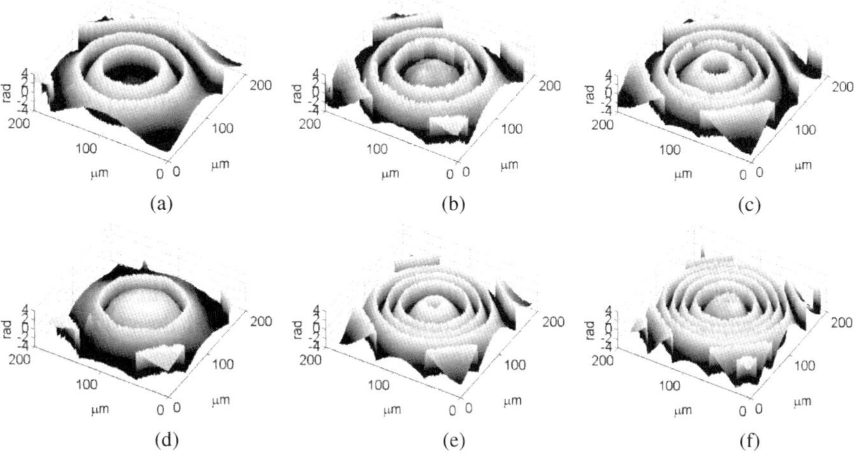

Fig. 5.11 Movie frames showing the evolution of the wrapped mod 2π phase map during the heating (**a**, **b**, **c**) and the cooling process (**d**, **e**, **f**) (from [43])

Table 5.1 List of coefficient values (from [43])

	Array 1			Array 2
	Lens A	Lens B	Lens C	Lens B
The constant term	3.21	3.23	3.96	2.75
Tilt about x (10^{-4})	25.0	8.54	21.0	−22.0
Tilt about y (10^{-4})	−82.0	−94	−45.0	−31.0
Astigmatism with axis at 45° (10^{-6})	−2.90	−3.88	0.84	−5.61
Focus shift (10^{-4})	−9.26	−7.23	−7.46	−6.91
Astigmatism with axis at 0° or 90° (10^{-5})	−5.57	1.46	−3.78	2.56
Triangular Astigmatism on x axis (10^{-8})	−0.29	6.50	−1.94	31.97
Triangular Astigmatism on y axis (10^{-7})	−4.27	−5.38	−0.62	−1.61
Third-order spherical aberration (10^{-8})	3.61	2.77	2.57	3.02

that our device displays the same properties for each microlens. Figure 5.14 shows the surface distribution of each polynomial that contributes to the fitted surface for a single lens, in order to compare the focus term to the other terms which describe the wavefront curvature. In particular, Fig. 5.14c displays the focus term, which is the main term contributing to the phase distribution. Figure 5.14d shows the distribution of the third-order spherical aberration that appears flat in the central region while diverging in the peripheral one. This divergence is due to the oil layer existing between adjacent lenses, in fact the oil layer redistributes its mass during the temperature variation but, in this case, without breaking up between adjacent lenses. Figure 5.14e–h shows the distributions of the astigmatism terms. The information provided by Figure 5.14d and Table 5.1 reveals that the spherical aberration term has a quite high value with the main contribution due to the intrinsic waviness at the air–liquid interface in between two adjacent microlenses where the curvature changes from convex to concave.

5.3.3 Characterization of SLR Tunable Lenses by Quantitative Phase Microscopy

A similar analysis was carried out in case of the *SLR*. Figure 5.1a–d shows the movie frames of the two-dimensional wrapped phase distribution corresponding to this configuration for a portion of 4 × 4 microlenses evolving during the heating and the cooling treatments, respectively. In Fig. 5.15a the temperature varies form 25.6 to 79.0°C. As can be noticed in the movie of Fig. 5.15a, while the temperature reaches 57°C (at about half of the duration of the movie) the liquid layer breaks up and the flat surface of the substrate becomes visible.

At the end of the movie, corresponding to about 79.0°C, the liquid droplets are clearly separated. Conversely, the movie frames in Fig. 5.15b show the behaviour of the open microfluidic system while the temperature changes between 79.3 and 34.4°C. When the temperature reaches 34.4°C the lens effect disappears completely and the liquid layer relaxes. The number of fringes is clearly higher compared to the case of the *WLR* and their density increases from the central to the side region of the lenses. This means that, in case of *SLR*, the slope is more pronounced in

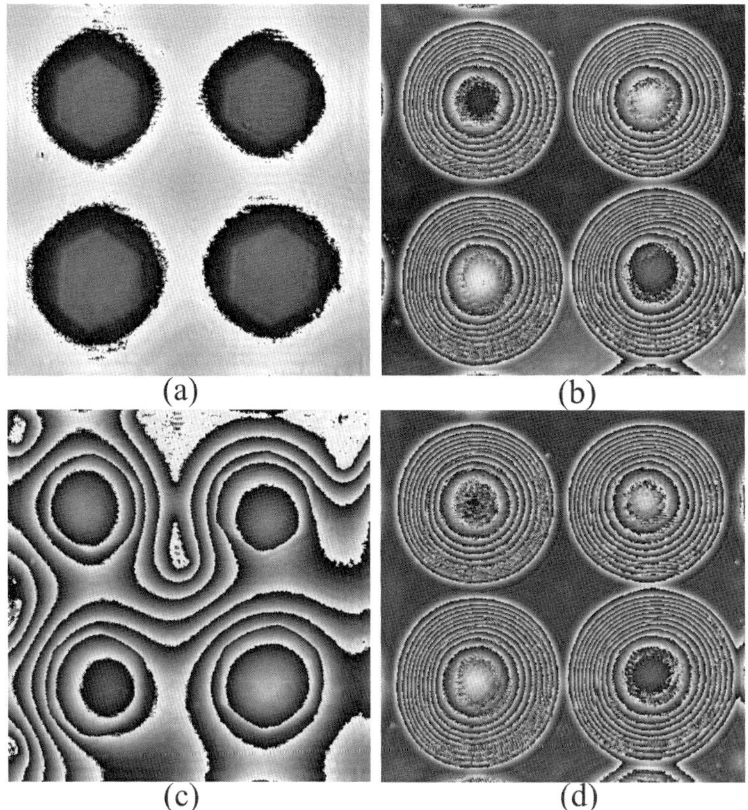

Fig. 5.15 Movie frames of the wrapped phase distribution evaluated for a portion of the lens array in case of separated lenses SRL (**a**, **b**) during heating and (**c**, **d**) during cooling (from [43])

correspondence of the peripheral regions and that the lenses are relatively flat in the centre. This behaviour is proved by the unwrapped phase profiles shown in Fig. 5.16a, b, where the two pictures refer to the heating and the cooling process, respectively.

As in the previous case, the cooling treatment leads to a more pronounced phase variation compared to that occurring during heating, while the focal length values range between around 30 and 10 mm in both processes (see Fig. 5.16c, d). The values of the focal length at the end of both processes are one order of magnitude higher than in case of *WLR*, due to the flatness of the oil profile in the centre of the lenses. The aberrations were evaluated by a one-dimensional fitting of the wavefronts, and Fig. 5.17a shows the measured and the fitted phase profile. Different from the analysis carried out for the *WLR*, the high frequency of the fringes makes the two-dimensional unwrapping procedure very difficult to be accomplished, so that the following equation was used for the fitting procedure:

$$W(x) = a\,(1)\,x^4 + a(2)x^2 + a(3)x + a(4) \tag{5.4}$$

where $\boldsymbol{W}(\boldsymbol{x})$ is the OPD. The calculated coefficients are reported in Fig. 5.17b.

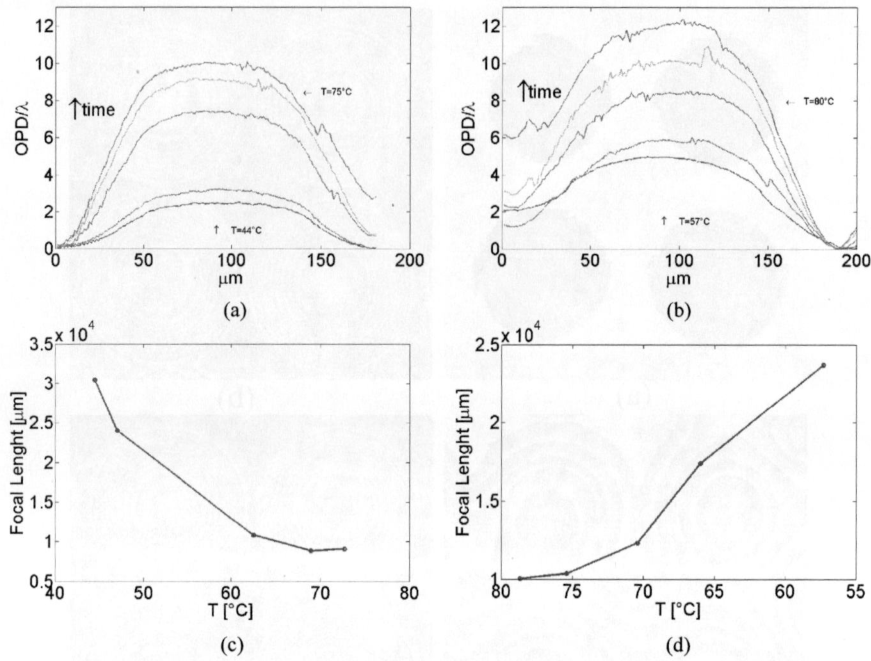

Fig. 5.16 Profile of the phase distribution during (**a**) heating and (**b**) cooling; (**c**),(**d**) temperature dependence of the focal length for the heating and the cooling process, respectively, in case of separated lenses (from [43])

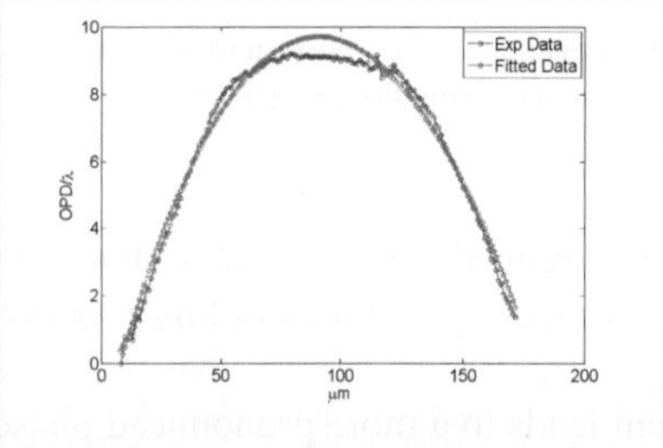

Coefficient	Value	Meaning
a(1)	-2.8×10^{-9}	Spherical Aberration Coefficient
a(2)	-6.2×10^{-4}	Defocusing Coefficient
a(3)	−0.0763	Tilt Coefficient
a(4)	2.7918	Constant Term

(a) (b)

Fig. 5.17 (**a**) Measured and fitted phase profile; (**b**) list of the coefficient values of the linear expansion resulting from the fitting process (from [43])

The typical appearance of a single microlens under the *SLR* is shown by the optical microscope frames in Fig. 5.18, where the oil breakup between adjacent lenses is clearly visible.

In this case the sample was observed under the microscope after treating thermally the microlens array already formed by the previous heating of the oil-coated PPLN substrate. The movie frames show the slight morphology variation

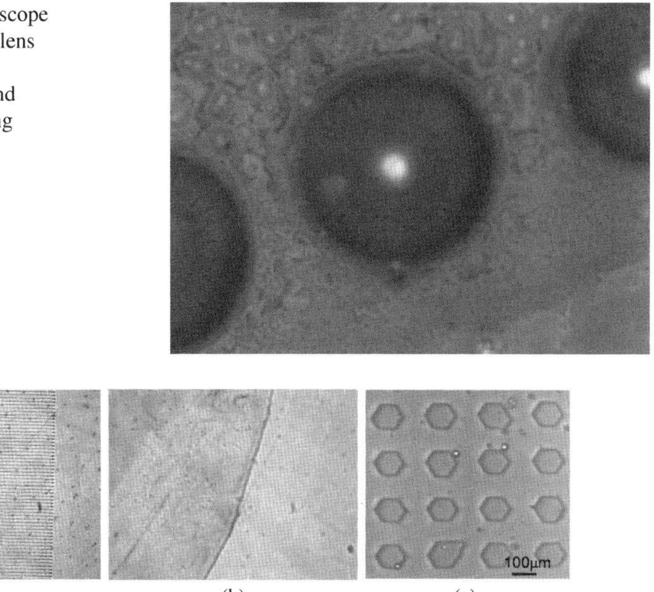

Fig. 5.18 Optical microscope image of a single liquid lens of an array of separated lenses after formation and successive thermal tuning (from [43])

(a) (b) (c)

Fig. 5.19 (**a**) Images of the 1D PPLN sample used for the cylindrical lens; (**b**) image of a part of the domain wall used for single toroidal lens; (**c**) typical periodic hexagonal structures used to form a micro-array of toroidal lenses (from [52])

experienced by the microlens during this experiment, thus demonstrating the possibility to achieve a focus tuning of the microlenses after their formation. This effect may be of great interest to the field of optics, as demonstrated by some works presented recently in literature on this subject [46]. In fact, microlenses with adjustable focal lengths may be useful to eliminate the need for mechanically moving parts, thus reducing the size of the optical systems while enabling precision focusing. Moreover, it is important to note that the frames in Fig. 5.19 show how, in case of the *SLR*, the evolution of the focal length is only slightly visible by the simple observation under the optical microscope. Conversely, the phase reconstruction in Fig. 5.15 clearly shows the curvature variation during the thermal process, thus demonstrating the importance of the QPM for a deep understanding of the optical behaviour of such liquid microlenses.

5.4 Hemicylindrical Liquid Microlenses

Even though many approaches have been developed for realizing liquid lenses, only a few works have been reported on the fabrication of cylindrical and toroidal microlenses [45, 49, 50]. Recently liquid droplets having hemicylindrical shape have been obtained by inducing strong anisotropic wetting onto nanopatterned surfaces [51]. This section shows how the PEW technique can be used for inducing the spontaneous formation of hemicylindrical or even hemitoroidal liquid structures onto PPLN substrates.

The crystal sample is coated with a thin film of an oily substance and then it is heated and cooled in order to modify the shape of the liquid layer. During the temperature change a fast reshaping of the oil film is observed. In the cylindrical case, the process tends to an equilibrium state where the oil is gathered up into two strips normal to the walls of the periodic domain pattern. These stripes exhibit lensing behaviour and we characterize their focal length by means of QPM. In literature many authors measure the focal length variations of liquid lenses but only by indirect calculations. Indeed, starting from optical microscope images, they calculate the height of the lens and then recover the focal length or measure the focal spot formation [1, 10, 12, 19, 46]. The QPM approach provides a direct measurement of the focal length from the computed complex wavefront at the lens exit pupil.

The cylindrical liquid lenses are generated by using a one-dimensional domain reversed grating, while the toroidal microlenses are obtained through a single large domain with a circular geometry or through an array of hexagonal domains. Figure 5.19 shows the optical microscope images of the PPLN samples. The 1D domain grating in Fig. 5.19a has a period around 19 μm and is about 1 mm wide and the schematic view of the domain structure is shown in Fig. 5.20a.

The samples were spin coated (3000 RPM) with the same oily substance as in the previous experiments (pentanoic acid) in order to get an oil film about 100 μm thick. The sample is positioned onto a thermo-controlled hotplate and the variation of the liquid morphology is recorded during the temperature rising from 30 to 100°C under an optical microscope. The thermo-controlled plate is made of a Peltier cell driven by a software. Moreover a NTC sensor measured the temperature of the sample. The device was able to furnish temperature ramps and cycles. Both the heating and the cooling processes are analysed here as in the previous experiments. The oil film appears to modify its shape in a regular and repeatable way during the cooling

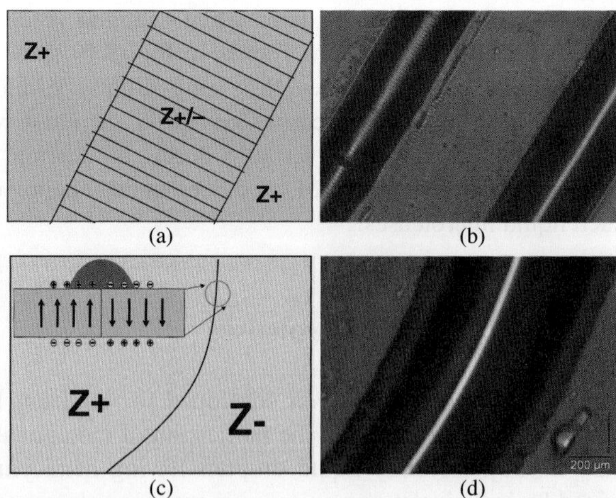

Fig. 5.20 Schematic pictures of the poled region for the PPLN sample (**a**) and the single domain wall (**c**); optical microscope image of cylindrical liquid lens (**b**) and toroidal lens (**d**) (from [52])

process. The final appearance of the oil film is shown in Fig. 5.20b, where two cylindrical profiles are clearly visible, with their principal axes normal to the edges of the periodic domain pattern. As described in the sketch of Fig. 5.20c, the presence of an electric surface charge, generated by the pyroelectric effect, lowers the surface tension due to the repulsion between like charges [17]. Therefore the liquid matter is accumulated at the domain boundaries in order to minimize the energy of the whole system. The details of the physics are outside the scope of this work and can be found in [17, 43]. It is important to note that, as shown in Fig. 5.20b, the fine PPLN structure does not affect the final overall behavior of the liquid that is collected along the main domain walls as for the case shown in Fig. 5.20d where only two domain walls exist. Figure 5.20d shows the microscope image of the toroidal lens. In this case the reversed domain grating has a diameter of 5 mm and a circular geometry. During the cooling process the oil film collects itself along the domain wall disappearing from the domain grating area.

5.4.1 Characterization of Hemicylindrical Lenses by Quantitative Phase Microscopy

An accurate QPM characterization of the cylindrical liquid structure is performed for retrieving the focal length variation. Furthermore we show that an array of liquid hemitoroidal lenses can be obtained on microscale dimension that could find application as resonant micro-cavities for whispering gallery modes [53, 54]. Other configurations of the oil distribution depend on the oil thickness and on the irregular poling. In the configuration shown in Fig. 5.21, a PPLN sample with a two-dimensional geometry is presented.

The oil clearly exhibits an high-fidelity accumulation along the hexagonal domain boundaries. Such micro-array of liquid toroidal structures can find application as resonant liquid micro-cavities [53, 54]. In order to characterize in-situ the liquid lenses, we adopted a DH interferometer in transmission configuration. The

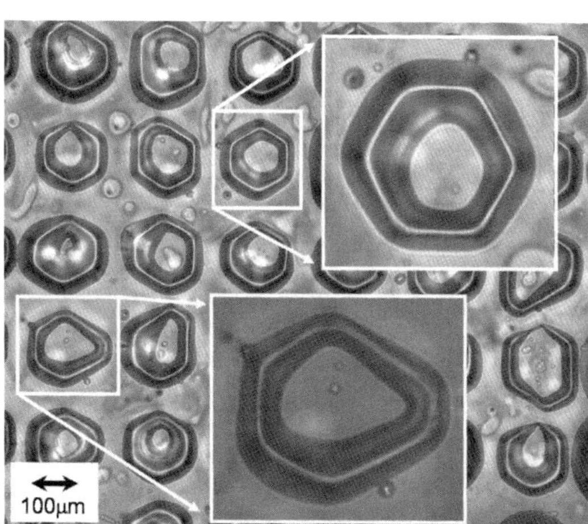

Fig. 5.21 Liquid micro-array of liquid toroidal structures. The liquid follows the non-uniform geometry of each hexagon (from [52])

sample was put on a hotplate and positioned in a DH setup. By DH we performed an interferometric analysis that was able to quantify the focal length for the liquid lens during its formation process. The setup was based on a Mach–Zehnder interferometer with a laser emitting at 532 nm (Fig. 5.5a). The object wave passed through the sample, then it is collected by a microscope objective and made interfere with the reference beam on a CCD camera placed, far from the objective image plane of a distance *d*, named reconstruction distance. The complex wavefield in the image plane is numerically calculated, starting from the interference pattern recorded by the CCD, through the scalar theory of diffraction [55] and allow to recover the intensity distribution and the phase retardation of the complex wavefield in the image plane. One hundred holograms were recorded during the cooling process when the covering substance modifies itself up to reach the final configuration showed in Fig. 5.20b. The hologram recording rate was 15 frame/s and the temperature was changed between 100 and 30°C. We computed the complex wavefield for each hologram in the image plane of the cylindrical lens exit pupil. We calculate the wrapped phase of the optical beam for each hologram and after a numerical unwrapping procedure we recovered the optical phase map of the wavefront retardation. The wrapped and unwrapped phase maps, corresponding to the final frame, are shown, respectively, in Fig. 5.22a, b.

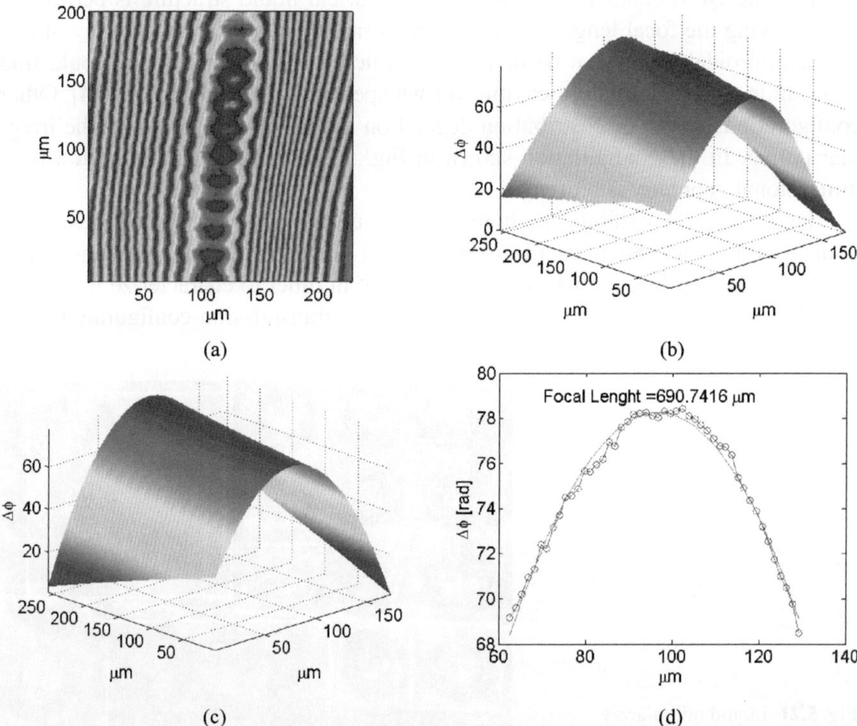

Fig. 5.22 (**a**) Wrapped phase distribution and (**b**) unwrapped phase distribution of the optical wavefront of the final frame at 30°. (**c**) Fitted surface obtained by the two-dimensional fitting procedure. (**d**) Focal length measurement at the stationary condition (from [52])

The two groups of movie frames show the temporal evolution of the phase map at the exit pupil face. We performed a two-dimensional fitting procedure to evaluate the focal length of the cylindrical lens. The equation used is $\Delta\varphi = \boldsymbol{a}\boldsymbol{x}^2 + \boldsymbol{b}\boldsymbol{x}\boldsymbol{y} + \boldsymbol{c}$. From the coefficient of the quadratic term we calculated the focal length f according to the equation $a = \pi/\lambda f$. The focal length varies from infinity to a very short value. The final focal length is $f = 670\,\mu\text{m}$. The fitted surface obtained is showed in Fig. 5.22a while in Fig. 5.22b the experimental data and fitting results are presented for a cylindrical lens cross section.

The PEW-driven cylindrical microlenses shown here could be useful in a wide variety of optics applications where the focusing light along one dimension is required. They can be used for inducing and/or correcting anamorphism in laser beam or images, correcting astigmatism in images, focusing light into a slit or for converging light in a line scan detector.

5.5 Arrays of Polymer-Based Microlenses

Solid microlenses are small lenses generally with diameters less than a millimeter and often down to $10\,\mu\text{m}$. Examples of such microlenses can be found in nature, ranging from simple structures to gather light for photosynthesis in leaves to compound eyes in insects. Nowadays, artificial solid single microlens and microlens arrays play a fundamental role in the photonic technology. Numerous materials and a variety of processes have been investigated for fabricating microlenses. In particular, most of solid microlens arrays are fabricated by means of moulding or embossing processes starting from an original matrix [56–58]. Differently from the liquid lenses, the solid microlenses, especially made up of glasses or polymers, have well-defined optical properties and are much less influenced by external agents. Consequently, the applications of these microlenses are different, as they are mainly used in the field of optical transmissions and photonics (fibre optics, CCD sensors, etc.). Even though a great interest has been focused on microlenses with changeable focal length, as explained in the previous sections, a great interest is still maintained in fabricating microlens arrays that are not tunable, by using different materials. In this way, polymeric materials such as PMMA [5] or PDMS [59] have been extensively investigated. PDMS is an elastomer material widely used in different applications, such as micro/nanofluidics, electrical insulation, micro/nanoelectromechanical systems (MEMS/NEMS), soft lithography, quantum dots and charge patterning in thin-film electrects [60]. PDMS offers many advantages for the fabrication of patterned structures. It is electrically insulating, mechanically elastic, gas-permeable and optically transparent. The latter property permits the study of this material through interferometric techniques, such as digital holography, treating the PDMS structures as phase objects. Moreover, PDMS is also biocompatible, thus finding application in the field of bioengineering, where the position of cells on a substrate is important for different purposes. These include biosensor fabrication for drug toxicity and environmental monitoring, tissue engineering, patterning of active proteins, patterning of animal cells and basic biology studies where the role of cell adhesion,

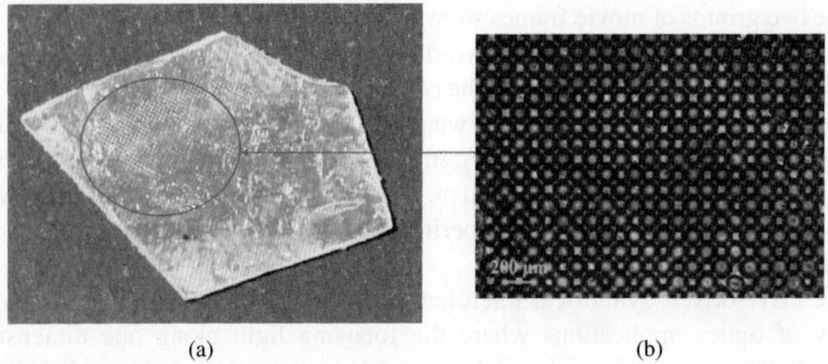

(a) (b)

Fig. 5.23 (**a**) Example of fabricated sample. (**b**) Optical microscope image of the PDMS microlens array contained in the circle (from [62])

shape, proliferation and differentiation are studied as a function of cell–cell and cell–extracellular matrix interactions. The ability to reliably pattern PDMS in the form of both thick substrates and thin membranes or films is critical for expanding the scope of its applications, especially in the fields of microfluidics and bioengineering. In particular, PDMS is specifically suitable because of its good optical quality and the simplicity of the lenses fabrication process known as "soft lithography," where it is only necessary to have a mould to obtain a negative replica of the structure [56, 57]. Fabrication of PDMS microlens arrays was demonstrated by surface wrinkling technique [61] or by using electromagnetic force-assisted UV printing. Moreover, patterning of arrayed PDMS structures through parylene C lift-off and capillary forming process [58] was demonstrated too. It is important to note that lenses made with polymeric materials such as PDMS can be also tuned thanks to the inherent elastic properties of the material. In fact, it has been demonstrated that a single PDMS microlens can be actuated and tuned thermically [63].

The PEW technique is used here for fabricating arrays of polydimethylsiloxane (PDMS)-based microlenses onto LN substrates. PDMS microlens arrays made of thousands of lenses are fabricated, as shown in Fig. 5.23, having micrometric size (100 μm of diameter) and focal length in the range of 300–1000 μm.

5.5.1 Fabrication of the Samples

The fabrication process is carried out onto a LN substrate, opportunely functionalized as described in Sect. 5.2. Three different samples are prepared, named A, B and C. Sample A is wet-etched in pure HF for 10 min (etching depth $\approx 12\,\mu$m) after the poling process with the aim at obtaining a structure similar to that shown in Fig. 5.24. Conversely, the samples B and C are not etched.

For samples A and B, a layer of PDMS polymer solution (Dow Corning Sylgard 184, 10:1 mixing ratio base to curing agent) is spin coated onto the z-face of the PPLN substrate at 6000 RPM for 2 min. The PDMS-coated samples (with a PDMS layer thickness of about 3 μm) are then placed onto a hotplate at a temperature of

Fig. 5.24 SEM image of a HF etched LN sample after photolithographic and poling processes, having a slightly different pitch and depth of etching of sample *A* (from [62])

170°C for 30 s, thus inducing rapid heating of the sample and generating a microlens array following the arrangement of the hexagonal domains [64]. The subsequent cooling from 170°C down to room temperature solidifies the PDMS lenses that are so ready for being characterized, as shown in Fig. 5.25a, b. Sample *C* is prepared in a different way. A layer of 1.3 μm thick photoresist is first spin coated on a LN

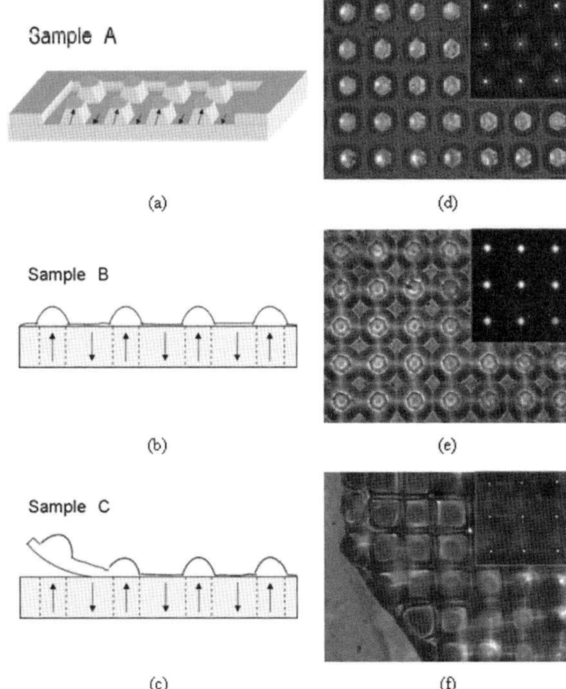

Fig. 5.25 Drawings (**a–c**) and optical microscope images (**d–f**) of fabricated samples. *Sample A* (**a**,**d**): etched PPLN with reversed hexagonal domains, on which a PDMS microlens square array is formed. *Sample B* (**b**,**e**): non-etched PPLN with PDMS microlenses. *Sample C* (**c**,**f**): PDMS microlenses alone, peeled off from a non-etched PPLN substrate. The pitch of the structures is about 200 μm (from [62])

z-face, at 3000 RPM for 1 min. Subsequently, the sample is placed onto a hotplate at a temperature of 115°C for 60 s in order to bake the resist. Then, it is covered with a drop of PDMS diluted with hexane (mixing ratio 3:1 PDMS to hexane), placed onto a hotplate at a temperature of 130°C for 60 s, chilled and finally peeled off. In this way the microlens array is detached from the substrate adopted to realize it, as displayed in Fig. 5.25c.

The photoresist (being a dielectric material) appears not to affect too much the formation of the PDMS microlenses, at least with the very thin layer used for this experiment. Figure 5.25d–f shows the optical microscope images corresponding, respectively, to the samples A, B and C in the focus plane of the substrate, the image in the focus plane of the lenses being displayed as insets.

5.5.2 Characterization of Polymer Microlenses by QPM

The three samples exhibit different optical properties and they have been characterized by a DH-based technique for accurate QPM analysis. The experimental setup consists in a typical Mach–Zehnder interferometer arranged in a transmission configuration, as depicted in Fig. 5.5a. The source is a continuous-wave solid state laser emitting light at 532 nm. The beam is divided in two by a beam splitter; the object beam passes through a thin PDMS microlens array and is imaged by a 5× microscope objective onto a CCD camera. The reference beam, opportunely expanded and with the same polarization of the object one, is recombined with it thanks to another beam splitter. The resulting interference pattern, i.e. the *digital hologram*, is recorded by the CCD camera (pixel size 4.4 μm), placed at a distance d from the image plane according to the holographic technique. The intensity and the phase of the complex wavefields passing through the PDMS sample are numerically calculated starting from these digitally recorded holograms that contain all the specimen's information. Thanks to the flexibility of DH, wavefields reconstruction is possible in different image planes without changing the setup, i.e. without moving the sample. In particular, an accurate quantitative phase analysis allows us to precisely characterize the curvature of each microlens with a resolution of about 1 μm.

Figure 5.26a, c, e displays the phase maps of a portion of each sample. The phase of three microlenses, each picked out from a different sample and indicated by a red frame, has been analysed in order to recover the focal length f. Figure 5.26b, d, f shows a parabolic fit along both dimensions of the unwrapped phase map $\varphi(x, y)$, according to (5.3). Here, we consider the parabolic approximation of the spherical wavefront of the beam that passed through the lens. The focal lengths recovered from the parabolic fits result to be 307 μm, 1010 μm and 394 μm for samples A, B and C, respectively [62], with a difference among lenses inside the same sample of few microns.

By the measurements it results that the smallest focal length is that of sample A. This effect is probably due to etching and thus reduction of the domain areas. For sample B the value of f is higher and the lenses are flatter – as a consequence of the

Fig. 5.26 (**a**),(**c**),(**e**) Phase map of sample A, B and C, respectively; (**b**), (**d**), (**f**) parabolic fit corresponding to the microlens in red frame (from [62])

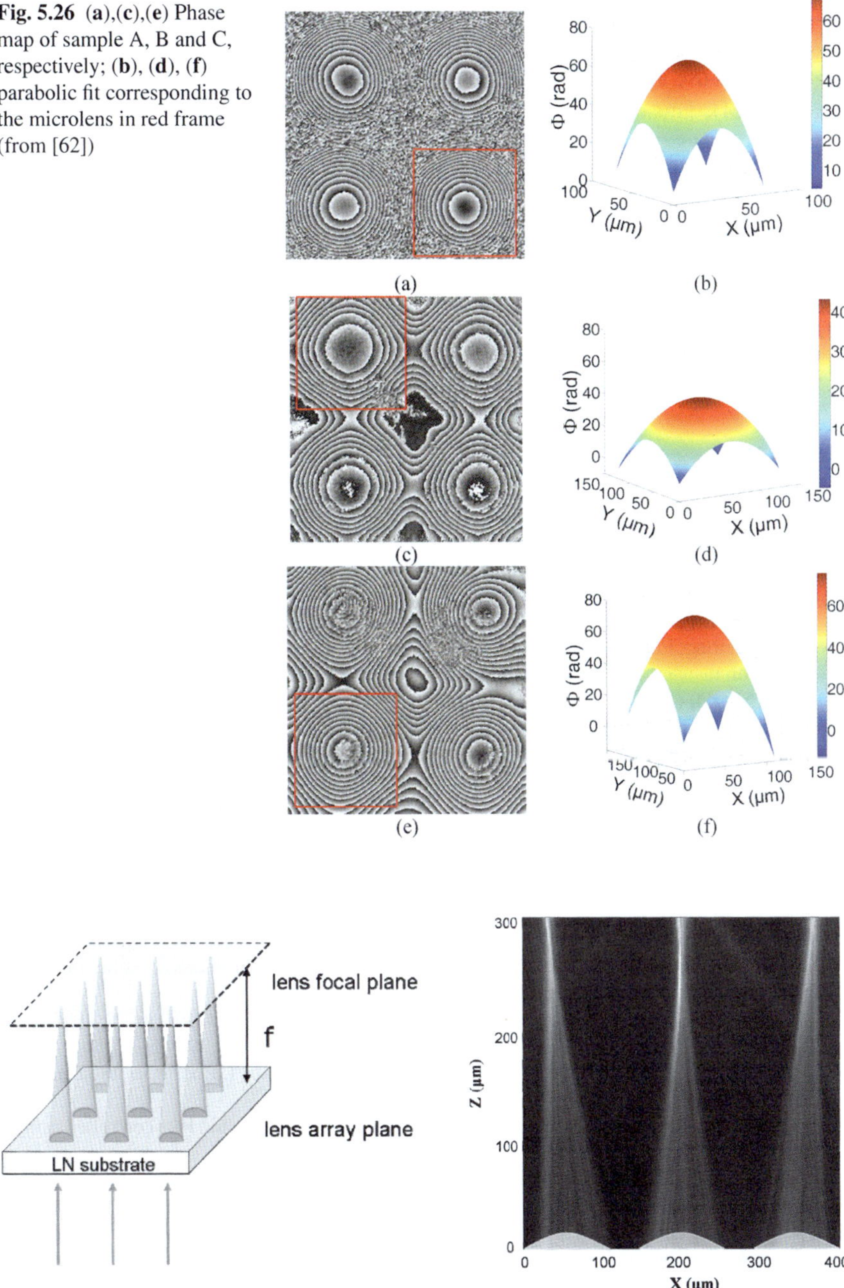

Fig. 5.27 (**a**) Scheme of the path of the rays passing through the sample. (**b**) Reconstructed amplitude of the beams exiting from the microlenses and focusing after about 0.3 mm (from [62])

fabrication process – as results from the fringes number and from the lower value of the phase (see Fig. 5.26c, d). As regards sample C, it seems that the process of peeling do not cause appreciable deformations in the lenses' shape or aberrations (see Fig. 5.26e, f). This could be very important for a realistic application of polymeric microlens arrays.

The evaluated focal length is of the same order of magnitude of sample A. For sample A an amplitude reconstruction varying the reconstruction distance d is also performed. By a single acquisition with the CCD it is possible to recover, by varying d, the amplitude of the beams exiting from the microlenses, at different steps, starting from the lens array plane up to the lens focus plane, as shown in Fig. 5.27a. The result is displayed in Fig. 5.27b, the resulting focal length being about 300 μm, in agreement with that obtained from the parabolic fit for sample A [62].

References

1. S. Kuiper, B.H.W. Hendriks, Variable- focus liquid lens for miniature cameras. Appl. Phys. Lett. **85**, 1128–1130 (2004)
2. L. Dong, A.K. Agarwal, D.J. David, J. Beebe, H. Jiang, Adaptive liquid microlenses activated by stimuli-responsive hydrogels. Nature **442**, 551–554 (2006)
3. B. Berge, J. Peseux, Variable focal lens controlled by an external voltage: an application of electrowetting. Eur. Phys. J. E **3**, 159–163 (2000)
4. L.G. Commander, S.E. Day, D.R. Selviah, Variable focal length microlenses. Opt. Commun. **177**, 157–170 (2000)
5. P.H. Huang, T.C. Huang, Y.T. Sun, S.Y. Yang, Fabrication of large area resin microlens arrays using gas-assisted ultraviolet embossing. Opt. Express **16**, 3041–3048 (2008)
6. A. Pikulin N. Bityurin, G. Langer, D. Brodoceanu, D. Bauerle, Hexagonal structures on metal-coated two-dimensional microlens arrays. Appl. Phys. Lett. **91**, 191106 (2007)
7. F. Krogmann, W. Monch, H. Zappe, A MEMS-based variable micro-lens system. J. Opt. A **8**, S330–S336 (2006)
8. C.C. Cheng, C.A. Chang, J.A. Yeh, Variable focus dielectric liquid droplet lens. Opt. Express **14**, 4101–4106 (2006)
9. C.C. Cheng, J.A. Yeh, Dielectrically actuated liquid lens. Opt. Express **15**, 7140–7145 (2007)
10. N. Chronis, G.L. Liu, K.H. Jeong, L.P. Lee, Tunable liquid-filled microlens array integrated with microfluidic network. Opt. Express **11**, 2370–2378 (2003)
11. D.Y. Zhang, N. Justis, Y.H. Lo, Integrated fluidic adaptive zoom lens. Opt. Lett. **29**, 2855–2857 (2004)
12. P.M. Moran, S. Dharmatilleke, A.H. Khaw, K.W. Tan, M.L. Chan, I. Rodriguez, Fluidic lenses with variable focal length. Appl. Phys. Lett. **88**, 041120 (2006)
13. H. Ren, D. Fox, P.A. Anderson, B. Wu, S.T. Wu, Tunable-focus liquid lens controlled using a servo motor. Opt. Express **14**, 8031–8036 (2006)
14. L. Hou, N. Smith, J. Heikenfeld, Electrowetting modulation of any flat optical film. Appl. Phys. Lett. **90**, 251114 (2007)
15. N. Smith, D. Abeysinghe, J. Heikenfeld, J.W. Haus, Agile wide-angle beam steering with electrowetting microprisms. Opt. Express **14**, 6557 (2006)
16. B. Sun, K. Zhou, Y. Lao, W. Cheng, J. Heikenfeld, Scalable fabrication of electrowetting pixel arrays with self-assembled oil dosing. Appl. Phys. Lett. **91**, 011106 (2007)
17. J.L. Lin, G.B. Lee, Y.H. Chang, K.Y. Lien, Model description of contact angles in electrowetting on dielectric layers. Langmuir **22**, 484–489 (2006)
18. W.H. Hsieh, J.H. Chen, Lens-profile control by electrowetting fabrication technique. IEEE Photon. Tech. Lett. **17**, 606–608 (2005)

19. H. Ren, S.T. Wu, Tunable-focus liquid microlens array using dielectrophoretic effect. Opt. Express **16**, 2646–2652 (2008)
20. K.H. Jeong, G.L. Liu, N. Chronis, L.P. Lee, Tunable microdoublet lens array. Opt. Express **12**, 2494–2500 (2004)
21. D. Grahan-Rowe, Liquid lenses make a splash. Nat. Photon. Volume sample, 2–4 (2006)
22. G. Beni, M.A. Tenan, Dynamics of electrowetting displays. J. Appl. Phys. **52**, 6011–6015 (1981)
23. R. Hayes, D.J. Feenstra, Video-speed electronic paper based on electrowetting. Nature **425**, 383–385 (2003)
24. D. Psaltis, S.R. Quache, C. Yang, Developing optofluidic technology through the fusion of microfluidics and optics. Nature **442**, 381–386 (2006)
25. F. Mugele, S. Herminghaus, Electrostatic stabilization of fluid microstructures. Appl. Phys. Lett. **81**, 2303–2305(2002)
26. E.L. Wooten et al., A review of lithium niobate modulators for fiber-optic communications systems. IEEE J. Sel. Top. Quantum Electron. **6**, 69–82 (2000)
27. R.L. Byer, Nonlinear optics and solid-state lasers: 2000. IEEE J. Sel. Top. Quantum Electron. **6**, 911–930 (2000)
28. M. Yamada, N. Nada, M. Saitoh, K. Watanabe, First-order quasi-phase matched $LiNbO_3$ waveguide periodically poled by applying an external field for efficient blue second-harmonic generation. Appl. Phys. Lett. **62**, 435–436 (1993)
29. S. Grilli, M. Paturzo, L. Miccio, P. Ferraro, *In situ* investigation of periodic poling in congruent $LiNbO_3$ by quantitative interference microscopy. Meas. Sci. Tech. **19**, 074008 (2008)
30. S. Grilli, L. Miccio, V. Vespini, A. Finizio, S. De Nicola, P. Ferraro, Liquid micro-lens array activated by selective electrowetting on lithium niobate substrates. Opt. Express **16**, 8084–8093 (2008)
31. K. Nassau, H.J. Levinstein, G.M. Loiacono, The domain structure and etching of ferroelectric lithium niobate. Appl. Phys. Lett. **6**, 228–229 (1965)
32. S. Grilli, P. Ferraro, P. De Natale, B. Tiribilli, M. Vassalli, Surface nanoscale periodic structures in congruent lithium niobate by domain reversal patterning and differential etching. Appl. Phys. Lett. **87**, 233106–3 (2005)
33. V. Gopalan, T.E. Mitchell, *In situ* video observation of 180° domain switching in $LiTaO_3$ by electro-optic imaging microscopy. J. Appl. Phys. **85**, 2304–2311 (1999)
34. R.S. Weis, T.K. Gaylord, Lithium niobate: Summary of physical properties and crystal structure. Appl. Phys. A **37**, 191–203 (1985)
35. E.M. Bourim, C.-W. Moon, S.-W. Lee, I.K. Yoo, Investigation of pyroelectric electron emission from monodomain lithium niobate single crystals. Phys. B **383**, 171–182 (2006)
36. B. Rosenblum, P. Bräunlich, J.P. Carrico, Thermally stimulated field emission from pyroelectric $LiNbO_3$. Appl. Phys. Lett. **25**, 17–19 (1974)
37. G. Rosenman, D. Shur, Y.E. Krasik, A. Dunaevsky, Electron emission from ferroelectrics. J. Appl. Phys. **88**, 6109–6161 (2000)
38. E. Colgate, H. Matsumoto, An investigation of electrowetting-based micro actuation. J. Vac. Sci. Technol. A **8**, 3625–3633 (1990)
39. M.G. Lippmann, Relations entre les phénomènes électrique et capillaries. Ann. Chim. Phys. **5**, 494 (1875)
40. F. Beunis, F. Strubbe, M. Marescaux, K. Neyts, A.R.M. Verschueren, Diffuse double layer charging in nonpolar liquids. Appl. Phys. Lett. **91**, 182911–182913 (2007)
41. F. Mugele, J.-C. Baret, Electrowetting: from basics to applications. J. Phys. Condens. Matt. **17**, R705–R774 (2005)
42. P. Ferraro, S. De Nicola, G. Coppola, in *Optical Imaging Sensors and Systems for Homeland Security Applications*, vol. 2, Series ed. by B. Javidi. Digital holography: recent advancements and prospective improvements for applications in microscopy. Advanced Sciences and Technologies for Security Applications. (Springer, Heidelberg, 2005), pp. 47–84

43. L. Miccio, A. Finizio, S. Grilli, V. Vespini, M. Paturzo, S. De Nicola, P. Ferraro, Tunable liquid microlens arrays in electrode-less configuration and their accurate characterization by interference microscopy. Opt. Express **17**, 2487–2499 (2009)
44. G. Milne, G.D.M. Jeffries, D.T. Chiu, Tunable generation of Bessel beams with a fluidic axicon. Appl. Phys. Lett. **92**, 261101 (2008)
45. X. Mao, J.R. Waldeisen, B.K. Juluri, T.J. Huang, Hydrodynamically tunable optofluidic cylindrical microlens. Lab. Chip **7**, 1303–1308 (2007)
46. X. Huang et al., Thermally tunable polymer microlenses. Appl. Phys. Lett. **92**, 251904 (2008)
47. Y. Choi, H.R. Kim, K.H. Lee, Y.M. Lee, J.H. Kim, A liquid crystalline polymer microlens array with tunable focal intensity by the polarization control of a liquid crystal layer. Appl. Phys. Lett. **91**, 221113 (2007)
48. H. Ren, Y.H. Fan, S.T. Wu, Liquid-crystal microlens arrays using patterned polymer networks. Opt. Lett. **29**, 1608–1610 (2004)
49. Y.-H. Lin et al., Tunable- focus cylindrical liquid crystal lenses. Jpn. J. Appl. Phys. **44**, 243–244 (2005)
50. J.-H Lee et al., Efficiency improvement and image quality of organic light-emitting display by attaching cylindrical microlens arrays. Opt. Express **16**, 21184–21190 (2008)
51. D. Xia, S.R.J. Brueck, Strongly anisotropic wetting on one-dimensional nanopatterned surfaces. Nano Lett. **8**, 2819–2824 (2008)
52. L. Miccio, M. Paturzo, S. Grilli, V. Vespini, P. Ferraro, Hemicylindrical and Toroidal Liquid Microlens formed by Pyro-Electro-Wetting (PEW). Opt. Lett. **34**, 1075–1077 (2009)
53. A. Kiraz, Y. Karadag, A.F. Coskun, Spectral tuning of liquid microdroplets standing on a superhydrophobic surface using electrowetting. Appl. Phys. Lett. **92**, 1911041–1911043 (2008)
54. S.I. Shopova, H. Zhou, X. Fan, P. Zhang, Optofluidic ring resonator based dye laser. Appl. Phys. Lett. **90**, 2211011–2211013 (2007)
55. P. Ferraro, S. Grilli, M. Paturzo, S. De Nicola, in *Ferroelectric Crystals For Photonic Applications*, eds. by P. Ferraro, S. Grilli, P. De Natale. Visual and quantitative characterization of ferroelectric crystals and related domain engineering processes by interferometric techniques. (Springer, Heidelberg, 2008), pp. 165–208
56. T.K. Shih, J.R. Ho, J.W.J. Cheng, A new approach to polymeric microlens array fabrication using soft replica molding. IEEE Phot. Tech. Lett. **16**, 2078 (2004)
57. T.K. Shih, C.F. Chen, J.R. Ho, F.T. Chuang, Fabrication of PDMS (polydimethylsiloxane) microlens and diffuser using replica molding. Microelectron. Eng. **83**, 2499 (2006)
58. C.Y. Chang, S.Y. Yang, L.S. Huang, K.H. Hsieh, Fabrication of polymer microlens arrays using capillary forming with a soft mold of micro-holes array and UV-curable polymer. Opt. Express **14**, 6253 (2006)
59. S. Grilli, V. Vespini, P. Ferraro, Surface-charge lithography for direct PDMS micro-patterning. Langmuir **24**, 13262 (2008)
60. H.O. Jacobs, G.M. Whitesides, Submicrometer patterning of charge in thin-film electrets. Science **291**, 1763–1766 (2001)
61. E.P. Chan, A.J. Crosby, Fabricating microlens arrays by surface wrinkling. Adv. Mater. **18**, 3238 (2006)
62. F. Merola, M. Paturzo, S. Coppola, V. Vespini, P. Ferraro, Self-patterning of a polydimethylsiloxane microlens array on functionalized substrates and characterization by digital holography. J. Micromech. Microeng. **19**, 125006 (5pp) (2009)
63. S.Y. Lee, H.W. Tung, W.C. Chen, W. Fang, Thermal actuated solid tunable lens. IEEE Phot. Tech. Lett. **18**, 2191 (2006)
64. P. Ferraro, S. Grilli, L. Miccio, V. Vespini, Wettability patterning of lithium niobate substrate by modulating pyroelectric effect to form microarray of sessile droplets. Appl. Phys. Lett. **92**, 213107 (2008)

Chapter 6
Quantitative Phase Imaging in Microscopy Using a Spatial Light Modulator

Vicente Micó, Javier García, Luis Camacho, and Zeev Zalevsky

Abstract In this chapter, we present a new method capable of recovery of the quantitative phase information of microscopic samples. Essentially, a spatial light modulator (SLM) and digital image processing are the basics to extract the sample's phase distribution. The SLM produces a set of misfocused images of the input sample at the CCD plane by displaying a set of lenses with different power at the SLM device. The recorded images are then numerically processed to retrieve phase information. Computations are based on the wave propagation equation and lead to a complex amplitude image containing information of both amplitude and phase distributions of the input sample diffracted wave front. The proposed configuration becomes a non-interferometric architecture (conventional transmission imaging mode) where no moving elements are included. Experimental results are provided in comparison with conventional digital holographic microscopy.

6.1 Introduction

Optical light microscopy [1] is one of, if not the most, enabling tools in many application fields such as semiconductor industry, material science, biomedicine, and micromechanical testing, just to cite a few. The role of optical microscopes is essentially to interpret 2D or, better still, the 3D structure of the specimen being studied. In particular and in biological research, there are two disciplines that are capturing the attention of the scientific community. On one hand, *fluorescence imaging* is becoming a standard tool at many fields in genetics, embryology, and cell biology for imaging of biomolecules in a labeled sample while being illuminated in the presence of synthetic fluorophores and/or fluorescent proteins. On the other hand, *contrast enhancing methods* allow the characterization of unstained biosamples and are widely used to examine dynamic events in live cell imaging.

Contrast enhancing techniques solve the question of imaging the specimens without altering them by killing or adding chemical dyes or fluorophores. Since

V. Micó (✉)
Departamento de Óptica, Universitat de Valencia, C/Doctor Moliner 50, 46100 Burjassot, Spain
e-mail: vicente.mico@uv.es

P. Ferraro et al. (eds.), *Coherent Light Microscopy*, Springer Series in Surface Sciences 46, DOI 10.1007/978-3-642-15813-1_6,

many specimens in microscopy are essentially transparent, that is, they are phase objects, image contrast provided by the microscope is extremely poor and the sample remains almost invisible. This reason encourages scientists to develop new methods to improve specimen visibility. Thus, darkfield, phase contrast (PC), differential interference contrast (DIC), oblique illumination, and polarized light microscopy have been successfully proposed and validated along the past century as imaging methods capable to bring images of phase objects with improved contrast [2]. Just as example, PC [3] and DIC [4] microscopy are two interference methods capable of converting the phase variation incoming from changes in optical path length (thickness and/or refractive index changes) of the specimen into amplitude variations allowing thus intensity fluctuations of the obtained image. In PC microscopy, the interference is caused by phase shifting the undiffracted (undeviated or zero order term) light passing through the sample with respect to the diffracted (deviated or non-zero order term) light scattered by the sample, or vice versa. On the contrary, DIC microscopy mixes two slightly shifted images of the same specimen in such a way that the final image becomes related with the optical path length gradient in the direction of the shear. However, in spite of their interferential nature, PC and DIC only provide visualization of phase objects, that is, both methods are qualitative from an optical path length measurement point of view with a nonlinear response between intensity and phase in the final specimen image [5].

Nowadays, classical PC [3] and DIC [4] microscopy have evolved to new methods [6–11], most of them capable of extracting quantitative phase information [7–11]. In that sense, optical phase measurement of specimens provides valuable information about morphology, refractometry, topography, and dynamics of cells and tissues at nanometer scale. One can divide the quantitative phase measurement methods in accordance with single-point and full-field techniques [12]. *Single-point phase measurement methods* allow ultrafast and ultrasmall changes in optical path length of a small area. Just as example of a single-point technique, *phase dispersion microscopy* (PDM) is based on the measurement of the phase difference between the fundamental and the second-harmonic non-scattered light transmitted through the sample in an interferometric configuration such as Michelson [13] and Mach–Zehnder [14]. PDM allows highly differential optical path sensitivity incoming from subtle refractive index differences. Large area phase measurement is possible by raster scanning the full sample field of view. However, this strategy is time consuming, and because recent developments in 2D array sensor devices permit high-speed imaging detection (thousands of frames per second), *full-field phase measurement methods* appear as more promising techniques.

6.2 Full-Field Quantitative Phase Imaging

We have divided this section into approaches that are based on holography as underlying principle for achieving full-field quantitative phase imaging and approaches that are not.

6.2.1 Holographic Approaches

In addition to PC and DIC microscopy, the third classical method for phase imaging is holography. Although PC and DIC techniques are also based on interferometry, essentially both methods provide visualization of the boundaries of a phase sample structure. On the contrary, holographic methods allow full-field quantitative measurement of the absolute phase sample information [15].

Imaging with holographic tools was established by Dennis Gabor in 1948 [16] when trying to overcome the limitations due to the use of lenses in electron microscopy. Essentially, the Gabor's setup implies an in-line configuration where the imaging wave (caused by sample's diffraction) and the reference wave (incoming from the non-diffracted light passing through the sample) interfere at the recording plane. If the sample is a weak diffractive sample, the interference process results with holographic recording where complete (amplitude and phase) information about the sample wave front becomes accessible. Otherwise, the sample excessively blocks the reference wave, and diffraction dominates the process preventing the accurate recovery of the sample's complex wave front.

Nowadays, the hologram is recorded by electronic devices (typically a CCD or CMOS camera) and digital Fourier filtering as well as numerical reconstruction algorithms are commonly used for quantitative phase imaging [17, 18]. In that sense, *digital in-line holographic microscopy* (DIHM) implements the Gabor basic setup with digital holographic recording provided by a CCD [19, 20], and it has been successfully implemented for a long range of applications including underwater observations, tracking of moving objects and particles, as well as the study of erosion processes in coastal sediments [21–24].

DIHM configuration has also been studied under phase-retrieval purposes [25–28]. The main objective is to reconstruct the complete complex amplitude of the original sample wave front, both real and phase distributions, avoiding the noise caused by the twin image of the holographic recording. This task can be tackled by means of non-iterative [25, 27, 28] as well as iterative methods [26] and using one [28] or more [25–27] holograms (recording planes). Essentially, all those methods are based on back propagation of the recorded diffracted wave front until other planes (including the input sample as well as other hologram planes) and then apply some kind of mathematical or physical constraints.

However, all those methods cannot work outside the Gabor assumption concerning weak diffractive samples. This restriction can be easily removed by inserting an external reference beam at the recording plane [15, 29]. In this case, the sample information is placed in one interferometric beam and the reference beam in a different one in such a way that the distorted wave front passing through the sample interferes with an undistorted reference beam. The resulting image contains quantitative information about the phase delay introduced by the sample, information that can be used to reconstruct a 3D quantitative microscopic image with nanometric resolution along the optical axis since phase delay is directly related with thickness and refractive index changes in the sample.

In that sense, *digital holographic microscopy* (DHM) combines high-quality imaging provided by microscopy, whole-object wave front recovery provided by holography, and numerical processing capabilities provided by computers [30–33]. As a result, DHM avoids the limited depth of focus in high NA lenses and allows visualization of phase samples that are not visible under conventional microscope imaging [34].

In DHM, the basic architecture is defined by an interferometric setup where the imaging system is placed on one branch (imaging arm) and a reference beam is reinserted at the CCD plane incoming from a second branch (reference arm). Owing to its interferometric underlying principle, different classical interferometric configurations can be employed [31, 35–39]. But in any case, DHM implies the mixing of an imaging and a reference beam in order to recover the complex amplitude distribution of the wave front diffracted from the input sample. Such phase and amplitude distribution extraction can be performed in off-axis configuration and by using Fourier filtering [15, 17] or in on-axis mode and by using phase-shifting process [15, 40, 41].

Over the last decade, a new generation of full-field phase imaging methods has been proposed showing a growing interest in quantitative phase imaging [42–62]. Zicha et al. proposed that *digitally recorded interference microscopy with automatic phase shifting* (DRIMAPS) combines phase-shifting interferometry with Horn microscopy for quantitative phase imaging in biology and it has been applied to measuring cell spreading, cell motility, cell growth, and dry mass [42, 43]. However, DRIMAPS is affected by phase noise, and this fact will prevent accurate long-term dynamic studies.

Spiral phase contrast microscopy (SPCM) proposes a Zernike-like setup with a spiral phase filter in the Fourier plane of the microscope. Such spiral phase filter produces boundary enhancement of the sample and can be used to extract its topographic information [44]. In addition, by computing a sequence of three images corresponding to different rotational orientations of the filter, SPCM is capable of retrieving full-field quantitative phase sample information [45, 46].

Fourier phase microscopy (FPM), a combination of PC microscopy with phase-shifting interferometry [47], is capable of retrieving quantitative phase images with subnanometer optical path stability due to its common-path interferometric architecture. In addition, FPM enables coherence noise reduction incoming from the use of partially coherent light sources and it has been applied to study cell dynamics over extended periods of time [48–50]. However, FPM has a limitation incoming from the use of phase-shifting procedure that prevents the use of FPM for very rapid (in the millisecond scale) phenomena such as cell membrane fluctuations and neural activity.

In that sense, *Hilbert phase microscopy* (HPM) applies mathematical Hilbert transform to the off-axis hologram recorded by using a regular Mach–Zehnder interferometric configuration in order to quantitatively extract the phase delay introduced by the specimen [51–53]. HPM is a single-shot technique capable of measuring events in the kilohertz range because it is only limited by the acquisition time of the recording device. However, HPM suffers from optical length path instabilities, due to mechanical and/or thermal changes on both optical paths, incoming from the Mach–Zehnder interferometric setup.

For this reason, *diffraction phase microscopy* (DPM) is proposed as a combination with the main advantages of FPM and HPM, that is, it provides full-field quantitative phase imaging with high phase stability over broad temporal scales while working in a single illumination shot [37, 54, 55]. DPM has also been combined with fluorescence [56], confocal [57], dynamic light scattering [58], and Fourier transform light scattering [59, 60] microscopy enabling multimodal microscopy techniques with different capabilities.

Recently proposed *spatial light interference microscopy* (SLIM) implies a combination between PC microscopy and Gabor holography [61]. SLIM completes a phase-shifting cycle needed to recover quantitative phase information in PC microscopy by using variable phase shift provided by an SLM. While conventional PC microscopy provides $\pi/2$ phase shift between undeviated and deviated light passing through the sample, the SLM permits the additional phase steps of π, $3\pi/2$, and 2π closing the full phase-shifting cycle. Thus, four images corresponding to each phase shift were recorded under white light illumination and combined to produce a full-field quantitative phase image. SLIM reveals the intrinsic contrast of biosample structures while renders quantitative optical path length maps across the specimen. SLIM has also been combined with DPM to enable single-shot operation principle [62].

6.2.2 *Non-Holographic Approaches*

However, phase retrieval can also be conducted without holography, that is, without the addition of a reference beam. The key point is to consider numerical reconstructions derived from the intensity variations of the input object's diffracted wave front [63–70]. All these methods are inherently stable against phase noise since they do not require the use of two separate beams for phase quantification as in typical interferometric setups. Also, those methods can be implemented with light sources having a partial degree of coherence in conventional microscopes, which makes simpler the experimental setup.

Barty et al. [65] proposed *quantitative optical phase* microscopy (QOPM) as a non-interferometric method for the extraction of full-field quantitative phase profile information based on the analysis of how the propagation beam is affected by the sample. They used an ordinary transmission microscope where different intensity images (in focus and defocused images) were measured by moving the sample in the z-axial direction using a stepper motor. The resulting intensity measurements acted as inputs of a deterministic phase-retrieval algorithm based on the transport of intensity equation [66–69].

Aimed in the same direction but using coherence illumination, Pedrini et al. [71–73] reported on a complex wave front reconstruction non-interferometric method based on the recording of the volume speckle field provided by an object under laser illumination and phase-retrieval algorithm. They validated the method working without lenses and displacing axially the CCD to provide the different intensity measurement planes that are needed for the computations. After that, reconstruction algorithms based on iteration of the wave propagation equation allow the recovery of the object's complex amplitude wave front. This *single-beam multiple-intensity reconstruction* (SBMR) method has also been validated in shape, deformation, and angular displacement measurements of 3D objects [74, 75].

But in the work reported on [26, 65, 71–75], the ability to defocus the input sample to measure several intensity distributions is obtained by either mechanical displacement of the sample [65] or the CCD camera [26, 71–75]. Thus, such approaches become extremely sensible to small misalignments due to non-orthogonalities and tilts in the experimental setup and to noise incoming from environmental disturbances between different recorded images.

However, all those drawbacks can be removed by considering a static experimental setup having no moving components. Thus, in order to improve the capabilities of previous approaches [71, 72], Bao et al. reported on a new method where the set of diffracted patterns is recorded by displacing neither the CCD nor the sample but by tuning the illumination wavelength [76]. Thus, the system setup becomes static where no moving components are needed. This approach improves the convergence of the phase-retrieval algorithm in comparison with previous works [26, 71, 72] while reduces the impact of noise incoming from the use of mechanical components.

In this chapter, a new approach for phase retrieval based on an SLM and where no moving elements are considered is presented in the field of digital microscopy [77]. Recently, several applications have been proposed in microscopy validating the use of an SLM as a programmable versatile element for quantitative phase [45, 48], qualitative [78] and quantitative [79] PC, depth of field extension [80], and DIC [81] microscopy. Here, the SLM is used to provide a set of different misfocused images of the sample without the need of mechanical moving neither the sample nor the CCD. Thus, the experimental setup becomes an in-line configuration (conventional imaging system in transmission mode) where a microscope lens magnifies the input sample onto a CCD placed at the output image plane. The misfocus is originated by displaying at the SLM a phase profile lens with finite focal length (or non-zero power). Thus, by varying the lens focal length that is being displayed at the SLM, it is possible to record a set of misfocused images in a way that is equivalent to the multiple intensity recordings provided by displacing the CCD in the SBMR method [26, 71–75] or by displacing the sample [65]. Once the whole set of misfocused images is stored in the computer's memory, they are digitally processed in a similar way as that in [71, 72] but taking into account that the addition of a lens in the SLM changes not only the imaging plane but also its magnification. This numerical process is carefully described in Sect. 6.3, where also a system setup description is provided and finally yields in-phase information recovery of the input sample. Experimental results are included both in Sect. 6.3 for a better understanding of the numerical computation process and for calibration purposes, and in Sect. 6.4 for a 3D biosample composed of swine sperm cells.

6.3 Full-Field Quantitative Phase Imaging in Microscopy by Means of Defocusing Using an SLM

The experimental setup used to validate our method is depicted in Fig. 6.1. A linearly polarized and collimated laser beam illuminates the input sample which is imaged by an infinity-corrected long working distance microscope objective in infinity imaging configuration. An additional tube lens brings the input sample into focus at its focal plane where a CCD is placed to capture imaging. Essentially, it is a conventional non-interferometric microscope configuration. However, an SLM in reflective configuration is placed between microscope and tube lenses. The SLM is controlled by a computer in order to display the different lenses. Finally, an additional polarizer placed after the SLM optimizes the SLM phase modulation process.

In this configuration and when no lens is displayed at the SLM, the CCD images a given 2D section of the 3D sample according to the depth of focus provided by the microscope lens. Thus, objects outside the depth of focus will appear blurred. This situation is represented in Fig. 6.2a where only the ray tracing in red is focused onto the CCD. Then, by varying the power of the lens displayed at the SLM from positive to negative in the range that the SLM is able to, different sections of the 3D sample will be in focus at the CCD. Obviously, the SLM lens power is very low in comparison with the one that is defined by the microscope lens, but it means a refocusing capability over different sections of the 3D input sample; that is, different transversal planes of the sample will be in focus at the CCD by varying the power

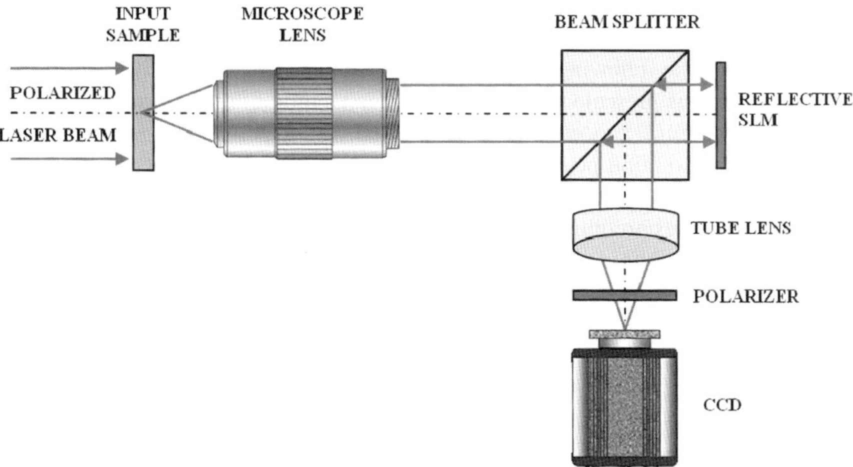

Fig. 6.1 Experimental setup drawing for phase retrieval in reflection mode

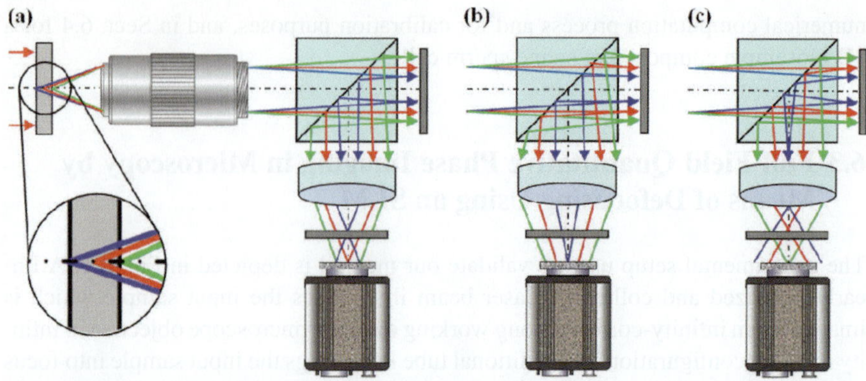

Fig. 6.2 Ray tracing for a 3D sample: (**a**–**c**) correspond to the central, left, and right parts of the 3D sample (red, blue, and green ray tracings), respectively, corresponding to no-lens, negative lens, and positive lens cases displayed at the SLM, respectively

of the SLM-displayed lens. When negative lenses are considered (Fig. 6.2b), we are imaging further sections of the sample in comparison to the previous case where no lens was displayed at the SLM. The contrary happens when using positive lenses (Fig. 6.2c). The magnified area of the 3D sample in Fig. 6.2a helps us to identify the color ray tracing with the imaged section of the sample.

This refocusing ability provides the system with an additional advantage in comparison with previous works [71–75]: since different transversal sections of the 3D input sample become imaged, the phase-retrieval algorithm uses real intensity imaging planes along the iteration procedure. Or in other words, some of the different amplitude distributions taken during the phase iteration process are real imaging planes instead of diffracted (or defocused) wave fronts. As we will see in the experimental section, this fact allows a better final image quality resulting from the whole process.

Now and in the following lines, the numerical manipulation involved in the proposed approach is presented. For a better understanding of the algorithm, the qualitative computational procedure is accompanied with experimental results obtained when a high-resolution negative USAF test target is used as input object. Moreover, aside of being useful for clarifying the numerical processing, the USAF test procedure provides preliminary calibration stage for the proposed setup when more complex samples (phase samples) are considered provided that the misfocus generated by the SLM will be the same.

Although the SLM can produce a larger set of images, we take one in-focus image of the USAF test obtained when no lens is considered at the SLM, four misfocused images using negative lenses, and four misfocused images using positive lenses. This set of nine intensity images corresponding to different sections of the input sample is stored in the computer's memory. Let us name such intensities as I_N, where N varies from 1 to 9. The whole set of images for the case of the USAF test target is represented in Fig. 6.3 where case (e) corresponds to the SLM

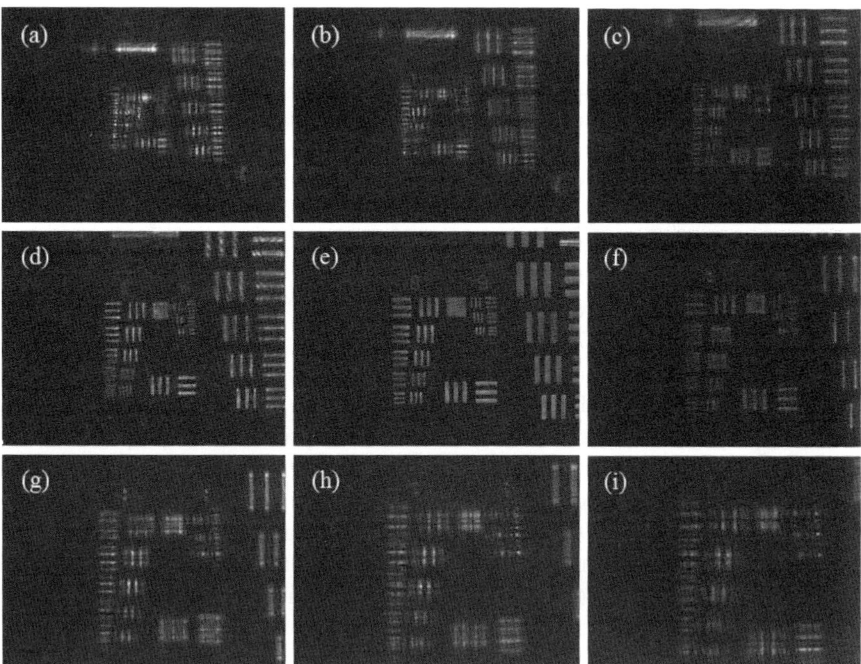

Fig. 6.3 Raw direct images of the USAF test central part obtained when varying the power of the lens displayed at the SLM

no-lens case. However, since the final image magnification in an infinity-corrected imaging configuration depends on the ratio between the focal lengths of the tube lens and the microscope objective, an increase in the power of the tube lens implies a reduction in the overall image magnification. Assuming that the SLM lens can be interpreted as a slight modification of the tube lens, positive lenses at the SLM will produce a reduction in the image magnification (Fig. 6.3a–d) while negative lenses will increase the magnification of the image (Fig. 6.3f–i) when comparing with the no-lens SLM imaging case (Fig. 6.3e).

For this reason and prior to applying the Rayleigh–Sommerfeld (RS)-based phase-retrieval algorithm on the digital propagation between the different intensity planes, the I_N inputs must be matched in magnification and transverse location. Otherwise, the numerical propagation will need magnification compensation and will cause costly computational difficulties. Moreover, a precise propagation distance between planes is also needed.

The procedure for accurately matching the magnification and lateral displacement, as well as the propagation distance between planes, implies a double-step algorithm based on the maximization of the correlation peak value. The first step is divided into two rounds. In the first one, different scaled versions of each misfocus image (I_N with $N \neq 5$) are correlated in intensity with the image obtained when no lens is displayed by the SLM (I_5 or Fig. 6.3e). Nevertheless and aside magnification change, the SLM lens also introduces a linear phase factor (or in other words, a shift) in the blur coming from misalignments in the experimental setup. Therefore, a control over the image displacement will also be needed. Thus, the maximum value and the displacement of the correlation peak from the image centre are indicative of both the best scale factor and the image displacement, respectively, to be applied for each I_N ($N \neq 5$) image in this first round of the adjusting procedure. Taking these values as inputs, we perform a second round where a finer adjustment of the scale factor and the displacement is achieved by polynomial adjustment around the maximum values of the correlation distributions.

However, all of the correlation operations performed in the first step are obtained between blurred images (since they are originated by varying the SLM lens power) and the in-focus image (no-lens imaging case). So, in order to get more accurate values of scale factors and displacements, we need to consider an additional variable: the propagation distance. Thus, the second step iterates the propagation distance with the scale factors and displacements. Numerical propagation is based on the well-known RS equation where the approximation based on convolution operation is applied [15, 18, 39]. Thus, the diffraction integral is calculated using three Fourier transformations through the convolution theorem, that is, $\mathrm{RS}(x, y; d) = \mathrm{FT}^{-1}\{\mathrm{FT}\{U(x, y)\} \cdot \mathrm{FT}\{h(x, y; d)\}\}$, where $\mathrm{RS}(x, y)$ is the propagated wave field, $U(x, y)$ is the recorded hologram, $h(x, y)$ is the impulse response, (x, y) are the spatial coordinates, FT is the Fourier transform operation (realized with the FFT algorithm), and d is the propagation distance.

In our approach, the recorded hologram is the amplitude distribution incoming from the square root of the in-focus image intensity (I_5). This input image is digitally propagated to different distances and correlated with scaled and shifted versions of the rest of the images (I_N with $N \neq 5$). Obviously, the starting values for the scale and shift are those obtained in the first step. After the first round of propagations, we store in the computer's memory the values of propagation distances that produce maximization in the correlation peak. With these best values, we refine the scale factor and displacement of each image (I_N with $N \neq 5$). Then, we re-scale and re-shift again each defocused image and compute again the correlation operation varying slightly the propagation distance. This process is repeated until the difference between new and previous values is lower than 10^{-3}.

As a consequence of this double-step process, a double result is obtained. On one hand, a new set of intensity images (named as I'_N) having both the same object size and the same lateral position is obtained. Figure 6.4 depicts this new set of I'_N images. On the other hand, the propagation distance between the different images is known in a precise way.

Taking into account the new set of I'_N images, phase-retrieval algorithm is applied to reconstruct the complex amplitude distribution of the diffracted object wave field. The phase iteration process starts with an initial complex amplitude distribution $U_1(x, y)$ coming from the square root of the first intensity image (I'_1 or Fig. 6.4a) multiplied by an initial constant phase: $U_1(x, y) = I'_1(x, y)\exp(\mathrm{i}\varphi_0(x, y))$, being $\varphi_0(x, y) = 0$. This initial complex amplitude distribution $U_1(x, y)$ is digitally propagated to the next measured plane, that is, to

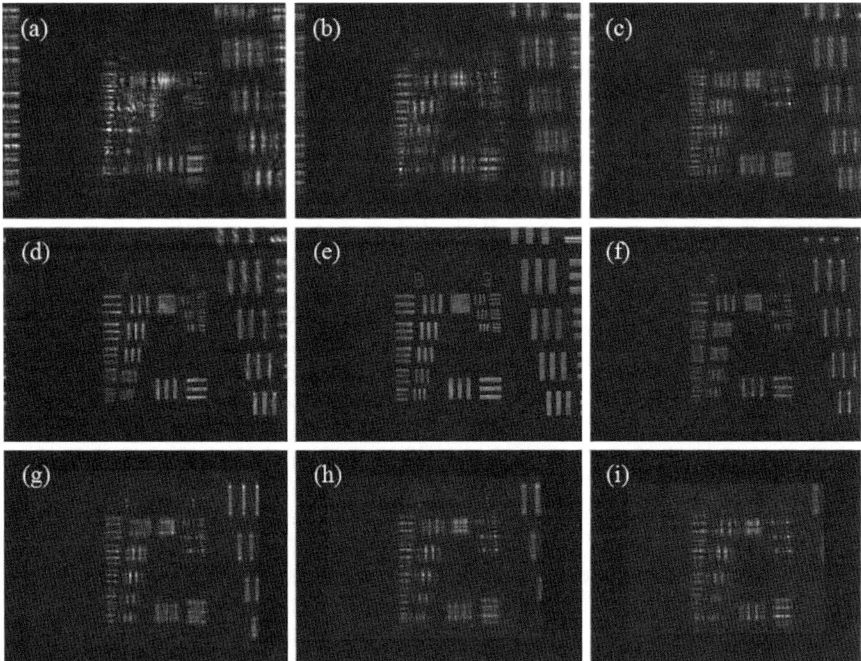

Fig. 6.4 Image compensation for magnification and lateral position introduced by the lens displayed at the SLM for the USAF test central part

the image represented by I_2' (or Fig. 6.4b). Once again, numerical computation of the RS equation by convolution approximation is considered, but now the recorded hologram is the complex amplitude distribution $U_1(x, y)$, and the Fresnel approximation is used in the calculation of the Fourier transform of the impulse response $H(u, v; d) = \mathrm{FT}\{h(x, y; d)\}$, where (u, v) are the spatial frequency coordinates. Then, the calculation of the propagated wave field from the first measured intensity to the second one separated by a distance of d_2 is simplified to $\mathrm{RS}_2(x, y; d_2) = \mathrm{FT}^{-1}\{T_1(u, v)\ H(u, v; d_2)\}$, where $T_1(u, v) = \mathrm{FT}\{U_1(x, y)\}$ is the Fourier transform of the initial complex wave field.

This procedure is repeated for every plane where measurements are performed, that is, $\mathrm{RS}_{N+1}(x, y; d_{N+1}) = \mathrm{FT}^{-1}\{T_N(u, v)\ H(u, v; d_{N+1})\}$. However, for subsequent propagations ($N \neq 1$), we retain the phase distribution $\varphi_N(x, y)$ incoming from the previous propagation and replace the obtained amplitude by the square root of the intensity measured at that plane: $T_N(u, v) = \mathrm{FT}\{U_N(x, y)\} = \mathrm{FT}\{(I_N'(x, y))^{1/2} \exp(\mathrm{i}\varphi_N(x, y))\}$. Let us name the iterative process from the first image (I_1) to the last one (I_9) step by step considering all the images as a *cycle*. Thus, once one cycle is performed, we propagate from I_9 to I_1 and the iterative process starts again; that is, a second cycle is considered. The iterative process is repeated until the quality of the reconstructed image at the imaging plane (I_5 or Fig. 6.3e) will be smaller than some predefined threshold. This threshold is obtained

Fig. 6.5 Representation of the normalized rms (*vertical coordinate*) versus the number of cycles (*horizontal coordinate*)

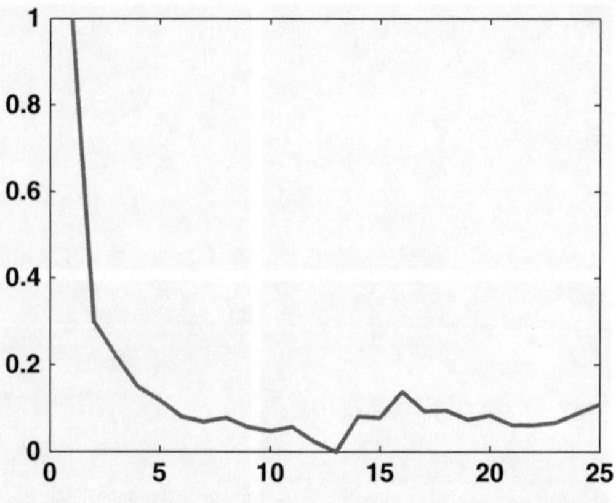

by computing the root mean square (rms) between the real image I'_5 and the one being obtained from the last image of the set (I'_9) propagated to plane number 5. Figure 6.5 plots the normalized variation of the rms as the number of cycles increase (from 1 to 25). We stop the iteration process at cycle number 6 when the rms value equals the background rms.

Finally, the whole process retrieves phase information about the input sample and, thus, the entire complex amplitude wave front reconstruction. Figure 6.6a depicts the case when the central part of the USAF test target is directly imaged

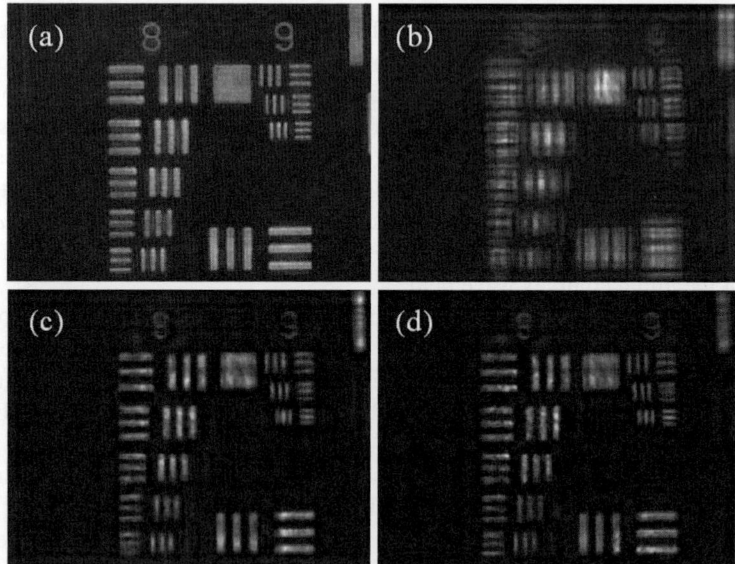

Fig. 6.6 (**a**) Direct imaging of the USAF test central part. (**b–d**) Propagated images resulting without cycle, with one cycle, and with six cycles in the iteration process, respectively

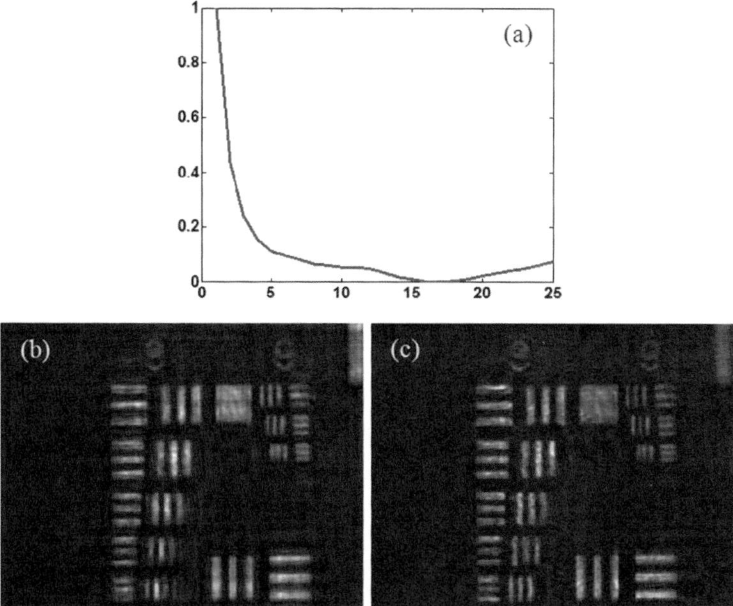

Fig. 6.7 (**a**) Representation of the normalized rms (*vertical coordinate*) versus the number of cycles (*horizontal coordinate*). (**b** and **c**) Propagated image resulting after 1 and 13 cycles, respectively, in the iteration process

over the CCD, case b corresponds to the direct propagation of image I'_9 (Fig. 6.4i) without applying the proposed approach, and cases c and d represent the resulting image obtained after one and six cycles, respectively, in the phase iterative process. We can see that no image reconstruction is possible when the proposed approach is not considered while a very good image quality is reconstructed by considering only one cycle in the iteration process.

As final step, we have checked the robustness and the capabilities of the proposed approach when considering less reconstruction planes in the iteration process. As first case, we have avoided the use of plane numbers 4, 5 (real imaging plane), and 6 corresponding to images d–e–f of Fig. 6.4, respectively, in the reconstruction process. That is, we are considering only misfocused images in the phase iteration reconstruction. The resulting reconstructions are depicted in Fig. 6.7 where the number of cycles increases from 6 to 13 in order to achieve a similar result as the one obtained in Fig. 6.6d.

And as second case, we have considered a less number of images in the reconstruction process: only image numbers 7–8–9 corresponding to images g–h–i in Fig. 6.4, respectively. The results are depicted in Fig. 6.8. Once again, an improvement in the number of cycles (from 6 to 9) is needed to achieve a similar result as the one obtained in Fig. 6.6d. Also as we can see, the reconstruction obtained with only one cycle (Fig. 6.8b) is significantly worse than the case depicted in Fig. 6.7b.

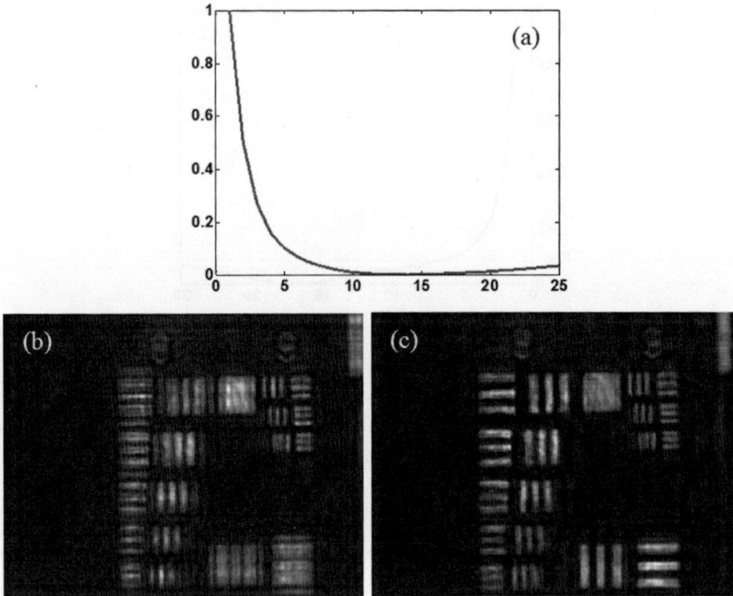

Fig. 6.8 (**a**) Representation of the normalized rms (*vertical coordinate*) versus the number of cycles (*horizontal coordinate*). (**b** and **c**) Propagated image resulting after one and nine cycles, respectively, in the iteration process

6.4 Experimental Validation of the Proposed Approach

In this section we present the experimental implementation of the reported approach when considering a complex sample composed by swine sperm cells enclosed in a counting chamber having a thickness of 20 μm. The sperm cells have an elliptical head shape of height and width around 6 × 9 μm, a total length of 55 μm, and a tail's width of 2 μm on the head side and below 1 μm on the end, approximately. It is an unstained sample that is dried up allowing fixed sperm cells for the experiments. Because of drying, the cells are fixed at different sections of the chamber allowing different sections in 3D.

The experimental setup assembled to validate the proposed approach is shown in Fig. 6.9. A collimated and horizontally polarized laser beam (Roithner Lasertechnik; green diode pumped laser modules, 532 nm wavelength) impinges over the input sample. A long working distance infinity-corrected microscope lens (Mitutoyo, M Plan Apo; 0.55 NA) in infinity imaging mode is used as imaging lens providing the image of the input sample at infinity. Imaging over the CCD (Basler A312f, 582 × 782 pixels, 8.3 μm pixel size, 12 bits/pixel) is achieved by a tube lens (doublet lens with 300 mm focal length). In this configuration, the resulting image has a magnification of 75×. A reflective SLM (Holoeye HEO 1080 P, 1920 × 1080 pixel resolution, 8 μm pixel pitch) that is placed between the microscope lens and the tube lens and a standard cube beam splitter (20 mm cube size) allows the reflective

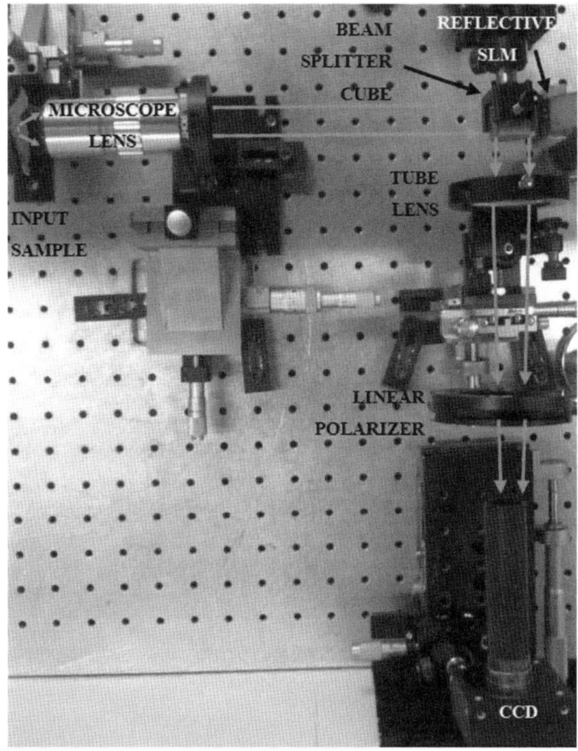

Fig. 6.9 Picture of the experimental setup for quantitative phase imaging by using an SLM in a reflective configuration

configuration. The SLM is connected to a computer where the different lenses are generated and sent to the SLM by displaying them in a figure that is transferred to the SLM. Matlab software is used to manage the whole process and for performing the required digital image processing. Two linear polarizers, one before the input sample and the other after the SLM, allow high-efficiency phase modulation in the SLM. Additional neutral density filters (not visible in the picture) are used to adjust the beam laser power.

Figure 6.10 shows the set of nine direct images recorded by the CCD when varying the SLM lens power in a similar way that Fig. 6.3 is for the USAF test case. Figure 6.11 shows the same set of images once the magnification and lateral displacement are compensated. Since the lenses displayed at the SLM are the same ones used in the USAF test case, the values obtained when studying the USAF case are applied in the sperm cell biosample case. Finally and for clarity, Fig. 6.12 magnifies the images corresponding to Fig. 6.11e, i. We can see as different sperm cells appear in focus, and since the sample is essentially a phase object, the cells appear invisible in conventional bright field imaging mode.

Then, we perform phase extraction iteration considering the whole set of nine images of the input sample (Fig. 6.11). In this case, the iteration cycle is repeated 15 times. The obtained results (real part and phase distribution) are depicted in

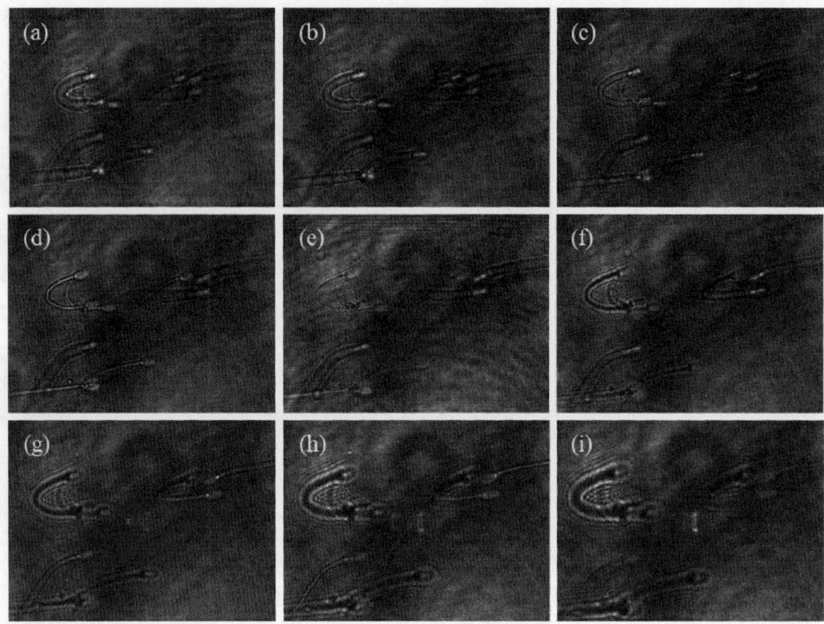

Fig. 6.10 Raw direct images of the sperm cell biosample obtained when varying the power of the lens displayed at the SLM

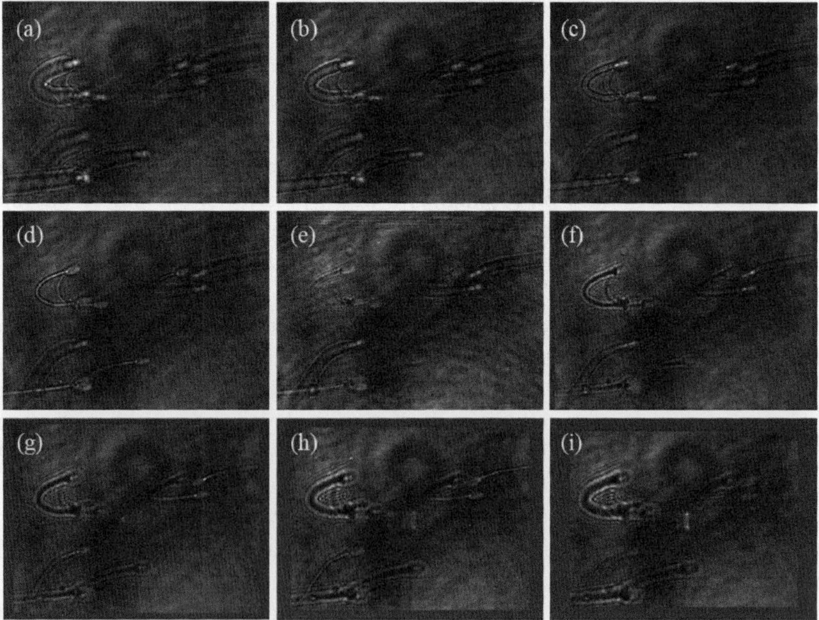

Fig. 6.11 Image compensation for magnification and lateral position introduced by the lens displayed at the SLM for the sperm cell biosample

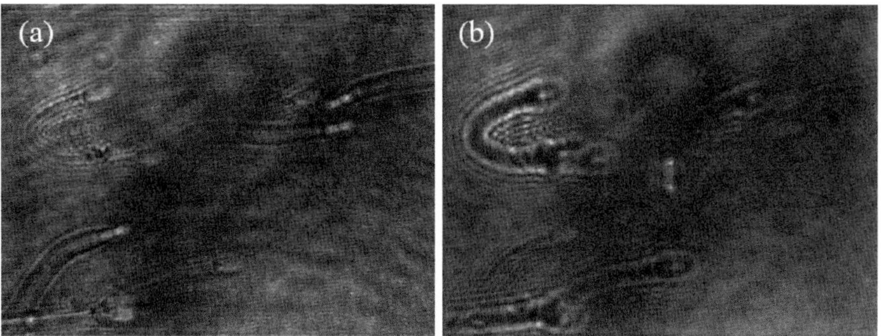

Fig. 6.12 Two sections of the 3D biosample where different sperm cells are in focus

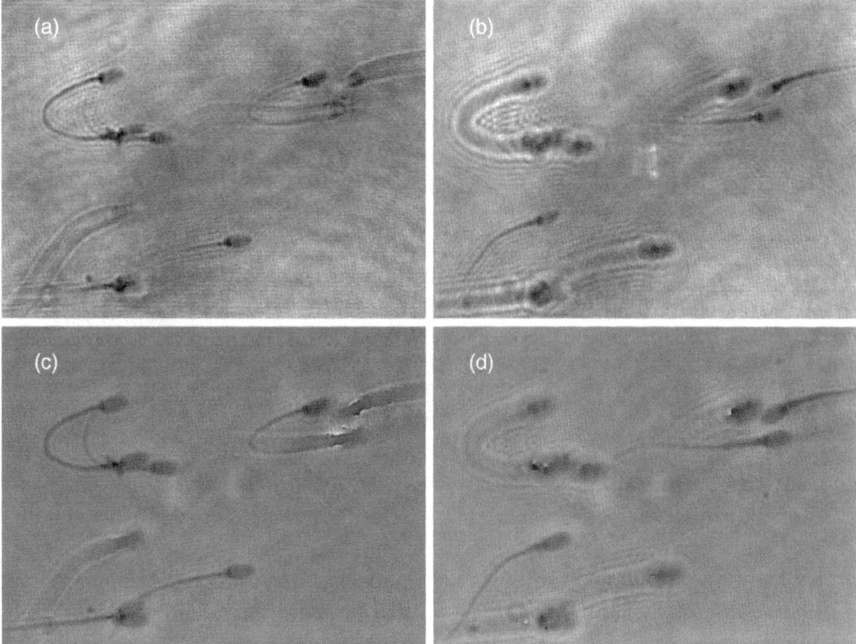

Fig. 6.13 (**a** and **b**) Real and (**c** and **d**) phase distributions of the retrieved complex amplitude images using the proposed approach and corresponding to the sections of Fig. 6.12

Fig. 6.13. We can see the cells that are basically invisible in Fig. 6.12a, b become now visible in Fig. 6.13a–c and b–d, respectively.

Finally, since phase distribution is related with 3D information of the sample, we can plot a 3D representation of the unwrapped phase distribution in order to obtain quantitative phase information of the sperm cell biosample. Figure 6.14 shows the 3D representations of both sections where the sperm cells are in focus (Fig. 6.13c, d).

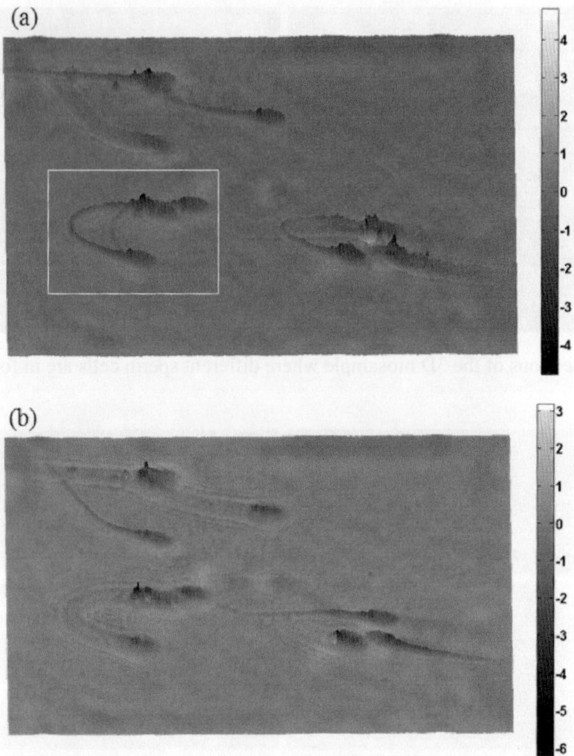

Fig. 6.14 (**a** and **b**) Three-dimensional representations of the in-focus sperm cells unwrapped phase distributions corresponding with Fig. 6.13c, d, respectively. Gray-level scale represents optical phase in radians

Now and in order to compare and validate the results obtained using the proposed method, we have assembled a conventional Mach–Zehnder digital holographic microscope configuration at the laboratory, and we have compared the results provided by both methods. In DHM experiment, the imaging arm contains the biosample and the microscope lens, and a reference beam is inserted in off-axis holographic configuration at the CCD plane. Off-axis holographic recording permits the recovery of the transmitted frequency band pass by Fourier transforming the recorded hologram and Fourier filtering of one of the hologram diffraction orders. Once the spectral filtered distribution is centered at the Fourier domain, inverse Fourier transformation retrieves complex amplitude imaging. Figure 6.15 shows the 3D plots of the unwrapped phase distributions in both considered cases: (a) corresponds to the iterated phase distribution obtained using the proposed approach (white rectangle depicted in Fig. 6.14a) and (b) corresponds to the phase provided by DHM. Although the cells are not the same ones in both figures, they come from the same swine sperm biosample allowing thus direct comparison. As it can be seen, the phase step between the background and the higher part of the sperm's

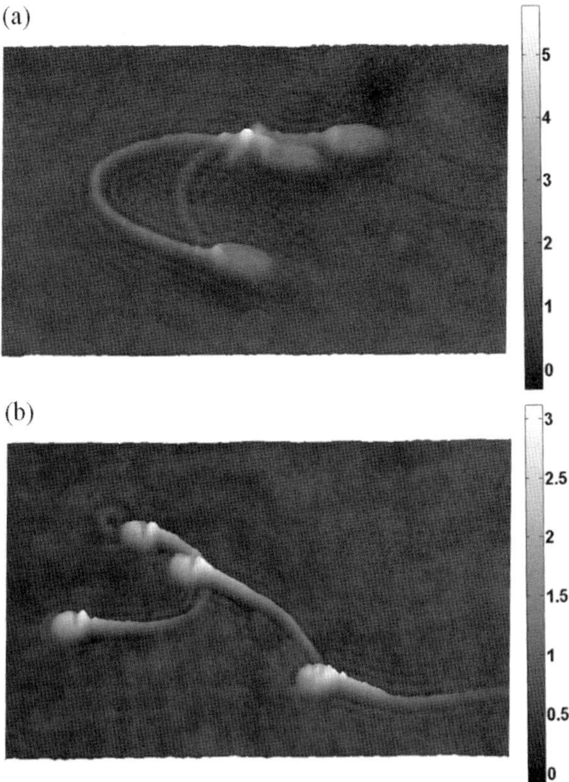

Fig. 6.15 Three-dimensional representations of sperm cells: (**a**) group of cells marked with a *solid white line rectangle* in Fig. 6.14a and (**b**) another group of sperm cells of the same biosample obtained with a conventional DHM configuration. Gray-level scale represents optical phase in radians

heads is around 2.5 rad in both figures. This fact shows a high degree of correlation between the unwrapped phase distributions provided by both methods. Note that the background phase value of both figures has been equalized for direct comparison.

Acknowledgments Part of this work was supported by the Spanish Ministerio de Educación y Ciencia under the project FIS2007-60626. Also, the authors want to thank Prof. Carles Soler and Paco Blasco from Proiser R+D S.L. (www.proiser.com) for providing the swine sperm biosample.

References

1. P. Torok, F.J. Kao, *Optical Imaging and Microscopy* (Springer, Heidelberg, 2003)
2. M.W. Davidson, M. Abramowitz, Optical microscopy. Encycl. Imaging Sci. Technol. **2**, 1106–1140 (2002)
3. F. Zernike, How I discovered phase contrast. Science **121**, 345–349 (1955)

4. M. Pluta, *Advanced Light Microscopy, Vol. 2: Specialized Methods* (Elsevier, Amsterdam, 1989)
5. F.H. Smith, Microscopic interferometry. Research **8**, 385–395 (1955)
6. A. Jesacher, S. Fürhapter, S. Bernet, M. Ritsch-Marte, Shadow effects in spiral phase contrast microscopy. Phys. Rev. Lett. **94**, 233902 (2005)
7. D.K. Hamilton, C.J.R. Sheppard, Differential phase contrast in scanning optical microscopy. J. Microsc. **133**, 27–39 (1984)
8. R. Liang, J.K. Erwin, M. Mansuripur, Variation on Zernike's phase contrast microscope. Appl. Opt. **39**, 2152–2158 (2000)
9. M.R. Arnison, K.G. Larkin, C.J.R. Sheppard, N.I. Smith, C.J. Cogswell, Linear phase imaging using differential interference contrast microscopy. J. Microsc. **214**, 7–12 (2004)
10. S.V. King, A. Libertun, R. Piestun, C.J. Cogswell, C. Preza, Quantitative phase microscopy through differential interference imaging. J. Biomed. Opt. **13**, 024020 (2008)
11. S.S. Kou, L. Waller, G. Barbastathis, C.J. Sheppard, Transport-of-intensity approach to differential interference contrast (TI-DIC) microscopy for quantitative phase imaging. Opt. Lett. **35**, 447–449 (2010)
12. G. Popescu, *Quantitative Phase Imaging of Nanoscale Cell Structure and Dynamics*, Chapter 5 in Methods Cell Biology (Elsevier, Amsterdam, 2008)
13. Ch Yang, A. Wax, I. Georgakoudi, E.B. Hanlon, K. Badizadegan, R.R. Dasari, M.S. Feld, Interferometric phase-dispersion microscopy. Opt. Lett. **25**, 1526–1528 (2000)
14. A. Ahn, C. Yang, A. Wax, G. Popescu, Ch. Fang-Yen, K. Badizadegan, R. R. Dasari, M.S. Feld, Harmonic phase-dispersion microscope with a Mach-Zehnder interferometer. Appl. Opt. **44**, 1188 (2005)
15. T. Kreis, *Handbook of Holographic Interferometry: Optical and Digital Methods* (Wiley-VCH, Weinheim, 2005)
16. D. Gabor, A new microscopic principle. Nature **161**, 777–778 (1948)
17. E. Cuche, F. Bevilacqua, C. Depeursinge, Digital holography for quantitative phase-contrast imaging. Opt. Lett. **24**, 291–293 (1999)
18. L.P. Yaroslavsky, *Digital Holography and Digital Image Processing: Principles, Methods, Algorithms* (Kluwer, New York, NY, 2003)
19. W. Xu, M.H. Jericho, I.A. Meinertzhagen, H.J. Kreuzer, Digital in-line holography for biological applications. Proc. Natl. Acad. Sci. USA **98**, 11301–11305 (2001)
20. L. Repetto, E. Piano, C. Pontiggia, Lensless digital holographic microscope with light-emitting diode illumination. Opt. Lett. **29**, 1132–1134 (2004)
21. P.R. Hobson, J. Watson, The principles and practice of holographic recording of plankton. J. Opt. A: Pure Appl. Opt. **4**, S34–S49 (2002)
22. W. Xu, M.H. Jericho, I.A. Meinertzhagen, H.J. Kreuzer Tracking particles in 4-D with in-line holographic microscopy. Opt. Lett. **28**, 164–166 (2003)
23. H. Sun, R.G. Perkins, J. Watson, M.A. Player, D.M. Paterson, Observations of coastal sediment erosion using in-line holography. J. Opt. A: Pure Appl. Opt. **6**, 703–710 (2004)
24. E. Malkiel, J.N. Abras, J. Katz, Automated scanning and measurements of particle distributions within a holographic reconstructed volume. Meas. Sci. Technol. **15**, 601–612 (2004)
25. Y. Zhang, X. Zhang, Reconstruction of a complex object from two in-line holograms. Opt. Express **11**, 572–578 (2003)
26. Y. Zhang, G. Pedrini, W. Osten, H. Tiziani, Whole optical wave field reconstruction from double or multi in-line holograms by phase retrieval algorithm. Opt. Express **11**, 3234–3241 (2003)
27. Y. Zhang, G. Pedrini, W. Osten, H.J. Tiziani, Reconstruction of in-line digital holograms from two intensity measurements. Opt. Lett. **29**, 1787–1789 (2004)
28. T. Latychevskaia, H.W. Fink, Simultaneous reconstruction of phase and amplitude contrast from a single holographic record. Opt. Express **17**, 10697–10705 (2009)
29. U. Schnars, W.P. Jueptner, *Digital Holography* (Springer, Heidelberg, 2005)
30. Ch. Mann, L. Yu, Ch. Lo, M. Kim, High-resolution quantitative phase-contrast microscopy by digital holography. Opt. Express **13**, 8693–8698 (2005)

31. P. Marquet, B. Rappaz, P.J. Magistretti, E. Cuche, Y. Emery, T. Colomb, Ch. Depeursinge, Digital holographic microscopy: A noninvasive contrast imaging technique allowing quantitative visualization of living cells with subwavelength axial accuracy. Opt. Lett. **30**, 468–470 (2005)
32. F. Charrière, F. Montfort, J. Kühn, T. Colomb, A. Marian, E. Cuche, P. Marquet, Ch. Depeursinge, Cell refractive index tomography by digital holographic microscopy. Opt. Lett. **31**, 178–180 (2006)
33. B. Kemper, G. von Bally, Digital holographic microscopy for live cell applications and technical inspection. Appl. Opt. **47**, A52–A61 (2008)
34. J. Lobera, J.M. Coupland, Contrast enhancing techniques in digital holographic microscopy. Meas. Sci. Technol. **19**, 025501 (2008)
35. H. Iwai, C. Fang-Yen, G. Popescu, A. Wax, K. Badizadegan, R.R. Dasari M.S. Feld, Quantitative phase imaging using actively stabilized phase-shifting low-coherence interferometry. Opt. Lett. **29**, 2399–2401 (2004)
36. S. Reichelt, H. Zappe, Combined Twyman-Green and Mach-Zehnder interferometer for microlens testing. Appl. Opt. **44**, 5786–5792 (2005)
37. G. Popescu, T. Ikeda, R.R. Dasari, M.S. Feld, Diffraction phase microscopy for quantifying cell structure and dynamics. Opt. Lett. **31**, 775–777 (2006)
38. V. Mico, Z. Zalevsky, J. Garcia, Common-path phase-shifting digital holographic microscopy: a way to quantitative phase imaging and superresolution. Opt. Commun. **281**, 4273–4281 (2008)
39. V. Mico, Z. Zalevsky, C. Ferreira, J. García, Superresolution digital holographic microscopy for three-dimensional samples. Opt. Express **16**, 19260–19270 (2008)
40. I. Yamaguchi, T. Zhang, Phase-shifting digital holography. Opt. Lett. **22**, 1268–1270 (1997)
41. I. Yamaguchi, J. Kato, S. Ohta, J. Mizuno, Image formation in phase-shifting digital holography and applications to microscopy. Appl. Opt. **40**, 6177–6185 (2001)
42. G.A. Dunn, D. Zicha, *Using DRIMAPS system of transmission interference microscopy to study cell behavior. in Cell biology: A laboratory handbook* (Academic, New York, NY, 1997), pp. 44–53
43. D. Zicha, E. Genot, G. A. Dunn, I.M. Kramer, TGF beta 1 induces a cell-cycle-dependent increase in motility of epithelial cells. J. Cell Sci. **112**, 447–454 (1999)
44. S. Fürhapter, A. Jesacher, S. Bernet, M. Ritsch-Marte, Spiral interferometry. Opt. Lett. **30**, 1953–1955 (2005)
45. S. Bernet, A. Jesacher, S. Fürhapter, C. Maurer, M. Ritsch-Marte, Quantitative imaging of complex samples by spiral phase contrast microscopy. Opt. Express **14**, 3792–3805 (2006)
46. Ch. Maurer, A. Jesacher, S. Fürhapter, S. Bernet, M. Ritsch-Marte, Upgrading a microscope with a spiral phase plate. J. Microsc. **230**, 134–142 (2007)
47. A.Y.M. NG, C.W. See, M.G. Somekh, Quantitative optical microscope with enhanced resolution using a pixelated liquid crystal spatial light modulator. J. Microsc **214**, 334–340 (2004)
48. G. Popescu, L.P. Deflores, J.C. Vaughan, K. Badizadegan, H. Iwai, R.R. Dasari, M.S. Feld, Fourier phase microscopy for investigation of biological structures and dynamics. Opt. Lett. **29**, 2503–2505 (2004)
49. G. Popescu, K. Badizadegan, R.R. Dasari, M.S. Feld, Observation of dynamic subdomains in red blood cells. J. Biomed. Opt. Lett. **11**, 040503 (2006)
50. N. Lue, W. Choi, G. Popescu, T. Ikeda, R.R. Dasari, K. Badizadegan, M.S. Feld, Quantitative phase imaging of live cells using fast Fourier phase microscopy. Appl. Opt. **32**, 1836–1842 (2007)
51. T. Ikeda, G. Popescu, R.R. Dasari, M.S. Feld, Hilbert phase microscopy for investigating fast dynamics in transparent systems. Opt. Lett. **30**, 1165–1167 (2005)
52. G. Popescu, T. Ikeda, K. Badizadegan, R.R. Dasari, M.S. Feld, Erythrocyte structure and dynamics quantified by Hilbert phase microscopy. J. Biomed. Opt. Lett. **10**, 060503 (2005)
53. G. Popescu, T. Ikeda, K. Goda, C.A. Best-Popescu, M.L. Laposata, S. Manley, R.R. Dasari, K. Badizadegan, M.S. Feld, Optical measurement of cell membrane tension. Phys. Rev. Lett. **97**, 218101 (2006)

54. Y.K. Park, G. Popescu, R.R. Dasari, K. Badizadegan, M.S. Feld, Fresnel particle tracking in three dimensions using diffraction phase microscopy. Opt. Lett. **32**, 811–813 (2007)
55. G. Popescu, Y.K. Park, R.R. Dasari, K. Badizadegan, M.S. Feld, Coherence properties of red blood cell membrane motions. Phys. Rev. E **76**, 021902 (2007)
56. Y.K. Park, G. Popescu, K. Badizadegan, R.R. Dasari, M.S. Feld, Diffraction phase and fluorescence microscopy. Opt. Express **14**, 8263–8268 (2006)
57. N. Lue, W. Choi, K. Badizadegan, R. R. Dasari, M.S. Feld, G. Popescu, Confocal diffraction phase microscopy of live cells. Opt. Lett. **33**, 2074–2076 (2008)
58. M.S. Amin, Y. Park, N. Lue, R.R. Dasari, K. Badizadegan, M.S. Feld, G. Popescu, Microrheology of red blood cell membranes using dynamic scattering microscopy. Opt. Express **15**, 17001–17009 (2007)
59. H. Ding, Z. Wang, F. Nguyen, S.A. Boppart, G. Popescu, Fourier transform light scattering of inhomogeneous and dynamic structures. Phys. Rev. Lett. **101**, 238102 (2008)
60. H. Ding, F. Nguyen, S.A. Boppart, G. Popescu, Optical properties of tissues quantified by Fourier transform light scattering. Opt. Lett. **34**, 1372 (2009)
61. H. Ding, G. Popescu, Instantaneous spatial light interference microscopy. Opt. Express **18**, 1569–1575 (2010)
62. Z. Wang, I.S. Chun, X. Li, Z.Y. Ong, E. Pop, L. Millet, M. Gillette, G. Popescu, Topography and refractometry of nanostructures using spatial light interference microscopy. Opt. Let. **35**, 208–210 (2010)
63. R.W. Gerchberg, W.O. Saxton, A practical algorithm for the determination of phase from image and diffraction plane pictures. Optik **35**, 237–246 (1978)
64. J.R. Fienup, Phase retrieval algorithms: a comparison. Appl. Opt. **21**, 2758–2769 (1982)
65. A. Barty, K.A. Nugent, D. Paganin, A. Roberts, Quantitative optical phase microscopy. Opt. Lett. **23**, 817–819 (1998)
66. M.R. Teague, Deterministic phase retrieval: A Green's function solution. J. Opt. Soc. Am. **73**, 1434–1441 (1983)
67. N. Streibl, Phase imaging by the transport equation of intensity. Opt. Commun. **49**, 6–10 (1984)
68. M.R. Teague, Image formation in terms of transport equation. J. Opt. Soc. Am. A **2**, 2019–2026 (1985)
69. T.E. Gureyev, A. Roberts, K.A. Nugent, Partially coherent fields, the transport-of-intensity equation, and phase uniqueness. J. Opt. Soc. Am. A **12**, 1942–1946 (1995)
70. G.Z. Yang, B.Z. Dong, B.Y. Gu, J. Zhuang, O.K. Ersoy, Gerchberg-Saxton and Yang-Gu algorithms for phase retrieval in a nonunitary transform system: a comparison. Appl. Opt. **33**, 209–218 (1994)
71. G. Pedrini, W. Osten, Y. Zhang, Wave-front reconstruction from a sequence of interferograms recorded at different planes. Opt. Lett. **30**, 833–835 (2005)
72. P. Almoro, G. Pedrini, W. Osten, Complete wavefront reconstruction using sequential intensity measurements of a volume speckle field. Appl. Opt. **45**, 8596–8605 (2006)
73. P. Almoro, G. Pedrini, W. Osten, Aperture synthesis in phase retrieval using a volume-speckle field. Opt. Lett. **32**, 733–735 (2007)
74. A. Anand, V. K Chhaniwal, P. Almoro, G. Pedrini, W. Osten, Shape and deformation measurements of 3D objects using volume speckle field and phase retrieval. Opt. Lett. **34**, 1522–1524 (2009)
75. P.F. Almoro, G. Pedrini, A. Anand, W. Osten, S.G. Hanson, Angular displacement and deformation analyses using a speckle-based wavefront sensor. Appl. Opt. **48**, 932–940 (2009)
76. P. Bao, F. Zhang, G. Pedrini, W. Osten, Phase retrieval using multiple illumination wavelengths. Opt. Lett. **33**, 309–311 (2008)
77. L. Camacho, V. Micó, Z. Zalevsky, J. García, Quantitative phase microscopy using defocusing by means of a spatial light modulator. Opt. Express **18**, 6755–6766 (2010)
78. Ch. Maurer, A. Jesacher, S. Bernet, M. Ritsch-Marte, Phase contrast microscopy with full numerical aperture illumination. Opt. Express **16**, 19821–19829 (2008)

79. S.B. Mehta, C.J.R. Sheppard, Quantitative phase-gradient imaging at high resolution with asymmetric illumination-based differential phase contrast. Opt. Lett. **34**, 1924–1926 (2009)
80. T.J. McIntyre, Ch. Maurer, S. Bernet, M. Ritsch-Marte, Differential interference contrast imaging using a spatial light modulator. Opt. Lett. **34**, 2988–2990 (2009)
81. Ch. Maurer, S. Khan, S. Fassl, S. Bernet, M. Ritsch-Marte, Depth of field multiplexing in microscopy. Opt. Express **18**, 3023–3034 (2010)

Chapter 7
Quantitative Phase Microscopy of Biological Cell Dynamics by Wide-Field Digital Interferometry

Natan T. Shaked, Matthew T. Rinehart, and Adam Wax

Abstract Interferometric phase measurements of wide-field images of biological cells provide a quantitative tool for cell biology, as well as for medical diagnosis and monitoring. Visualizing rapid dynamic cell phenomena by interferometric phase microscopy can be performed at very fast rates of up to several thousands of full frames per second, while retaining high resolution and contrast to enable measurements of fine cellular features. With this approach, no special sample preparation, staining, or fluorescent labeling is required, and the resulting phase profiles yield the optical path delay profile of the cell with sub-nanometer accuracy. In spite of these unique advantages, interferometric phase microscopy has not been widely applied for recording the dynamic behavior of live cells compared to other traditional phase microscopy methods such as phase contrast and differential interference contrast (DIC) microscopy, which are label free but inherently qualitative. Recent developments in the field of interferometric phase microscopy are likely to result in a change in this situation in the near future. Through careful consideration of the capabilities and limitations of interferometric phase microscopy, important new contributions in the fields of cell biology and biomedicine will be realized. This chapter presents the current state of the art of interferometric phase microscopy of biological cell dynamics, the open questions in this area, and specific solutions developed in our laboratory.

N.T. Shaked (✉)
Department of Biomedical Engineering, Fitzpatrick Institute for Photonics, Duke University, Durham, NC 27708, USA
e-mail: natan.shaked@duke.edu

A. Wax (✉)
Department of Biomedical Engineering, Fitzpatrick Institute for Photonics, Duke University, Durham, NC 27708, USA
e-mail: a.wax@duke.edu

P. Ferraro et al. (eds.), *Coherent Light Microscopy*, Springer Series in Surface Sciences 46, DOI 10.1007/978-3-642-15813-1_7,

7.1 Introduction

Visualizing dynamic cell phenomena that occur at second to millisecond time scales, such as membrane fluctuations, cell swelling, and action potential activation, requires development of wide-field microscopy techniques that can achieve high data acquisition rates while retaining sufficient resolution and contrast to enable measurements of fine cellular features. Biological cells are, however, mostly transparent three-dimensional objects that are very similar to their surrounding in terms of absorbance and reflection, and thus conventional intensity-based light microscopy techniques lack the required contrast to achieve this goal. Exogenous contrast agents such as fluorescent dyes are frequently used to solve the contrast problem. However, fluorescent contrast agents tend to photobleach, reducing the available imaging time. Other concerns include the potential for cytotoxicity and the possibility that the agents themselves will influence cellular behavior. As an alternative, phase microscopy can provide label-free information on cellular structure and dynamics. Traditional phase microscopy methods, such as phase contrast (PhC) and differential interference contrast (DIC) microscopy, are widely used today [1]. However, these approaches are not inherently quantitative, and each method presents its own distinct imaging artifacts. On the other hand, wide-field digital interferometry (WFDI) has the potential to provide a powerful, label-free method for quantitative phase measurements of biological cell dynamics [2–8]. WFDI is based on measuring an interference pattern composed of a superposition of the light field which has interacted with the sample and a mutually coherent reference field. With this approach, the entire complex wavefront describing the sample is captured. From the recorded complex field, it is possible to digitally reconstruct the quasi-three-dimensional distribution of the sample field without the need for mechanical scanning.

Chapter 9 presents an introduction to the field of quantitative phase microscopy of biological cells using digital interferometry, as well as reviews several specific methods and related applications. The current chapter mainly focuses on the utility of WFDI for quantitative phase microscopy of cell dynamics.

In practice, WFDI has been less commonly used for recording dynamic behaviors of live cells compared to other label-free phase microscopy methods such as PhC and DIC, in spite of its many attractive advantages. However, we believe WFDI becomes more widely used in the near future by careful analysis of its limitations and identification of possible solutions. Specific areas which need to be considered are (**a**) time resolution – camera bandwidth consumption tradeoff; (**b**) phase noise, stability, and sensitivity of the interferometric system; (**c**) phase unwrapping problem; (**d**) lack of specificity inside cells; and (**e**) utilizing the visualized phase profile for quantitative–functional analysis of the cells.

Each section in this chapter analyzes a different limitation from the list above and its impact on imaging live biological cells. Then, various approaches for coping with these limitations are presented in each section. Several specific experimental demonstrations of imaging of different dynamic behaviors of live cells are presented throughout the chapter to illustrate these points. These demonstrations include contracting cardiomyocytes, dynamic cytology of articular chondrocyte transient behavior, microscopic unicellular organism movement, and neuronal dynamics.

7.2 Time Resolution – Camera Bandwidth Consumption Tradeoff

WFDI is based on recording an interference pattern by a digital camera. This interference pattern is composed of light that has interacted with a sample and a mutually coherent light field that is derived directly from the light source, typically without interacting with the sample. With this approach, the complex sample field, which includes the phase profile of the sample, is recorded by the camera, even though the camera is only sensitive to intensity. The problem with this approach is that the intensity interferometric signal contains unwanted zero-order and twin image diffracted waves, in addition to the desired complex sample field. In order to extract the sample phase profile, these unwanted components must be eliminated during the digital processing stage. Off-axis interferometry [9–17] deals with this problem by imposing a large angle between the reference and the sample beams, creating a spatial separation between the desired and the undesired field components. However, this approach comes at the expense of ineffective use of the camera spatial frequency bandwidth, which means that high spatial frequencies (or alternatively the captured field of view (FOV)) of the sample might be reduced or lost entirely. When acquiring the phase profiles of dynamic biological samples, the requirement for the digital camera frame rate might be demanding, especially when one considers that high frame rates are frequently obtained by reducing the number of camera pixels per frame, which further narrows the camera spatial bandwidth. Thus, when camera bandwidth is intentionally limited in order to obtain high frame rates (e.g., by pixel binning in the camera hardware level), camera bandwidth consumption is an important consideration. Therefore, even though traditional off-axis interferometry only requires a single exposure for acquiring the sample field, it might not always be the best choice for recording cell dynamics due to the fact that it does not effectively use the camera spatial bandwidth.

An alternative approach to off-axis interferometry which more effectively uses the camera spatial bandwidth is on-axis interferometry [18–26]. In this approach, the angle between the sample and the reference beams is set to zero. This results in a required camera bandwidth that is the same as that needed for acquiring the sample intensity image alone; however, this approach also causes the undesired diffracted waves to occlude the desired sample field. The traditional solution to this problem is to acquire three or four on-axis, phase-shifted interferograms of the same sample and to digitally separate the sample field through signal processing using all acquired interferograms. This approach may not be practical for dynamic processes, where the sample may change between the several frames of acquisition. In addition, phase noise may increase due to system fluctuations between the frames [27].

Hence, there is a clear tradeoff between time resolution and camera spatial bandwidth, and both approaches, on-axis interferometry and off-axis interferometry, have their own disadvantages for phase microscopy of cell dynamics. These disadvantages have to be carefully considered when designing a WFDI system for this purpose. Finding the optimal working point, in which time resolution and camera bandwidth consumption are optimized, is a requirement for rapid phase imaging. The next two sections present two approaches developed in our laboratory for optimizing time resolution and camera bandwidth consumption in WFDI.

7.2.1 Slightly Off-Axis Interferometry (SOFFI)

In [28], we introduced the slightly off-axis interferometry (SOFFI) technique for measuring dynamic processes. This method does not require separate measurements of the reference or sample fields, as required in on-axis interferometry. Furthermore, it does not require a detector with as much spatial frequency bandwidth as that needed for traditional off-axis interferometry. This method is, in fact, an intermediate solution between on-axis and off-axis interferometry. In the proposed method, the sample phase is obtained by using only two slightly off-axis phase-shifted interferograms, without the need for additional off-line measurements or estimations, and a simple digital process that can be implemented in real time. We first introduce this digital process mathematically; then we explain the conditions of the angular separation between the reference and the sample beams, as well as propose an optical system for implementing this approach.

As stated above, according to the SOFFI approach, only two slightly off-axis interferograms $I_k, k = 1, 2$, are acquired. These interferograms can be mathematically expressed as follows:

$$I_k = |E_{\mathrm{r}}|^2 + |E_{\mathrm{s}}|^2 + |E_{\mathrm{s}}|\left|E_{\mathrm{r}}^*\right| \exp[j(\phi_{\mathrm{OBJ}} + qx + \alpha_k)] + |E_{\mathrm{r}}|\left|E_{\mathrm{s}}^*\right| \times \exp[-j(\phi_{\mathrm{OBJ}} + qx + \alpha_k)] \quad (7.1)$$

where E_{r} and E_{s} are the reference and sample field distributions, respectively, ϕ_{OBJ} is the spatially varying phase associated with the object, q is the fringe frequency due to the angular shift between the sample and the reference fields, x is the direction of the angular shift (assuming straight fringes), and $\alpha_1 = 0$, $\alpha_2 = \beta$ define a phase shift which is intentionally induced between the interferograms by the use of wave plates. From (7.1), it can be seen that incorrect estimation of β or q will add constant or modulated phase errors to the calculated object phase profile. Thus, even though q and β can be measured or calculated in advance, we propose to measure these parameters digitally after each frame of acquisition. By doing this, the method avoids variations in these parameters that might occur across repeated measurements of a dynamic process. Assuming a horizontal fringe pattern, these parameters can be calculated by summing the fringe pattern columns and fitting the resulting vector to a sine wave. In practice, this process allows us to find q and β with high enough accuracy to prevent degradation of the phase modulation induced by the sample.

Once the two interferograms have been acquired, they can be digitally processed to recover the recorded sample's wrapped phase distribution as follows:

$$F = \left[I_1 - I_2 + j \cdot \mathrm{HT}\{I_1 - I_2\}\right] \cdot \frac{\exp(-jqx)}{[1 - \exp(j\beta)]} \quad (7.2)$$
$$\tilde{\phi}_{\mathrm{OBJ}} = \arctan\{\mathrm{Im}F/\mathrm{Re}F\}$$

where HT denotes a Hilbert transform.

The spatial frequency representation of each of the two interferograms includes two autocorrelation terms (ACTs), both located at the origin of the spatial frequency domain, and two crosscorrelation terms (CCTs), each located on a different side of the spatial frequency domain. Assuming that the CCTs are spatially separated, (7.2) first isolates the positive-frequency CCT from both ACTs and the negative-frequency CCT. To explain this, let us also recall that the Fourier transform (FT) of a HT of a signal g obeys the following rule: $\mathrm{FT}\{\mathrm{HT}\{g\}\}(\omega) = -j\cdot\mathrm{sign}(\omega)\cdot\mathrm{FT}\{g\}(\omega)$, where $\mathrm{sign}(\omega)$ is the signum function applied to the frequency axis ω. Therefore, multiplying the imaginary unit j by the Hilbert transform of each interferogram negates the negative frequencies (including the negative-frequency CCT and the negative-frequency half of the ACTs), while leaving the positive frequencies untouched (including the positive-frequency CCT and the positive-frequency half of the ACTs). Thus, by taking $I_k + j\cdot\mathrm{HT}\{I_k\}, k = 1, 2$, we have twice the positive-frequency CCT and twice the positive-frequency half of the ACTs. Then, by the subtraction of $I_1 + j\cdot\mathrm{HT}\{I_1\}$ and $I_2 + j\cdot\mathrm{HT}\{I_2\}$ (where I_1 and I_2 are phase shifted), the ACT parts left are eliminated as well.

Next, (7.2) centers the remaining positive-frequency CCT at the origin of the spatial frequency domain by multiplication with $\exp(-jqx)$. The arctan function in (7.2) yields the wrapped phase profile of the object, and an unwrapping algorithm, as reviewed in Sect. 7.4, is then used on $\tilde{\phi}_{\mathrm{OBJ}}$ to remove 2π phase ambiguities, yielding ϕ_{OBJ}. To compensate for stationary phase aberrations [29], it is recommended to subtract a reference phase image with no sample present. This can be achieved by capturing an additional pair of interferograms off-line, before recording the dynamic process and without the presence of the sample and calculating the object phase according to the procedure described above. Then, this sample-less phase image can be subtracted from the real-time sample object phase images to obtain the final object phase image.

To illustrate the advantage of the proposed approach, we theoretically compare the spatial frequency spectra of three methods: traditional off-axis interferometry, SOFFI, and traditional on-axis interferometry. These spectra are illustrated in Fig. 7.1, where for simplicity only one spatial frequency axis is shown. In all cases the ACTs, located around the origin of the spatial spectrum, are composed of the reference-field ACT and the sample-field ACT. Let us assume that the reference is a constant plane wave over the camera illumination area. Then, the total width of the ACTs is solely dominated by the spatial frequency bandwidth of the sample-field ACT, given by four times the highest spatial frequency ω_0 of the object, as imaged

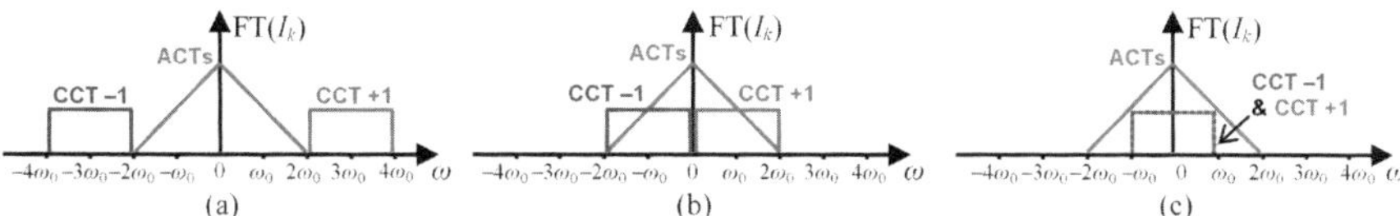

Fig. 7.1 Schematic comparison between the spatial frequency domains of (**a**) off-axis interferometry, (**b**) SOFFI, and (**c**) on-axis interferometry. For simplicity, only one spatial frequency axis is shown

on the camera. In comparison, the width of each CCT is only $2\omega_0$. As shown in Fig. 7.1a, traditional off-axis interferometry requires that the angle between the reference and the object beams is large enough to cause a complete separation of the CCTs from the ACTs. This occurs when $q \geq 3\omega_0$, resulting in a highest spatial frequency bandwidth of at least $4\omega_0$. Note that ω_0 can be experimentally measured a single time before recording the dynamics of the object of interest, under the reasonable assumption that this bandwidth stays constant during the multiple measurements of the object. Alternatively, for biological cell microscopy, ω_0 is usually inversely proportional to the diffraction-limited spot of the optics used for imaging.

For SOFFI, the CCTs should not overlap, but an overlap of the ACTs with each of the CCTs is allowed. As shown in Fig. 7.1b, this occurs if $q = \omega_0$. Therefore, the highest spatial frequency needed per exposure is only $2\omega_0$, half of that needed in traditional off-axis interferometry. Thus, when the maximum fringe frequency allowed by the camera does not create enough separation between the ACTs and the CCTs, an advantage can be gained by acquiring two off-axis interferograms using the SOFFI mathematical process described by (7.2), instead of using a single off-axis interferogram.

In the case of on-axis interferometry, shown in Fig. 7.1c, all CCTs and ACTs are centered at the origin. Therefore, the highest spatial frequency per exposure that is necessary is only ω_0. However, in this case three or four interferograms (depending on the technique chosen) are typically required to fully eliminate both ACTs and one of the CCTs.

For observing dynamic processes, if more than one interferogram is needed (on-axis interferometry and SOFFI), it is better to acquire the data with as few measurements as possible to minimize reconstruction errors and data acquisition times. In this aspect, SOFFI is superior to on-axis interferometry. In addition, traditional on-axis interferometry requires strict control of the phase shifts between the interferograms since slight variations can introduce an error in the reconstruction [27]. SOFFI avoids this type of error by digitally measuring the exact phase shift between the interferograms after the acquisition of each sample image by fitting the background fringe pattern in each interferogram to sine waves, as explained above.

To experimentally demonstrate the proposed method, we have constructed the optical system shown in Fig. 7.2. This interferometric setup is capable of sequentially acquiring two or more phase-shifted interferograms. Linearly polarized light from a coherent laser source is split into reference and sample beams using a modified Mach–Zehnder interferometer. To image the phase and amplitude information, lenses L_1 and L_3, as well as lenses L_2 and L_3, are positioned in 4f configurations. By varying the orientations of the two wave plates positioned in the reference arm, different phase shifts between the interferograms generated on the camera can be created [30]. Other phase-shifting configurations, such as the one presented in [31], are also possible for implementing this method.

Using the optical system shown in Fig. 7.2, we have performed two sets of experiments. In the first set of experiments, a static, water-immersed polystyrene microsphere (Duke Scientific Corporation; 12 μm diameter) on a microscope coverslip was used as the sample. Figure 7.3a demonstrates the low visibility of a simple

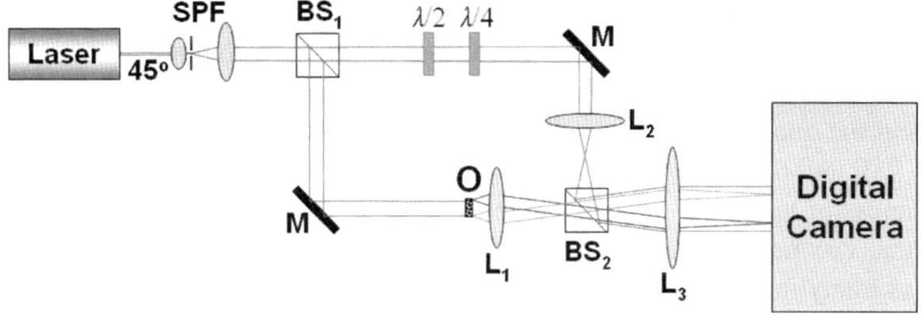

Fig. 7.2 Possible experimental setup for phase-shifting interferometry (sequential acquisition). SPF = spatial filter (beam expander including a confocally positioned pinhole); BS_1, BS_2 = beam splitters; M = mirror; $\lambda/2$ = half wave plate; $\lambda/4$ = quarter wave plate; O = object/sample; L_1, L_2, L_3 = lenses [28]

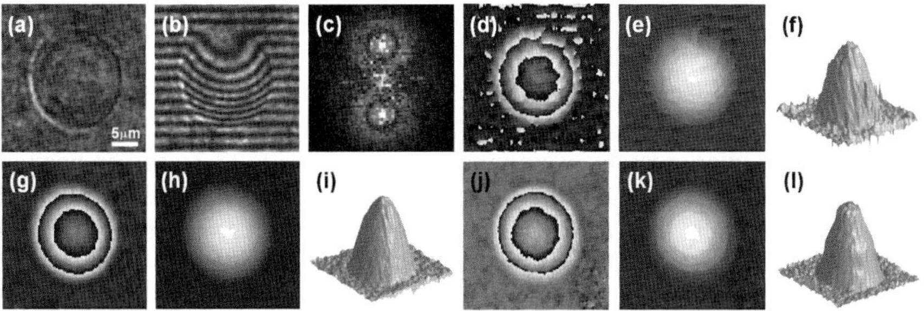

Fig. 7.3 Static polymer microsphere (12 μm in diameter): (**a**) Regular intensity image through the system; (**b**) single off-axis interferogram; (**c**) the middle part of the spatial spectrum of this interferogram. Digitally reconstructed phase obtained by (**d–f**) traditional off-axis interferometry (using a single interferogram), (**g–i**) SOFFI (using two phase-shifted interferograms), and (**j–l**) on-axis interferometry (using four phase-shifted interferograms). (**d**, **g**, **j**) Wrapped phase; (**e**, **h**, **k**) final unwrapped phase; (**f**, **i**, **l**) surface plot of the final unwrapped phase presented in (**e**, **h**, **k**), respectively [28]

intensity image of the sample. Phase reconstructions using off-axis, slightly off-axis, and on-axis geometries were executed. For the off-axis case, the best spatial resolution can be achieved when the fringe frequency on the camera is set to match the maximum fringe frequency allowed in this demonstration with the resulting interferogram shown in Fig. 7.3b. However, as shown in the spatial frequency spectrum of the interferogram (Fig. 7.3c), this fringe frequency was not high enough to separate the ACTs and CCTs. Figure 7.3d shows the corresponding reconstructed wrapped phase, and Fig. 7.3e, f shows the final unwrapped phase, demonstrating the low-quality images obtained by traditional off-axis interferometry under a restricted detector spatial frequency bandwidth.

To demonstrate the utility of SOFFI, another interferogram was acquired with the same fringe frequency but with a 90° phase shift introduced by rotating the quarter wave plate (see Fig. 7.1). The two interferograms were processed according to (7.2), yielding the wrapped phase shown in Fig. 7.3g and the unwrapped phase shown in Fig. 7.3h, i, demonstrating the improvement that was obtained by the SOFFI technique. For comparison, four on-axis interferograms (with phase shifts of 0°, 90°, 180°, and 270°) were obtained by varying the relative orientations of the half wave plate and the quarter wave plate [30], with the reconstructed wrapped phase shown in Fig. 7.3j and the corresponding unwrapped phase shown in Fig. 7.3k, l. From the smoothness of the shape and the flatness of the background, the traditional on-axis technique was performed comparably to the SOFFI technique, although it required twice as many image acquisitions. To quantify these results, fitting Fig. 7.3f, i, and l to a simulated phase profile of the microsphere with a flat background yielded root-squared-value accuracies of 91% for off-axis, 98% for slightly off-axis, and 99% for on-axis, respectively.

In the second set of experiments, we visualized a human skin cancer cell (A431, epithelial carcinoma) in growth media. The direct image of this sample through the optical system is shown in Fig. 7.4a. A single off-axis interferogram of the sample and its spatial frequency domain representation are shown in Fig. 7.4b and c, respectively, where again there was not enough separation between the ACTs and the CCTs, and thus the final unwrapped phase (Fig. 7.4d, g) is highly degraded. For the SOFFI case, an additional phase-shifted interferogram of the same sample was acquired. Both interferograms were processed according to (7.2), and an unwrapping algorithm was applied to the result, yielding the final unwrapped phase shown in Fig. 7.4e, h. Again, from observing the flatness of the background and the texture of the cell, it can be seen that substantial improvement is obtained by the SOFFI approach compared to the traditional off-axis approach.

For comparison, four on-axis interferograms, each phase shifted by 90°, were acquired for the same sample. However, for cancer cells in growth media, the final unwrapped phase for the on-axis case (Fig. 7.4f, i) was worse compared to the slightly off-axis case (Fig. 7.4e, h), especially at the cell boundaries. One cause for the degraded image in Fig. 7.4f, i was sample motion due to vibrations and Brownian motion of the cell media. Consequently, it was much harder to obtain the accurate phase shifts required for the traditional on-axis case. On the other hand, for

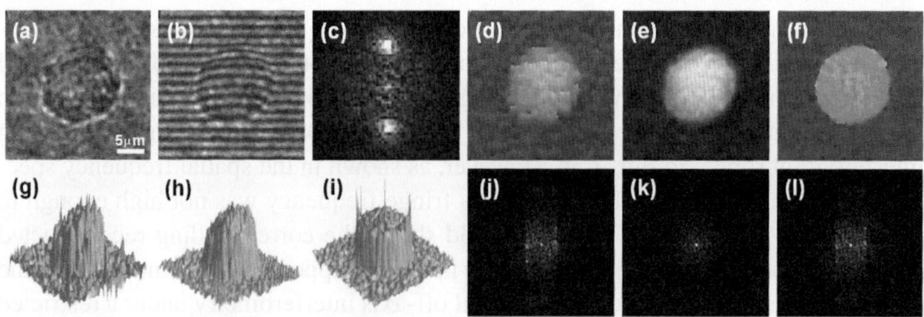

Fig. 7.4 Live human skin cancer (A431, epithelial carcinoma) cell in growth media: (**a**) Regular intensity image through the system; (**b**) single off-axis interferogram; (**c**) the middle part of the spatial spectrum of this interferogram. Final unwrapped phase obtained by (**d**) traditional off-axis interferometry (using a single interferogram), (**e**) SOFFI (using two phase-shifted interferograms), and (**f**) on-axis interferometry (using four phase-shifted interferograms); (**g–i**) surface plots corresponding to (**d–f**), respectively; and (**j–l**) MTF images of the background of (**d–f**), respectively [28]

the slightly off-axis case, the exact phase shifts were easily found by digitally fitting the background fringes to sine waves, and thus temporal phase shift errors were minimized. In addition, SOFFI required only two measurements, rather than four, which made the phase profile less vulnerable to sample changes arising from motion between measurements and resulted in a higher image throughput. To quantify the improvement of the proposed method for the cell measurement as well, Fig. 7.4j–l shows the modulation transfer function (MTF) of the image background (simply growth media with no cell present) of Fig. 7.4d–f, respectively. Excluding the zero-order point at the center, the half-power bandwidth in the vertical dimension for the slightly off-axis case is 0.31 and 0.34 of that for the two other cases. The fact that the slightly off-axis MTF (Fig. 7.4k) is narrower than in the two other cases indicates a flatter background and consequently produces a smoother overall phase profile for the SOFFI case.

7.2.2 Parallel On-Axis Interferometry (PONI)

In [7], we presented an optical setup that could acquire two slightly off-axis phase-shifted interferograms in a single camera exposure, using the same HT-based digital processing elaborated in the previous subsection. This method avoids temporal phase noise that can arise between successive phase-shifted frames.

Experimental results from this system are shown in Fig. 7.5. The phase imaging of a live human cancer cell and a beating rat myocardial cell (cardiomyocyte) demonstrate that sub-nanometer optical path delay stability can be obtained at millisecond time scales when acquiring the two phase-shifted interferograms in parallel.

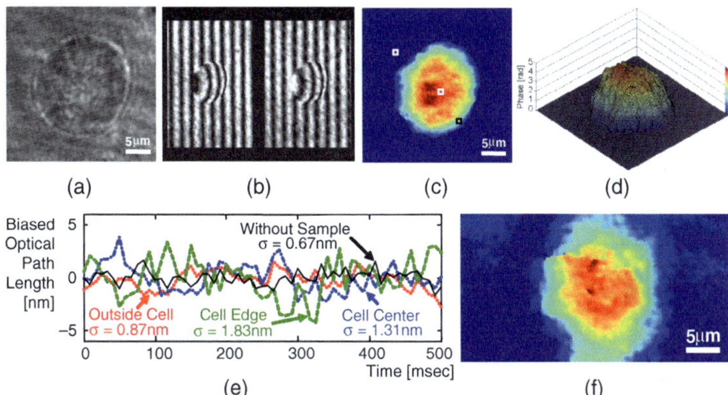

Fig. 7.5 (**a**–**e**) WFDI quantitative phase microscopy of human breast cancer (MDA-MB-468) cell in growth media: (**a**) Intensity image through the system (low visibility); (**b**) two phase-shifted interferograms of the sample captured in a single camera exposure; (**c**) final unwrapped phase profile; (**d**) surface plot of the phase profile shown in (**c**); (**e**) temporal phase stability without the sample and with the sample in the three *points* marked in (**c**); and (**f**) the final unwrapped phase profile of a rat beating myocardial cell (Video 1) [7]

In [32], a similar parallel optical system was used to acquire two *on-axis* interferograms in a single camera exposure, which is more advantageous in the detector bandwidth consumption than traditional approaches. As explained above, the bandwidth consumption of a single on-axis hologram is the same as that needed for capturing the intensity image of the sample. However, traditionally on-axis interferometry requires at least three interferograms to separate the wanted CCT from the unwanted CCT and ACTs. Alternative methods for acquiring three or four on-axis interferograms in a single camera exposure have been proposed (e.g., [20, 21] and others). However, since at least three interferograms are needed by these methods, the camera bandwidth is still not utilized optimally. Two-step on-axis WFDI, in which two (rather than three) on-axis interferograms are required, has been suggested in [19]. However, this technique requires off-line measurements of both the reference and the sample fields, which may be unsuitable for acquiring dynamic processes. Meng et al. [22] have suggested a two-step phase-shifting on-axis WFDI approach that requires only two interferograms and an off-line reference wave intensity measurement. Awatsuji et al. [24] have demonstrated how to parallelize this technique for non-biological, non-microscopic, and amplitude objects.

In [32], we proposed a parallel two-step on-axis WFDI method [32], called parallel on-axis interferometry (PONI), which is suitable for measuring the quantitative phase profile of cell dynamics since only a single exposure is required for each interferometric measurement of the dynamic process. The PONI method is more cost-effective in utilizing the camera bandwidth than off-axis interferometry, maximizing the amount of spatial information that can be captured from the sample, given a finite spatial camera bandwidth. Figure 7.6 shows the PONI scheme. The system is composed of a modified Mach–Zehnder interferometer, followed by an image/polarization splitter. A 45° linearly polarized laser light is splitted by beam splitter BS_1 into two beams. One beam is transmitted through the biological sample and magnified by the microscope objective. The other beam serves as a reference beam and is transmitted through a quarter wave plate, creating circular polarization, and then magnified by a similar objective lens. The beams are combined by beam splitter BS_2, where there is no angular offset between the beams. Lens L_0 is in a 4f

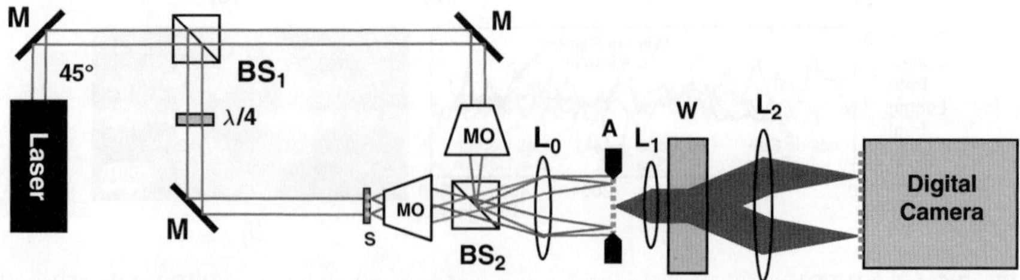

Fig. 7.6 PONI microscope for quantitative phase measurements of live cell and microorganism dynamics. BS_1, BS_2 = beam splitters; M = mirror; $\lambda/4$ = quarter wave plate; S = sample; A = rectangular aperture; MO = microscope objective; L_0, L_1, L_2, = lenses; WP = Wollaston prism [32]

configuration with both microscope objectives, imaging the phase and amplitude of the sample field in the aperture plane (A). This rectangular aperture ensures that only half of the digital camera plane is illuminated by each interferogram. The image/polarization splitter, lying beyond the aperture, includes an additional 4f subsystem that images the field at the aperture plane onto the camera sensor. A Wollaston prism, positioned in the confocal plane of lenses L_1 and L_2, is used to split the beam into two perpendicularly polarized beams. These two beams create two on-axis interferograms at the camera, phase shifted by 90° compared to each other.

The two 90° phase-shifted on-axis interferograms, which are acquired in a single camera exposure, can be mathematically expressed as follows:

$$\begin{aligned} I_1 &= I_R + I_S + 2\sqrt{I_R I_S}\cos(\phi_{OBJ} + \phi_C) \\ I_2 &= I_R + I_S + 2\sqrt{I_R I_S}\sin(\phi_{OBJ} + \phi_C) \end{aligned} \tag{7.3}$$

where I_R and I_S are the reference and sample intensity distributions, respectively; ϕ_{OBJ} is the spatially varying phase of the sample; and ϕ_C is the spatially varying background phase, without the presence of the sample. Let us define $I_0 = I_R + I_S$ and the overall phase as $\varphi = \phi_{OBJ} + \phi_C$. From (7.3), we get $2\sqrt{I_R I_S}\cos\varphi = I_1 - I_0$ and $2\sqrt{I_R I_S}\sin\varphi = I_2 - I_0$. Using the following trigonometric relation $\cos^2\varphi + \sin^2\varphi = 1$, we get the following quadratic equation:

$$2I_0^2 - 2I_0(I_1 + I_2 + 2I_R) + (I_1^2 + I_2^2 + 4{I_R}^2) = 0 \tag{7.4}$$

This quadratic equation yields the following solutions for I_0:

$$I_0 = \frac{I_1 + I_2 + 2I_R \pm \sqrt{(I_1 + I_2 + 2I_R)^2 - 2(I_1^2 + I_2^2 + 4I_R^2)}}{2} \tag{7.5}$$

It can be shown that if the reference field is strong enough compared to the sample field, the negative sign in (7.5) should always be chosen [22, 23] and thus I_0 can be found.

Note that in (7.5) I_0 can be obtained without direct knowledge of the sample intensity I_S, and thus this procedure is valid even if I_s changes during observation of the dynamic process. The only a priori knowledge that is needed is an off-line measurement of the reference intensity I_R, which is assumed to stay constant during the observation period. The dynamic process measurements are then acquired by recording of two 90° phase-shifted on-axis interferograms (I_1, I_2), using the system illustrated in Fig. 7.6.

In practice, the phase retrieval procedure should include a stationary phase-referencing step where the background phase is measured. The complete phase retrieval procedure is defined as follows. In the first step, three off-line measurements are taken without the presence of the sample: the sample arm intensity I'_S, the reference arm intensity I'_R, and two interferograms (I'_1, I'_2). Using these data, the wrapped background phase $\tilde{\phi}_C$ can be calculated as follows:

$$F' = \left[\left(I_1' - I_S' - I_R'\right) + j\left(I_2' - I_S' - I_R'\right)\right] / \sqrt{I_R'}; \quad \tilde{\phi}_C = \arctan(\text{Im } F' / \text{Re } F') \tag{7.6}$$

Then, an unwrapping algorithm, as reviewed in Sect. 7.4, is applied to solve 2π ambiguities in $\tilde{\phi}_C$, which yields ϕ_C.

During the recording of sample dynamics and at each sample observation time point, a pair of interferograms (I_1, I_2) is continuously recorded, where each pair is recorded in a single camera exposure. The overall wrapped phase is calculated as follows:

$$F = \left[(I_1 - I_0) + j\,(I_2 - I_0)\right] / \sqrt{I_R}; \quad \tilde{\varphi} = \arctan(\text{Im } F / \text{Re } F) \tag{7.7}$$

where $I_R = I_R'$ and thus it is already known, and I_0 is calculated by (7.5). An unwrapping algorithm, as reviewed in Sect. 7.4, is applied to solve 2π ambiguities in $\tilde{\varphi}$, which yields φ. Finally, the sample phase is calculated as follows:

$$\phi_{\text{OBJ}} = \varphi - \phi_C \tag{7.8}$$

Since the information for each observation of the sample is acquired in a single camera exposure, the method itself is not limited to the rate in which the dynamic process changes, and the only limiting factor here is the true frame rate of the digital camera used.

As illustrated in Fig. 7.2a, the requirement for the spatial frequency separation for an off-axis image interferogram causes the highest spatial frequency required on the camera to be four times the highest frequency of the sample ω_0. Despite this disadvantage, off-axis interferometry is considered as the leading technique for interferometric recording of dynamic processes, since the required information can be acquired in a single camera exposure. In the on-axis geometry, on the other hand, there is no angle between the reference and the sample beams, and thus the camera bandwidth consumption is more effective since all four terms are centered at the origin of the spatial frequency domain (Fig. 7.2c). However, the problem of separating the desired terms from the unwanted terms is solved by acquiring multiple phase-shifted interferograms of the same sample. If these interferograms are acquired in sequence, the sample or the system noise might change during the acquisition time points.

Projecting several phase-shifted on-axis interferograms on the same camera can be used to overcome this drawback. However, since three or four phase-shifted interferograms are needed in the traditional on-axis approaches, the highest spatial frequency required from the camera is equivalent to at least $3\omega_0$ or $4\omega_0$, respectively. In the proposed PONI method, only two phase-shifted interferograms acquired simultaneously, and therefore the highest frequency on camera is equivalent to $2\omega_0$, half of that required for the off-axis case.

To experimentally demonstrate the PONI method, we have implemented the optical system shown in Fig. 7.6. For comparison purposes, we have also implemented a typical off-axis setup within the same system. This was accomplished by removing the wave plate and Wollaston prism (see Fig. 7.6), inducing a large angle between the reference and the sample beams, changing lens L_2 to another lens with a double focal length, and shifting the camera accordingly to retain the 4f configuration of lenses L_1 and L_2. The replacement of lens L_2, which results in a doubling of the total magnification of the optical system, allows full utilization of the digital camera imaging area, since for the off-axis case only one interferogram is needed. To make this comparison valid, the same sample focus conditions were kept, and for the off-axis case, the angle between the reference and the sample beams was made large enough to take advantage of the full available camera spatial bandwidth.

We have first experimentally demonstrated the PONI method by imaging a fixed rat hippocampal neuron in phosphate-buffered saline (PBS). This sample was chosen since it contains small spatial details outside of the cell body, the neuronal axons and the dendrites. Figure 7.7a shows two 90° phase-shifted on-axis interferograms of the neuron acquired in a single camera exposure. Figure 7.7b shows the phase profile of this neuron, obtained by processing the two interferograms shown in Fig. 7.7a according to the digital process explained above. Figure 7.7c shows the phase profile obtained by the off-axis technique. Despite the fact that this phase image is double in size (since in the off-axis case the entire sensor area is used for each interferogram), there is a loss of spatial resolution compared to the proposed method (Fig. 7.7b).

For the second experimental demonstration of the PONI method, we have imaged the dynamics of *Euglena gracilis*, a unicellular protist, in water as the sample. This sample was chosen due to the fine details of the euglena's flagellum, the thin tail used for propulsion, which is located outside the euglena body. The thin flagellum

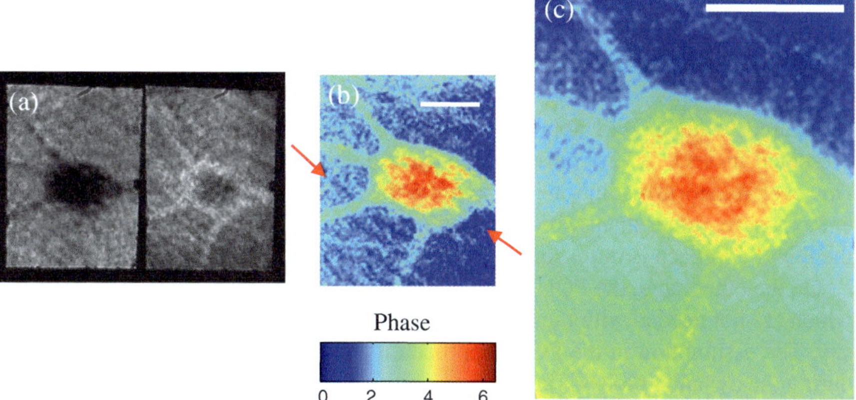

Fig. 7.7 Rat hippocampal neuron: (**a**) Two phase-shifted on-axis interferograms acquired in a single camera exposure; (**b**) unwrapped phase profile obtained by the suggested parallel on-axis method, 33× magnification, no binning; (**c**) unwrapped phase profile obtained by the traditional off-axis method, 66× magnification, 2× vertical binning. The *white scale bars* indicates 5 μm. Phase color bar is valid for both (**b**) and (**c**). In spite of being half the size compared to (**c**), finer details can be detected in (**b**) compared to (**c**). *Red arrows* in (**b**) indicate two examples for that [32]

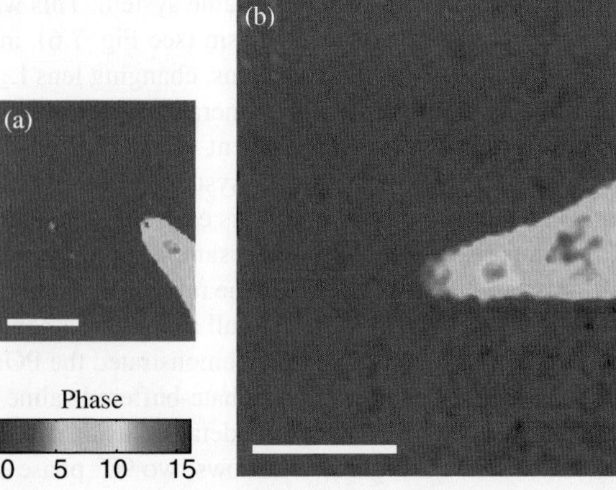

Fig. 7.8 *Upper part* of *Euglena gracilis* (a unicellular protist). (**a**) Unwrapped phase profile obtained by the suggested parallel on-axis method, 33× magnification, 2×2 binning (dynamic behavior: Video 2); (**b**) unwrapped phase profile obtained by the traditional off-axis method, 66× magnification, 4×2 binning (dynamic behavior: Video 3). The *white scale bars* indicates 10 μm. Phase color bar is valid for both (**a**) and (**b**) [32]

of the euglena moves at the millisecond time scale and thus should be observed with a WFDI technique that loses neither spatial nor time resolution. The final phase images are shown in Fig. 7.8a, b for the proposed on-axis and for the traditional off-axis methods, respectively. The flagellum dynamics are presented in Video 2 and Video 3, demonstrating the high temporal acquisition rate, as well as the spatial resolution improvement obtained by the proposed method compared to the traditional off-axis case.

Using the PONI approach, it is possible to optimize the consumption of the spatial bandwidth of the camera (number of pixels required for each camera exposure), while not sacrificing time resolution by acquiring the required information in a single exposure.

7.2.3 SOFFI Versus PONI

Compared to the SOFFI approach, presented in Sect. 7.2.1, the PONI approach, presented in Sect. 7.2.2, consumes the same camera spatial bandwidth per frame. However, temporal resolution is superior to the PONI approach since only one camera exposure, rather than two, is needed to acquire the required information. Thus, the PONI approach is more suitable for observing rapid dynamic processes.

On the other hand, the PONI approach requires additional off-line measurements which are not needed with the SOFFI approach, including the reference field intensity. The assumption in the PONI approach is that reference field does not change during the dynamic process (including its noise characteristics), which might not be valid for all types of measurements (practically for those with low signal-to-noise ratio). In addition, when using the PONI approach, one has to assume that the reference field intensity is large compared to the sample field.

Another inherent disadvantage of the PONI approach is that it requires equalization of intensity variations and accurate digital image registration when dividing the single camera image into two different digital interferograms. These problems are avoided when using the SOFFI approach, but come at the cost of sacrificing temporal resolution since two exposures per sample observation are needed.

7.3 Phase Noise and Stability and Sensitivity of the Interferometric System

Interferometric optical systems typically contain beam splitters that divide the beam into reference and sample arms and then combine these two beams at the detector to generate interference patterns. The two beams usually pass through different paths and thus experience different environmental influences, giving rise to noise features. For example, air perturbations can be different in the sample and reference arm paths causing differential noise. Furthermore, this noise might change temporally, which negates the possibility of measuring it off-line prior to the recording period in an effort to cancel it later via subtraction. Speckle noise, laser power stability, detector noise, and mechanical stability of the optical system are additional problems that can arise when imaging dynamic processes by WFDI. In addition, when imaging fast dynamic phenomena, the camera exposure time might need to be required to be short, yielding low intensities at the detector plane and a low phase profile signal-to-noise ratio (SNR). It is not always possible to simply increase the laser power to solve this problem since this might induce damage to the biological sample. The goal is thus to design the interferometric system so that the signal level will always be kept above the noise level.

Various papers have analyzed different noise factors that are relevant for digital interferometry. For example, Charrière et al. [33] analyzed the influence of shot noise on phase imaging SNR in the framework of digital holographic microscopy. Using a framework based upon statistical decision theory, the SNR of an arbitrary phase image can be calculated; however, this requires knowledge of an ideal image signal in order to assess the difference between the reconstructed image signal and the ideal signal. Although this method is suitable for simulations where the image signal is known beforehand, it is almost impossible to determine the ideal image signal for a real measurement, especially for biological samples. Therefore, Charrière et al. also defined the SNR of the hologram as the intensity of the interference cross-terms over the summed intensity of the reference and object waves. After establishing a theoretically based, shot noise-limited SNR expression, plots of the SNR at various reference/signal intensity ratios as well as varying total collected photon counts were compared. The approach was validated experimentally by analysis of a holographically imaged quartz–chrome binary grating that was reconstructed with varying ratios of reference and signal intensities.

A further study by Charrière et al. [34] examined the influence of shot noise on phase measurement accuracy. Recognizing that there is no precise and universally acknowledged method for quantifying phase accuracy, this group defines a parameter based on the standard deviation (STD) of the phase profile across a region of interest obtained by subtracting 10 averaged blank phase profiles from an individual blank phase profile. In this configuration, static optical aberrations and misalignment errors are removed. To characterize quantization error, perfect holograms were quantized to levels ranging from 1 to 16 bits. The resulting SNR in reconstruction was measured using the STD method. As expected, the graphs of phase STD appear to be inversely proportional to the bit depth used. SNR measurements of holograms with simulated shot noise showed a plateau at quantization levels above 6 bits. Therefore, the experimental effects of quantization are disregarded in further characterization of shot noise influence on SNR. Simulations that include shot noise ranging from photon counts of 10^0–10^4 show an inversely proportional relationship between the photon count and the phase measurement STD. At a level of 100 photons per pixel, the phase accuracy is limited to $\sim 2.25°$, while at 10^4 photons/pixel, the shot noise-limited phase accuracy is $\sim 0.25°$.

Brophy [27] describes the relationship between intensity error and phase variance in phase-shifting interferometry. Brophy's statistical treatment of noise sources yields an expression for phase variance based on the degree of correlation between intensity errors. Intensity error in this context can be any variation of the measured intensity at a single point from the expected measured intensity. The methods developed can be used to model both correlated and uncorrelated sources of noise, including shot noise, thermal noise, quantization error, temporal noise due to vibrations, temporal laser intensity noise, etc. In this paper, Brophy primarily examines two interesting cases: uncorrelated noise with equal variance from frame to frame and intensity quantization error with specific structure of correlation between images. In phase-shifting interferometry, multiple projections with specific phase shifts are summed in different combinations to produce x-axis and y-axis projections in phase space in order to compute the phase at each point by taking the arctangent of the constructed argument (y-axis projection over x-axis projection). Using a Taylor expansion of the phase, the variance is approximated by an expression that is dependent on the structure of phase image combinations used to construct the argument and also the phase at which the phase error is measured. It is particularly useful to examine the phase error averaged over all possible phases, as noise sources commonly have random phase. In the case of uncorrelated intensity noise that changes approximately uniformly from frame to frame, the average phase variance becomes proportional to the relative variance of the modulated portion of the intensity. An increase in the number of frames, N, used by a given phase-shifting algorithm results in a proportional decrease in phase variance of $\sim 1/N$, which is expected for statistically uncorrelated noise sources. Brophy next examines quantization error to demonstrate the effects of intensity noise with structured correlation between frames. In 90° four-step phase-shifting algorithms, the quantization error of every fourth frame is positively correlated and every second frame is negatively correlated. Successive frames are taken to have no correlation when averaged over a small region of the phase. Using these correlation assumptions, the phase variance is found to be approximately proportional to $1/3Q^2$, where Q is the number of quantization levels, and the factor of 3 is specific to 90° phase-shifting algorithms. While the quantization error is not dependent on the number of frames, choosing a phase step other than 90° will change the scaling factor. For example, choosing a step that yields less phase space symmetry and thus less overall correlation between frames will decrease the variance for a given quantization level.

When considering sequential phase-shifting interferometry, specifically for recording dynamic processes, vibrations play an important role in the noise characteristics. Phase-shifting interferometry is typically performed in the time domain by physically shifting the phase of the reference field by carefully controlled increments. These systems tend to suffer from noise introduced by mechanical vibrations that differ from system to system. de Groot [35] developed a transfer function that can be applied to a system with a well-characterized complex noise spectrum to determine the maximum phase error due to vibrations and also the typical performance of an instrument in the presence of uncorrelated and randomly phased vibrations. This analysis primarily shows that different methods of phase shifting can be applied to a system with a characteristic vibration spectrum in order to reduce the magnitude of phase error. This is a particularly useful result for applications of repeated static object testing. A more recent study [36] makes use of active feedback in the form of a piezoelectric modulator that is controlled by a photodiode detector to lock the phase step of the reference arm. With this method, the phase drift of a test sample was reduced by two orders of magnitude when actively stabilized.

Building on the error analysis presented by Brophy in [27], the effects of phase step choice in phase-shifting interferometry are analyzed and optimized by Remmersmann et al. [37] to reduce overall noise. Phase error is defined here to be the STD of intensity fluctuations across frames. The digitization error is again recognized to be independent of the object's actual phase profile. It is shown analytically that uncorrelated noise sources contribute the least amount of error when the phase steps are chosen to be $2\pi/3$. Errors due to misaligned reference phase shifts are not uncorrelated and thus are dealt with separately. Defining an expression for the phase error in terms of the phase shift, β, and the misalignment, $\beta - \Delta\beta$, reveals two expressions for β in terms of $\Delta\beta$ that minimize the error due to misaligned phase steps. When an upper bound on $\Delta\beta$ from piezo actuators is applied, angular steps within the range of 75–90° all increase the experimental robustness against inaccurate phase shifting.

In live cell samples, however, vibrations in a system can propagate through the sample growth medium (i.e., aqueous material) and even cause slight deformations within the sample itself. Because of this, it is difficult to characterize the noise structure of a given system and subsequently tune a phase-shifting algorithm to minimize noise levels. Therefore, it is preferable to capture all phase-shifted interferograms in a single exposure if possible to minimize the effects of vibration in reconstructing a given phase profile. Depending on the delay between acquisition time points, multiple acquisitions will typically result in higher variation due to vibration and mechanical fluctuation of media perturbations.

Remmersmann et al. [37] used modified Michelson and Linnik interferometers to experimentally examine the effects of partial coherence on phase errors. A range of red, blue, green, and yellow LEDs, as well as a superluminescent diode (SLD), a HeNe laser, and a laser diode are used to create holograms of the illumination plane wave. These experiments confirm that digitization beyond 256 levels (8 bits) does not significantly increase system performance since other noise sources become dominant. Furthermore, phase steps of $2\pi/3$ are confirmed to minimize the phase error contributions from statistically uncorrelated processes. Comparing light sources reveals the red LEDs to have a lower magnitude of noise, presumably due to the reduction in speckle noise and multiple reflection interference (1.3° for red LED, a 0.8° reduction in standard deviation compared to that for a HeNe laser). The phase noise also exhibits a $1/V$ proportionality, where V is the fringe visibility. Again the red LED yields the lowest phase noise in these measurements. Finally, the dependence of contrast on phase shift speed is investigated. At higher phase-shifting speeds, the fringe visibility for a long integration time decreases by $\sim$10%, but does not significantly contribute to phase error. Remmersmann et al. also examined fixed pancreas tumor cells that were sandwiched between two glass coverslips. When imaging with a fully coherent laser source (HeNe), circular noise rings appear as a result of multiple reflections. These rings cause significant distortions in the final phase images. However, the use of LED illumination results in interference patterns and phase images that are completely free from this source of noise. This imaging example clearly illustrates the utility of partially coherent illumination in phase-shifting interferometry and other holographic imaging modalities.

The use of partially coherent sources in digital holography introduces a limit to the range over which the object wave will interfere with the reference wave. This limit in turn affects the distance over which a hologram can be propagated in digital reconstruction. Spatial coherence function expressions are used by Dubois et al. [38] to derive the maximum refocusing distance allowed, based on the geometry of the presented holographic system. In spite of this limit, there are several advantages of using an illumination scheme with partial spatial coherence that are discussed. (1) Partially coherent illumination suppresses speckle noise. (2) While the coherence length limits the maximum refocusing distance, any perturbations in the optical system beyond this limit will only weakly affect the reconstruction. (3) With reduced temporal coherence, any spurious interference arising from multiple reflections (as is common in biological specimen imaging when multiple glass coverslips may be present) does not contribute to the interferometric signal. Dubois's system consists of an incoherent LED from which light is collimated, passed through an aperture, and magnified in order to increase spatial coherence. The choice of magnification optics and adjustment of the aperture size govern the spatial coherence area. Thus it is possible to adjust these elements to optimize the degree of coherence in the imaging system. With this system, two of the three advantages of holography using a partially coherent source are demonstrated. The high image quality obtained is attributed to a lack of speckle noise. Computer reconstructions of the same image at four different ranges demonstrate the loss of resolution as refocus distance increases. It is important to note that this study examined amplitude reconstruction of images, not phase image reconstruction. However, the sample principles can be applied to phase imaging.

In [39], illumination with partial spatial coherence is produced for a holographic microscope using a rotating ground glass diffuser. Moving this glass along the optical axis of the presented system adjusts the spot size of the laser beam on the glass diffuser and also adjusts the average speckle size. At longer refocusing distances, the reconstruction of holograms recorded with lower spatial coherence has significantly reduced resolution. Examining a cross section of an onion peel using this system reveals that the lower spatial coherence not only reduces apparent background speckle but also reduces the amplitude of multiply reflected interference patterns.

7.4 Phase Unwrapping Problem

Direct detection of phase by digital interferometry methods relies on the use of an arctangent operation to calculate phase. Since the signed two-argument arctangent function is inherently limited to a range of $-\pi$ to π, it is impossible to unambiguously measure the phase directly using this method. As a solution, many algorithms have been developed to detect discontinuities of 2π in phase maps and remove them to create a smooth and slowly varying true phase profile. This process is widely known as *phase unwrapping*. There are three major complications in phase unwrapping: (1) Noise in wrapped phase profiles can introduce discontinuities in the true phase profile that cause algorithms to incorrectly unwrap phase profiles; (2) any true phase changes larger than π between two adjacent points break the underlying assumption of a slowly varying phase function, which is necessary for classic two-dimensional phase unwrapping; and (3) the computational expense of an algorithm must be low enough to allow rapid unwrapping when working with large images and three-dimensional data sets of time-resolved phase changes.

Various digital two-dimensional phase unwrapping algorithms have been developed with the goal of reducing the effects of these three complications. The classic algorithms aim to minimize computational time while maximizing robustness to noisy data [40–42]. It is also possible to unwrap phase changes over time, which is of particular interest for applications where the absolute phase is of less interest than the deformations produced as a result of a dynamic sample [43]. A comprehensive overview on digital phase unwrapping algorithms including software codes can be found in the book authored by Ghiglia and Pritt [44].

An alternative approach to post-processing digital unwrapping is to modify the experimental setup to acquire additional information that can enable one to detect and then correct unwrapping problems. For example, the use of two or more illumination wavelengths has been proposed as an alternative approach to the unwrapping digital algorithms mentioned above. This method has been demonstrated as a highly effective tool for extending the unambiguous phase measurement range significantly while maintaining high measurement sensitivity, by creating a synthetic "beat" wavelength between the two illumination sources [45–52]. These multiple-wavelength systems can be designed to acquire holograms at several wavelengths simultaneously, which is well suited for dynamic measurements of both mechanical and biological samples [50–52].

Another example of an experimental approach to solving the phase unwrapping problem involves differential low-coherence phase measurements with a dispersive element included in the optical scheme to introduce a quadratic term into the phase function [53]. The obtained phase profile can then be integrated to produce a relative height profile free from 2π ambiguities. Another approach, recently presented, measures the phase profile as a function of spectral wave number without 2π ambiguities as a means to overcome the phase unwrapping problem [54].

7.5 Lack of Specificity inside Cells

Many biological cells are at the range of 5–20 micron in thickness. Cells contain internal organelles that might be of specific interest for medical and biomedical studies. For example, the characteristics of the cell nucleus have been recognized as histological markers of genetic and epigenetic changes leading to cancer [55]. However, WFDI provides whole-cell information that lacks the specificity for identifying subcellular components. Solving this limitation can offer additional information that is currently lacking with WFDI techniques.

One approach for addressing the inner specificity problem is to use exogenous tags, and functionalize them to attach to cell organelles of interest, which can be detected by WFDI. For example, Amin et al. [56] have suggested a method for tracking microbeads that are targeted to attach to the cell surface, in order to enable analysis of the cytoskeleton dynamics. Alternative techniques use nanoparticles (NPs) as labeling tags. Particularly, metal NPs are considered to be attractive contrast agents for biomedical applications due to their biocompatibility and because they are tunable through a broad range of wavelengths including the visible and near-infrared regions. In addition, they do not exhibit photobleaching or cause cytotoxicity [57, 58] as might happen in other contrast agents such as fluorescence dyes. In photothermal molecular imaging, the optical absorption of the NPs at their plasmon resonance results in a sharp temperature rise in the vicinity of the particle due to the photothermal effect. This change of temperature leads to a variation in the local refraction index on nanometer scales which can be detected optically [59]. Our group, in cooperation with the research group of Dr. J. Izatt, has demonstrated a photothermal molecular-imaging interferometric system with high spatial resolution that uses immunolabeled plasmonic NPs. This optical coherence tomography (OCT) system can obtain deeper penetration depths than that possible by regular light microscopy and yet avoids the photobleaching or cytotoxicity concerns of other contrast agents [60]. In the experiment, photothermal heating of gold nanospheres is achieved by a local, selective illumination using a green laser (532 nm) to excite the plasmonic resonance of 60 nm gold nanosphere. The local heating induced a phase signature which was then detected using a phase-sensitive spectral domain OCT scheme. The system could distinguish the presence of gold NPs with a measured sensitivity of 14 ppm (weight/weight). The system was used to characterize the epidermal growth factor receptor (EGFR) expression of different types of cultured cancer cells. Using plasmonic metal NPs to tag specific subcellular components might also be possible by WFDI, as currently explored by our group. In contrast to OCT, WFDI will provide the advantage of acquiring the three-dimensional spatial distribution of the cells in a single camera exposure, which is more suitable for recording fast cell dynamics than OCT, since OCT typically employs point or line scanning.

However, any technique using exogenous tags in WFDI loses at least part of the advantage of WFDI as being a label-free technique. This includes a potential modification of the normal behavior of the cells during the measurement, since it can be hard to definitively state that cellular behavior will be completely unchanged when these exogenous tags are present.

An alternative approach to solve the specificity problem inside of cells is based on acquiring three-dimensional spatial information using many camera exposures. This includes topographic-based methods [61, 62], confocal scanning methods [63], and various OCT methods (see Chap. 8). However, acquiring information across multiple exposures is not inherently suitable for recording fast cell dynamics. Although current technologies can enable high scanning and data acquisition rates that may be suitable for acquiring images of some dynamic cell phenomena, they have inherent temporal resolution limitations which might be critical barriers for studying certain dynamic phenomena.

Performing digital sectioning ('slicing') inside of cells based on the use of digital algorithms might also be possible for imaging certain types of cells. Brady et al. [64] have shown that for objects that are sparse enough and are imaged with high SNR, it is possible to reject out-of-focus haze digitally by using an iterative algorithm. Our laboratory is now working on using a similar method for performing sectioning inside phase objects, including several layers of cells and interiors of cells. Although this technique may work well for certain types of cells, it is not a complete solution to the WFDI specificity problem due to its assumption that the objects are sparsely distributed.

To conclude, as reviewed in this section, all of the solutions to the WFDI problem of specificity inside cells described above are only partial, rather than general, and each of them has its own disadvantages for recording fast cell dynamics. Yet, careful selection of one of these solutions, taking into account its unique inherent limitations, might provide a fit for particular applications.

7.6 From Phase Profiles to Quantitative–Functional Analysis

WFDI is inherently a quantitative recording technique. However, simple quasi-three-dimensional visualization of samples of interest need not be the end of the investigative process. A quantitative analysis should permit extraction of numerical parameters of merit which are useful for cell biology studies or medical diagnosis. The phase profiles acquired with WFDI can be interpreted to provide specific information for these purposes. When implemented as a transmission mode interferometric system, WFDI can measure optical thickness which is the multiplication between the index of refraction differences and the geometrical thickness. Separating these two factors is important in contributing an incisive analysis tool. For example, local changes in the index of refraction may occur during action potential or whenever there is an ion flux inside the cell. Independently or not, geometrical path changes can also occur due to movement of intracellular components. Thus, these conjugated parameters, the index of refraction differences and the geometrical thickness, may not be distinct when acquiring the phase profile of a dynamic cell. This fact should be considered during the system development and especially in the subsequent data analysis.

WFDI has been applied to examine various types of biological cells. However, many publications in this area have focused on the development of optical interferometric methods and simply present visual phase images, rather than insightful analysis of the images to obtain quantitative and functional information that is useful for cell biology research. Contrary to this trend, Popescu et al. [4, 5, 63–68], Kemmler et al. [69], and others have shown that digital interferometry can be used to quantitatively analyze red blood cells, cancer cells, and kidney cells for relatively slow cell processes.

In [8], we have used WFDI for quantitative analysis of the relatively fast dynamics of chondrocytes. These are the single type of cells that compose articular cartilage, the connective tissue that distributes mechanical loads between bones. The phenotypic expression and metabolic activity of these cells are strongly influenced by shape and volume changes occurring due to mechanical and osmotic stresses [70]. Chondrocytes are a particularly interesting model system for studying physical signal transduction and cellular injury due to their sensitivity to physiologic changes in osmotic stress [71]. Depending on the environmental stimuli, chondrocyte swelling can happen within a period of several seconds, with meaningful intermediate events that potentially occur within milliseconds [72]. Important calculations that are associated with these rapid intermediate events include accurate volume evaluation at the time of maximal swelling (e.g., just before cell rupture), as well as dry mass calculations of single cells and cell populations in a monolayer during swelling, where other cells and intracellular parts rapidly move across the FOV. Label-free cell visualization of these rapid dynamic phenomena, with the goal of obtaining quantitative and functional volumetric data, requires microscopy techniques with fast three-dimensional acquisition rates that are beyond the scope of most conventional scanning microscopy methods.

The off-axis WFDI setup shown in Fig. 7.9 was used for recording the chondrocyte rapid transient during swelling due to hypo-osmotic pressure. Light from a laser source is split into reference and object beams. The object beam is transmitted through the sample and magnified by a microscope objective. The reference beam is transmitted through a compensating microscope objective and then combined with the object beam at an angle. The combined beams are projected onto a digital camera, creating an off-axis interferogram of the sample. As explained in Sect. 7.2.1, a single interferogram can be used to retrieve both phase and amplitude of the sample field, provided that there is a sufficient angle between the reference and the object beams and that the spatial bandwidth of the detector is sufficiently wide.

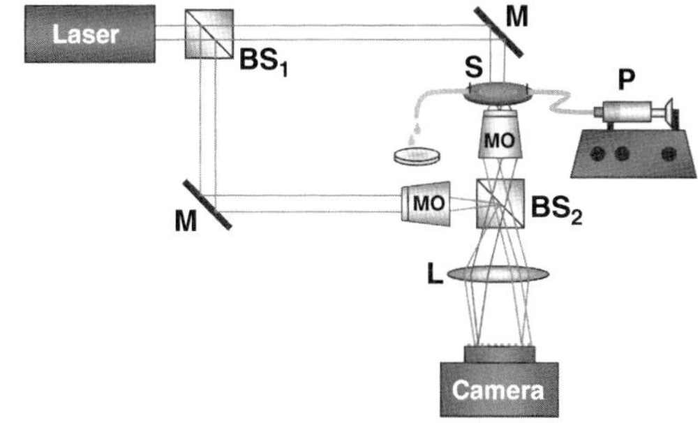

Fig. 7.9 Off-axis WFDI phase microscopy system for recording isolated chondrocyte dynamics under extreme hypo-osmotic pressure. BS_1, BS_2 = beam splitters; M = mirror; S = sample; P = controlled syringe pump; MO = microscope objective; L = lens

Small changes in this angle during the measurement of dynamic processes can cause potential error in the retrieved sample field. To avoid this issue, we digitally measure the fringe frequency q separately for each acquired interferogram (similar to what was performed for the slightly off-axis case presented in Sect. 7.2.1, but using a single off-axis interferogram this time). Assuming vertically aligned fringes, summing the background interference vertically around the sample yields a relatively noise-free sine wave, which then can be digitally fit to an ideal sine wave to estimate the ideal fringe frequency q. The interferogram is multiplied by $\exp(jqx)$ to center the useful image CCT at the origin of the spatial frequency domain. Afterward, the remaining unwanted terms (the twin-image CCT and the two ACTs) are spatially filtered by gradually discarding high spatial frequencies. The phase argument of the resulting complex matrix is the wrapped phase profile of the sample field, with an unwrapping algorithm used to remove 2π ambiguities in this profile.

We have used the system presented in Fig. 7.9 to acquire 120 frames per second (fps) data of single chondrocyte and chondrocyte monolayer behavior due to extreme osmolarity reduction. The osmolarity change was induced by using a syringe pump that quickly replaced the culture cell media with deionized water, resulting in a rapid cell swelling. Figure 7.10a shows the surface plot of the chondrocyte phase profile during swelling at three chosen time points. As expected, during swelling, the phase profile height decreases and the two-dimensional area of the cell image increases [5]. Figure 7.10b shows an image from Video 4, presenting the same cell swelling, whereas Fig. 7.10c show an image from Video 5, presenting another chondrocyte swelling that terminates with cell rupture. An image from Video 6, presenting chondrocyte monolayer dynamics during swelling is shown in Fig. 7.10d, illustrating that chondrocyte swelling and bursting dynamics vary from cell to cell. Note that these videos were down sampled from the original videos to reduce their file sizes. However, the quantitative analysis results presented below were obtained from the images obtained at the full 120 fps rate.

For comparison purposes, we have collected similar data using a DIC microscope (LSM 510; Carl Zeiss, Thornwood, NY) at the maximum frame rate possible by this specific device (10 fps for single-cell imaging (Video 7) and 0.75 fps for cell monolayer imaging (Video 8)). Although the two experiments (DIC and WFDI)

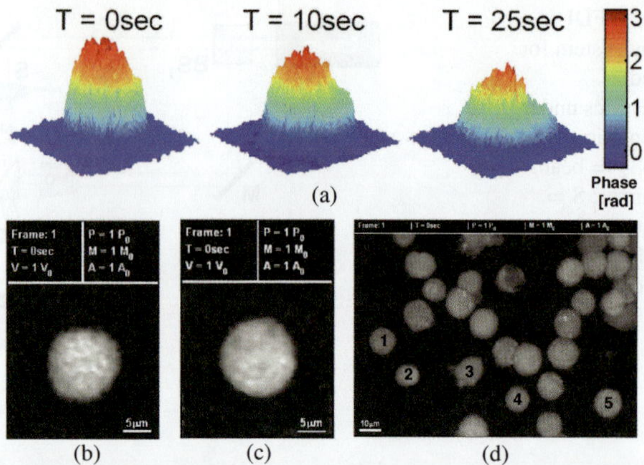

Fig. 7.10 Final unwrapped phase profiles of live chondrocyte dynamics obtained by WFDI: (**a**) Surface plots of single-cell swelling in three chosen time points; (**b–d**) two-dimensional view videos of (**b**) single-cell swelling, 60 fps (Video 4), (**c**) single-cell bursting, 60 fps (Video 5), and (**d**) cell monolayer dynamics, 7.5 fps (Video 6). Note that the original quality of the presented videos is decreased to reduce the file size [8]

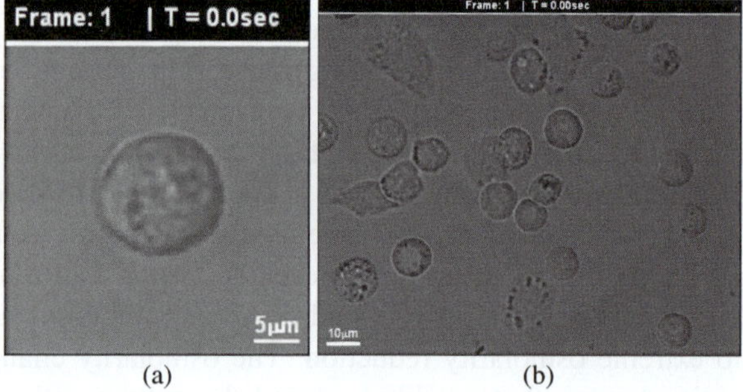

Fig. 7.11 Videos of live chondrocyte dynamics phase profiles obtained by DIC microscopy: (**a**) Single-cell swelling and bursting, 10fps (Video 7). (**b**) Cell monolayer dynamics, 0.75fps (Video 8) [8]

have been performed using separate systems, the sample and imaging conditions were made as similar as possible.

Figure 7.11a, b (Video 7 and 8) show the phase images obtained with the DIC microscope for the cases of single chondrocyte and chondrocyte monolayer dynamics, respectively. As expected from DIC microscopy, only the cell edges are seen, and the characteristic shading artifact is present. As also shown in these videos, after several seconds, the chondrocytes are no longer in focus due to changes in cell volumes. Since the DIC data collection is two dimensional and non-quantitative in nature, it is impossible to bring the sample back into focus without losing important data during the transient. Therefore, we see that obtaining accurate quantitative volumetric data is impractical using DIC microscopy.

On the other hand, in Fig. 7.11 since the entire wavefront is captured by WFDI, it is possible to refocus the sample image digitally; however, the amount of refocusing that is needed must be determined at each time point during the dynamic process.

In our case, since biological cells are phase objects, digital propagation along the axial direction was performed according to the phase-object focus criterion defined in [73].

The quantitative analysis performed on the final phase profiles obtained by WFDI (Fig. 7.10) includes monitoring the temporal changes in the chondrocyte relative area (A/A_0), volume (V/V_0), dry mass (M/M_0), and average phase (P/P_0). Figure 7.12a presents graphs of the temporal dependence of these parameters during the transient visualized in Fig. 7.10b (Video 4). As can be seen from these graphs, the chondrocyte gained about 50% in volume and more than 50% in area during swelling, while retaining its dry mass at approximately the same level. Figure 7.12b shows the parameter graphs for the transient visualized in Fig. 7.10c (Video 5). In this case, the chondrocyte started swelling, gaining volume and area, but then burst and lost dry mass. This finding provides experimental support of the dry mass calculations that are based on the chondrocyte phase profile. Another validation for this calculation is the small jumps that can be seen on the graphs before the chondrocyte bursts. These jumps exactly occur at the time points at which intercellular parts from other, already burst chondrocytes have entered the FOV (see the related frames in Video 5). This demonstration validates the approach of assessing dry mass loss-induced cell bursting using WFDI-based calculations. Note that the chondrocyte volume was calculated up until the time point of cell bursting (and not beyond) due to the fact that the volume calculation assumes isotropic volume change [9]. Based on the high temporal resolution obtained, we calculate the chondrocyte volume just prior to bursting as $V_L = 1.28$ times the initial cell volume V_0. Figure 7.12c shows graphs of the time dependence of the relative chondrocyte area, dry mass, and average phase of the cell monolayer visualized in Fig. 7.10d (Video 6). The graphs illustrate the tradeoffs in these parameters that occur during the dynamic response of the monolayer. Different chondrocytes start swelling at different time points, swell to various extents, and burst at different time points. Individual cell swelling and

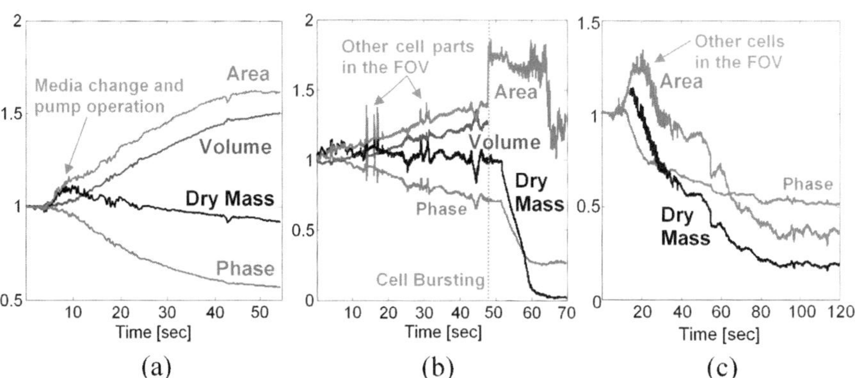

Fig. 7.12 WFDI-based graphs of the relative change in the values of the volume, dry mass, area, and average phase of the chondrocytes during: (**a**) single-cell swelling (visualized in Fig. 7.10b and Video 4), (**b**) single-cell bursting (visualized in Fig. 7.10c and Video 5), (**c**) cell monolayer dynamics (visualized in Fig. 7.10d and Video 6) [8]

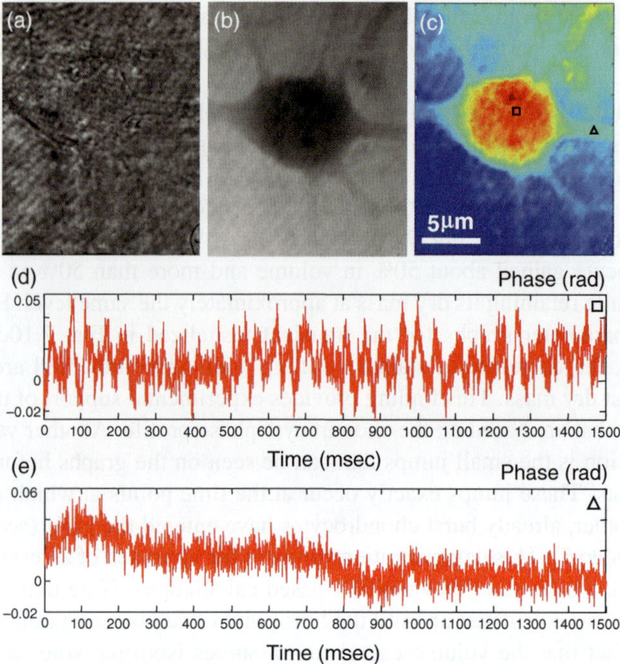

Fig. 7.13 WFDI quantitative phase microscopy of hippocampal neuron dynamics, recorded at 2,000 frames per second: (**a**) regular intensity image through the system (low visibility, only edges are seen); (**b**) final unwrapped phase (optical path delay) profile obtained by the digital interferometric processing; (**c**) the phase profile with digital coloring. No exogenous contrast agents were used; (**d**) neuronal phase dynamics on the neuron body (marked by a *square* in (**c**)); (**e**) neuronal phase dynamics on the dendrites (marked by a *triangle* in (**c**)). different dynamic behaviors are seen. Similar results were obtained by checking many other similar points on the neuron body and dendrites

bursting result in a decrease in the average phase value. The rupture of an individual cell is characterized by a loss of dry mass and in a temporal increase of viewable area (until the burst chondrocyte intracellular parts leaves the FOV). On the other hand, other chondrocytes and chondrocyte intracellular parts entering the FOV result in an increase in dry mass and area. As time passes, it is demonstrated that the values of all three parameters decrease due to rupture of most chondrocytes in the monolayer, which results in an approximately uniform distribution of the intracellular parts in the chamber. Using the monolayer WFDI results, it is also possible to calculate individual chondrocyte maximum volume V_L before cell rupture in response to the same monolayer osmotic conditions. For example, the chondrocytes marked by 1–5 in Fig. 7.10d have yielded $V_{L,1} = 1.58V_0$, $V_{L,2} = 1.81V_0$, $V_{L,3} = 1.37V_0$, $V_{L,4} = 1.35V_0$, and $V_{L,5} = 1.22V_0$.

These experimental demonstrations show that WFDI can provide a powerful research tool for observing the dynamics of articular chondrocytes and other types of biological cells, as well as for performing various cytological measurements in vitro including cell viability assays.

The quantitative phase profiles obtained by WFDI can also be used to assess the local dynamics characterizing various subcellular components. Figure 7.13 shows images of a rat hippocampal neuron acquired in 2000 fps using a similar optical setup to that shown in Fig. 7.9. Different dynamic behaviors have been quantitatively observed at different locations on the neuron body and the dendrites. Detecting the local changes in the index of refraction during the spatial propagation of a millisecond rate neuronal action potential might also be possible by WFDI, a goal of our future research.

7.7 Conclusion

We have presented an overview of WFDI phase microscopy techniques which enable visualization of the dynamics of three-dimensional biological cells and permit extraction of quantitative morphological parameters. These techniques can provide phase measurements of dynamic cell processes with sub-nanometer optical path delay accuracy over sub-millisecond time scales. By understanding the WFDI limitations reviewed in the chapter, and applying the suggested solutions which frequently depend on the specific application and the sample of interest, we believe that WFDI will become a common tool for quantitative analysis of cell dynamics.

Acknowledgments This work was supported by grants from the National Science Foundation (BES 03-48204, CBET-0651622). N.T.S. greatly acknowledges the support of the Bikura Postdoctoral Fellowship from Israel.

References

1. M. Pluta, *Advanced Light Microscopy*, vol. 2. (Elsevier, Amsterdam, 1988)
2. F. Dubois, C. Yourassowsky, O. Monnom et al., Digital holographic microscopy for the three-dimensional dynamic analysis of *in vitro* cancer cell migration. J. Biomed. Opt. **11**, 054032 (2006)
3. G. Popescu, in *Quantitative Phase Imaging of Nanoscale Cell Structure and Dynamics*, ed. by B. Jena. Methods in Cell Biology. (Elsevier, Amsterdam, 2008)
4. N. Lue, W. Choi, G. Popescu, K. Badizadegan, R.R. Dasari, M.S. Feld, Synthetic aperture tomographic phase microscopy for 3D imaging of live cells in translational motion. Opt. Express **20**, 16240 (2008)
5. G. Popescu, Y.K. Park, N. Lue, C.A. Best-Popescu, L. Deflores, R.R. Dasari, M.S. Feld, K. Badizadegan, Optical imaging of cell mass and growth dynamics. Am. J. Physiol.-Cell Physiol. **295**, C538 (2008)
6. G. Di Caprio, G. Coppola, S. Grilli et al., Microfluidic system based on the digital holography microscope for analysis of motile sperm. Proc. SPIE **7389**, 738907 (2009)
7. N.T. Shaked, M.T. Rinehart, A. Wax, Dual-interference-channel quantitative-phase microscopy of live cell dynamics. Opt. Lett. **34**, 767–769 (2009)
8. N.T. Shaked, J.D. Finan, F. Guilak, A. Wax, Quantitative phase microscopy of articular chondrocyte dynamics by wide-field digital interferometry. J. Biomed. Opt. Lett. **15**, 010505 (2010)

9. E. Cuche, P. Marquet, C. Depeursinge, Simultaneous amplitude-contrast and quantitative phase-contrast microscopy by numerical reconstruction of Fresnel off-axis holograms. Appl. Opt. **38**, 6994–7001 (1999)
10. E. Cuche, P. Marquet, C. Depeursinge, Spatial filtering for zero-order and twin-image elimination in digital off-axis interferometry. Appl. Opt. **39**, 4070–4075 (2000)
11. Y. Zhang, Q. Lu, B. Ge, Elimination of zero-order diffraction in digital off-axis interferometry. Opt. Commun. **240**, 261–267 (2004)
12. T.M. Kreis, W.P.P. Jupiter, Suppression of the dc term in digital interferometry. Opt. Eng. **36**, 2357–2360 (1997)
13. Y. Takaki, H. Kawai, H. Ohzu, Hybrid-interferometric microscopy free of conjugate and zero-order images. Appl. Opt. **38**, 4990–4996 (1999)
14. P. Marquet, B. Rappaz, P.J. Magistretti, E. Cuche, Y. Emery, T. Colomb, C. Depeursinge, Digital holographic microscopy: A noninvasive contrast imaging technique allowing quantitative visualization of living cells with subwavelength axial accuracy. Opt. Lett. **30**, 468–470 (2005)
15. B. Rappaz, P. Marquet, E. Cuche, Y. Emery, C. Depeursinge, P. Magistretti, Measurement of the integral refractive index and dynamic cell morphometry of living cells with digital holographic microscopy. Opt. Express **13**, 9361–9373 (2005)
16. T. Ikeda, G. Popescu, R.R. Dasari, M.S. Feld, Hilbert phase microscopy for investigating fast dynamics in transparent systems. Opt. Lett. **30**, 1165 (2005)
17. G. Popescu, T. Ikeda, R.R. Dasari, M.S. Feld, Diffraction phase microscopy for quantifying cell structure and dynamics. Opt. Lett. **31**, 775 (2006)
18. I. Yamaguchi, T. Zhang, Phase-shifting digital interferometry. Opt. Lett. **22**, 1268–1270 (1997)
19. P. Guo, A.J. Devaney, Digital microscopy using phase-shifting digital holography with two reference waves. Opt. Lett. **29**, 857–859 (2004)
20. T. Kiire, S. Nakadate, M. Shibuya, Simultaneous formation of four fringes by using a polarization quadrature phase-shifting interferometer with wave plates and a diffraction grating. Appl. Opt. **47**, 4787–4792 (2008)
21. G. Rodriguez-Zurita, N. Toto-Arellano, C. Meneses-Fabian, J.F. Vázquez-Castillo, One-shot phase-shifting interferometry: Five, seven, and nine interferograms. Opt. Lett. **33**, 2788–2790 (2008)
22. X.F. Meng, L.Z. Cai, X.F. Xu, X.L. Yang, X.X. Shen, G.Y. Dong, Y.R. Wang, Two-step phase-shifting interferometry and its application in image encryption. Opt. Lett. **31**, 1414–1416 (2006)
23. J.P. Liu, T.C. Poon, Two-step-only quadrature phase-shifting digital interferometry. Opt. Lett. **34**, 250–252 (2009)
24. Y. Awatsuji, T. Tahara, A. Kaneko, T. Koyama, K. Nishio, S. Ura, T. Kubota, O. Matoba, Parallel two-step phase-shifting digital holography. Appl. Opt. **47**, D183–D189 (2008)
25. G. Popescu, L.P. Deflores, K. Badizadegan et al., Fourier phase microscopy for investigation of biological structure and dynamics. Opt. Lett. **29**, 2503 (2004)
26. N. Lue, W. Choi, G. Popescu, T. Ikeda, R.R. Dasari, K. Badizadegan, M.S. Feld, Quantitative phase imaging of live cells using fast Fourier phase microscopy. Appl. Opt. **32**, 811 (2007)
27. C.P. Brophy, Effect of intensity error correlation on the computed phase of phase-shifting interferometry. J. Opt. Soc. Am. A **7**, 537–541 (1990)
28. N.T. Shaked, Y. Zhu, M.T. Rinehart, A. Wax, Two-step-only phase-shifting interferometry with optimized detector bandwidth for microscopy of live cells. Opt. Express **17**, 15585–15591 (2009)
29. P. Ferraro, S. De Nicola, A. Finizio, G. Coppola, S. Grilli, C. Magro, G. Pierattini, Compensation of the inherent wave front curvature in digital holographic coherent microscopy for quantitative phase-contrast imaging. App. Opt. **42**, 1938–1946 (2003)
30. E. Tajahuerce, O. Matoba, S.C. Verrall, B. Javidi, Optoelectronic information encryption with phase-shifting interferometry. Appl. Opt. **39**, 2313–2320 (2000)
31. K.J. Chalut, W.J. Brown, A. Wax, Quantitative phase microscopy with asynchronous digital interferometry. Opt. Express **15**, 3047–3052 (2007)

32. N.T. Shaked, T. Newpher, M.D. Ehlers, A. Wax, Parallel on-axis holographic phase microscopy of biological cells and unicellular microorganism dynamics. Appl. Opt. **49**, 2872–2878 (2010)
33. F. Charrière, T. Colomb, F. Montfort, E. Cuche, P. Marquet, C. Depeursinge, Shot-noise influence on the reconstructed phase image signal-to-noise ratio in digital holographic microscopy. Appl Opt. **45**, 7667–7673 (2006)
34. F. Charrière, B. Rappaz, J. Kühn, T. Colomb, P. Marquet, C. Depeursinge, Influence of shot noise on phase measurement accuracy in digital holographic microscopy. Opt. Express **15**, 8818–8831 (2007)
35. P.J. de Groot, Vibration in phase-shifting interferometry. J Opt. Soc. Am. A **12**, 2212 (1995)
36. H. Iwai, C. Fang-Yen, G. Popescu, A. Wax, K. Badizadegan, R.R. Dasari, M.S. Feld, Quantitative phase imaging using actively stabilized phase-shifting low-coherence interferometry. Opt. Lett. **29**, 2399–2401 (2004)
37. C. Remmersmann, S. Stürwald, B. Kemper, P. Langehanenberg, G. von Bally, Phase noise optimization in temporal phase-shifting digital holography with partial coherence light sources and its application in quantitative cell imaging. Appl Opt. **48**, 1463–1472 (2009)
38. F. Dubois, L. Joannes, J.C. Legros, Improved three-dimensional imaging with a digital holography microscope with a source of partial spatial coherence. Appl. Opt. **38**, 7085–7094 (1999)
39. F. Dubois, M.N. Requena, C. Minetti, O. Monnom, E. Istasse, Partial spatial coherence effects in digital holographic microscopy with a laser source. Appl. Opt. **43**, 1131–1139 (2004)
40. M.A. Schofield, Y. Zhu, Fast phase unwrapping algorithm for interferometric applications. Opt. Lett. **28**, 1194–1196 (2003)
41. J. Huntley, Noise-immune phase unwrapping algorithm. Appl. Opt. **28**, 3268–3270 (1989)
42. T.J. Flynn, Two-dimensional phase unwrapping with minimum weighted discontinuity. J. Opt. Soc. Am. A **14**, 2692 (1997)
43. J. M. Huntley, H. Saldner, Temporal phase-unwrapping algorithm for automated interferogram analysis. Appl. Opt. **32**, 3047–3052 (1993)
44. D.C. Ghiglia, M.D. Pritt, *Two-Dimensional Phase Unwrapping: Theory, Algorithms, and Software* (Wiley, New York, NY, 1998)
45. Y.Y. Cheng, J.C. Wyant, Two-wavelength phase shifting interferometry. Appl. Opt. **23**, 4539 (1984)
46. R. Onodera, Y. Ishii, Two-wavelength interferometry that uses a Fourier-transform method. Appl. Opt. **37**, 7988–7994 (1998)
47. M.T. Rinehart, N.T. Shaked, N.J. Jenness, R.L. Clark, A. Wax, Simultaneous two-wavelength transmission quantitative phase microscopy with a color camera. Opt. Lett. **35**, 2612–2614 (2010)
48. C.J. Mann, P.R. Bingham, V.C. Paquit, K.W. Tobin, Quantitative phase imaging by three-wavelength digital holography. Opt. Express **16**, 9753–9764 (2008)
49. J. Gass, A. Dakoff, M.K. Kim, Phase imaging without 2π ambiguity by multiwavelength digital holography. Opt. Lett. **28**, 1141–1143 (2003)
50. J. Kühn, T. Colomb, F. Montfort, F. Charrière, Y. Emery, E. Cuche, P. Marquet, C. Depeursinge, Real-time dual-wavelength digital holographic microscopy with a single hologram acquisition. Opt. Express **15**, 7231–7242 (2007)
51. A. Khmaladze, M. Kim, C. Lo, Phase imaging of cells by simultaneous dual-wavelength reflection digital holography. Opt. Express **16**, 10900–10911 (2008)
52. Y. Fu, G. Pedrini, B.M. Hennelly, R.M. Groves, W. Osten, Dual-wavelength image-plane digital holography for dynamic measurement. Opt. Lasers Eng. **47**, 552–557 (2009)
53. C.K. Hitzenberger, M. Sticker, R. Leitgeb, A.F. Fercher, Differential phase measurements in low-coherence interferometry without 2pi ambiguity. Opt. Lett. **26**, 1864–1866 (2001)
54. J. Zhang, B. Rao, L. Yu, Z. Chen, High-dynamic-range quantitative phase imaging with spectral domain phase microscopy. Opt. Lett. **34**, 3442–3444 (2009)
55. A. Wax, C. Yang, M. Müller, R. Nines, C. W. Boone, V. E. Steele, G. D. Stoner, R. R. Dasari, M.S. Feld, In situ detection of neoplastic transformation and chemopreventive

effects in rat esophagus epithelium using angle-resolved low-coherence interferometry. Cancer Res. **63**, 3556–3559 (2003)
56. M.S. Amin, Y.K. Park, N. Lue, R.R. Dasari, K. Badizadegan, M.S. Feld, G. Popescu, Microrheology of red blood cell membranes using dynamic scattering microscopy. Opt. Express **15**, 17001 (2007)
57. A. Wax, K. Sokolov, Molecular imaging and darkfield microspectroscopy of live cells using gold plasmonic nanoparticles. Laser Photon Rev **3**, 146–158 (2009)
58. L. Cognet et al., Single metallic nanoparticle imaging for protein detection in cells. Proc. Natl. Acad. Sci. USA. (PNAS) **100**, 11350–11355 (2003)
59. D. Boyer et al., Photothermal imaging of nanometer-sized metal particles among scatterers. Science **297**, 1160–1163 (2002)
60. M.C. Skala et al., Photothermal optical coherence tomography of epidermal growth factor receptor in live cells using immunotargeted gold nanosphere. Nano Lett. **8**, 3461–3467 (2008)
61. F. Charrière, A. Marian, F. Montfort, J. Kuehn, T. Colomb, E. Cuche, P. Marquet, C. Depeursinge, Cell refractive index tomography by digital holographic microscopy. Opt. Lett. **31**, 178–180 (2006)
62. M.S. Feld, Tomographic phase microscopy. Nat. Methods **4**, 717–719 (2007)
63. N. Lue, W. Choi, K. Badizadegan, R.R. Dasari, M.S. Feld, G. Popescu, Confocal diffraction phase microscopy of live cells. Opt. Lett. **33**, 2074–2076 (2008)
64. D.J. Brady, K. Choi, D.L. Marks, R. Horisaki, S. Lim, Compressive holography. Opt. Express **17**, 13040–13049 (2009)
65. G. Popescu, K. Badizadegan, R.R. Dasari, M.S. Feld, Observation of dynamic subdomains in red blood cells. J. Biomed. Opt. Lett. **11**, 040503 (2006)
66. G. Popescu, Y.K. Park, R.R. Dasari, K. Badizadegan, M.S. Feld, Coherence properties of red blood cell membrane motions. Phys. Rev. E. **76**, 021902 (2007)
67. G. Popescu, Y.K. Park, W. Choi, R.R. Dasari, M.S. Feld, K. Badizadegan, Imaging red blood cell dynamics by quantitative phase microscopy. Blood Cells Mol. Dis. (2008)
68. M. Mir, Z. Wang, K. Tangella, G. Popescu, Diffraction phase cytometry: Blood on a CD-Rom. Opt. Express **17**, 2579–2585 (2009)
69. M. Kemmler, M. Fratz, D. Giel, N. Saum, A. Brandenburg, C. Hoffmann, Noninvasive time-dependent cytometry monitoring by digital holography. J. Biomed. Opt. **12** (2007)
70. J.P. Urban, A.C. Hall, K.A. Gehl, Regulation of matrix synthesis rates by the ionic and osmotic environment of articular chondrocytes. J. Cell Physiol. **154**, 262–270 (1993)
71. F. Guilak, G.R. Erickson, H.P. Ting-Beall, The effects of osmotic stress on the viscoelastic and physical properties of articular chondrocytes. Biophys. J. **82**, 720–727 (2002)
72. W.R. Trickey, F.P.T. Baaijens, T.A. Laursen, L.G. Alexopoulos, F. Guilak, Determination of the Poisson's ratio of the cell: Recovery properties of chondrocytes after release from complete micropipette aspiration. J. Biomech. **39**, 78–87 (2006)
73. F. Dubois, C. Schockaert, N. Callens, C. Yourassowsky, Focus plane detection criteria in digital holography microscopy by amplitude analysis. Opt. Express **14**, 5895–5908 (2006)

Chapter 8
Spectral Domain Phase Microscopy

Hansford C. Hendargo, Audrey K. Ellerbee, and Joseph A. Izatt

Abstract Spectral domain phase microscopy (SDPM) is a functional extension of optical coherence tomography (OCT) using common-path interferometry to produce phase-referenced images of dynamic samples. Like OCT, axial resolution in SDPM is determined by the source coherence length, while lateral resolution is limited by diffraction in the microscope optics. However, the quantitative phase information SDPM generates is sensitive to nanometer-scale displacements of scattering structures. The use of a common-path optical geometry yields an imaging system with high phase stability. Due to coherence gating, SDPM can achieve full depth discrimination, allowing for independent motion resolution of subcellular structures throughout the sample volume. Here we review the basic theory of OCT and SDPM along with applications of SDPM in cellular imaging to measure topology, Doppler flow in single-celled organisms, time-resolved motions, rheological information of the cytoskeleton, and optical signaling of neural activation. Phase imaging limitations, artifacts, and sensitivity considerations are discussed.

8.1 Introduction

Phase-based microscopy has enabled observation and measurement of biological processes and dynamics occurring on a sub-wavelength scale. The idea of using phase microscopy to enhance image contrast has existed since the early twentieth century. Phase contrast [1, 2] and differential interference contrast microscopy [3] are long-standing techniques that convert the phase changes of light passing through a sample into intensity changes, thereby allowing for visualization of transparent objects. These techniques have dramatically changed the field of cellular imaging, in which many samples of interest, while thin and optically transparent, lack the intrinsic contrast necessary for bright-field visualization of internal structures. Such

J.A. Izatt (✉)
Department of Biomedical Engineering, Duke University, 136 Hudson Hall, Box 90281, Durham, NC 27708, USA
e-mail: jizatt@duke.edu

P. Ferraro et al. (eds.), *Coherent Light Microscopy*, Springer Series in Surface Sciences 46, DOI 10.1007/978-3-642-15813-1_8,

objects alter the phase of reflected or transmitted light based on local changes in the optical path length (either due to differences in the refractive index or in the physical path length). While these early-phase microscopy techniques enabled qualitative observations of underlying sample structures, they were unable to provide quantitative information due to the nonlinear relations between the phase and intensity changes.

In recent years, there has been considerable interest in developing quantitative phase imaging modalities for application in cellular imaging. Phase shifting interferometry [4], digital holography [5, 6], Fourier phase microscopy [7, 8], Hilbert phase microscopy [9], and diffraction phase microscopy [10] have all demonstrated the ability to translate phase changes into quantitative optical path length changes that arise due to either index of refraction or thickness changes throughout a given sample. Phase-sensitive derivatives of optical coherence tomography (OCT) have also been used for quantitative phase imaging. OCT is an imaging modality that uses low-coherence interferometry to generate depth-resolved reflectivity maps of a given sample [11]. This technique has most often been used for imaging the scattering structure of biological samples to provide morphological data [12, 13]. However, phase maps corresponding to minute changes in refractive index of the tissue medium or the position of sample scatterers may also be extracted from the interferometric signal generated during OCT imaging. This chapter focuses on the theory, limitations, and applications of one phase-sensitive derivative of OCT: spectral domain phase microscopy (SDPM). The remainder of this section includes a brief overview of OCT implementation and theory, which forms the basis of SDPM theory; more complete treatments on this subject are presented elsewhere [14, 15].

8.1.1 Optical Coherence Tomography (OCT)

A typical OCT setup is shown in Fig. 8.1 using a standard Michelson interferometer configuration. Light from a broadband source is split (generally in a 50:50 ratio) between a reference arm containing a mirror and a sample arm. The light reflected from each arm recombines at the beam splitter and generates an interference pattern at the detector. Due to the low-coherence nature of the source, interference for a given sample reflector is observable only when the optical path length between the reference and sample arms is matched to within the coherence length of the source (i.e., coherence gating). A one-dimensional, depth-resolved reflectivity profile of the structure (e.g., position of optical interfaces) of the sample may be generated either by axially scanning the reference mirror, as in time domain OCT (TDOCT), or through Fourier processing of a spectrally encoded interference signal (also called an interferogram), as in Fourier domain OCT (FDOCT). Coherence gating, therefore, provides the ability to resolve sample structure along the axis of the incident probing beam; the reflectivity profile thus created is known as an A-scan. The sample beam may be scanned laterally across the sample to build a two-dimensional image (B-scan) or a three-dimensional volume data set. Time domain and Fourier domain systems represent the two major subdivisions of OCT systems.

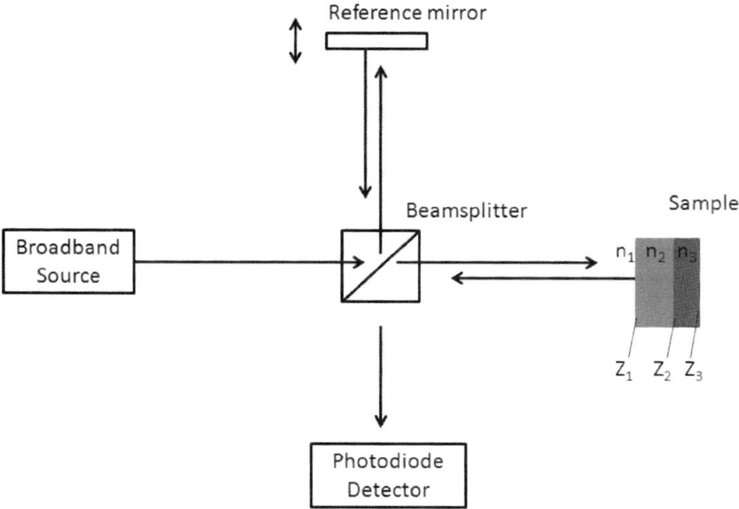

Fig. 8.1 Michelson interferometer used in a time domain OCT setup. Images are acquired by modulating the reference mirror along the path of the optical beam in time. A sample with two layers (three distinct interfaces) at positions Z_1, Z_2, and Z_3 with indices of refraction n_1, n_2, and n_3, respectively, is shown. Figure 8.2 gives the time-encoded signal incident on the detector as a function of the position of the reference mirror

8.1.1.1 Time Domain OCT (TDOCT)

Time domain systems were the first generation of OCT devices developed. To acquire an A-scan, one uses a Michelson interferometer, broadband source, and photodiode detector, as shown in Fig. 8.1. The position of the reference mirror is modulated along the axis of the beam, and the readout of the photodiode detector consists of a time-encoded "fringe burst" whose peaks correspond with those positions in depth where the reference arm path length matches that of an individual scatterer in the sample arm. The example shows a sample with three distinct interfaces: the interfaces correspond to locations where the local refractive index changes, thus causing some of the incident light to be reflected back toward the detector. Back-reflected photons from the sample arm interfere with those reflected from the reference mirror. Detection of the envelope of the interference pattern, whose width is proportional to the coherence length of the source, allows for the localization of scatterers in the sample (to within a coherence length), as shown in Fig. 8.2. Observe that the resulting envelope profile resolves the position of all interfaces through the depth of the sample relative to one another. The coherence length of the source determines the axial resolution, which is given by the full width at half maximum intensity of the interference envelope.

8.1.1.2 Fourier Domain OCT (FDOCT)

In contrast to TDOCT, FDOCT employs the same Michelson interferometer setup but uses a stationary reference mirror to encode the depth information of a given

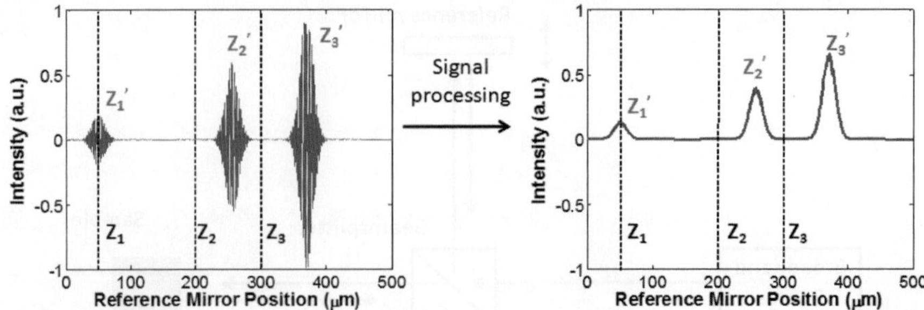

Fig. 8.2 Results of TDOCT data acquisition and processing from Fig. 8.1. The interference signal is collected over time as a function of the reference mirror position. A fringe burst pattern results when the optical path length of the reference arm matches that of light reflected from each of the three distinct sample layers. Software processing yields the envelope of each fringe burst, showing the locations of the three reflective interfaces as a depth-resolved reflectivity profile of the sample. The *dashed lines* indicate the geometrical distance between each reflector. The detected profile (*blue trace*) shows the optical path length difference between each surface, which differs from the geometrical distance according to the index of refraction of each layer. The full width at half maximum of each peak gives the axial resolution of the system, determined by the source bandwidth

sample as a function of wavenumber. Two different implementations of FDOCT have been introduced: spectral domain OCT (SDOCT) and swept-source OCT (SSOCT), diagrammed in Fig. 8.3. A broadband source along with a spectrometer is used in the former case, and a tunable wavelength swept source and photodiode detector are used in the latter. FDOCT implementations have been demonstrated to have a sensitivity advantage of at least 20 dB over TDOCT systems and also can achieve faster data acquisition rates [16–18]. The example in Fig. 8.3 shows the imaging configuration for a sample with three distinct layers, as before; Fig. 8.4 shows the basic method for signal detection and processing.

In FDOCT, the interferogram is sampled as a function of the wavenumber of the source. Spectrometer-based systems typically sample linearly in wavelength, but

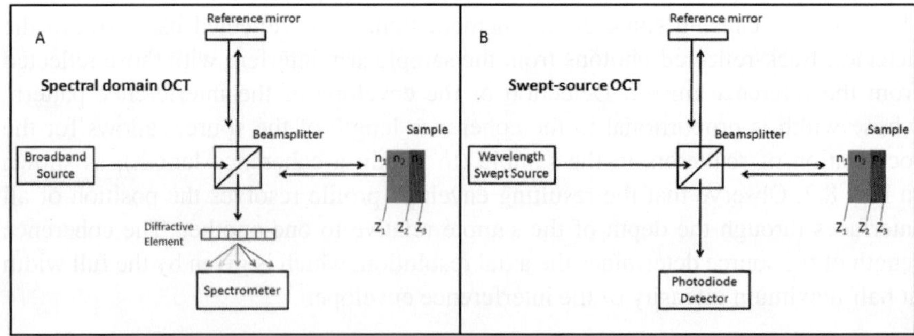

Fig. 8.3 Implementations of FDOCT. (**a**) SDOCT setup using a broadband source and spectrometer-based detection. (**b**) SSOCT setup with a narrow linewidth wavelength swept source and photodiode detector. The same two-layer sample as in Fig. 8.1 is presented

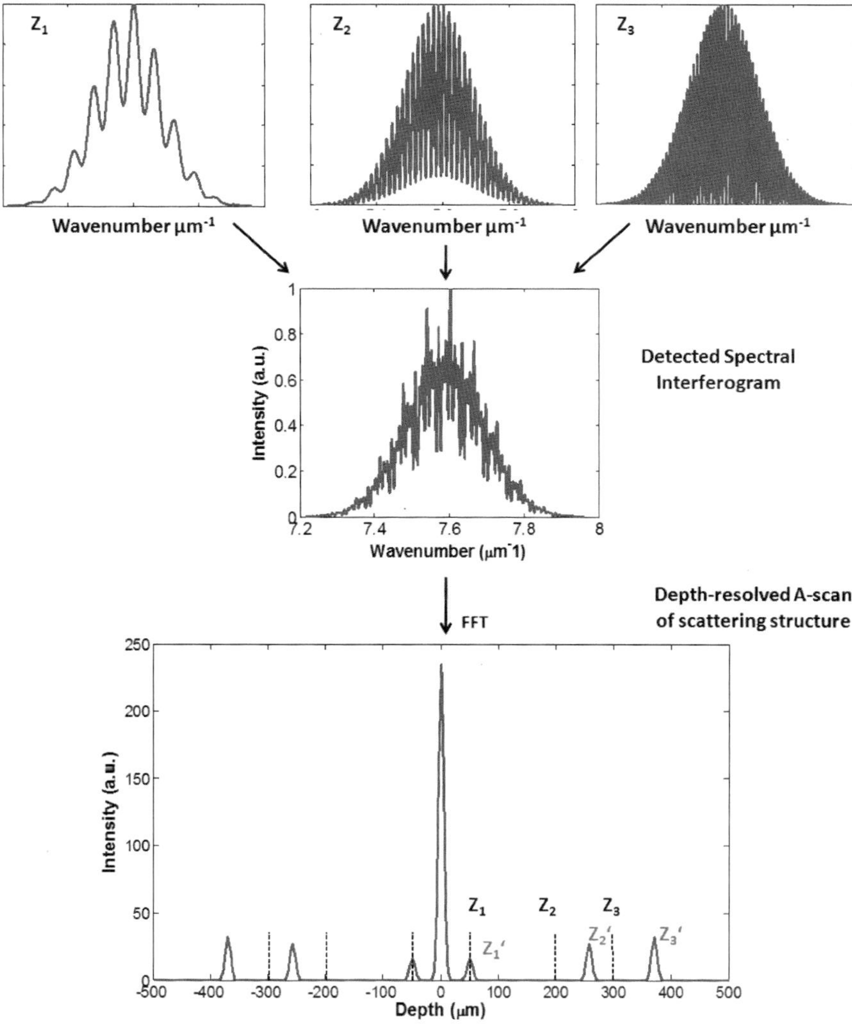

Fig. 8.4 Basic signal detection process in FDOCT. Each distinct sample reflector generates an interference pattern that overlays the source spectrum. The frequency of the interference fringes increases with the optical path length mismatch between the reference and sample arms. The detected spectral interferogram is the summation of the interference patterns generated from each sample reflector. Taking the magnitude of the Fourier transform of the interferogram yields a depth-resolved A-scan. A large peak centered at zero path length difference, known as the DC peak, results from non-interfering photons. Because the detected interferogram is real valued, the Fourier transform results in a complex conjugate artifact, or mirror image of each reflector on the opposite side of the DC peak. Using only half of the A-scan profile yields the same information as in TDOCT. *Black dashed lines* indicate geometrical positions of each reflector relative to DC. The *blue profile* shows the detected reflector positions, which are shifted from those in *black* due to differences in the index of refraction of each layer

processing is used to convert the wavelength sampling into samples that are linearly spaced in wavenumber. A few groups have demonstrated spectrometer designs and swept-source systems that directly sample linearly in wavenumber [19, 20]. The frequency of the interference pattern varies with the path length mismatch between the reference and sample arms. The detected spectral interferogram is a summation of interference patterns generated by photons backscattered from each sample reflector and mixed with light from the reference mirror (i.e., the interference patterns of all samples in depth for a given wavenumber overlap). Fourier transformation of the signal yields information about the reflectivity of each scatterer and can be used to obtain a reflectivity map of sample scatterers. A large peak centered at zero path length difference is generated by non-interfering light reflected from the reference arm and sample reflectors. This constant peak, known as the DC peak, is collocated with the position of the reference mirror. The position of each individual reflector is thus shifted away from DC by the optical path difference between the reference mirror and sample reflector.

As seen in Fig. 8.4, Fourier processing of the resulting interferogram yields a twin-sided image of the sample structure with identical structural features occurring on both the negative and positive sides of DC. This is because the detected signal is real valued, so the Fourier transform operation results in an artifact profile (known as the complex conjugate) on the opposite side of the DC peak from the real position of the sample scatterer. It is possible to ignore this artifact by confining the sample structure to one side of the DC peak, though a scattering profile sufficiently extensive in depth may necessitate the use of techniques to remove the complex conjugate image [21–25]. Such methods are collectively known as complex conjugate resolution techniques.

8.1.2 Spectral Domain Phase Microscopy (SDPM)

The Fourier transform of the spectral interferogram from FDOCT results in a complex-valued data set. The magnitude of the signal reveals the position and reflectivity of sample reflectors (A-scan), but phase information is also obtained simultaneously without additional processing. Because FDOCT has been used largely as a morphological imaging technique, the phase information is typically ignored when generating images. However, the phase can be exploited to give functional information about the sample and has been used in low-coherence interferometry setups to detect and measure flow [26, 27], birefringence [28], and sub-wavelength optical path length changes in a sample [29]. Because the phase allows for highly sensitive measurements of minute structure and dynamics and can be easily collected in OCT, there has been much interest recently in developing phase-based OCT for applications in cellular imaging.

Common-path interferometers offer enhanced phase stability over two-path setups due to the reduced noise threshold that results from the reference and sample light sharing the same optical path. Such systems have been demonstrated to have a phase stability on the order of tens of picometers [30, 31]. One offshoot of FDOCT that uses a common-path configuration has been termed spectral domain phase microscopy (SDPM) [30], or alternatively spectral domain optical coherence phase microscopy (SD-OCPM) [31]. SDPM has been applied to cellular imaging of features and dynamics that occur on the sub-micron scale. Such features are not easily resolved using amplitude information alone. The remainder of this chapter discusses the theory of SDPM as well as its limitations, artifacts, and applications in cellular and functional imaging.

8.2 Spectral Domain Phase Microscopy Theory and Implementation

The principal theory of spectral domain phase microscopy follows closely that of SDOCT. Here, attention is given to analysis of the phase signal and its relationship to the signal-to-noise ratio. Explanations on the relation between phase and structural variation are also provided. Basic conceptual designs for experimental implementation of such systems are also discussed.

8.2.1 SDPM Theoretical Analysis

8.2.1.1 SDPM Phase Analysis

For the case of a single reflector, the SDOCT interferometric signal at the detector as a function of wavenumber is given by

$$i(k) = \frac{1}{2}\rho S(k)\delta k \left\{ R_{\mathrm{R}} + R_{\mathrm{S}} + \sqrt{R_{\mathrm{R}} R_{\mathrm{S}}} \cos(2kn(x + \Delta x)) \right\}, \tag{8.1}$$

where $S(k)$ is the spectral density of the source, ρ is the detector responsivity, δk is the wavenumber spacing per detector pixel, R_{R} and R_{S} are the reference and sample reflectivities, respectively, n is the index of refraction, and $x + \Delta x$ is the distance between the reference and sample reflectors. Here, x accounts for the discrete sampling of the detector in the spatial domain while Δx represents sub-resolution differences in the sample position from x [32]. The first two terms in (8.1) are so-called DC terms that get mapped to the position of the reference reflector while the third term contains the interferometric data of interest in OCT and SDPM. Taking the Fourier transform of (8.1) and ignoring the DC terms yields a depth-dependent reflectivity profile of the sample, known as an A-scan, with peaks located at $\pm 2nx$. These peaks can be described as

$$I_{\mathrm{signal}}(\pm 2nx) = \frac{\rho S \Delta t \sqrt{R_{\mathrm{R}} R_{\mathrm{S}}}}{2e} E(2nx) \exp(\pm j 2k_{\mathrm{o}} n \Delta x), \tag{8.2}$$

where S is the total source power, Δt is the detector integration time, e is the electronic charge, $E(2nx)$ is the coherence envelope function, j is the imaginary number, and k_o is the center wavenumber of the source. The phase of the detected signal during the jth measurement can be used to track sub-wavelength deviations in the sample with respect to a reference phase measurement, ϕ_o. These deviations are related to the detected phase by

$$\Delta x = \frac{\lambda_o \Delta\phi}{4\pi n} + m\frac{\lambda_o}{2n}, \tag{8.3}$$

where Δx is the sub-resolution feature of the sample reflector, λ_o is the center wavelength of the source, $\Delta\phi = \phi_j - \phi_o$, and m is an integer number of half wavelengths. A factor of 2 in the denominator accounts for the double pass in the optical path length due to the reflection geometry of the optical setup traditionally used in FDOCT. It should be noted that the phase reference may be a point in time or a spatial location.

When (8.3) assumes a constant index of refraction within a sample, the phase difference can be used to calculate a change in sample thickness. However, changes in the index of sample structure affect the phase expression in (8.2) as do physical changes, and the two are inherently coupled. Assuming a trivial contribution from one allows for calculation of changes in the other; however, validation of such an assumption is critical for correct quantitative assessment.

A given phase value is not an absolute determination of Δx, but can potentially represent Δx plus any integer number of half wavelengths. Without a priori knowledge of the phase topography of the sample, there is no way to know the exact value of m in (8.3), as any displacement that is a multiple of $\lambda_o/2$ will give the same value. Such phase ambiguity can lead to a problem known as phase wrapping. If $m = \pm 1$, simply adding or subtracting 2π to the phase can correctly unwrap the artifact. However, if $|m| \geq 2$, it is impossible to unwrap the phase accurately using this simple method without prior sample information. Techniques used to overcome this problem are discussed in Sect. 8.3.

8.2.1.2 Theoretical Phase Sensitivity and Its Dependence on Signal to Noise

The number of photons generated by a source will vary from one point in time to the next. This time-varying number of photons is known as shot noise. Shot noise presents a fundamental limit on the optical noise level for a given detection system. Here, we present the analysis of the effects of shot noise on the phase sensitivity of SDPM, which has been discussed much in past literature [16–18, 30, 32]. Equation (8.2) gives the interferometric component of the SDPM signal after Fourier transformation of the detected interferogram. The noise associated with this signal in the shot noise limit can be described as

$$I_{\text{noise}}(\pm 2xn) = \left(\frac{\rho S \Delta t R_{\text{R}}}{2e}\right)^{1/2} \exp(j\phi_{\text{rand}}), \tag{8.4}$$

where ϕ_{rand} is the random phase contribution from shot noise and S is the total source power. The signal-to-noise ratio of a measurement is thus given by

$$SNR = \left(\frac{|I_{\text{signal}}|}{|I_{\text{noise}}|}\right)^2 = \frac{\rho S \Delta t R_S}{2e}. \tag{8.5}$$

The relation between I_{signal} and I_{noise} is illustrated in the vector map in Fig. 8.5. For a given measurement, the phase noise, $\delta\phi$, is a measure of the difference in angle (phase) between the actual sample signal and the measured signal. Note that this phase difference, which depends on ϕ_{rand}, is a maximum when I_{noise} is orthogonal to I_{signal}. The parallel component of I_{noise} affects the sensitivity of the amplitude measurement but does not affect the phase. The phase resolution, $\delta\tilde{\phi}$, is typically defined as the average value of $\delta\phi$ using the following equation:

$$\delta\tilde{\phi} = \frac{2}{\pi}\int_0^{\pi/2} \arctan\left(\frac{|I_{\text{noise}}|}{|I_{\text{signal}}|}\sin\phi_{\text{rand}}\right) d\phi_{\text{rand}}, \tag{8.6}$$

where $\delta\tilde{\phi}$ is averaged over a quarter period with the assumption that ϕ_{rand} is uniformly distributed over 2π. Integrating over a single quadrant gives an approximation of the phase noise at any given time, since the effect of ϕ_{rand} in the second quadrant is redundant with ϕ_{rand} in the first quadrant from the perspective of the measured phase noise. Also, the third and fourth quadrant values of ϕ_{rand} have the same absolute value of phase noise as the second and first quadrants, respectively.

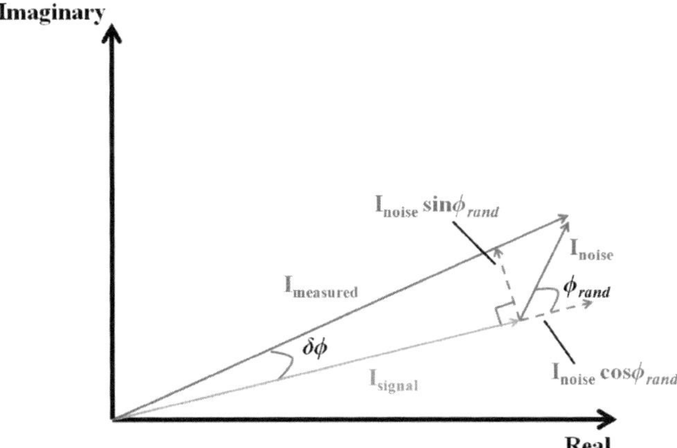

Fig. 8.5 Effects of random phase contributions from pure shot noise on measured phase. $\delta\phi$ is defined as the angle between the actual signal vector, I_{signal}, and the measured signal vector, I_{measured}. The phase noise vector, I_{noise}, is randomly oriented according to a random phase, ϕ_{rand}. Only the component of I_{noise} that is orthogonal to I_{signal} contributes to phase noise in the measurement. The phase resolution, $\delta\tilde{\phi}$, is an averaged value of $\delta\phi$

The measured intensity is $I_{\text{signal}} + I_{\text{noise}}$. and I_{noise} influences both the phase and amplitude sensitivity. For the case where $|I_{\text{signal}}| >> |I_{\text{noise}}|$ (small angle approximation), (8.6) can be approximated using

$$\arctan\left(\frac{|I_{\text{noise}}|}{|I_{\text{signal}}|}\sin\phi_{\text{rand}}\right) \approx \frac{|I_{\text{noise}}|}{|I_{\text{signal}}|}\sin\phi_{\text{rand}}. \tag{8.7}$$

Combining (8.2) and (8.4) into (8.6) and using the approximation from (8.7) yields

$$\delta\tilde{\phi} = \frac{2}{\pi}\frac{|I_{\text{noise}}|}{|I_{\text{signal}}|} = \frac{2}{\pi}\left(\frac{2e}{\rho S \Delta t R_{\text{S}}}\right)^{1/2} = \frac{2}{\pi}\left(\frac{1}{SNR}\right)^{1/2}. \tag{8.8}$$

The phase resolution, $\delta\tilde{\phi}$, for a given measurement thus depends upon the inverse square root of the SNR and is considered the smallest reliable phase change that can be detected by a given system.

The phase resolution may also be converted into physical displacement resolution by using (8.8) in the relation expressed in (8.3):

$$\delta\tilde{x} = \frac{\lambda_{\text{o}}\delta\tilde{\phi}}{4\pi n}. \tag{8.9}$$

When expressed in this way, $\delta\tilde{x}$ is the smallest difference in sample thickness that can be reliably detected for two relative phase measurements, assuming a constant index of refraction.

8.2.1.3 Velocity Resolution

The phase information can also be used to obtain data on the velocity of moving scatterers in a sample. We have already shown how the phase can be used to track the sub-wavelength changes in the position of a sample reflector. By taking the phase difference between two phase measurements in time and dividing by the time in between acquisitions, the instantaneous velocity of a sample reflector at that position can be measured [26, 32]. The time derivative of the phase leads to a Doppler shift in the interference frequency. The measured velocity, $v(t)$, depends upon the angle between the probing beam and the direction of motion of a moving scatterer and is given by

$$v(t) = \frac{\lambda_{\text{o}}}{4\pi\cos\theta}\frac{(\phi_i - \phi_{i-1})}{\Delta t} = \frac{\lambda_{\text{o}}}{4\pi\cos\theta}\frac{\Delta\phi}{\Delta t}, \tag{8.10}$$

where θ is the Doppler angle between the optical axis and the direction of scatterer motion, and the index i indicates the ith measurement in time. The factor $\Delta\phi/\Delta t$ is the slope of the phase between sequential measurements and is related to the Doppler shift frequency, f_{Dopp}, as

$$\frac{\Delta\phi}{\Delta t} = 2\pi f_{\mathrm{Dopp}}. \tag{8.11}$$

The velocity thus depends on a difference operation, which reduces the velocity resolution (increases the noise floor) by a factor of $\sqrt{2}$ over the phase resolution due to propagation of error. We assume there is no uncertainty in the time measurement, and that the integration time of the detector is short enough that motion of the sample reflector can be considered of constant velocity during sampling. The velocity resolution is thus given by

$$\delta\tilde{v} = \frac{\lambda_{\mathrm{o}}}{4\pi\,\Delta t\cos\theta}\sqrt{2}\delta\tilde{\phi} = \frac{\lambda_{\mathrm{o}}}{\sqrt{2}\pi^2\Delta t\cos\theta}\frac{1}{\mathrm{SNR}^{1/2}}. \tag{8.12}$$

The SNR advantage of FDOCT clearly benefits SDPM by leading to improved phase stability, which in turn yields greater structural and motion sensitivity.

8.2.2 SDPM Implementation

8.2.2.1 SDPM System Design

To date, SDPM has been demonstrated using two types of common-path interferometer setups [30, 33]. These systems allow for cancellation of common mode noise and elimination of noise due to mechanical instability of the reference mirror, since light from the sample and reference arms travels the same optical path. Thus, the phase of the sample relative to the reference will be stable save for changes induced by the sample itself. Both spectrometer and swept-source designs have been demonstrated, and configurations capable of raster scanning have been utilized for imaging in three dimensions. Figure 8.6a shows a schematic of an SDOCT system that has been rendered common path by blocking the reference arm. Typically, a glass coverslip is placed in the sample arm, with the sample of interest (cells in this case) resting on top of the coverslip. The reflection from the bottom coverslip surface effectively provides a suitable alternative reference reflection to the separate reference mirror used in standard OCT. An objective lens with a sufficiently high NA can be used to visualize the cells. A similar SDPM design using a swept-source configuration is shown in Fig. 8.6b. Both setups yield high phase stability, as any vibrations to the optical setup generally affect both the reference and sample reflectors to the same degree.

8.2.2.2 Self-Phase Referencing

Another possible technique to achieve highly stable phase measurements uses a separate reference arm along with a reflector in the sample arm that acts as a phase reference. For a standard SDOCT system, phase noise created due to small vibrations in the reference arm can lead to large fluctuations in the phase measurement. However, changes to the phase caused by reference mirror instability affect all sample

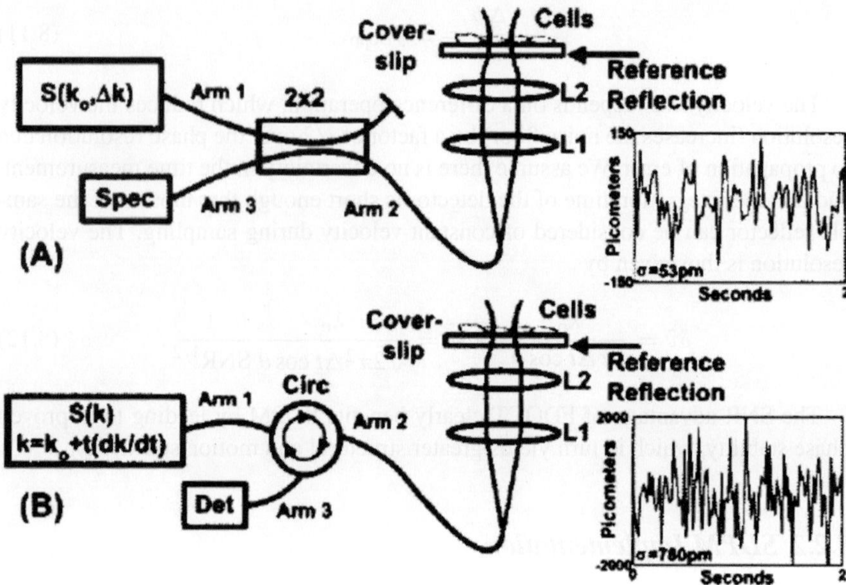

Fig. 8.6 Schematics of an SDPM setup. (**a**) SDPM is performed using a fiber-based SDOCT setup with the reference arm blocked. (**b**) A SS-SDPM setup using a wavelength swept laser and optical circulator. Plot insets show the phase stability of measurements from a clean glass coverslip. Images from [30]

reflectors simultaneously. Thus, by taking the difference of the phases detected from two different sample reflectors, the noise introduced by the reference arm mirror can be removed. By using a reflector in the sample as a phase reference, which may be a reference from either an axial depth location or a lateral location, the relative motion of other reflectors can be detected with reduced noise [34, 35].

8.3 Limitations and Artifacts

Though SDPM offers many advantages in terms of phase stability, there are also several important disadvantages that must be considered when implementing this method. We discuss here the effects of the reduced SNR typical in common-path systems, the tradeoff between lateral resolution and depth of focus, as well as the consequences of phase corruption and phase wrapping artifacts. This section also presents possible solutions to some of these problems.

8.3.1 SNR Reduction

Due to the common-path design, the SNR of images produced by SDPM systems is typically 30 dB less than similar images obtained using a standard FDOCT setup.

This occurs because the reference reflection provided by the air–glass interface of the bottom surface of a glass only reflects 4% of the incident light. Moreover, unless an optical circulator is used, the light that would be typically used to illuminate the reference arm (50% of the source light) is wasted. Because the reflectivity of the coverslip surface still dominates the reflectivity of the sample, the SNR expression in (8.5) still holds, but the dynamic range of the system is reduced. Thus, low SNR reflectors normally detected by standard FDOCT are difficult to observe using common-path SDPM. The self-phase referencing method described earlier can be used to recover the full dynamic range of FDOCT to be utilized, though the phase stability tends to degrade compared to common-path systems [30, 35].

8.3.2 Lateral Resolution vs Depth of Focus

Another limitation of SDPM is a more general limitation of OCT techniques: the tradeoff between the lateral resolution and depth of focus inherent to all objectives. High NA objectives are typically necessary to image at the cellular level. However, the depth of focus scales inversely as NA^2. In SDPM, however, the problem is more critical. Since SDPM relies upon a reference reflection from the bottom of a glass coverslip, typically 150–200 μm thick, it is important to consider the effects of a reduced depth of focus on the strength of the interferometric signal detected. To ensure adequate signal from the reference reflection, it should ideally be kept within the depth of focus of the imaging beam. Because the sample arm optics comprise a confocal detection scheme, the detected intensity along the axis of the sample beam can be described, as in [36], by

$$I(u) = \left(\frac{\sin(u/2)}{u/2}\right)^2, \tag{8.13}$$

where u, defined as the normalized axial range parameter, is related to the axial distance, z, of a given imaging point from the central focus position and the maximum collection angle of the objective, α, by

$$u(z) = \frac{8\pi}{\lambda} z \sin^2(\alpha/2). \tag{8.14}$$

Figure 8.7 shows the falloff in SNR over depth for an SDPM system using various NA objectives. A typical coverslip has a thickness of about 150–200 μm. Assuming the focus of the sample arm beam is placed at the sample (i.e., near the top surface of the coverslip), the reference surface may not fall within the depth of focus, thus less light will be collected upon reflection. For high NA objectives, the large SNR decrease caused by the reference reflection being located outside the depth of focus can be compensated by increasing the dynamic range of the system. This can be accomplished most easily by increasing the sample arm power or the detector integration time. Thus, it is possible in SDPM to detect samples outside of the depth

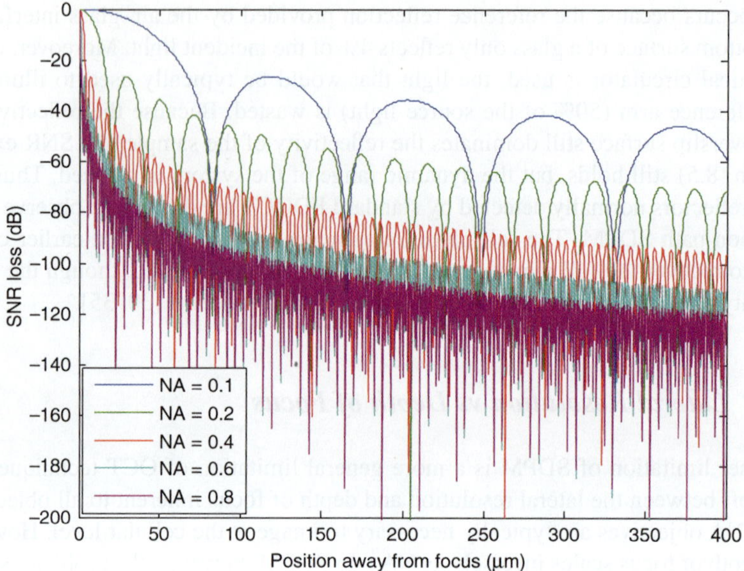

Fig. 8.7 Simulated effects on SNR falloff using high NA objectives for a single reflector. The loss in signal power results from the increasing distance of the reflector from the center of the focus. As the NA of the objective increases, the depth of focus decreases. Thus, the signal of reflectors far from the central focal spot suffers greater degradation with higher NA objectives

of focus, though the strength of such signals is reduced compared to detection by standard FDOCT.

Recently, a line-scan, self-phase-referenced SDPM system for cellular imaging was developed to utilize high NA objectives [35], circumventing the SNR loss typically associated with common-path systems. In this system a separate reference arm was used, which allowed for independent focusing of the reference and sample arms. A cylindrical lens focused light into a line incident in both the reference and sample arms. HeLa cells resting on a glass coverslip were imaged. A portion of the line illumination was incident upon the cells and another portion was incident upon a coverslip, simultaneously acquiring the phase from the sample of interest and the coverslip surface. Because an entire lateral image was acquired over a single detector integration period, the coverslip surface could be used as a phase reference to account for reference arm motion. The use of a separate reference arm provided a sufficiently high reference signal to enable use of a high NA objective (NA = 1.2), thus improving the lateral resolution of the system while still providing a stable phase reference in the plane of the sample.

8.3.3 Phase Corruption

The phase of adjacent reflectors in a given sample may influence the phase of nearby scatterers. This problem, known as phase corruption, can cause erroneous phase measurements for samples of weak reflectivity positioned near strong reflectors (particularly stationary reflectors like the surface of a coverslip). A full treatment of this problem is given in [37]. Briefly, the measured phase at a given depth, x, is influenced by the presence of nearby reflectors according to

$$\phi(x) = \tan^{-1}\left(\frac{\sum_{m=1}^{M} P_m(x)\sin\left[2n_m k_o(x - x_m)\right]}{\sum_{m=1}^{M} P_m(x)\cos\left[2n_m k_o(x - x_m)\right]}\right) \tag{8.15}$$

where $P_m(x)$ is the convolution of the source spectral density function and the point spread function of the mth reflector, x_m is the axial depth of the mth reflector, k_o is the center wavenumber of the source, n_m is the refractive index of the mth reflector, and M is the total number of reflectors in the sample. If we let $P_m(x) = A_m \exp[-2(x - x_m)^2/2\sigma^2]$, an ideal Gaussian, the phase and slope of the phase in the dx neighborhood of two isolated reflectors ($M = 2$) separated by distance y are given by the following equations:

$$\phi(x_m + \mathrm{d}x) = \tan^{-1}\left\{\frac{\beta\exp(-\mathrm{d}x^2/\sigma^2)\sin(2k_o\mathrm{d}x) + \exp\left[-(y+\mathrm{d}x)^2/\sigma^2\right]\sin(2k_o(y+\mathrm{d}x))}{\beta\exp(-\mathrm{d}x^2/\sigma^2)\cos(2k_o\mathrm{d}x) + \exp\left[-(y+\mathrm{d}x)^2/\sigma^2\right]\cos(2k_o(y+\mathrm{d}x))}\right\}, \tag{8.16}$$

$$\frac{\partial\phi(x_m + \mathrm{d}x)}{\partial x} = 2k_o - \frac{2y\beta\sin(2k_o y)}{\sigma^2\left\{\exp[-(y^2+2y\mathrm{d}x)/\sigma^2] + \beta^2\exp[(y^2+2y\mathrm{d}x)/\sigma^2] + 2\beta\cos(2k_o y)\right\}}, \tag{8.17}$$

where $y = x_2 - x_1$ and $\beta = A_2/A_1$. Clearly a linear slope for the phase is expected for reflectors separated by large distances ($y \to \infty$), when one reflector is much stronger than the other ($\beta \to \infty$), or when the system has a high axial resolution (e.g., achieved through a large source bandwidth $\sigma \to \infty$). However, for realistic scenarios, the measured phase for a single reflector is corrupted by information about the phase of closely spaced reflectors.

This problem can be mitigated through software processing or design of an appropriate sample arm. The most straightforward method to correct the problem is to obtain a clean background signal without the sample. By subtracting the background from the sample data, phase corruption due to the coverslip reflection, which typically poses the most significant problem, can be avoided. Figure 8.8 gives an example of a spontaneously beating cardiomyocyte on a coverslip. The presence of the coverslip reflection corrupts the phase data measured from the beating cell. The phase-corrupted data do not reveal the motion of the cell surface; the phase-corrected data, however, reveal phase fluctuations corresponding to the beating of the cell after the background interferometric signal was removed.

8.3.4 Phase Wrapping

The change in phase of the OCT signal is linearly related to axial motion of the sample over time. However, the measured phase is limited to a range of $-\pi$ to $+\pi$. A problem known as *phase wrapping* occurs when a change in phase between

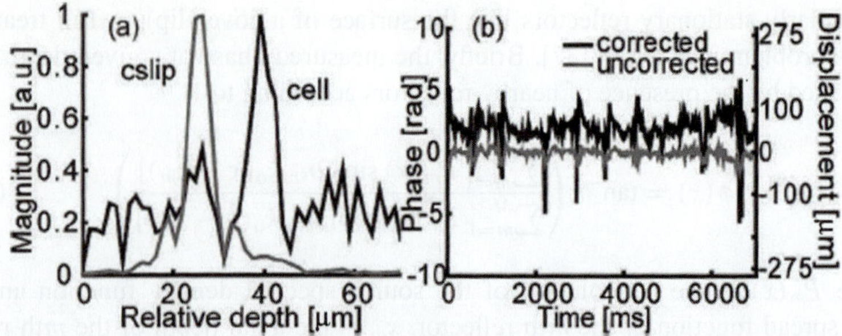

Fig. 8.8 SDPM data from a spontaneously beating cardiomyocyte resting on a coverslip surface. (**a**) Magnitude information showing the position of the cell and coverslip both before (*gray*) and after (*black*) correction for phase corruption. (**b**) Phase information corresponding to the surface motion effects of the beating cell. The beating pattern of the cell is obscured in the corrupted data. Software correction allows the beating to be observed. Image from [37]

consecutive measurements is such that the total phase change falls outside the $[-\pi, \pi]$ interval and thus yields an ambiguous result, as illustrated in Fig. 8.9 for a linearly sloping phase surface. When the optical path length exceeds the 2π detection bandwidth, the detected phase wraps to the opposite limit of the 2π interval. Because SDPM operates in a reflective geometry, variations in the position of a sample reflector in space or time will cause phase wrapping whenever the cumulative path length change exceeds half of the source center wavelength.

Singly wrapped phases correlate identically with the actual phase change of the object when slow phase variations are present, as in Fig. 8.9. Such artifacts can be removed readily by simple software methods that add 2π to the phase at relevant locations to smooth the phase profile [38]. However, such techniques fail when multiple phase wrapping occurs, as may be the case for fast sample motion or sharp

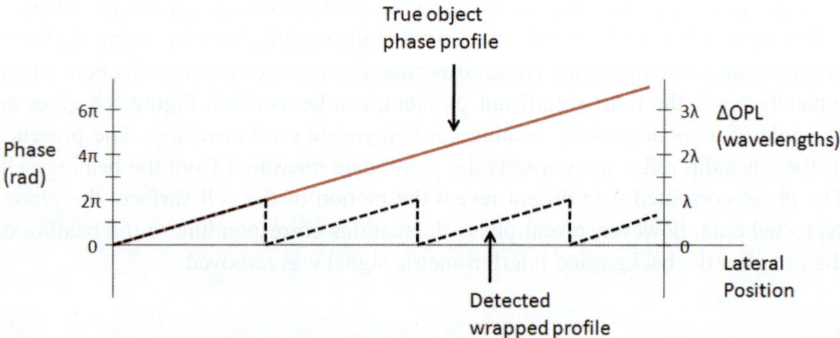

Fig. 8.9 Example of a phase wrapping artifact. Measurements from a linearly sloped phase profile result in the detected phase being "wrapped" into a 2π bandwidth. The measured phase actually falls between an interval from $-\pi$ to $+\pi$, but here has been shifted to be between 0 and 2π for illustrative purposes

spatial variations. Multiple wrapping can be avoided temporally with rapid acquisition times or spatially with fine lateral sampling. Unfortunately, these solutions may not be possible in some cases due to hardware or data acquisition limitations; in these cases, software-based solutions can be implemented to correctly unwrap such phase images.

Many software algorithms exist to correct for phase wrapping [38]. These methods typically rely upon calculation and comparison of the phase gradient throughout a given image and detection of phase discontinuities. These algorithms vary in complexity, speed, and utility. An alternative technique that relies upon a hardware-based solution is multi-wavelength phase unwrapping, in which the phase information at two or more wavelengths is used for robust phase unwrapping [39, 40]. This method has been implemented in FDOCT by taking advantage of broadband source light to yield phase information from many wavelengths simultaneously [41], as briefly described in Fig. 8.10.

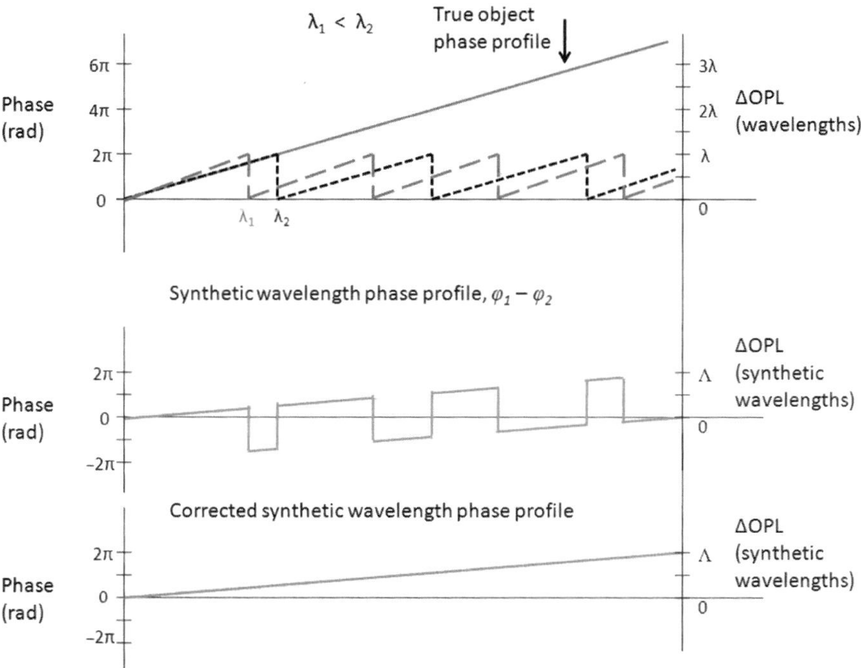

Fig. 8.10 Illustration of synthetic wavelength phase unwrapping. (*Top*) The linearly sloped phase profile is measured with two different wavelengths, each resulting in a slightly shifted wrapped phase profile. (*Middle*) The phases of the two single wavelength profiles are subtracted from each other, yielding the synthetic wavelength phase profile. (*Bottom*) A phase value of 2π is added to locations on the synthetic profile that are less than 0, generating a correctly unwrapped profile. Note that the newly unwrapped profile describes the true phase profile, but in terms of the synthetic wavelength Λ

A broadband source captures a phase profile at two or more different wavelengths shown. The difference in the phase information obtained at the two wavelengths can be recast in terms of an equivalent phase associated with a longer synthetic wavelength, Λ, which is a function of each of the imaging wavelengths:

$$\Lambda = \frac{\lambda_1 \lambda_2}{|\lambda_1 - \lambda_2|}. \tag{8.18}$$

This recast expression allows for wrap-free measurements of changes in optical path length less than $\Lambda/2$, which can be significantly larger than any of the single imaging wavelengths alone.

Synthetic wavelength phase unwrapping may be applied to OCT data to resolve correctly sample motions that are larger than $\lambda_o/2$. Image processing in OCT uses the Fourier transform of a broadband spectrum. By windowing the signal spectrum to isolate information for a given wavelength before applying the Fourier transform, phase information at multiple "center" wavelengths may be obtained. This method has an advantage over other software techniques in that it is not necessary to rely upon phase gradients across image pixels to unwrap the phase at a given point. Rather, two phase maps for a given sample depth can be generated via Fourier transformation of the two windowed spectra. The phase maps are designed to have corresponding image pixels, the difference of which generates the synthetic wavelength phase profile. Each image pixel can be unwrapped without need for calculation of phase gradients across the image.

Synthetic wavelength unwrapping was applied to phase measurements from human epithelial cheek cells imaged with SDPM. The bright-field image of a group of cells is shown in Fig. 8.11a. The wrapped phase profile was initially obtained in Fig. 8.11b and was later unwrapped using the synthetic wavelength method. As can be seen from the bright-field microscope image, single cells as well as a cluster of cells stacked together are present in the field. The cell cluster is expected to

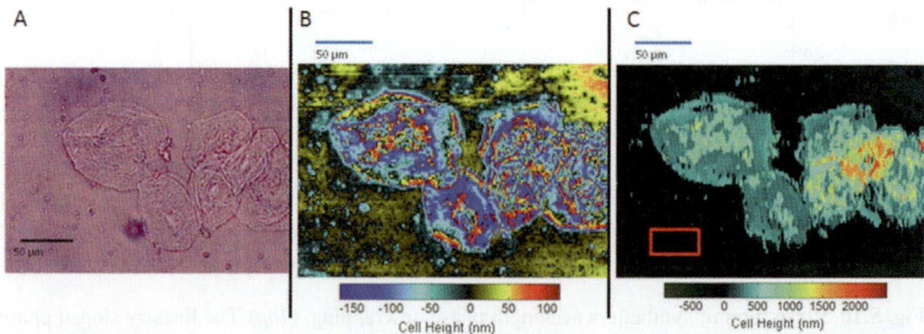

Fig. 8.11 Phase images of human epithelial cheek cells. (**a**) Bright-field microscopy image of a single cell and cluster of cells. (**b**) Phase image produced from initial wrapped phase measurements. (**c**) Synthetic wavelength unwrapped phase map. Note the difference in the color bar scales for (**b**) and (**c**). The region in the *red box* indicates the area used as the phase reference. Image from [41]

introduce multiple wrapping artifacts at areas where its thickness exceeds that for which a single cell would produce only a single wrap. A clear picture of the cell height above the coverslip surface is obtained in Fig. 8.11c. The heights of both single cells and the cell cluster are correctly unwrapped and resolved.

8.4 Applications

Phase microscopy finds great utility in the realm of cellular imaging. The sensitivity of phase measurements to small motion or changes in cell structure allows for the detection and measurement of dynamics and features on the micron and sub-micron scale. To this end, SDPM lends itself as a useful tool for cellular biologists studying cellular motility or behavior.

Some of the earliest studies demonstrating SDPM used temporal changes in phase to monitor cell contractility, as shown in Fig. 8.12, or spatial phase changes across the cell surface to map the cellular topology [30, 31]. The phase sensitivity advantage of SDPM was also extended to measure (very slow) cytoplasmic flow [32] as well as the mechanical properties of cellular membranes [42]. Multi-dimensional measurements of the phase were demonstrated across contracting cardiomyocytes [43]. SDPM has been combined with multi-photon microscopy to provide both highly sensitive functional and structural imaging of fibroblast cells [44]. A self-phase-referenced SDOCT setup was used to measure phase changes during action potential propagation in a crustacean nerve [34].

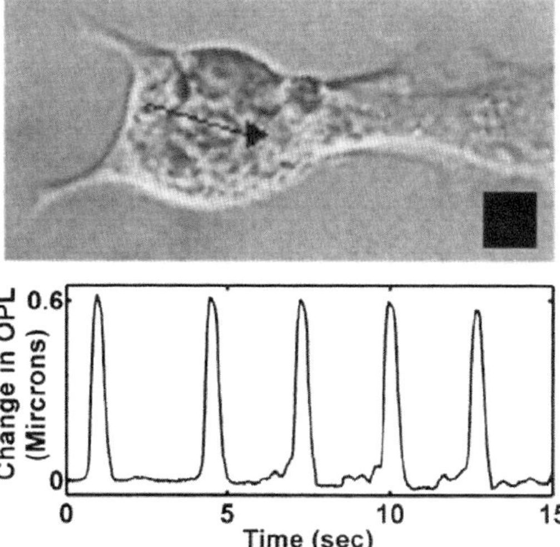

Fig. 8.12 SDPM used to monitor the spontaneous beating of a cardiomyocyte from a chick embryo. *Top*: bright-field image of the cell. *Arrow* indicates location of the SDPM measurements. *Bottom*: SDPM phase trace measuring changes in optical path length of the reflected light from the cell at the depth corresponding to the cell surface. Image from [30]

8.4.1 Three-Dimensional Cellular Phase Microscopy

Three-dimensional phase imaging of cellular structure and dynamics is critical for studying the complete response of a cell to external stimuli. The ability to detect phase changes throughout the volume of the cell holds potential for studying translation and signaling processes. Multi-dimensional SDPM has been demonstrated using raster scanning [31] and full-field implementations [33]. In the former case, which the authors referred to as spectral domain optical coherence phase microscopy (SD-OCPM), galvanometer mirrors were used to scan the OCT probing beam in the lateral dimensions. The latter system utilized a two-dimensional CCD camera and a wavelength swept source to capture a full two-dimensional field in a single detector snapshot as shown in Fig. 8.13. Both techniques are capable of revealing the surface topology of cells.

While several initial demonstrations of multi-dimensional SDPM used static samples, one of the benefits of these systems is the ability to monitor dynamic processes occurring throughout in the cell over time. A demonstration of this capability was carried out by monitoring the contractions of spontaneously beating cardiomyocytes [43]. A raster scanning implementation was used to obtain cross-sectional images of beating cells, and the depth-resolved phase was monitored across the cell, shown in Fig. 8.14. The beating pattern at multiple locations in single cells as well as for clusters of cells was measured. The ability to measure localized Doppler flow from cytoplasmic streaming in *Amoeba proteus* was also demonstrated, shown in Fig. 8.15. Such studies demonstrate the potential to use this technique to detect the location of origin of biological activity in dynamic cellular processes.

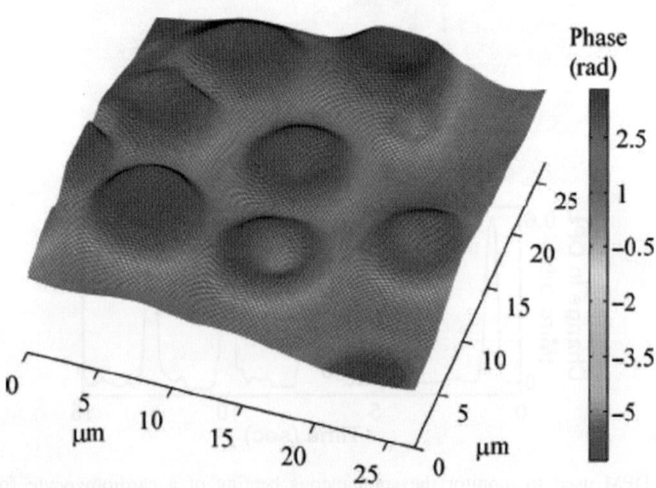

Fig. 8.13 Full-field SDPM image of red blood cells from [33]. Note the concave shape of the blood cells is clearly resolved

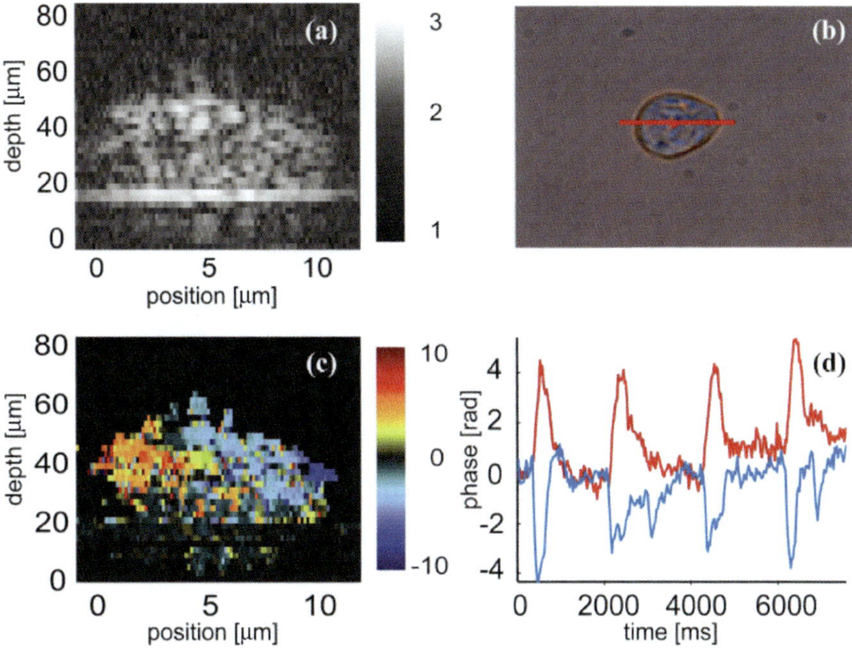

Fig. 8.14 Image of contracting cardiomyocyte from a chick embryo. (**a**) OCT intensity image of the cell resting on the surface of a coverslip. (**b**) Bright-field image of an isolated cell. *Red* line indicates location of the cross-sectional images. (**c**) SDPM phase image of the cell during a contraction. Negative phase (*blue*) relates to motion of the cell toward the coverslip surface; positive phase (*yellow* and *red*) indicates motion away from the surface. (**d**) Phase trace over time showing the biphasic motion at two different sites within the cell. Image from [43]

8.4.2 Doppler Velocimetry of Cytoplasmic Streaming

Cytoplasmic streaming plays an important role in transport mechanisms on the cellular level. Movement of cytoplasm controls the motility of cells and is one mechanism by which metastasis occurs. By understanding the process of cell motility and studying the ability of cells to spread, one can gain insight into the progression of certain diseases. Nutrient transport across cell membranes is also an important process and can result in fluid flow across the cell membrane. Doppler velocimetry measurements may be made using SDPM to quantify the motion of cellular phenomena. This was demonstrated in *A. proteus* by monitoring cytoplasmic streaming in extending pseudopods of the organism [32], shown in Fig. 8.16.

An amoeba was monitored using bright-field microscopy concurrently with an SDPM microscope. The probing beam was directed onto the cytoplasmic channel of the organism and the phase at all depths was measured over time. The measured phase changes from sequential A-scans revealed the axial component of the velocity of the flow. As a pseudopod extended from the amoeba, a positive velocity was measured. Between 18 and 20 s after the start of data acquisition, a drop of $CaCl_2$

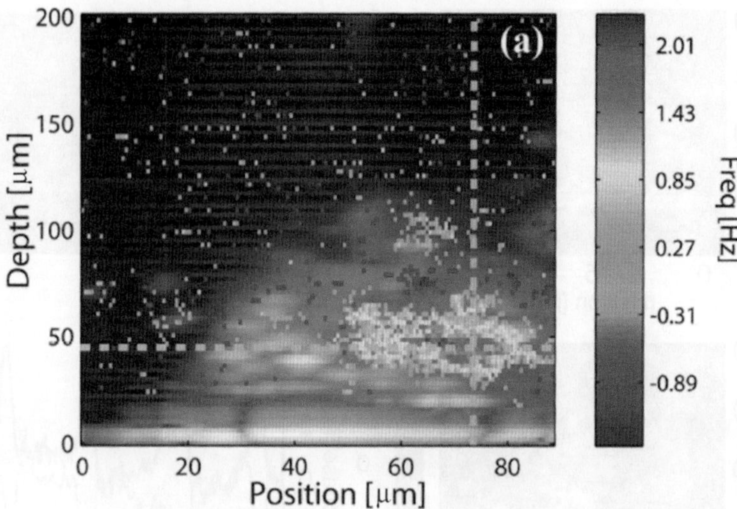

Fig. 8.15 Cross-sectional image of amoeba with the overlaid Doppler frequency shift. The gray-scale intensity image reveals the structure of the amoeba resting on the coverslip surface. The colored phase data correlate with the velocity of the cytoplasm flowing within the cell. Image from [43]

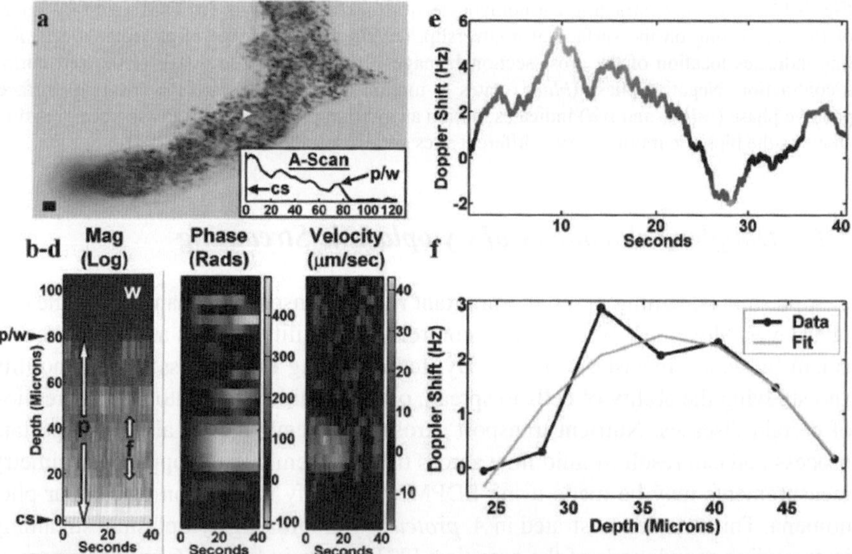

Fig. 8.16 Results from demonstration of SDPM Doppler flow in amoeba. (**a**) Bright-field image of an amoeboid. (**b–d**) Intensity, phase, and velocity maps of SDPM data taken at the point indicated by the *white triangle* in (**a**). Data reveal the velocity of the cytoplasmic flow. (**e**) Trace of the induced Doppler shift over time showing a directional change in cytoplasmic flow in response to the addition of salt to the environment. (**f**) Parabolic fit of the Doppler data obtained through the depth of the cell, suggesting a laminar flow profile as expected. Image from [32]

was added to the solution containing the amoeba that caused the pseudopod to retract and the phase to become negative. Thus, the velocity of the fluid flow was measured with high sensitivity, and changes in the directionality of the flow in response to environmental conditions were also detected. Figure 8.16f indicates that a laminar flow profile was detected within the cytoplasmic channel.

8.4.3 Cellular Cytoskeletal Rheology

Phase information has also been used to measure the mechanical properties of cells. Describing the behavior of the cytoskeletal structure poses an important biological problem for understanding the basic functionality of the cell. Mechanical modeling of cellular structure has traditionally been difficult due to its complex biological properties as well as differences between cell types. SDPM has been used to quantify the mechanical strength of the cellular membrane in single cells and could be useful for further studies on cellular rheology [42]. The cellular response was fitted to exponential and power law models to describe the behavior of the membrane under applied force.

Figure 8.17 shows results from the experiment. Magnetic beads were attached to the membrane of MCF-7 human breast cancer cells. The cells were placed onto a

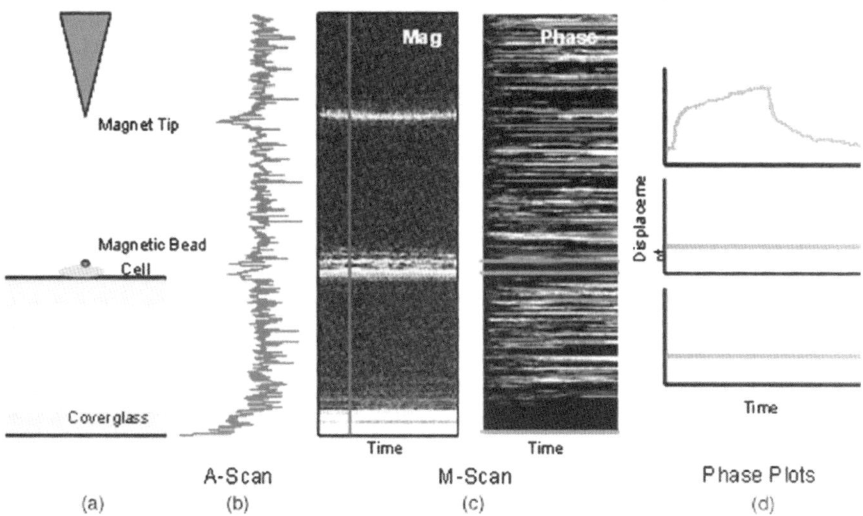

Fig. 8.17 Scheme and sample data from a cytoskeletal strength experiment. (**a**) The cell of interest with an attached magnetic bead rests on the top surface of a glass coverslip. A magnetic tip is suspended above the cell and used to pull on the bead. The SDPM probing beam was incident on the cell from below; the bottom coverslip surface provided the reference reflection. (**b**) Sample A-scan data showing peaks corresponding to the location of the top surface of the coverslip and the cell as well as the tip of the magnet. (**c**) Sample A-scan data acquired over time (M-scan) showing the cross-sectional magnitude and phase data acquired at a single lateral position on the cell. (**d**) Phase plots showing depth-resolved phase changes at depths corresponding to the cell, the top coverslip surface, and the bottom coverslip surface (from *top* to *bottom*). Image from [42]

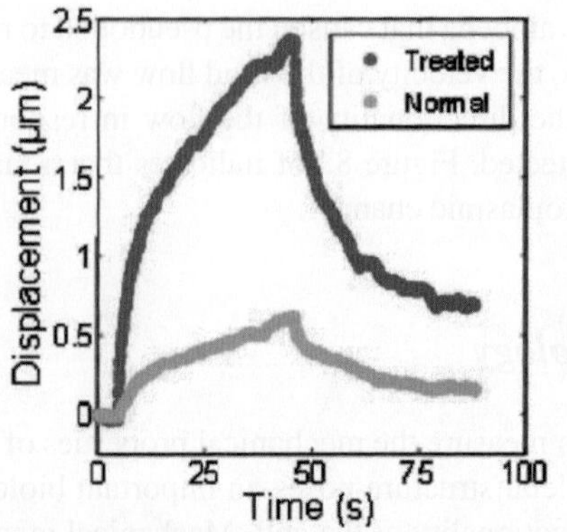

Displacement after 40 s

	Normal	Treated
Avg.	0.57 μm	2.40 μm
Std. Dev.	0.35 μm	1.86 μm

Fig. 8.18 Cell surface displacement measured from phase changes comparing normal MCF-7 cells to cells treated with cytochalasin D. Treated cells show greater displacement due to the disruption of their actin fiber networks and reduced cytoskeletal elasticity. Image from [42]

glass coverslip, and an electromagnet was used to pull the attached beads. SDPM was used to monitor the change in phase at a constant location through the cell as the beads were pulled. The measured changes in the optical phase captured with SDPM correlated to physical displacement of the membrane surface. Cells were then treated with cytochalasin D (Cyto-D), which caused the actin fibers throughout the cell to depolymerize, weakening the structural integrity of the cell membrane. The cell surface notably displaced more, demonstrating reduced cytoskeletal elasticity. Results in Fig. 8.18 from SDPM measurements revealed an increase in the distance moved by the membranes of the Cyto-D-treated cells when a constant force was applied through the electromagnet.

8.4.4 Multi-modal Phase Imaging with Molecular Specificity

Phase-based OCT can also be combined with other imaging techniques to obtain information with enhanced contrast and specificity. Phase changes in OCT arise due to changes in either the index of refraction or the structural size of a given sample. The molecular origins of these signal changes, however, cannot be detected as the resolution of OCT systems is inadequate for monitoring the activity of such particles. A combined SDPM (also called SD-OCPM) and multi-photon microscope was developed to add the ability to distinguish specific molecular targets within a cell [44]. Green fluorescent protein was used to label the actin filaments in skin fibroblast cells. Simultaneous imaging with two-photon fluorescence and phase imaging was carried out. While the fluorescence revealed the structure of the actin filaments, correlating with the detected phase in the OCT signal, the phase measurement also revealed the location of the cell nucleus, which could not be observed in the multiphoton image. This system demonstrated the ability to combine imaging modalities to obtain complementary information.

8.4.5 Action Potential Propagation

Increasing interest has grown in the use of optical signals to detect neural activity in biological specimens. Optical signaling provides an intrinsic contrast mechanism through which dynamic electrical activity can be recorded. When an action potential occurs, changes to the neuron structure due to ionic flux and membrane depolarization may cause cell swelling or localized refractive index changes. These changes yield optical signaling changes that may correspond to electrical activity in the cell. Such optical signals give rise to detectable optical path length changes on the order of nanometers for light reflected or transmitted through the nerve. Non-invasive detection of electrical activity holds great potential for studying nervous activity without the need for exogenous contrast agents or contact methods that could affect biological behavior.

The phase sensitivity of common-path techniques enables their use in monitoring optical responses during electrical nervous activity. Furthermore, FDOCT setups detect light reflected from a sample, which holds potential for in vivo applications. The standard FDOCT setup in Fig. 8.19 was used to investigate the potential of using

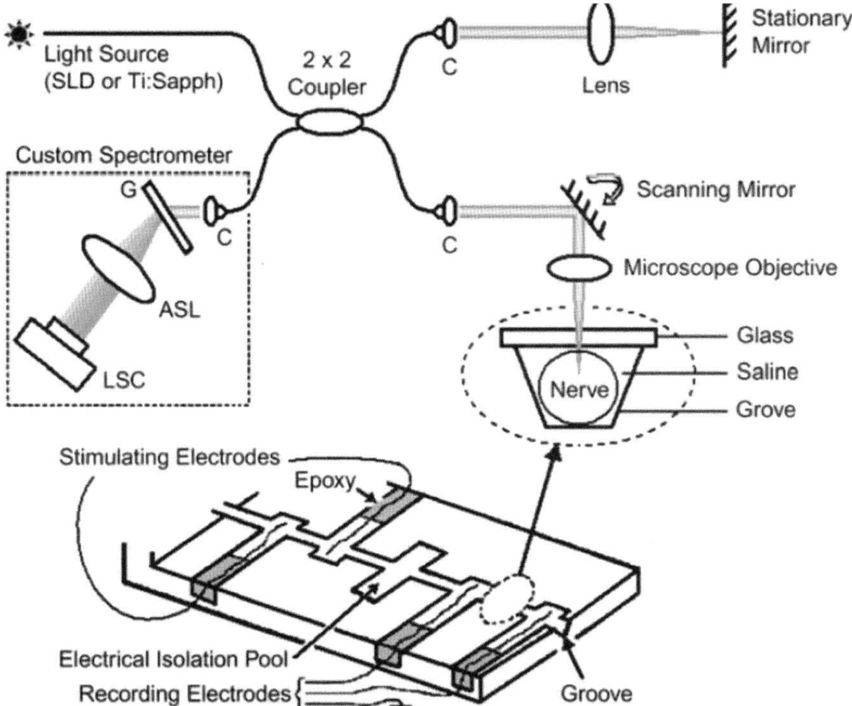

Fig. 8.19 SDOCT system for imaging action potential propagation in isolated nerves. The nerve chamber is shown with stimulation and recording electrodes. Stable phase measurements are made by using the phase of the glass/saline interface as a reference. C: collimating lens, ASL: air-spaced lenses, LSC: line-scan camera, SLD: superluminescent diode. Image from [34]

self-phase referencing for studying the propagation of action potentials in nervous tissue [34]. The reflection from the glass saline interface provided a stable phase reference that could be subtracted from the phase measured at other depths through the nerve. The authors applied electrical stimulation to an isolated crayfish leg nerve as well as lobster leg nerve bundles to induce an action potential. Recordings from electrodes revealed the electrical activity of the action potential.

Results in Fig. 8.20 show that optical path length changes between 1 and 2 nm were measured on a millisecond timescale, corresponding well to values obtained using other optical techniques [45, 46]. Slight differences were present between the electrical and optical recordings. Such discrepancies may be attributed to differences in the nature of the signals detected optically and electrically. While the electrical signal is the summed response of the depolarization of all fibers in a nerve bundle,

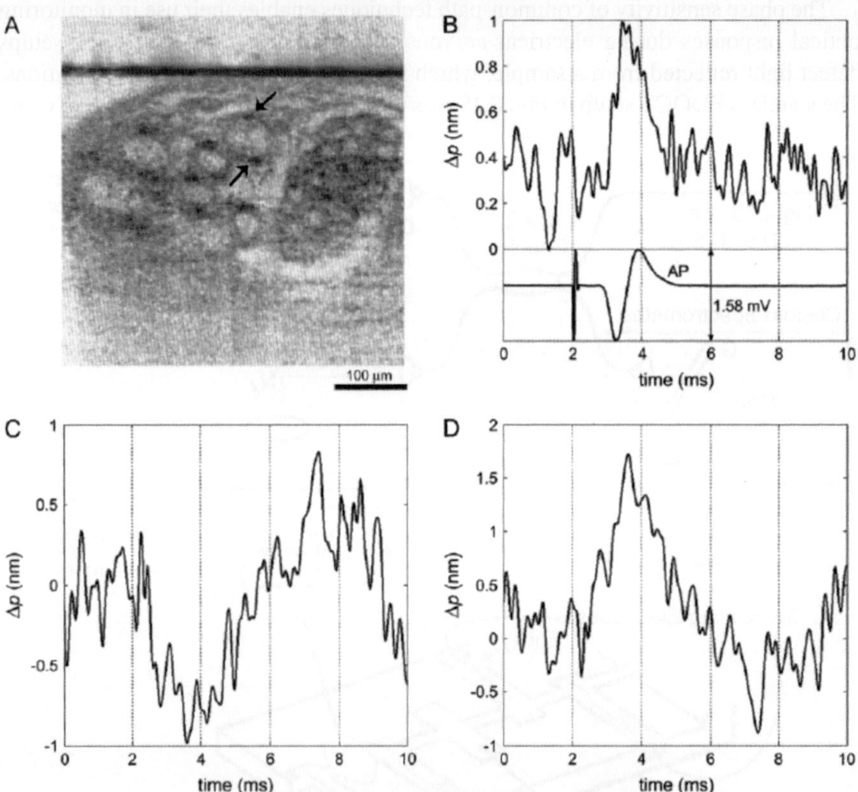

Fig. 8.20 Action potential measurement from lobster leg nerve. (**a**) Magnitude cross-sectional image of a nerve bundle. Dark *horizontal line* indicates interface of coverslide glass and saline solution. *Arrows* indicate positions at which phase measurements were made. (**b**) Optical path length change as calculated from the phase of the *top arrow* in (**a**). Electrical recording shown below. (**c**) Optical path length change calculated from the phase profile of the *bottom arrow* in (**a**). (**d**) Difference in optical path length between (**b**) and (**c**). All traces are averaged over 100 stimulation trials. Image from [34]

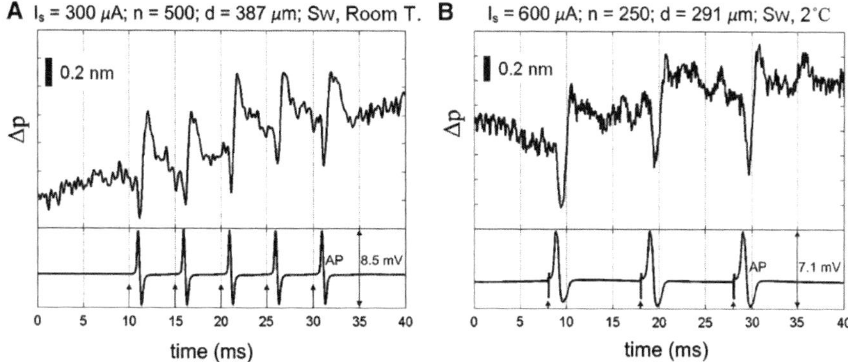

Fig. 8.21 Results from self-phase-referenced SDOCT measurements from a squid giant axon. (**a**) Measured displacement of nerve tissue during repeated stimulation. (**b**) Measured displacement for an axon when cooled to 2°C and repeatedly stimulated. Data from [47]

the optical signal originates from a single location within the nerve and is sensitive to changes in optical path length.

A further study revealed the ability to detect the dependence of action potential propagation on environmental conditions [47]. A similar setup to that shown above was used to detect optical signaling from action potential propagation in a squid giant axon. Based on the phase difference between the top and bottom surface of the axon membrane, the authors observed changes in membrane thickness that varied from 1 to 20 nm depending on temperature and ionic concentrations of the solution in which the axon was placed. Increasing salt concentration was found to increase the magnitude of the optical response. Cooling the sample led to an increase in the magnitude of the signal measured relating to the change in membrane size caused by cell swelling. Example results are shown in Fig. 8.21. These studies demonstrate the possibility of using phase-based OCT implementations for performing detection of fast, nanoscale biological processes.

8.5 Conclusion

Phase microscopy techniques have grown in importance as biological studies have evolved to probe cellular dynamics and structure. SDPM is one technique that has been developed to enhance these studies, providing non-invasive, depth-resolved, real-time phase measurements that can quantitatively be correlated to changes in the index of refraction or the structural properties of the cell.

Future work on SDPM will involve the application of faster data acquisition techniques to image phenomena occurring on faster time scales. Faster imaging will also help to eliminate phase wrapping artifacts and will enable volume acquisitions to capture complete cellular dynamics. Techniques to enhance lateral resolution with higher NA objectives will be helpful for increasing the utility of nanometer-scale phase measurements.

Acknowledgments We would like to acknowledge the work and contributions of past and present graduate students and post-doctoral members of the Izatt Biophotonics Laboratory at Duke University. We would like to acknowledge Brad Bower, Michael Choma, Al-Hafeez Dhalla, Ryan McNabb, Justin Migacz, Neal Shepherd, Melissa Skala, Yuankai Tao, and Mingtao Zhao. Portions of this work were supported by NIH grants R01 HL715015, P01 HL36059, R21-RR019769, R21-EB006338, R24-EB00243, RR019769 and the Center for Biomolecular and Tissue Engineering at Duke University.

References

1. O.W. Richards, Phase difference microscopy. Nature **154**, 672 (1944)
2. F. Zernike, How I discovered phase contrast. Science **121**, 345–349 (1955)
3. C. Preza, D.L. Snyder, J.-A. Conchello, Theoretical development and experimental evaluation of imaging models for differential-interference contrast microscopy. J. Opt. Soc. Am. A **16**, 2185–2199 (1999)
4. K. Creath, Phase-shifting speckle interferometry. Appl. Opt. **24**, 3053–3058 (1985)
5. E. Cuche, F. Bevilacqua, C. Depeursinge, Digital holography for quantitative phase-contrast imaging. Opt. Lett. **24**, 291–293 (1999)
6. C.J. Mann, L. Yu, C.-M. Lo, M.K. Kim, High-resolution quantitative phase-contrast microscopy by digital holography. Opt. Express **13**, 8693–8698 (2005)
7. G. Popescu, L.P. Deflores, J.C. Vaughan, K. Badizadegan, H. Iwai, R.R. Dasari, M.S. Feld, Fourier phase microscopy for investigation of biological structures and dynamics. Opt. Lett. **29**, 2503–2505 (2004)
8. N. Lue, W. Choi, G. Popescu, T. Ikeda, R.R. Dasari, K. Badizadegan, M.S. Feld, Quantitative phase imaging of live cells using fast Fourier phase microscopy. Appl. Opt. **46**, 1836–1842 (2007)
9. T. Ikeda, G. Popescu, R.R. Dasari, M.S. Feld, Hilbert phase microscopy for investigating fast dynamics in transparent systems. Opt. Lett. **30**, 1165–1167 (2005)
10. N. Lue, W. Choi, K. Badizadegan, R.R. Dasari, M.S. Feld, G. Popescu, Confocal diffraction phase microscopy of live cells. Opt. Lett. **33**, 2074–2076 (2008)
11. D. Huang, E.A. Swanson, C.P. Lin, J.S. Schuman, W.G. Stinson, W. Chang, M.R. Hee, T. Flotte, K. Gregory, C.A. Puliafito, J.G. Fujimoto, Opt. Coherence Tomogr. Sci. **254**, 1178–1181 (1991)
12. C.A. Toth, D.G. Narayan, S.A. Boppart, M.R. Hee, J.G. Fujimoto, R. Birngruber, C.P. Cain, C.D. DiCarlo, W.P. Roach, A comparison of retinal morphology viewed by optical coherence tomography and by light microscopy. Arch. Ophthalmol. **115**, 1425–1428 (1997)
13. W. Drexler, J.G. Fujimoto, State-of-the-art retinal optical coherence tomography. Prog. Retin. Eye Res. **27**, 45–88 (2008)
14. A.F. Fercher, W. Drexler, C.K. Hitzenberger, T. Lasser, Optical coherence tomography – principles and applications. Rep. Prog. Phys. **66**, 239–303 (2003)
15. J.A. Izatt, M.A. Choma, in *Optical Coherence Tomography: Technology and Applications*, eds. by W. Drexler, J.G. Fujimoto. Theory of optical coherence tomography (Springer, Heidelberg, 2008)
16. R. Leitgeb, C.K. Hitzenberger, A.F. Fercher, Performance of Fourier domain vs. time domain optical coherence tomography. Opt. Express **11**, 889–894 (2003)
17. M.A. Choma, M.V. Sarunic, C. Yang, J.A. Izatt, Sensitivity advantage of swept source and Fourier domain optical coherence tomography. Opt. Express **11**, 2183–2189 (2003)
18. J.F.d. Boer, B. Cense, B.H. Park, M.C. Pierce, G.J. Tearney, B.B. Bouma, Improved signal-to-noise ratio in spectral-domain compared with time-domain optical coherence tomography. Opt. Lett. **28**, 2067–2069 (2003)
19. Z. Hu, A.M. Rollins, Fourier domain optical coherence tomography with a linear-in-wavenumber spectrometer. Opt. Lett. **32**, 3525–3527 (2007)

20. C.M. Eigenwillig, B.R. Biedermann, G. Palte, R. Huber, K-space linear Fourier domain mode locked laser and applications for optical coherence tomography. Opt. Express **16**, 8916–8937 (2008)
21. Y.K. Tao, M. Zhao, J.A. Izatt, High-speed complex conjugate resolved retinal spectral domain optical coherence tomography using sinusoidal phase modulation. Opt. Lett. **32**, 2918–2920 (2007)
22. Y. Yasuno, S. Makita, T. Endo, G. Aoki, M. Itoh, T. Yatagai, Simultaneous B-M-mode scanning method for real-time full-range Fourier domain optical coherence tomography. Appl. Opt. **45**, 1861–1865 (2006)
23. M.V. Sarunic, B.E. Applegate, J.A. Izatt, Real-time quadrature projection complex conjugate resolved Fourier domain optical coherence tomography. Opt. Lett. **31**, 2426–2428 (2006)
24. A.M. Davis, M.A. Choma, J.A. Izatt, Heterodyne swept-source optical coherence tomography for complete complex conjugate ambiguity removal. J. Biomed. Opt. **10**(6), 064005 (2005)
25. R.K. Wang, In vivo full range complex Fourier domain optical coherence tomography. Appl. Phys. Lett. **90**(5), 054103 (2007)
26. L. Wang, Y. Wang, S. Guo, J. Zhang, M. Bachman, G.P. Li, Z. Chen, Frequency domain phase-resolved optical Doppler and Doppler variance tomography. Opt. Commun. **242**, 345–350 (2004)
27. J.A. Izatt, M.D. Kulkarni, S. Yazdanfar, J.K. Barton, A.J. Welch, In vivo bidirectional color Doppler flow imaging of picoliter blood volumes using optical coherence tomography. Opt. Lett. **22**, 1439–1441 (1997)
28. J.F.d. Boer, T.E. Milner, M.J.C.v. Gemert, J.S. Nelson, Two-dimensional birefringence imaging in biological tissue by polarization-sensitive optical coherence tomography. Opt. Lett. **22**, 934–936 (1997)
29. C. Yang, A. Wax, M.S. Hahn, K. Badizadegan, R.R. Dasari, M.S. Feld, Phase-referenced interferometer with subwavelength and subhertz sensitivity applied to the study of cell membrane dynamics. Opt. Lett. **26**, 1271–1273 (2001)
30. M.A. Choma, A.K. Ellerbee, C. Yang, T.L. Creazzo, J.A. Izatt, Spectral-domain phase microscopy. Opt. Lett. **30**, 1162–1164 (2005)
31. C. Joo, T. Akkin, B. Cense, B.H. Park, J.F.d. Boer, Spectral-domain optical coherence phase microscopy for quantitative phase-contrast imaging. Opt. Lett. **30**, 2131–2133 (2005)
32. M.A. Choma, A.K. Ellerbee, S. Yazdanfar, J.A. Izatt, Doppler flow imaging of cytoplasmic streaming using spectral domain phase microscopy. J. Biomed. Opt. **11**(2), 024014 (2006)
33. M.V. Sarunic, S. Weinberg, J.A. Izatt, Full-field swept source phase microscopy. Opt. Lett. **31**, 1462–1464 (2006)
34. T. Akkin, C. Joo, J.F.d. Boer, Depth-resolved measurement of transient structural changes during action potential propagation. Biophys. J. **93**, 1347–1353 (2007)
35. Z. Yaqoob, W. Choi, S. Oh, N. Lue, Y. Park, C. Fang-Yen, R.R. Dasari, K. Badizadegan, M.S. Feld, Improved phase sensitivity in spectral domain phase microscopy using line-field illumination and self phase-referencing. Opt. Express **17**, 10681–10687 (2009)
36. M. Gu, C.J.R. Sheppard, X. Gan, Image formation in a fiber-optical confocal scanning microscope. J. Opt. Soc. Am. A **8**, 1755–1761 (1991)
37. A.K. Ellerbee, J.A. Izatt, Phase retrieval in low-coherence interferometric microscopy. Opt. Lett. **32**, 388–390 (2007)
38. D.C. Ghiglia, M.D. Pritt, *Two-Dimensional Phase Unwrapping: Theory, Algorithms, and Software* (Wiley, New York, NY, 1998)
39. C.R. Tilford, Analytical procedure for determining lengths from fractional fringes. Appl. Opt. **16**, 1857–1860 (1977)
40. Y.-Y. Cheng, J.C. Wyant, Two-wavelength phase shifting interferometry. Appl. Opt. **23**, 4539–4543 (1984)
41. H.C. Hendargo, M. Zhao, N. Shepherd, J.A. Izatt, Synthetic wavelength based phase unwrapping in spectral domain optical coherence tomography. Opt. Express **17**, 5039–5051 (2009)

42. E.J. McDowell, A.K. Ellerbee, M.A. Choma, B.E. Applegate, J.A. Izatt, Spectral domain phase microscopy for local measurements of cytoskeletal rheology in single cells. J. Biomed. Opt. **12**(4), 044008 (2007)
43. A.K. Ellerbee, T.L. Creazzo, J.A. Izatt, Investigating nanoscale cellular dynamics with cross-sectional spectral domain phase microscopy. Opt. Express **15**, 8115–8124 (2007)
44. C. Joo, K.H. Kim, J.F.d. Boer, Spectral-domain optical coherence phase and multiphoton microscopy. Opt. Lett. **32**, 623–625 (2007)
45. T. Akkin, D.P. Dave, T.E. Milner, H.G. Rylander III, Detection of neural activity using phase-sensitive optical low-coherence reflectometry. Opt. Express **12**, 2377–2386 (2004)
46. C. Fang-Yen, M.C. Chu, H.S. Seung, R.R. Dasari, M.S. Feld, Noncontact measurement of nerve displacement during action potential with a dual-beam low-coherence interferometer. Opt. Lett. **29**, 2028–2030 (2004)
47. T. Akkin, D. Landowne, A. Sivaprakasam, Optical coherence tomography phase measurement of transient changes in squid giant axons during activity. J. Membr. Biol. **231**, 35–46 (2009)

Chapter 9
Coherent Light Imaging and Scattering for Biological Investigations

Huafeng Ding and Gabriel Popescu

Abstract Quantitative phase imaging (QPI) of live cells has received significant scientific interest over the past decade or so, mainly because it offers structure and dynamics information at the *nanometer scale* in a completely noninvasive manner. Fourier transform light scattering (FTLS) relies on quantifying the optical phase and amplitude associated with a coherent image field and propagating it numerically to the scattering plane. It combines optical microscopy, holography, and light scattering for studying inhomogeneous and dynamic media. We present recent developments of QPI technology and FTLS for biological system structure and dynamics study. Their applications are classified into *static* and *dynamic* according to their temporal selectivity. Several promising prospects are discussed in the summary section.

9.1 Introduction

Phase contrast (PC) and differential interference contrast (DIC) microscopy have been used extensively to infer morphometric features of live cells without the need for exogenous contrast agents [1]. These techniques transfer the information encoded in the phase of the imaging field into the intensity distribution of the final image. Thus, the optical phase shift through a given sample can be regarded as a powerful endogenous contrast agent, as it contains information about both the thickness and the refractive index of the sample. However, both PC and DIC are *qualitative* in terms of optical path length measurement, i.e., the relationship between the irradiance and the phase of the image field is generally nonlinear [2, 3].

Quantifying the optical phase shifts associated with cells gives access to information about morphology and dynamics at the *nanometer scale*. Over the past decade, the development of quantitative phase imaging techniques has received increased scientific interest. Full-field phase measurement techniques provide simultaneous

G. Popescu (✉)
Quantitative Light Imaging Laboratory, Department of Electrical and Computer Engineering, Beckman Institute for Advanced Science and Technology, University of Illinois at Urbana-Champaign, Urbana, IL 61801, USA
e-mail: gpopescu@uiuc.edu

P. Ferraro et al. (eds.), *Coherent Light Microscopy*, Springer Series in Surface Sciences 46, DOI 10.1007/978-3-642-15813-1_9,

information from the whole image field on the sample, which has the benefit of studying both the temporal and the spatial behavior of the biological system under investigation [4–22]. With the recent advances in two-dimensional array detectors, full-field phase images can now be acquired at high speeds (i.e., thousands of frames per second). We review the recent developments in *quantitative phase imaging* (*QPI*) and its applications to studying biological structures and dynamics.

Light scattering has emerged as an important approach in the field for studying biological samples, as it is noninvasive, requires minimum sample preparation, and extracts rich information about morphology and dynamic activity [23–31]. In light scattering studies, by measuring the angular distribution of the scattered field, one can infer quantitative information about the sample structure (i.e., its spatial distribution of refractive index). Light scattering by cells and tissues evolved as a dynamic area of study, especially because this type of investigation can potentially offer a noninvasive window into function and pathology [32–35]. We show that because of the phase information, QPI is equivalent to a very sensitive light scattering measurement, thus bridging the fields of imaging and scattering. This new approach to light scattering, called Fourier transform light scattering (FTLS), is the *spatial* analog to Fourier transform spectroscopy.

In this chapter, we review the main QPI methods and scattering measurements by FTLS, which have been recently reported in the literature.

9.2 Quantitative Phase Imaging (QPI)

Recently, new quantitative phase imaging techniques have been recently developed for spatially resolved investigation of biological structures. Combining phase-shifting interferometry with Horn microscopy, DRIMAPS (digitally recorded interference microscopy with automatic phase shifting) has been proposed as a new technique for quantitative biology [4, 36]. This *quantitative phase imaging* technique has been successfully used for measuring cell spreading [5], cell motility [6], and cell growth and dry mass [37]. A full-field quantitative phase microscopy method was developed also by using the transport-of-irradiance equation [38, 39]. The technique is inherently stable against phase noise because it does not require using two separate beams as in typical interferometry experiments. This approach requires, however, recording images of the sample displaced through the focus and subsequently solving numerically partial differential equations.

Digital holography has been developed a few decades ago [40] as a technique that combines digital recording with traditional holography [41]. Typically, the phase and amplitude of the imaging field are measured at an out-of-focus plane. By solving numerically the Fresnel propagation equation, one can determine the field distribution at various planes. For optically thin objects, this method allows for reconstructing the in-focus field and, thus, retrieving the phase map characterizing the sample under investigation. This method has been implemented in combination with phase-shifting interferometry [42]. More recently, digital holography has been adapted for quantitative phase imaging of cells [11, 43, 44].

In recent years, new full-field quantitative phase imaging techniques have been developed for studying live cells. The advance of Fourier phase microscopy (FPM) [13, 45], Hilbert phase microscopy (HPM) [14, 15], and diffraction phase microscopy (DPM) [16, 46] came in response to the need for high phase stability over broad temporal scales. The principles of operation of these techniques and their applications for cell biology are described below.

9.2.1 Fourier Phase Microscopy (FPM)

FPM combines the principles of phase contrast microscopy and phase-shifting interferometry, such that the scattered and unscattered light from a sample are used as the object and reference fields of an interferometer, respectively. More details of this experiment are presented elsewhere [13]. Here we present a brief description of the experimental setup depicted in Fig. 9.1. The collimated low coherence field from a superluminescent diode (SLD, center wavelength 809 nm and bandwidth 20 nm) is used as the illumination source for a typical inverted microscope. Through the video port, the microscope produces a magnified image positioned at the image plane IP. The lens L_1 is positioned at the same plane IP and has a focal length such that it collimates the zero spatial frequency field. The Fourier transform of the image field is projected by the lens L_2 (50 cm focal distance) onto the surface of a programmable phase modulator (PPM; Hamamatsu Photonics, model X8267). This PPM consists of an optically addressed, two-dimensional liquid crystal array with 768 × 768 active pixels. The polarizer P adjusts the field polarization in a direction parallel to the axis of the liquid crystal. In this configuration, the PPM produces precise control over the phase of the light reflected by its surface. The PPM pixel size is $26 \times 26\,\mu\text{m}^2$, whereas the dynamic range of the phase control is 8 bits over 2π. In the absence of PPM modulation, an exact phase and amplitude replica of the image field is formed at the CCD plane, via the beam splitter BS_1. For alignment purposes, a camera is used to image the surface of the PPM via the beam splitter BS_2. The PPM is used to controllably shift the phase of the scattered field component U_1 (dotted line) in four successive increments of $\pi/2$ with respect to the average field U_0 (solid line), as in typical phase-shifting interferometry measurements [47]. The phase difference between U_1 and U_0 is obtained by combining four recorded interferograms as follows:

$$\Delta\varphi(x, y) = \tan^{-1}\left[\frac{I(x, y; 3\pi/2) - I(x, y; \pi/2)}{I(x, y; 0) - I(x, y; \pi)}\right] \tag{9.1}$$

where $I(x, y; \alpha)$ represents the irradiance distribution of the interferogram corresponding to the phase shift α. If we define $\beta(x, y) = |U_1(x, y)| / |U_0|$, then the phase associated with the image field $U(x, y)$ can be determined

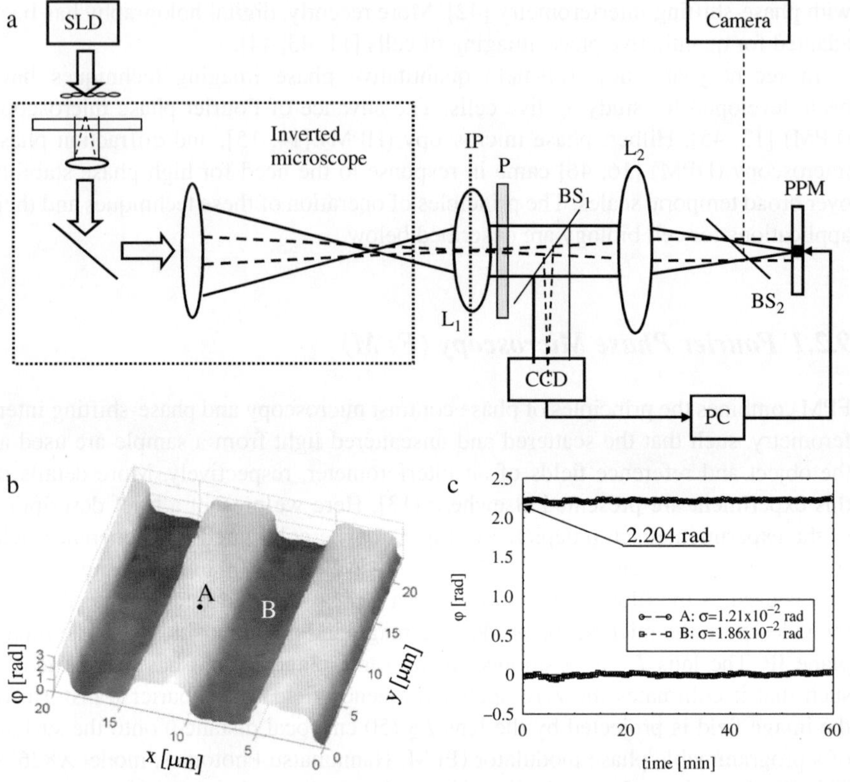

Fig. 9.1 (**a**) FPM experimental setup. (**b**) Quantitative phase image of a phase grating. (**c**) Temporal fluctuations of the path lengths associated with points *A* and *B* on the grating in (**b**)

$$\phi(x, y) = \tan^{-1}\left[\frac{\beta(x, y)\sin(\Delta\phi(x, y))}{1 + \beta(x, y)\cos(\Delta\phi(x, y))}\right] \tag{9.2}$$

The amplitude ratio β contained in (9.2) can be obtained from the four frames, taking into account that $\beta_{\phi\to 0} = 0$ [13]. The phase image retrieval rate is limited by the refresh rate of the liquid crystal PPM, which in our case is 8 Hz. However, this acquisition rate is not limited in principle and can be further improved using a faster phase shifter. In fact, we recently improved the data acquisition by approximately two orders of magnitude [45].

We employed the procedure presented here to experimentally determine the spatial phase modifications of a field propagating through various transparent media. Figure 9.1b shows an example of such measurement, obtained for a transmission phase grating. Using a 40 × (NA = 0.65) microscope objective, we retrieved the spatially varying phase delay induced by this grating, which is made of glass with the refractive index $n = 1.51$. The profile of the grating was measured by stylus profilometry, and the height was found to be 570 ± 10 nm while its pitch had a value of 4 μm. This corresponds to a phase profile of height $\varphi = 2.217 \pm 0.039$ rad. As can be seen in Fig. 9.1b, the measurement correctly recovers the expected phase distribution. Figure 9.1c shows the values of the reconstructed phase associated with the points A and B indicated in Fig. 9.1b, as a function of time. The phase values are averaged over an area that corresponds to $0.6 \times 0.6\,\mu\text{m}^2$ in the sample plane, which is approximately the diffraction limit of the microscope. The values of the standard deviation associated with the two points are 18 and 12 mrad, respectively, which demonstrate the significant stability of the technique in the absence of active stabilization. Interestingly, the phase stability of the measurement is actually better when wet samples are studied [48].

9.2.2 Hilbert Phase Microscopy (HPM)

Hilbert phase microscopy (HPM) extends the concept of complex analytic signals to the spatial domain and measures quantitative phase images from only one spatial interferogram recording [14, 15]. Due to its single-shot nature, the HPM acquisition time is limited only by the recording device and thus can be used to accurately quantify nanometer-level path length shifts at the millisecond time scales or less, where many relevant biological phenomena develop. The experimental setup is shown in Fig. 9.2a. A HeNe laser ($\lambda = 632$ nm) is coupled into a 1×2 single mode fiber optic coupler and collimated on each of the two outputs. One output field acts as the illumination field for an inverted microscope equipped with a 100× objective. All the optical fibers are fixed to minimize phase noise. The tube lens is such that the image of the sample is formed at the CCD plane via the beam splitter cube. The second fiber coupler output is collimated and expanded by a telescopic system consisting of another microscope objective and the tube lens. This reference beam can be approximated by a plane wave, which interferes with the image field. The reference field is tilted with respect to the sample field such that uniform fringes are created at an angle of 45° with respect to x- and y-axes. The CCD used (C7770; Hamamatsu Photonics) has an acquisition rate of 291 frames/s at the full resolution of 640×480 pixels, at 1–1.5 ms exposure time. The fringes are sampled by 6 pixels per period. The spatial irradiance associated with the interferogram across one direction is given by

$$I(x) = I_{\mathrm{R}} + I_{\mathrm{S}}(x) + 2\sqrt{I_{\mathrm{R}} I_{\mathrm{S}}(x)} \cos\left[qx + \phi(x)\right] \tag{9.3a}$$

where I_{R} and I_{s} are, respectively, the reference and sample irradiance distributions, q is the spatial frequency of the fringes, and ϕ is the spatially varying phase associated with the object, the quantity of interest in our experiments. Using high-pass spatial filtering and Hilbert transformation, the quantity ϕ is retrieved in each point of the single-exposure image [14].

To exemplify the ability of the new instrument to perform live-cell dynamic morphometry at the millisecond and nanometer scales, we obtained time-resolved HPM

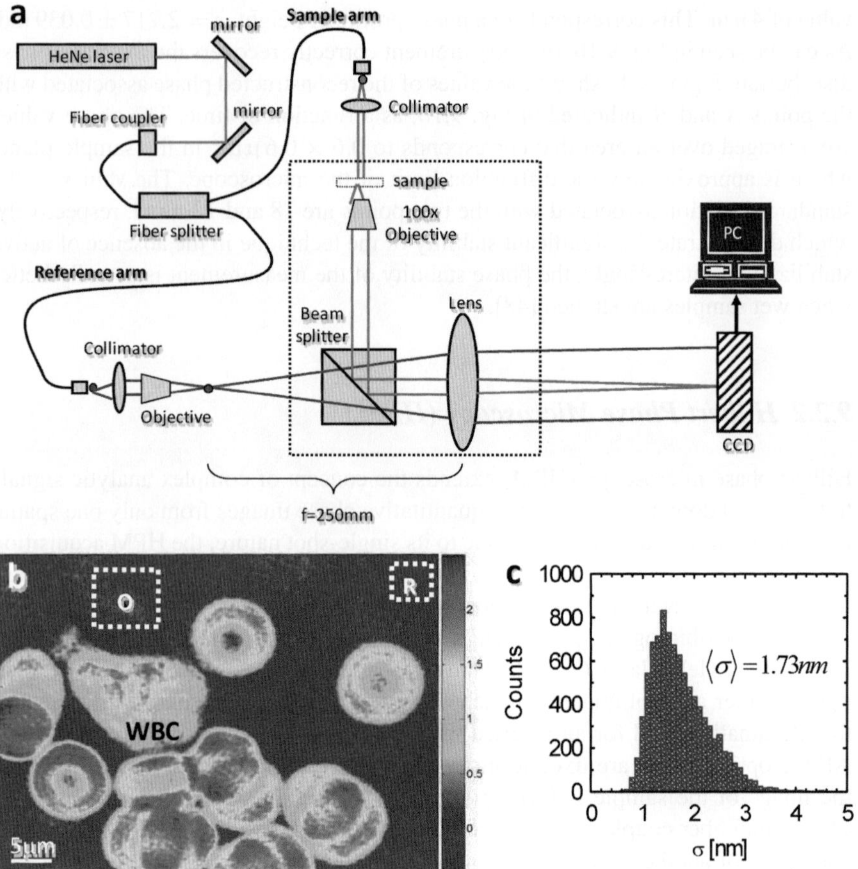

Fig. 9.2 (**a**) HPM experimental setup. (**b**) HPM image of a droplet of blood. (**c**) The histogram of standard deviations associated with a region in the field of view containing no cells. Adapted with permission from [15] © SPIE 2005

images of red blood cells (RBCs). Droplets of whole blood were simply sandwiched between coverslips, with no additional preparation. Figure 9.2b shows a quantitative phase image of live blood cells; both isolated and agglomerated erythrocytes are easily identifiable. A white blood cell (WBC) is also present in the field of view. Using the refractive index of the cell and surrounding plasma of 1.40 and 1.34, respectively [49], the phase information associated with the RBCs is translated into nanometer scale image of the cell topography. The assumption of optical homogeneity of RBC is commonly used [50, 51] and justified by the knowledge that cellular content consists mainly of hemoglobin solution. In order to eliminate the longitudinal noise between successive frames, each phase image was referenced to the average value across an area in the field of view containing no cells (denoted in Fig. 9.2b by R). To quantify the residual noise of the instrument in a spatially relevant way, we recorded sets of 1000 images, acquired at 10.3 ms each and analyzed the path length fluctuations of individual points within a 100 × 100 pixel area (denoted in Fig. 9.2b by O). The path length associated with each point in O was averaged over 5 × 5 pixels, which approximately corresponds to the dimensions of the diffraction limit spot. The histogram of the standard deviations associated with all the spots within region O is shown in Fig. 9.2c. The average value of this histogram is indicated. This noise assessment demonstrates that our HPM instrument is capable of providing quantitative information about structure and dynamics of biological systems, such as RBCs, at the nanometer scale. Recently, an active feedback loop has been added to the HPM system, which further improved the stability of the instrument [52].

9.2.3 Diffraction Phase Microscopy (DPM)

Diffraction phase microscopy (DPM) is a novel quantitative phase imaging technique that combines the single-shot feature of HPM with the common path geometry associated with Fourier phase microscopy [13]. As a result, DPM is characterized by the significant stability of the common path interferometers, while operating at high acquisition speeds, limited only by the detector. The experimental setup is shown in Fig. 9.3a. The second harmonic radiation of a Nd:YAG laser ($\lambda = 532$ nm) was used as illumination for an inverted microscope (Axiovert 35; Carl Zeiss Inc.), which produces the magnified image of the sample at the output port. The microscope image appears to be illuminated by a virtual source point VPS. A relay lens RL was used to collimate the light originating at VPS and replicate the microscope image at the plane IP. A phase grating G is placed at this image plane, which generates multiple diffraction orders containing full spatial information about the image. The goal is to select two diffraction orders (0th and 1st) that can be further used as reference and sample fields, respectively, as in Mach–Zehnder interferometer geometries. In order to accomplish this, a standard spatial filtering lens system L_1–L_2 is used to select the two diffraction orders and generate the final interferogram at the CCD plane. The 0th order beam is low-pass filtered using the spatial filter SF positioned in the Fourier plane of L_1, such that at the CCD plane it approaches a uniform field. The spatial filter allows passing the entire frequency content of the 1st diffraction order beam and blocks all the other orders. The 1st order is thus the imaging field, and the 0th order plays the role of the reference field. The two beams traverse the same optical components, i.e., they propagate along a common optical path, thus significantly reducing the longitudinal phase noise. The direction of the spatial modulation was chosen at an angle of 45° with respect to the x- and y-axes of the CCD, such that the total field at the CCD plane has the form

$$E(x, y) = |E_0|\, e^{i[\phi_0+\beta(x+y)]} + |E_1(x, y)|\, e^{i\phi_1(x,y)} \tag{9.3b}$$

In (9.3b), $\left|E_{0,1}\right|$ and $\varphi_{0,1}$ are the amplitudes and the phase, respectively, of the orders of diffraction 0, 1, while β represents the spatial frequency shift induced by the grating to the 0th order. Note that, as a consequence of the central ordinate theorem, the reference field is proportional to the spatial average of the microscope image field

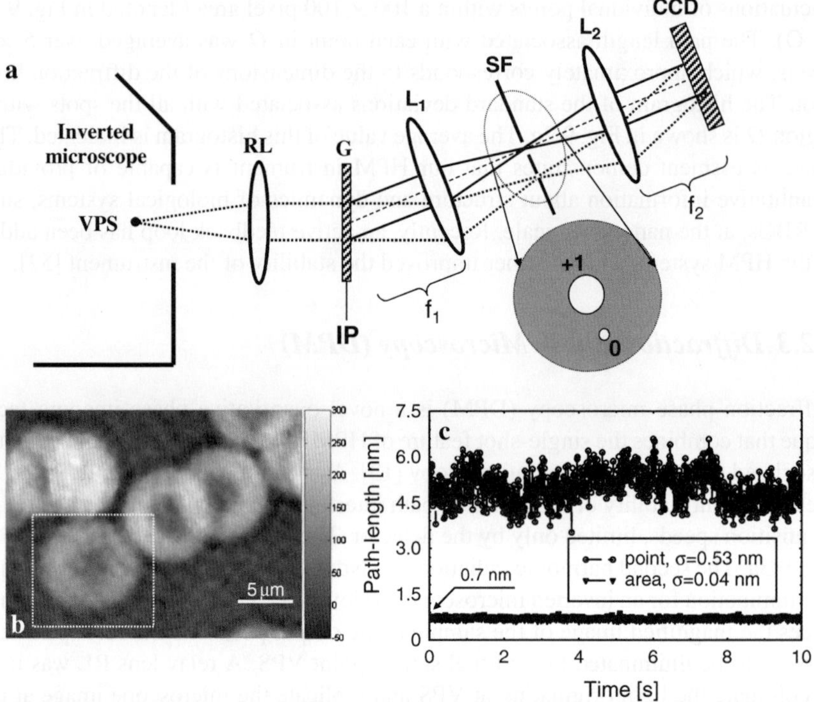

Fig. 9.3 (**a**) DPM experimental setup. (**b**) DPM image of a blood droplet. (**c**) Temporal path length fluctuations associated with a point and area. Adapted with permission from [16] © OSA 2006

$$|E_0|\,\mathrm{e}^{\mathrm{i}\phi_0} \propto \frac{1}{A}\int |E(x,y)|\,\mathrm{e}^{\mathrm{i}\phi(x,y)}\mathrm{d}x\mathrm{d}y \tag{9.4}$$

where A is the total image area. The spatial average of an image field has been successfully used before as a stable reference for extracting spatially resolved phase information [13].

The CCD (C7770; Hamamatsu Photonics) has an acquisition rate of 291 frames/s at the full resolution of 640×480 pixels. To preserve the transverse resolution of the microscope, the spatial frequency β is chosen to match or exceed the maximum frequency allowed by the numerical aperture of the instrument. Throughout our experiments, the microscope was equipped with a 40× (0.65 NA) objective, which is characterized by a diffraction-limited resolution of 0.4 μm. The microscope–relay lens combination produces a magnification of about 100, thus the diffraction spot at the grating plane has a size of approximately 40 μm. The grating pitch is 20 μm, which allows taking advantage of the full resolution given by the microscope objective. The L_1–L_2 lens system has an additional magnification of $f_2/f_1 = 3$, such that the sinusoidal modulation of the image is sampled by 6 CCD pixels per period. The spatially resolved quantitative phase image associated with the sample is retrieved from a single CCD recording via a spatial Hilbert transform, as described in Sect. 9.2.2. and in [14].

In order to demonstrate the inherent stability of the system and ability to image live cells, we imaged droplets of whole blood sandwiched between coverslips, with no additional preparation. Figure 9.3b shows a quantitative phase image of live blood cells, where the normal, discocyte shape can be observed. To quantify the stability of the DPM instrument and thus the sensitivity of cell topography to dynamical changes, we recorded sets of 1000 *no-sample* images, acquired at 10.3 ms each and performed noise analysis on both single points and entire field of view. The *spatial* standard deviation of the path length associated with the full field of view had a *temporal* average of 0.7 nm and a *temporal* standard deviation of 0.04 nm, as shown in Fig. 9.3c. Also shown in Fig. 9.3c is the temporal path length trace of an arbitrary point (3×3 pixel average), characterized by a standard deviation of 0.53 nm. Thus, DPM provides quantitative phase images which are inherently stable to the level of sub-nanometer optical path length and at an acquisition speed limited only by the detector. Recently, DPM has been combined with epifluorescence microscopy to simultaneously image, for the first time, both the nanoscale structure and dynamics and the specific functional information in live cells [46].

9.2.4 *Applications of QPI*

The full-field quantitative phase imaging techniques presented here are suitable for all cell visualization and morphometric applications associated with traditional microscopy techniques, e.g., bright field, dark field, phase contrast and Nomarski/DIC. In the following, we present several cell biology applications that are specific to quantitative phase imaging methods. The classification is with respect to the time scale of cell imaging.

9.2.4.1 Red Blood Cell Volumetry

Phase contrast microscopy [50] and reflection interference contrast microscopy [51] have been used previously to measure dynamic changes in RBC shape. Such nanoscale topographic information offers insight into the biophysical properties and health state of the cell. However, these methods are not inherently *quantitative*. Thus, 3D quantitative erythrocyte shape measurements have been limited to atomic force and scanning electron microscopy. Nevertheless, due to the heavy sample preparation required by these methods prior to imaging, their applicability to studying cells in physiological conditions has been limited [53, 54].

Mature erythrocytes represent a very particular type of structure; they lack nuclei and organelles and thus can be modeled as optically homogeneous objects, i.e., they produce local optical phase shifts that are proportional to their thickness. Therefore, measuring quantitative phase images of red blood cells provides cell thickness profiles with an accuracy that corresponds to a very small fraction of the optical wavelength. Using the refractive index of the cell and surrounding plasma of 1.40

and 1.34, respectively [49], the phase information associated with the RBCs can be easily translated into nanometer scale image of the cell topography. An example of nanometer scale topography of red blood cell quantified by Hilbert phase microscopy is shown in Fig. 9.4a. The thickness profile of the cell, $h(x, y)$, relates to the measured phase, $\phi(x, y)$, as $h(x, y) = (\lambda/2\pi\Delta n)\phi(x, y)$, where Δn is the refractive index contrast between the hemoglobin contained in the cell and the surrounding fluid (plasma). The cell thickness profile obtained by this method is shown in Fig 9.4b. The volume of individual cells can be measured from the HPM data as $V = \int h(x, y)\mathrm{d}x\mathrm{d}y$. Figure 9.4c depicts the volume of RBC measured by HPM during spontaneous hemolysis, i.e., after the membrane ruptured and the cell started to lose hemoglobin. This result demonstrates the ability of quantitative phase imaging to provide RBC volumetry without the need for preparation. It represents a significant advance with respect to current techniques that require the cells to be

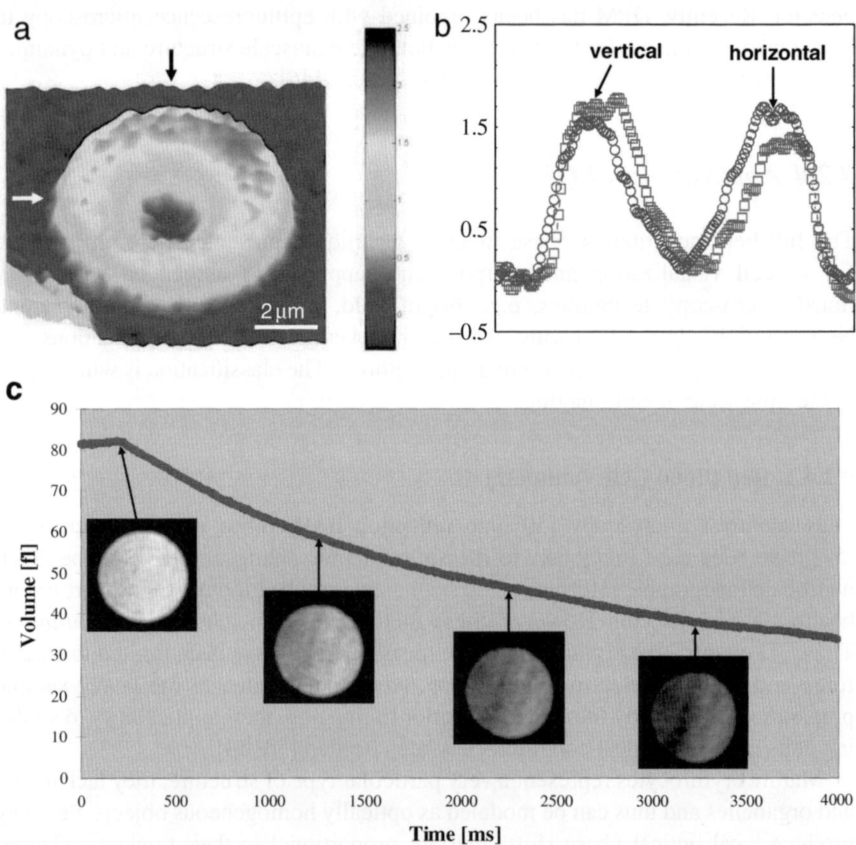

Fig. 9.4 (**a**) HPM image of a normal RBC. (**b**) Cell profile across two directions. (**c**) Cell volume as a function of time during hemolysis that starts at approximately $t = 300$ ms. Adapted with permission from [15] © SPIE 2005

prepared such that they assume spherical shapes [55]. The errors in the volume measurement due to changes in refractive index are not likely to be significant. Spatial inhomogeneities of the imaging fields may affect the accuracy of the volume measurement, and this influence was minimized by accurately spatially filtering and collimating the beams.

9.2.4.2 Cell Dry Mass

Several decades ago, it has been shown that the optical phase shift through the cells is a measure of the cellular *dry mass* content [56, 57]. Optical interferometry provides access to the phase information of a given transparent sample; the main challenge is to suppress the environmental noise, which hinders the ability to measure optical path length shifts quantitatively.

DRIMAPS employs Horn microscopy and phase-shifting interferometry to measure phase images from biological samples. The potential of DRIMAPS for studying cell growth has been demonstrated [6]. This technique, however, is not stable against phase noise, which limits its applicability to studying cell dynamics. Although other quantitative phase imaging techniques have been reported [7, 58], their potential for analysis of cellular dry mass has not been evaluated, to our knowledge.

In order to quantify cell dry mass, we employed Fourier phase microscopy described in Sect. 9.2.1. It was shown that FPM provides quantitative phase images of live cells with high transverse resolution and low noise over extended periods of time. The general expression for the spatially resolved quantitative phase images obtained from a cell sample is given by

$$\varphi(x, y) = \frac{2\pi}{\lambda} \int_0^{h(x,y)} \left[n_c^z(x, y, z) - n_0\right] \mathrm{d}z \tag{9.5}$$

In (9.5), λ is the wavelength of light, h is the local thickness of the cell, and n_0 is the refractive index of the surrounding liquid. The quantity n_c^z is the refractive index of cellular material, which is generally an inhomogeneous function in all three dimensions. Without loss of generality, (9.5) can be rewritten in terms of an axially averaged refractive index n_c, as

$$\varphi(x, y) = \frac{2\pi}{\lambda} \left[n_c(x, y) - n_0\right] h(x, y) \tag{9.6}$$

However, it has been shown that the refractive properties of a cell composed mainly of protein has, to a good approximation, a simple dependence on protein concentration [56, 57]

$$n_c(x, y) = n_0 + \alpha C(x, y) \tag{9.7}$$

In (9.7), α is referred to as the refraction increment (units of ml/g) and C is the concentration of dry protein in the solution (in g/ml). Using this relationship, the dry mass surface density σ of the cellular matter is obtained from the measured phase map as

$$\sigma(x, y) = \frac{\lambda}{2\pi\alpha}\varphi(x, y) \tag{9.8}$$

In order to illustrate the potential of FPM for measuring the dry mass distribution of live cells, we used the FPM instrument for imaging confluent monolayers of HeLa cells [59]. Figure 9.5 shows an example of the dry mass density distribution σ (units of pg/μm^2) obtained from nearly confluent HeLa cells. In applying (9.8), we used $\alpha = 0.2\,\text{ml/g}$ for the refraction increment, which corresponds to an average of reported values [56]. Quantitative information about the dry mass of the cells thus allows investigation of cell movement, growth, or shape change in a totally noninvasive manner. In order to quantify the phase stability of the instrument, we recorded 240 phase images over 2 h from a cell sample that contained regions with no cells. We measured the path length standard deviation of each pixel within a $15 \times 15\,\mu\text{m}^2$ region. The average of these standard deviations had a value of 0.75 nm, which indicates that the sensitivity for changes in dry mass surface density has a value of $3.75\,\text{fg}/\mu\text{m}^2$.

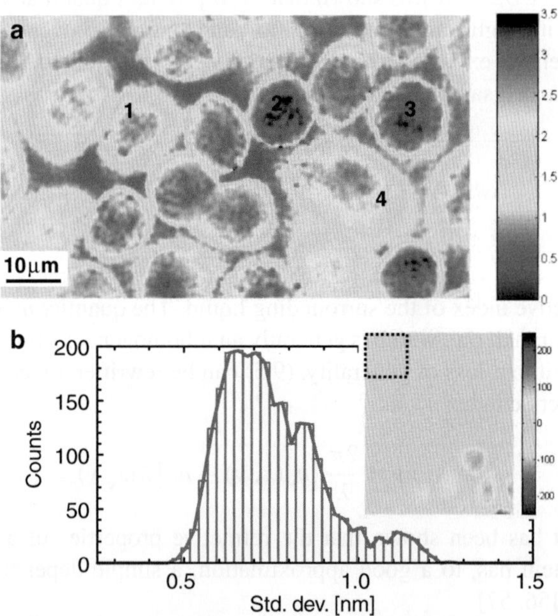

Fig. 9.5 (**a**) Dry mass density distribution $\sigma(x, y)$ obtained using FPM. The *color bar* has units of picograms/μm^2. (**b**) Histogram of the path length standard deviation corresponding to the pixels within the $15 \times 15\,\mu\text{m}^2$ selected area shown in *inset*. The *color bar* of the *inset* indicates optical path length in units of nanometers, which sets the ultimate sensitivity to dry mass changes to 4 fg/μm^2. Adapted with permission from [59] © APS 2008

9.2.4.3 Cell Growth

Quantitative knowledge about this phenomenon can provide information about cell cycling, functioning, and disease [60, 61]. We employed Fourier phase microscopy (FPM) to quantify the changes in dry mass of HeLa cells in culture. Data containing time-resolved FPM images were acquired at a rate of four frames a minute over periods of up to 12 h, and the dry mass surface density information was extracted as presented above. Each cell fully contained in the field of view was segmented using a MATLAB program based on iterative image thresholding and binary dilation, which was developed in our laboratory. Figure 9.6a shows the segmented images of the four cells shown in Fig. 9.5. We monitored the total dry mass of each cell over a period of 2 h. The results are summarized in Fig. 9.6b. Cell 4 exhibits linear growth, as does cell 1, although it is reduced by a factor of almost 5. In contrast, cell 2 shows virtually no change in mass, while cell 3 appears to exhibit a slight oscillatory behavior, the origin of which is not clearly understood. These results demonstrate the capability of FPM to quantify small changes in cell mass and therefore monitor in detail the evolution of cell cycle and its relationship with function.

9.2.4.4 Cell Membrane Fluctuations

Because RBCs have a relatively simple structure [62, 63], they represent a convenient model for studying cell membranes, which have broad applications in both science and technology [64, 65]. The lipid bilayer is 4–5 nm thick and exhibits

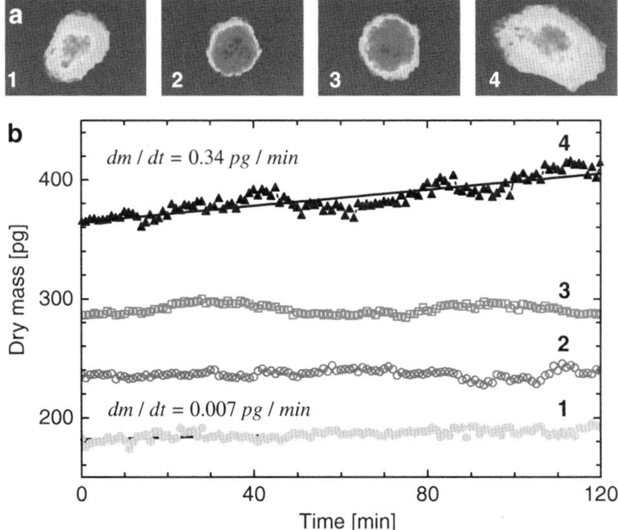

Fig. 9.6 (**a**) Images of the cells segmented from Fig. 9.1, as indicated. (**b**) Temporal evolution of the total dry mass content for each cell. The *solid lines* for cells 1 and 4 indicate fits with a linear function. The values of the slope obtained from the fit are indicated. Adapted with permission from [59] © APS 2008

fluid-like behavior, characterized by a finite bending modulus κ and a vanishing shear modulus $\mu \approx 0$. The resistance to shear, crucial for RBC function, is provided by the spectrin network, which has a mesh size of ~80 nm. Spontaneous membrane fluctuations, or "flickering," have been modeled theoretically under both static and dynamic conditions in an attempt to connect the statistical properties of the membrane displacements to relevant mechanical properties of the cell [50, 66–69]. These thermally induced membrane motions exhibit 100 nm scale amplitudes at frequencies of tens of Hertz. In past studies, measurements of the membrane mean squared displacement vs. spatial wave vector, $\Delta u^2(q)$, revealed a q^{-4} dependence predicted by the equipartition theorem, which is indicative of fluid-like behavior [50, 51, 70–72]. These results conflict with the static deformation measurements provided by micropipette aspiration [73, 74], high-frequency electric fields [75, 76], and, more recently, optical tweezers [77], which indicate an average value for the shear elasticity of the order of $\mu \sim 10^{-6}\,\mathrm{J/m^2}$. Gov et al. predicted that the cytoskeleton pinning of the membrane has an overall effect of confining the fluctuations and, thus, gives rise to superficial tension much larger than in the case of free bilayers [66]. This confinement model may offer new insight into the cytoskeleton–bilayer interaction that determines the morphology and physiology of the cell [78].

Existing optical methods for studying RBC dynamics, including phase contrast microscopy (PCM) [50], reflection interference contrast microscopy (RICM) [51], and fluorescence interference contrast (FLIC) [79], are limited in their ability to measure cell membrane displacements. It is well known that PCM provides phase shifts quantitatively only for samples that are optically much thinner than the wavelength of light, which is a condition hardly satisfied by any cell type. Similarly, a single RICM measurement cannot provide the absolute cell thickness unless additional measurements or approximations are made [80]. FLIC relies on inferring the absolute position of fluorescent dye molecules attached to the membrane from the absolute fluorescence intensity, which may limit both the sensitivity and the acquisition rate of the technique [79]. Thus, none of these techniques is suitable for making spatially resolved measurements of the dynamics of cell membrane fluctuations and testing the hypothesis of Gov et al.

We performed highly sensitive experimental measurements of thermal fluctuations associated with RBCs under different morphological conditions. The results reveal the effect of the cytoskeleton on the RBC fluctuations and support the model proposed by Gov et al. In order to quantify membrane fluctuations at the nanometer and millisecond scales with high transverse resolution, we developed a new quantitative phase imaging technique. The method combines Hilbert phase microscopy (HPM) [14, 15] with an electronic stabilization feedback loop and is referred to as the stabilized Hilbert phase microscopy (sHPM) [52].

Our samples were primarily not only composed of RBCs with typical discocytic shapes but also contained cells with abnormal morphology which formed spontaneously in the suspension, such as echinocytes, with a spiculated shape, and spherocytes, approaching a spherical shape. By taking into account the free energy contributions of both the bilayer and the cytoskeleton, these morphological changes have been successfully modeled [78]. Figure. 9.7a–c shows typical sHPM

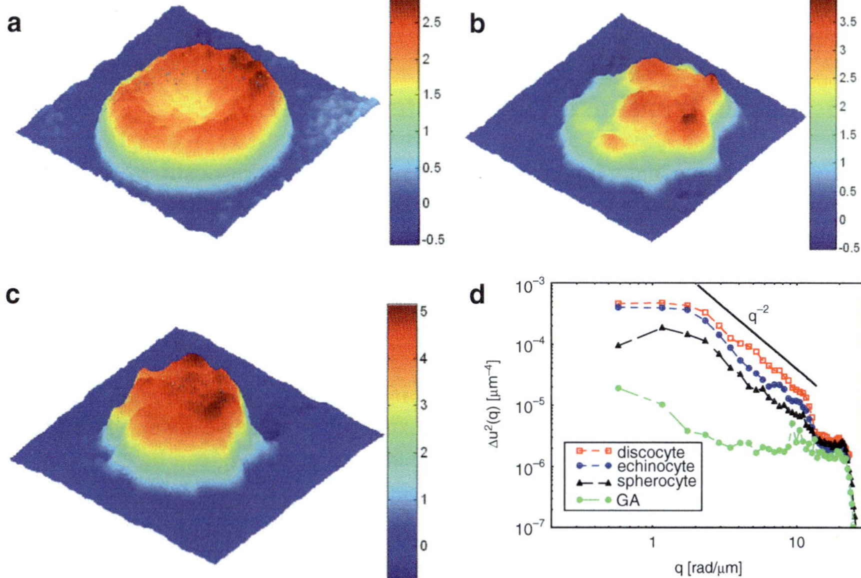

Fig. 9.7 sHPM images of a discocyte (**a**), echinocyte (**b**), and spherocyte (**c**). The *color bar* shows thickness in microns. (**d**) Mean squared displacements for the three RBC groups and for the gluteraldehyde (GA)-fixed cells. Adapted with permission from [52] © APS 2006

images of cells in these three groups. For comparison, we also analyzed the motions of RBCs fixed with 40 μM gluteraldehyde, using a standard procedure [81]. The resultant mean squared displacements, $\Delta u^2(q)$, for each group of four to five cells are summarized in Fig. 9.7d. The fixed cells show significantly diminished fluctuations, as expected. The curves associated with the three untreated RBC groups exhibit a power law behavior with an exponent $\alpha = 2$. As in the case of vesicles, this dependence is an indication of tension; however, the RBC tension is determined by the confinement of the bilayer by the cytoskeleton [66, 82]. Based on this model, we fitted the data to extract the tension coefficient for each individual cell. The average values obtained for the discocytes, echinocytes, and spherocytes are, respectively, $\sigma = (1.5 \pm 0.2) \cdot 10^{-6}\mathrm{J / m^2}$, $\sigma = (4.05 \pm 1.1) \cdot 10^{-6}\mathrm{J / m^2}$, and $\sigma = (8.25 \pm 1.6) \cdot 10^{-6}\mathrm{J / m^2}$. The tension coefficient of red blood cells is 4–24 times larger than what we measured for vesicles, which suggests that the contribution of the cytoskeleton might be responsible for this enhancement. Further, it is known that the cytoskeleton plays a role in the transitions from a normal red blood cell shape to abnormal morphology, such as echinocyte and spherocyte [78]. Therefore, the consistent increase in tension we measured for the discocyte–echinocyte–spherocyte transition can be explained by changes in the cytoskeleton, which pins the bilayer. These findings support the hypothesis that the fluctuations are laterally confined by a characteristic length, $\xi_0 = 2\pi\sqrt{\kappa/\sigma}$, which is much smaller than the cell size [66]. Compared to other optical techniques used for studying membrane fluctuations, the sHPM technique used here is quantitative in terms of membrane topography and displacements, highly sensitive to the nanoscale membrane motions, and provides high transverse resolution.

9.3 Fourier Transform Light Scattering (FTLS)

Light scattering techniques provide information that is intrinsically averaged over the measurement volume. Thus, the spatial resolution is compromised, and the scattering contributions from individual component are averaged. Particle tracking microrheology has been recently proposed to measure the particle displacements in the imaging (rather than scattering) plane [83, 84], in which the spatial resolution is reserved. However, the drawback is that relatively large particles are needed such that they can be tracked individually, which also limits the throughput required for significant statistical average. As discussed in the previous section, *phase-sensitive* methods have been employed to directly extract the refractive index of cells and tissues [85, 86]. From this information, the angular scattering can be achieved via the Born approximation [87].

With consideration of the limitations of the techniques above, Fourier transform light scattering (FTLS) was recently developed as an approach to study light scattering from biological samples based on DPM [16]. FTLS combines the high spatial resolution associated with optical microscopy and intrinsic averaging of light scattering techniques [88]. DPM is stable in optical path length to the sub-nanometer level due to its common path interferometric geometry. This feature allows FTLS perform studies on static and dynamic samples with extraordinary sensitivity.

9.3.1 Principle of FTLS

FTLS system requires accurate phase retrieval for elastic light scattering (ELS) measurements and, in addition, fast acquisition speed for DLS studies [88]. Figure 9.8

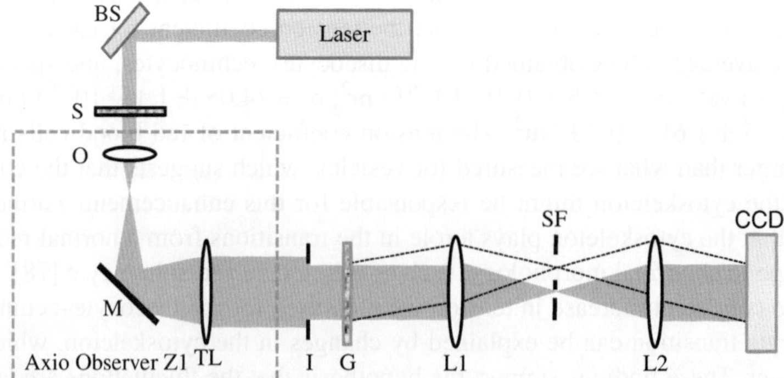

Fig. 9.8 FTLS experimental setup. BS, beam splitter; *S*, sample; *O*, objective lens; *M*, mirror; TL, tube lens; *I*, iris; *G*, grating; SF, spatial filter; L_1 and L_2, lenses. Adapted with permission from [35] © OSA 2009

depicts our experimental setup that satisfies these requirements by incorporating a common path interferometer (DPM) with a commercial computer-controlled microscope. The phase retrieval principle and operation for DPM were described in Sect. 9.2.3. In this FTLS setup, the two beams propagate along a common optical path, thus significantly reducing the longitudinal phase noise. The direction of the spatial modulation is along the x-axis, such that the total field at the CCD plane has the form [16]

$$U(x, y) = |U_0|\, e^{i(\phi_0+\beta x)} + |U_1(x, y)|\, e^{i\phi_1(x,y)} \tag{9.9}$$

In (9.9), $|U_{0,1}|$ and $\phi_{0,1}$ are the amplitudes and the phases, respectively, of the orders of diffraction 0, 1, while β represents the spatial frequency shift induced by the grating to the 0th order. To preserve the transverse resolution of the microscope, the spatial frequency β exceeds the maximum frequency allowed by the numerical aperture of the instrument. The L_1–L_2 lens system has an additional magnification of $f_2/f_1 = 5$, such that the sinusoidal modulation of the image is sampled by 4 CCD pixels per period. The obtained interferograms were used to calculate the phase information of the objects. The interferogram is spatially high-pass filtered to isolate the cross term

$$|U_0|\, |U_1(x, y)| \cos\left[\phi_1(x, y) - \phi_0 - \beta x\right] \tag{9.10}$$

which can be regarded as the real part of a complex analytic signal. The imaginary component, $\sin\left[\phi(x, y) - \phi_0 - \beta x\right]$, is obtained via a spatial Hilbert transform [14, 16, 52]. Thus, from a single CCD exposure, we obtain the spatially resolved phase and amplitude associated with the image field. From this image field information $\widetilde{U}$, the complex field can be numerically propagated at arbitrary planes; in particular, the far-field angular scattering distribution $\widetilde{U}$ can be obtained simply via a Fourier transformation [88]

$$\widetilde{U}(\mathbf{q}, t) = \int U(\mathbf{r}, t) e^{-i\mathbf{q}\cdot\mathbf{r}} d^2\mathbf{r} \tag{9.11}$$

With time lapse image acquisition, the temporal scattering signals are recorded and the sampling frequency is only limited by the speed of the camera. The power spectrum is obtained through Fourier transform of this time-resolved scattering signals.

9.3.2 Applications of FTLS

FTLS provides high-sensitivity light scattering study by taking optical phase and amplitude measurements at the image plane. Due to the interferometric experimental geometry and the reliable phase retrieval, spatial resolution of the scatterer positions is well preserved for FTLS. FTLS has been applied to study the tissue optical properties, cell type characterization, dynamic fluctuations of cell membrane, and cell actin dynamics.

9.3.2.1 Elastic (Static) FTLS Tissues

Upon propagation through inhomogeneous media such as tissues, optical fields suffer modifications in terms of irradiance, phase, spectrum, direction, polarization, and coherence, which can reveal information about the sample of interest. We use FTLS to extract quantitatively the scattering mean free path l_s and anisotropy factor g from tissue slices of different rat organs [35]. This direct measurement of tissue scattering parameters allows predicting the wave transport phenomena within the organ of interest at a multitude of scales. The scattering mean free path l_s was measured by quantifying the attenuation due to scattering for each slice via the Lambert–Beer law, $l_s = -d/\ln[I(d)/I_0]$, where d is the thickness of the tissue, $I(d)$ is the irradiance of the unscattered light after transmission through the tissue, and I_0 is the total irradiance, i.e., the sum of the scattered and unscattered components. The unscattered intensity $I(d)$, i.e., the spatial DC component, is evaluated by integrating the angular scattering over the diffraction spot around the origin. The resulting l_s values for 20 samples for each organ, from the same rat, are summarized in Fig. 9.9a.

The anisotropy factor g is defined as the average cosine of the scattering angle

$$g = \int_{-1}^{1} \cos(\theta) p[\cos(\theta)] d[\cos(\theta)] / \int_{-1}^{1} p[\cos(\theta)] d[\cos(\theta)] \tag{9.12}$$

where p is the normalized angular scattering, i.e., the phase function. Note that, since (9.11) applies to tissue slices of thickness $d < l_s$, it cannot be used directly in (9.12) to extract g since g values in this case will be thickness dependent. This is so because the calculation in (9.12) is defined over tissue of thickness $d = l_s$, which describes the average scattering properties of the tissue (i.e., independent of how the tissue is cut). Under the weakly scattering regime of interest here, this angular scattering distribution p is obtained by propagating the complex field numerically through $N = l_s/d$ layers of $d = 5\,\mu\text{m}$ thickness [35],

$$p(\mathbf{q}) \propto \left| \iint [U(\mathbf{r})]^N e^{i\mathbf{q}\cdot\mathbf{r}} d^2\mathbf{r} \right|^2 \tag{9.13}$$

Equation (9.13) applies to a slice of thickness l_s. It reflects that, by propagating through N weakly scattering layers of tissue, the total phase accumulation is the sum of the phase shifts from each layer, as is typically assumed in phase imaging of transparent structures [89]. The angular scattering distribution, or phase function, $p(\theta)$ is obtained by performing azimuthal averaging of the scattering map, $p(\mathbf{q})$, associated with each tissue sample. The maximum scattering angle was determined by the numeric aperture of the objective lens, and it is about 18° for our current

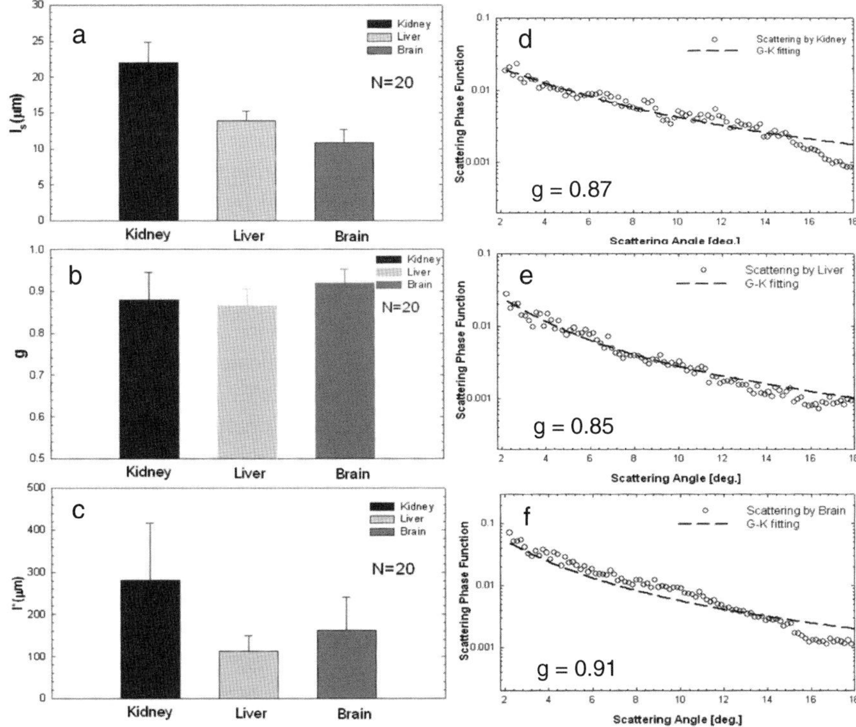

Fig. 9.9 FTLS measurements of the scattering mean free path l_s (**a**), anisotropy factors (**b**), and transport mean free path (**c**) for the three rat organs with 20 samples per group. The *error bars* correspond to the standard deviations ($N = 20$). (**d–f**) The typical angular scattering plots of different rat organs. The *dashed lines* indicate fits with the G–K phase function. Adapted with permission from [35] © OSA 2009

setup (10× objective applied for tissue study). The angular scattering data were further fitted with Gegenbauer Kernel (GK) phase function [90]

$$P(\theta) = ag \cdot \frac{(1-g^2)^{2a}}{\pi\left[1+g^2-2g\cos(\theta)\right]^{(a+1)}\left[(1+g)^{2a}-(1-g)^{2a}\right]} \tag{9.14}$$

Note that g can be estimated directly from the angular scattering data via its definition (9.12). However, because of the limited angular range measured, g tends to be overestimated by this method, and, thus, the GK fit offers a more reliable alternative than the widely used Henyey–Greenstein (HG) phase function with the parameter $a = 1/2$. The representative fitting plots for each sample are shown in Fig. 9.9d–f. The final values of g are included in Fig. 9.9b and agree very well with previous reports in the literature [91]. From these measurements of thin, singly scattering slices, we inferred the behavior of light transport in thick, strongly scattering tissue. Thus the transport mean free path, which is the renormalized scattering length to account for the anisotropic phase function, can be obtained as $l^* = l_s/(1-g)$. The l^* values for 20 samples from each organ are shown in Fig. 9.9c.

In order to extend the FTLS measurement toward extremely low scattering angles, we scanned large fields of view by tiling numerous high-resolution microscope images [88]. Figure 9.10a presents a quantitative phase map of a 5-μm-thick tissue slice obtained from the breast of a rat model by tiling ~1000 independent images. This 0.3-giga pixel composite image is rendered by scanning the sample with a 20 nm precision computerized translation stage. The phase function associated with this sample is shown in Fig. 9.10b. We believe that such a broad angular range, of almost three decades, is measured here for the first time and cannot be achieved via any single measurement. Notably, the behavior of the angular scattering follows power laws with different exponents, as indicated by the two dashed lines. This type of measurements over broad spatial scales may bring new light into unanswered questions, such as tissue architectural organization and possible self-similar behavior [92].

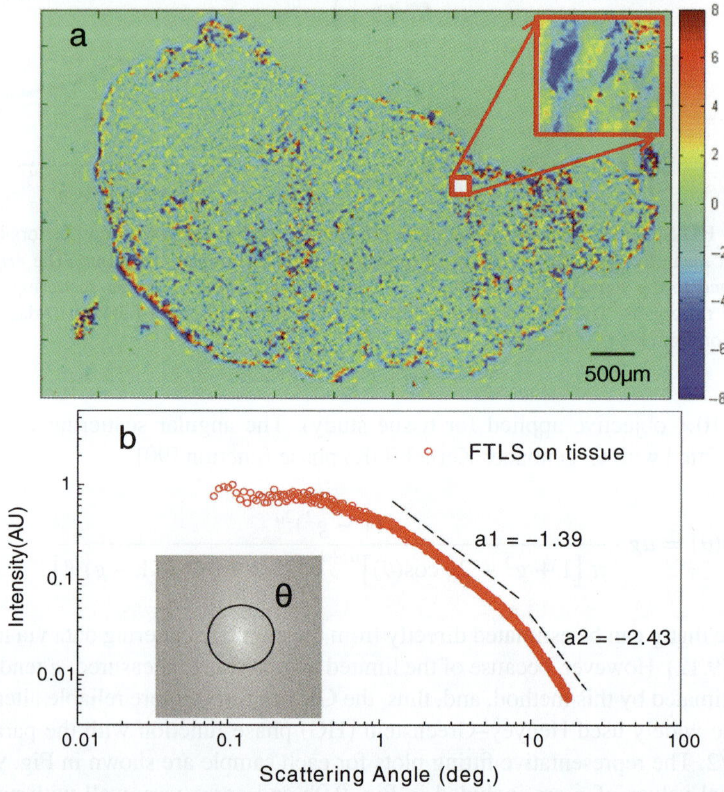

Fig. 9.10 **(a)** Terra-pixel quantitative phase image of a mouse breast tissue slice. *Color bar* indicates phase shift in radians. **(b)** Angular scattering from the tissue in a. The *inset* shows the 2D scattering map, where the average over each ring corresponds to a point in the angular scattering curve. The *dashed lines* indicate power laws of different exponents. Reprinted figure with permission from [88] © 2008 by the American Physical Society. http://prl.aps.org/abstract/PRL/v101/i23/e238102

The results above showed that FTLS can quantify the angular scattering properties of thin tissues, which thus provides the scattering mean free path l_s and anisotropy factor g for the macroscopic (bulk) organ. We note that, based on the knowledge of l_s, g, and l^*, one can predict the outcome of a broad range of scattering experiments on large samples (size $>> l^*$), via numerical solutions to the transport equation or analytical solutions to the diffusion equation. We envision that the FTLS measurements of unstained tissue biopsies, which are broadly available, will provide not only diagnosis value but also possibly the premise for a large scattering database, where various tissue types, healthy and diseased, will be fully characterized in terms of their scattering properties.

9.3.2.2 Elastic (Static) FTLS Cells

Light scattering investigations can noninvasively reveal subtle details about the structural organization of cells [29, 30, 93–96]. We employed FTLS to measure scattering phase functions of different cell types and demonstrate its capability as a new modality for cell characterization [97]. In order to demonstrate the potential of FTLS for cell sorting based on the rich scattering signals that it provides, we retrieved the scattering phase functions from three cell groups (Fig. 9.11a–c): red blood cells, myoblasts (C2C12), and neurons. Figure 9.11d–f shows the angular scattering distributions associated with these samples. For each group, we performed measurements on different fields of view. Remarkably, FTLS provides these scattering signals over approximately 35° (40× objective applied for cell study) in scattering angle and several decades in intensity. For comparison, we also measured the scattering signature of the background (i.e., culture medium with no cells in the field of view by using threshold), which incorporates noise contributions from the beam inhomogeneities, impurities on optics, and residues in the culture medium. These measurements demonstrate that FTLS is sensitive to the scattering signals from single cells, which contrast to previous measurements on cells in suspensions. Subtle details of the cell structures may be washed in studies on suspensions since the signals are averaged over various cell orientations.

We analyzed our FTLS data with a statistical algorithm based on the principal component analysis (PCA) aimed at maximizing the differences among the cell groups and providing an automatic means for cell sorting [98]. This statistical method mathematically transforms the data to a new coordinate system to illustrate the maximum variance by multiplying the data with the chosen individual vectors. Our procedure can be summarized as follows. First, we average the $n(n = 1, \ldots, 45)$ measurements for the three cell types (15 measurements per group), to obtain the average scattered intensity, $\overline{I(\theta_m)} = \frac{1}{45} \sum_{n=1,\ldots,45} I_n(\theta_m)$, with $m = 1, \ldots, 35$ denoting the number of scattering angles. Second, we generate a matrix ΔY_{nm} of variances, where n indexes the different measurements and m the scattering angles. The covariance matrix associated with ΔY, $\mathrm{Cov}(\Delta Y)$, is calculated and its eigenvalues and eigenvectors extracted. The three principal components are obtained by retaining three eigenvectors corresponding to the largest

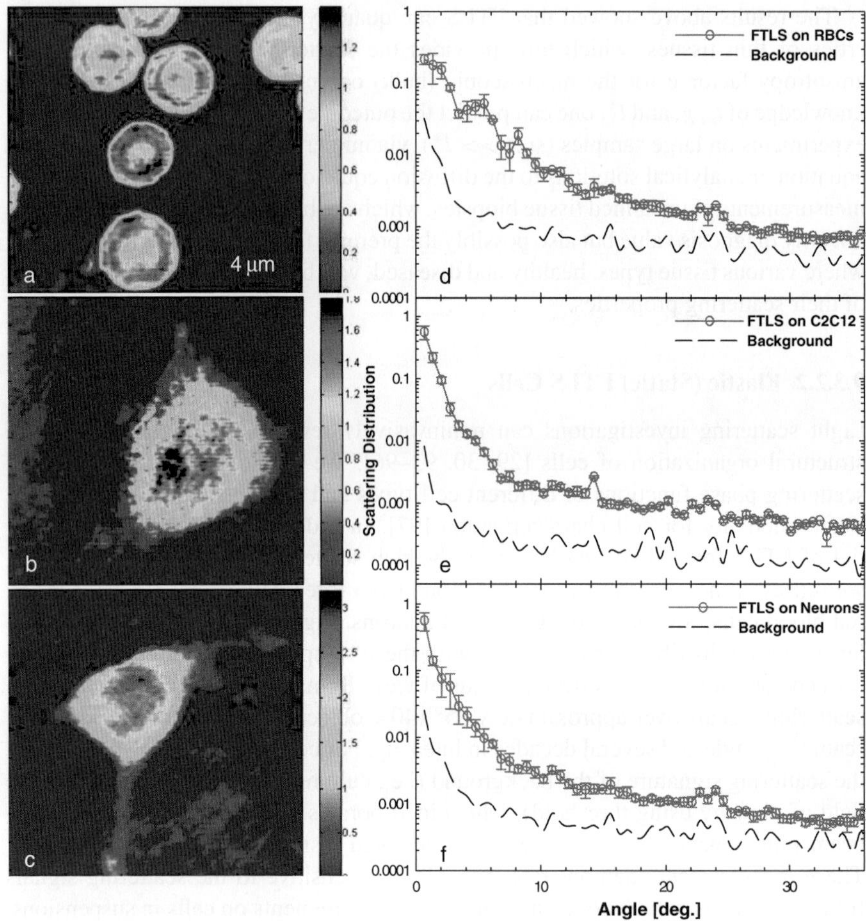

Fig. 9.11 Quantitative phase images of red blood cells (**a**), C2C12 cell (**b**), and neuron (**c**); the scale bar is 4 μm and the *color bar* indicates phase shift in radians. (**d–f**) Respective scattering phase functions measured by FTLS [97] © ASP 2010

eigenvalues. In order to build the training set, 45 measurements (i.e., 15 per cell type) were taken and processed following the procedures described above.

Figure 9.12 shows a representation of the data where each point in the plot is associated with a particular FTLS measurement. In addition to the 15 measurements per group for the training sets, we performed, respectively, 15, 15, and 10 test measurements for neurons, RBCs, and C2C12 cells. The additional test measurements allowed us to evaluate the sensitivity and specificity of assigning a given cell to the correct group [99]. We obtained sensitivity values of 100, 100, and 70% and specificities of 100, 88, and 100% for RBCs, neurons, and C2C12 cells, respectively.

We demonstrated here that FTLS can be used to differentiate between various cell types. Due to the particular imaging geometry used, scattering phase functions

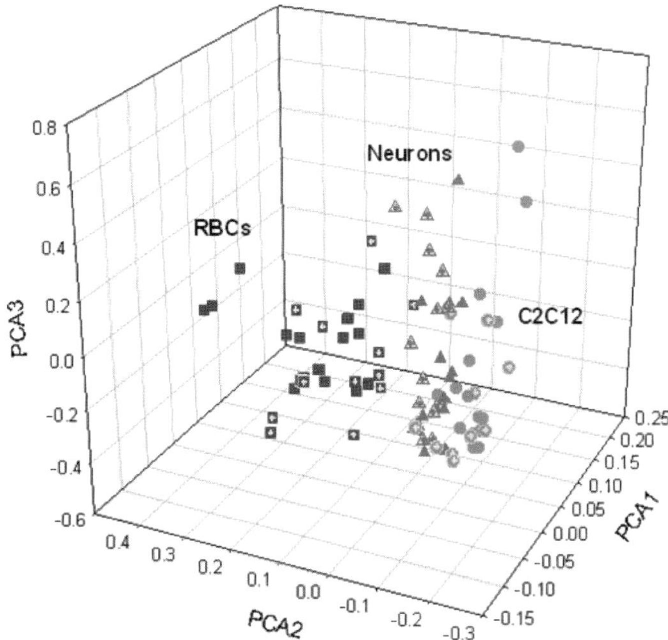

Fig. 9.12 PCA of the experimental data for the three cell types, as indicated. *Solid filled symbols* are the training sets of these three different biological samples included inside three ellipses. The symbols with "+" sign in the middle are the testing measurements for each sample [97] © ASP 2010

associated with single cells can be retrieved over a broad range of angles. This remarkable sensitivity to weak scattering signals may set the basis for a new generation of cytometry technology, which, in addition to the intensity information, will extract the structural details encoded in the phase of the optical field. FTLS may improve on fluorescence-based flow cytometry as it operates without the need for exogenous tags.

9.3.2.3 Quasi-elastic (Dynamic) FTLS of Cell Membrane Fluctuations

Quasi-elastic (dynamic) light scattering is the extension of ELS to dynamic inhomogeneous systems [100]. The temporal fluctuations of the optical field scattered at a particular angle by an ensemble of particles under Brownian motion relate to the diffusion coefficient of the particles. More recently, *microrheology* retrieves viscoelastic properties of complex fluids over various temporal and length scales, which is subject to sustained current research especially in the context of cell mechanics [101–105]. We employed FTLS to study red blood cell membrane fluctuations. To our knowledge, prior experimental investigations of light scattering by RBCs have been limited to measurements in suspension, where the signal is averaged over many cells and orientations. In this section, we briefly discussed the application of FTLS to study the fluctuating membranes of RBCs [88]. To determine how the cell

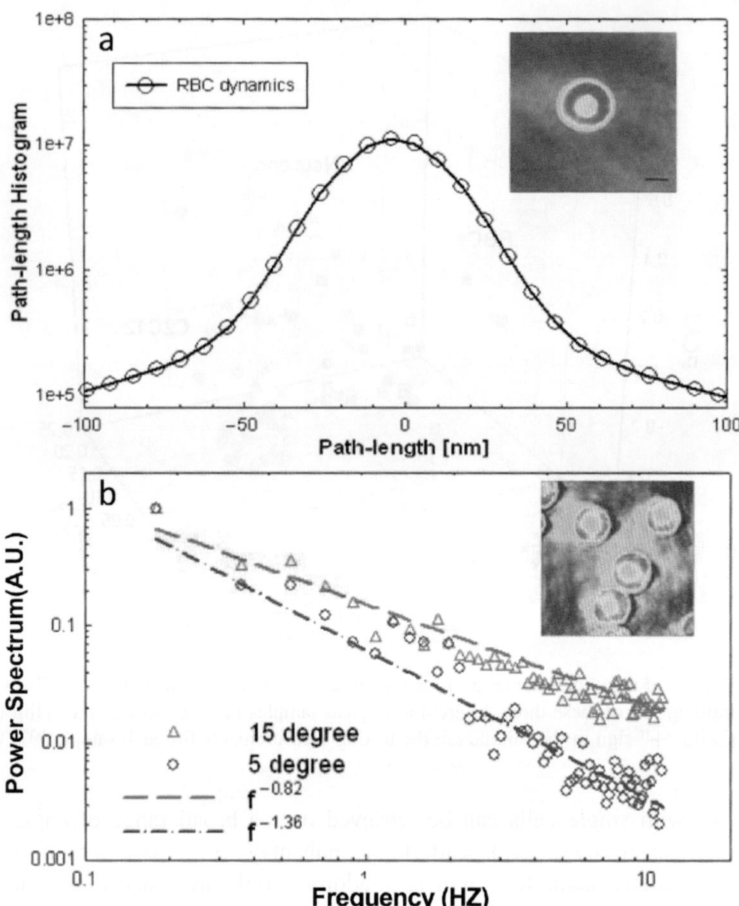

Fig. 9.13 (**a**) Histogram of the path length displacements of a RBC. The *inset* is the phase image. (**b**) Dynamics FTLS of RBCs: log–log power spectra at 5° and 15° with the respective power law fits, as indicated. The *inset* shows one RBC phase image from the time sequence. Reprinted figure with permission from [88] © 2008 by the American Physical Society. http://prl.aps.org/abstract/PRL/v101/i23/e238102

membrane flickering contributes to the dynamic light scattering of cells, RBCs from healthy volunteer sandwiched with two glass coverslips were imaged via DPM by acquiring 256 frames, at 20 frames per second, about 14 s. Figure 9.13a shows the membrane displacement histograms of a RBC. The power spectrum in Fig. 9.13b follows power laws with different exponents in time for all scattering angles (or, equivalently, wave vectors). As expected, the slower frequency decay at larger q values indicates a more solid behavior, i.e., the cell is more compliant at longer spatial wavelengths. Notably, the exponent of -1.36 of the longer wavelength (5° angle) is compatible with the -1.33 value predicted by Brochard et al. for the fluctuations at each point on the cell [50]. This is expected, as at each point the motions are dominated by long wavelengths [66]. The dynamic FTLS studies of RBC rheology can be performed on many cells simultaneously, which is an advantage over the previous flickering studies [50, 52, 71]. We believe that these initial results are extremely promising and that the label-free approach proposed here for studying cell dynamics will complement very well the existing fluorescence studies.

9.3.2.4 Quasi-elastic (Dynamic) FTLS of Cytoskeleton Dynamics

Recently, dynamic properties of cytoskeleton have been the subject of intense scientific interest [106–116]. In particular, it has been shown that actin filaments play an important role in various aspects of cell dynamics, including cell motility [107–109]. Previously, actin polymerization has been studied in real time by total reflection fluorescence microscopy [110, 111]. In this section, we demonstrate that FTLS is capable of sensing the spatiotemporal behavior of active (ATP-consuming) dynamics due to f-actin in single glial cells as addressed previously [117]. This activity is mediated by motor protein Myosin II and underlies diverse cellular processes, including cell division, developmental polarity, cell migration, filopodial extension, and intracellular transport.

We used FTLS to study the slow active dynamics of enteric glial cytoskeleton. During the FTLS measurement, the cells were maintained under constant temperature at 37°C via the incubation system that equips the microscope. The sensitivity of FTLS to actin dynamics was tested by controlling its polymerization activity. In order to inhibit actin polymerization, Cytochalasin-D (Cyto-D), approximately 5 μM in Hibernate-A, was added to the sample dishes. Cyto-D is a naturally occurring fungal metabolite known to have potent inhibitory action on actin filaments by capping and preventing polymerization and depolymerization at the rapidly elongating end of the filament. By capping this "barbed" end, the increased dissociation at the pointed end continues to shorten the actin filament.

The established single-particle tracking method [118] was first applied to test the efficacy of Cyto-D as actin inhibitor. With 1-μm diameter beads attached to the cell membrane as probes, the dynamic property of the cell membrane was investigated. Sets of 512 quantitative phase images of cells with beads attached were acquired at 1 frame/5 s, with a total acquisition time of 45 min. The x- and y- coordinates of the tracked beads were recorded as a function of time, and the trajectories were used to calculate the mean squared displacement (MSD) [119]

$$\mathrm{MSD}(\Delta t) = \left\langle [x(t+\Delta t) - x(t)]^2 + [y(t+\Delta t) - y(t)]^2 \right\rangle \tag{9.15}$$

where $< \ldots >$ indicates the average over time and also over all the tracked particles. We recorded the displacements of the attached beads before and after treatment with Cyto-D. The MSD results are summarized in Fig. 9.14. Each curve is the result of an average over all the beads ($N > 10$) tracked under the same experimental condition. The data for normal (before treatment) cells exhibit a power law trend over two distinct temporal regions, separated at approximately $\tau_0 = 215$ s. This change in slope reflects a characteristic lifetime τ_0 associated with actin polymerization, which is known to be in the minute range [120]. For both curves, the fit at $t > \tau_0$ gives a

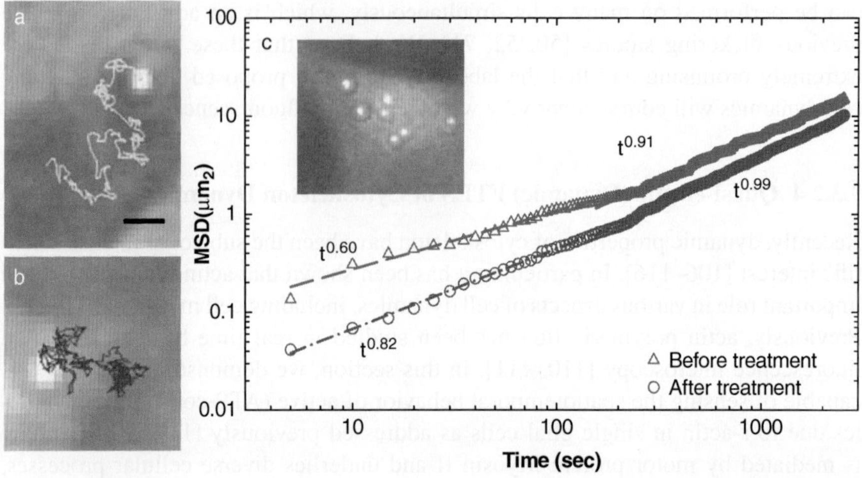

Fig. 9.14 Tracking beads attached to the cell membrane: (**a**) bead trajectories before Cyto-D treatment and (**b**) bead trajectories after drug treatment. (**c**) Corresponding mean square displacement of the tracked beads before (*red*) and after (*blue*) treatment. The fitting with a power law function over two different time windows is indicated. The *inset* shows a quantitative phase image of a EGC with 1-μm beads attached to the membrane. Adapted with permission from [117] © OSA 2010

power law of exponent approximately 1, indicating diffusive motion. However, the membrane displacements for the untreated cells are a factor of 1.3 larger, consistent with the qualitative observation of the particle trajectories. Thus, at long time, the effect of actin on membrane dynamics is to enhance its motions without changing its temporal statistics.

To determine how the actin cytoskeleton contributes to the dynamic light scattering of cells alone, cells without beads were imaged via DPM by acquiring 512 frames, at 0.2 frames per second, over ∼45 min, prior to and after Cyto-D application, respectively (Fig. 9.15a, b). Figure 9.15c shows a comparison between the membrane displacement histograms of a cell before and after the actin inhibitor. It is evident from this result that the polymerization phenomenon is a significant contributor to the overall cell membrane dynamics, as indicated by the broader histogram distribution. Further, both curves exhibit non-Gaussian shapes at displacements larger than 10 nm, which suggests that the cell motions both before and after actin inhibition are characterized by nonequilibrium dynamics. Figure 9.15d presents the comparison of the spatially averaged power spectra associated with the FTLS signal for a single cell, before and after treatment with the actin-blocking drug. The broader power spectrum of the untreated cell membrane motions is consistent with the histogram distribution in Fig. 9.14c and also the particle tracking results. Further, both frequency-averaged (statics) curves shown in Fig. 9.15e indicate similar functional dependence on the wave vector q, but with enhanced fluctuations for the normal cell, by a factor of ∼3.4. One key feature of Fourier transform light scattering is its ability to render simultaneous angular scattering from an entire range of angles. Figure 9.16a–d shows the power spectrum of the fluctuations for the same cell (shown in Fig. 9.15a, b) before and after the actin inhibition, as function of both

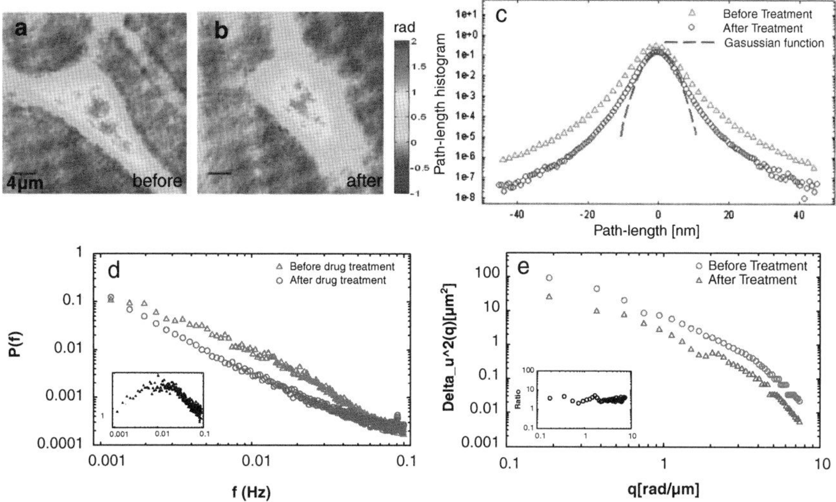

Fig. 9.15 (**a, b**) Quantitative phase images of glial cell before and after Cyto-D treatment; (**c**) histogram of the path-length displacements of a glial cell before and after drug treatment, as indicated. The *blue dashed line* indicates the fit with a Gaussian function; (**d**) spatially averaged power spectrum of glial cells before and after drug treatment, as indicated; (**e**) temporally averaged power spectrum before and after drug treatment. The *inset* shows the ratio of the two spectra. Adapted with permission from [117] © OSA 2010

frequency f (at two particular q-values) and wave vector q (at two particular frequencies ω). After actin inhibition, the functional dependence of $\Delta u^2(f)$ assumes a Lorentzian shape and does not change notably with varying q, which contrasts with the situation where the cell cytoskeleton is intact. This interesting behavior can be seen in Fig. 9.16a, b, where the temporal power spectra at two particular scattering angles (or wave vectors) are shown. These findings suggest that, as the actin is disrupted, the dynamic scattering signal is most likely due to Brownian-like membrane fluctuations. At frequencies $f > 1/\tau_0$ (Fig. 9.16c), there is a significant mismatch between the "before" and the "after" $\Delta u^2(q)$ curves. On the other hand, for $f < 1/\tau_0$ (Fig. 9.16d), the two dependencies look similar, with the normal cell exhibiting consistently higher fluctuations.

The trends shown in Figs. 9.14, 9.15, and 9.16 support the idea of actin-mediated membrane dynamics that is characterized by a time constant τ_0. Thus, we propose a physical model which captures the essence of our experimental results. We assume that the cell membrane is under the influence of both Brownian motion and actively driven displacements due to ATP-consuming actin polymerization activity. The membrane is under tension and connected to springs (actin filaments) that undergo continuous remodeling. This picture has been applied successfully to the active motions in red blood cell membranes [67, 121]. The Langevin equation for the membrane fluctuations is [122]

$$\dot{u}(q,t) + \omega(q)u(q,t) = \Lambda(q)f(q,t) \tag{9.16}$$

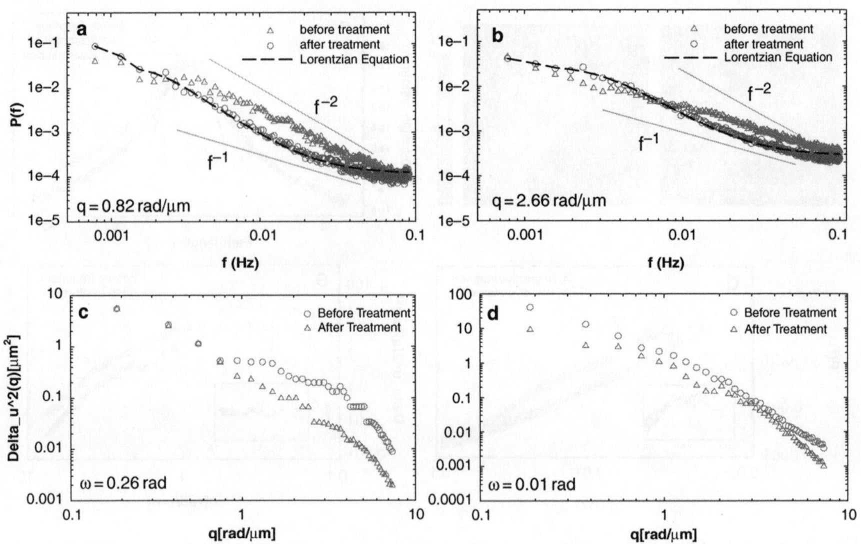

Fig. 9.16 (**a, b**) Temporal power spectrum at two different scattering angles, as indicated. The f^{-1} and f^{-2} dependences are shown as reference; (**c, d**) spatial power spectrum at two different frequencies, as indicated. Adapted with permission from [117] © OSA 2010

where u is the displacement of the membrane element, q the spatial wave vector, $\omega(q) = \left(\kappa q^3 + \sigma q + \gamma/q\right)/4\eta$, κ the bending modulus, σ the tension modulus, γ the confinement factor, η the viscosity of the surrounding medium, f the random force, and $\Lambda(q) = 1/(4\eta q)$ is the Oseen hydrodynamic factor. The displacement power spectrum can be obtained by Fourier transforming equation (9.16)

$$|u(q,\omega)|^2 = \frac{\Lambda^2(q)\,|f(q,\omega)|^2}{\omega^2 + \omega(q)^2} \tag{9.17}$$

The instantaneous force has both Brownian and direct components, characterized by their respective temporal correlations

$$\langle f_{\mathrm{B}}(q,t)\, f_{\mathrm{B}}(q,t+\tau)\rangle = 2k_{\mathrm{B}}T\delta(\tau)/\Lambda(q) \tag{9.18}$$

$$\langle f_{\mathrm{D}}(q,t)\, f_{\mathrm{D}}(q,t+\tau)\rangle = 2\alpha k_{\mathrm{B}}T\left(\mathrm{e}^{-|\tau|/\tau_0}/2\tau_0\right)/\Lambda(q) \tag{9.19}$$

In (9.18) and (9.19), f_{B} is the Brownian (thermal) and f_{D} the directed force, assumed to exhibit an exponential temporal correlation, with the decay time τ_0 given by the actin polymerization lifetime. The angular brackets denote temporal average such that $< f_{\mathrm{B}} > = < f_{\mathrm{D}} > = 0$. The factor α establishes the mean square force fluctuations, i.e., $\alpha = 0$ for Cyto-D-treated cells. Combining (9.17), (9.18), and (9.19), we obtain for the displacement power spectrum

$$|u(q,\omega)|^2 = \frac{2k_B T \Lambda(q)}{\omega^2 + \omega(q)^2}\left[1 + \alpha \frac{{\omega_0}^2}{\omega^2 + {\omega_0}^2}\right] \quad (9.20)$$

Remarkably, this simple model captures some essential features of the experimental results as follows. First, the active term in (9.20) captures the increase in the power spectrum at high frequencies (Fig. 9.15d). This is also evidenced in Fig. 9.16c, where at a fixed (high) frequency we found enhanced fluctuation spectrum vs. q. Second, upon blocking actin polymerization with Cyto-D, i.e., $\alpha \to 0$, $|u(q)|^2$ remains Lorentzian in shape for all q's, as found in Fig. 9.16a, b. Third, at long times ($\omega \ll \omega_0$) the active term is featureless, i.e., the actin contributions are Brownian like and can be described by an increased effective temperature, $T_{\text{eff}} = \alpha T$ [67]. In this case, the static behavior is obtained integrating (9.20) over ω, which yields

$$|u(q)|^2 = k_B T\,(1+\alpha)\,/\left(\kappa q^4 + \sigma q^2 + \gamma\right) \quad (9.21)$$

This explains the similar functional dependence for the two curves in Fig. 9.15e, resulting in a ratio $\alpha + 1 = 3.4$. A similar trend was found at a fixed (low) frequency, as shown in Fig. 9.16d. We fitted the two curves in Fig. 9.15e with (9.21) and obtained a good agreement in the low q, tension-dominated region. The values for σ obtained from the fit are 5.18e–4 and 1.85e–3, respectively. Note that both these values are approximately one to two orders of magnitude larger than those measured in red blood cells [52]. Thus, the membrane fluctuation noise produced by actin is equivalent to a decrease in surface tension or, equivalently, an increase in effective temperature $T_{\text{eff}} = 2.4\,T$.

We believe that these initial results are extremely promising and that the label-free approach proposed here for studying cytoskeleton dynamics will complement very well the existing fluorescence studies. However, the interpretation of these FTLS measurements requires improvement, as evidenced, for instance, in the poor fit of the curves in Fig. 9.15e at large q. It is likely that the simple model proposed here needs further refinements to include contributions from particle transport and polymer dynamics within the cell.

9.4 Summary and Outlook

In summary, we reviewed a number of reported techniques for quantitative phase imaging and scattering measurement. The extensive applications of QPI and FTLS were discussed for both biological system structure and dynamics. Further, the QPI capability can be judged by two main parameters: *phase sensitivity*, which defines the smallest phase change that the instrument can capture (i.e., at signal to noise of 1), and *acquisition speed*, which determines the fastest phenomena that can be studied.

Current applications of both static and dynamic QPI were discussed. In the case of red blood cells, which have an optically homogeneous structure, cell volumetry and membrane fluctuations have been demonstrated. In eukaryotic cells, the phase imaging problem is more challenging, as the cell topography and refractive index distribution contribute in a nontrivial manner to the measured path length map. Thus, decoupling the refractive index contribution from the thickness was discussed. Further, it has been shown that the phase shift map associated with eukaryotic cells can be translated into dry mass density. The sub-nanometer scale accuracy in optical path length allows for measurements of dry mass at the femtogram scale. With this approach, cell growth and motility can be monitored without the need for contrast agents.

FTLS was applied as a new approach for studying static and dynamic light scattering with unprecedented sensitivity, granted by the image plane measurement of the optical phase and amplitude. In FTLS, spatial resolution of the scatterer positions is well preserved. These remarkable features of FTLS are due to the interferometric experimental geometry and the reliable phase retrieval. FTLS has been applied to study the tissue optical properties, cell type characterization, and dynamic structure of cell membrane. We anticipate that this type of measurement will enable new advances in life sciences, due to the ability to detect weak scattering signals over broad temporal (milliseconds to days) and spatial (fraction of microns to centimeters) scales.

Recent developments in QPI and FTLS expanded its applicability to new areas. Here we briefly describe two directions which, in authors' opinion, promise to make a significant impact in medicine.

9.4.1 QPI for Blood Screening

Blood smear analysis has remained a crucial diagnostic tool for pathologists despite the advent of automatic analyzers such as flow cytometers and impedance counters. Though these current methods have proven to be indispensible tools for physicians and researchers alike, they provide limited information on the detailed morphology of individual cells and merely alert the operator to manually examine a blood smear by raising flags when abnormalities are detected. We have recently demonstrated an automatic *interferometry-based* smear analysis technique known as diffraction phase cytometry (DPC) [123], which is capable of providing the same information on red blood cells as is provided by current clinical analyzers, while rendering additional, currently unavailable parameters on the 2D and 3D morphology of individual red blood cells (RBCs). To validate the utility of our technique in a clinical setting, we performed a comparison between tests generated from 32 patients by a state-of-the-art clinical impedance counter and DPC [124].

Extending this principle of investigation to clinical-relevant problems, DPM was used to quantify membrane fluctuations of human RBCs parasitized by *Plasmodium falciparum* at 37 and 41°C [125]. Alterations of structure cause changes in RBC membrane fluctuations, with stronger effects seen in later stage parasitization and at febrile temperature. These findings offer potential new avenues for identifying, through cell membrane dynamics, pathological states that affect cell function and could result in human diseases [126].

9.4.2 FTLS of Cancerous Tissues

We believe that QPI can become a powerful tool for label-free tissue diagnosis, where the refractive index plays the role of cancer marker. In our initial studies, FTLS has been applied to image unstained 5-μm-thick tissue slices of rat breast tumors and various organs [35, 88]. The FTLS measurements of unstained tissue biopsies, which are broadly available, will provide a set of scattering signatures for automatic cancer detection and grading. Further, our measurements will set the stage for a large scattering database, where various tissue types, healthy and diseased, will be fully characterized in terms of their scattering properties without staining [35]. Finally, we will develop a fiber optics-based instrument, which will allow for in vivo imaging via a needle. Thus, we will essentially replace tissue biopsies which require laboratory preparation and human inspection — with "optical biopsies," with real-time image acquisition and largely computer-assisted analysis. FTLS provides an effective and staining-free modality for studying tissue biopsies by deriving the optical properties from "organelle-to-organ" scale.

Acknowledgments G.P. is grateful to the current and previous colleagues for their contributions to QPI: Zhuo Wang, Mustafa Mir, and Ru Wang (Quantitative Light Imaging Laboratory, Department of Electrical and Computer Engineering, University of Illinois at Urbana-Champaign); YoungKeun Park, Niyom Lue, Shahrooz Amin, Lauren Deflores, Seungeun Oh, Christopher Fang-Yen, Won-shik Choi, Kamran Badizadegan, Ramachandra Dasari, and Michael Feld (Spectroscopy Laboratory, Massachusetts Institute of Technology); and Takahiro Ikeda and Hidenao Iwai (Hamamatsu Photonics KK).

H.D. would like to acknowledge contributions from collaborators Catherine Best (College of Medicine, University of Illinois at Urbana-Champaign), Martha Gillette and Larry J. Millet (Department of Cell and Developmental Biology, University of Illinois at Urbana-Champaign), Alex Levine (University of California at Los Angeles), Marni Boppart and Freddy Nguyen (Department of Kinesiology and Community Health, Beckman Institute for Advanced Science and Technology, University of Illinois at Urbana-Champaign), Stephen A. Boppart and Jiangming Liu (Biophotonics Imaging Laboratory, Department of Electrical and Computer Engineering, Beckman Institute for Advanced Science and Technology, University of Illinois at Urbana-Champaign), Michael Laposata (Division of Laboratory Medicine and Clinical Laboratories, Vanderbilt University Medical Center), and Carlo Brugnara (Department of Laboratory Medicine, Children's Hospital Boston).

References

1. D.J. Stephens, V.J. Allan, Light microscopy techniques for live cell imaging. Science **300**, 82–86 (2003)
2. F. Zernike, How I discovered phase contrast. Science **121**, 345 (1955)
3. F.H. Smith, Microscopic interferometry. Research (London) **8**, 385 (1955)
4. D. Zicha, G.A. Dunn, An image-processing system for cell behavior studies in subconfluent cultures. J. Microsc **179**, 11–21 (1995)
5. G.A. Dunn, D. Zicha, P.E. Fraylich, Rapid, microtubule-dependent fluctuations of the cell margin. J. Cell Sci. **110**, 3091–3098 (1997)
6. D. Zicha, E. Genot, G.A. Dunn, I.M. Kramer, TGF beta 1 induces a cell-cycle-dependent increase in motility of epithelial cells. J. Cell Sci. **112**, 447–454 (1999)

7. D. Paganin, K.A. Nugent, Noninterferometric phase imaging with partially coherent light. Phys. Rev. Lett. **80**, 2586–2589 (1998)
8. B.E. Allman, P.J. McMahon, J.B. Tiller, K.A. Nugent, D. Paganin, A. Barty, I. McNulty, S.P. Frigo, Y.X. Wang, C.C. Retsch, Noninterferometric quantitative phase imaging with soft x rays. J. Opt. Soc. Am. A-Opt. Image Sci. Vis. **17**, 1732–1743 (2000)
9. S. Bajt, A. Barty, K.A. Nugent, M. McCartney, M. Wall, D. Paganin, 'Quantitative phase-sensitive imaging in a transmission electron microscope. Ultramicroscopy **83**, 67–73 (2000)
10. C.J. Mann, L.F. Yu, C.M. Lo, M.K. Kim, High-resolution quantitative phase-contrast microscopy by digital holography. Opt. Expr. **13**, 8693–8698 (2005)
11. P. Marquet, B. Rappaz, P.J. Magistretti, E. Cuche, Y. Emery, T. Colomb, C. Depeursinge, Digital holographic microscopy: a noninvasive contrast imaging technique allowing quantitative visualization of living cells with subwavelength axial accuracy. Opt. Lett. **30**, 468–470 (2005)
12. H. Iwai, C. Fang-Yen, G. Popescu, A. Wax, K. Badizadegan, R.R. Dasari, M.S. Feld, Quantitative phase imaging using actively stabilized phase-shifting low-coherence interferometry. Opt. Lett. **29**, 2399–2401 (2004)
13. G. Popescu, L.P. Deflores, J.C. Vaughan, K. Badizadegan, H. Iwai, R.R. Dasari, M.S. Feld, Fourier phase microscopy for investigation of biological structures and dynamics. Opt. Lett. **29**, 2503–2505 (2004)
14. T. Ikeda, G. Popescu, R.R. Dasari, M.S. Feld, Hilbert phase microscopy for investigating fast dynamics in transparent systems. Opt. Lett. **30**, 1165–1168 (2005)
15. G. Popescu, T. Ikeda, C.A. Best, K. Badizadegan, R.R. Dasari, M.S. Feld, Erythrocyte structure and dynamics quantified by Hilbert phase microscopy. J. Biomed. Opt. Lett. **10**, 060503 (2005)
16. G. Popescu, T. Ikeda, R.R. Dasari, M.S. Feld, Diffraction phase microscopy for quantifying cell structure and dynamics. Opt. Lett. **31**, 775–777 (2006)
17. J.W. Pyhtila, A. Wax, Improved interferometric detection of scattered light with a 4f imaging system. Appl. Opt. **44**, 1785–1791 (2005)
18. J.W. Pyhtila, A. Wax, Rapid, depth-resolved light scattering measurements using Fourier domain, angle-resolved low coherence interferometry. Opt. Expr. **12**, 6178–6183 (2004)
19. M.A. Choma, A.K. Ellerbee, C.H. Yang, T.L. Creazzo, J.A. Izatt, Spectral-domain phase microscopy. Opt. Lett. **30**, 1162–1164 (2005)
20. C. Fang-Yen, S. Oh, Y. Park, W. Choi, S. Song, H.S. Seung, R.R. Dasari, M.S. Feld, Imaging voltage-dependent cell motions with heterodyne Mach-Zehnder phase microscopy. Opt. Lett. **32**, 1572–1574 (2007)
21. C. Joo, T. Akkin, B. Cense, B.H. Park, J.E. de Boer, Spectral-domain optical coherence phase microscopy for quantitative phase-contrast imaging. Opt. Lett. **30**, 2131–2133 (2005)
22. W.S. Rockward, A.L. Thomas, B. Zhao, C.A. DiMarzio, Quantitative phase measurements using optical quadrature microscopy. Appl. Opt. **47**, 1684–1696 (2008)
23. R. Drezek, A. Dunn, R. Richards-Kortum, Light scattering from cells: finite-difference time-domain simulations and goniometric measurements. Appl. Opt. **38**, 3651–3661 (1999)
24. J.R. Mourant, M. Canpolat, C. Brocker, O. Esponda-Ramos, T.M. Johnson, A. Matanock, K. Stetter, J.P. Freyer, Light scattering from cells: the contribution of the nucleus and the effects of proliferative status. J. Biomed. Opt. **5**, 131–137 (2000)
25. C.S. Mulvey, A.L. Curtis, S.K. Singh, I.J. Bigio, Elastic scattering spectroscopy as a diagnostic tool for apoptosis in cell cultures. IEEE J. Sel. Top. Quant. Electron. **13**, 1663–1670 (2007)
26. H. Ding, J.Q. Lu, R.S. Brock, T.J. McConnell, J.F. Ojeda, K.M. Jacobs, X.H. Hu, Angle-resolved Mueller matrix study of light scattering by R-cells at three wavelengths of 442, 633, and 850 nm. J. Biomed. Opt. **12**, 034032 (2007)
27. M.T. Valentine, A.K. Popp, D.A. Weitz, P.D. Kaplan, Microscope-based static light-scattering instrument. Opt. Lett. **26**, 890–892 (2001)
28. W.J. Cottrell, J.D. Wilson, T.H. Foster, Microscope enabling multimodality imaging, angle-resolved scattering, and scattering spectroscopy. Opt. Lett. **32**, 2348–2350 (2007)

29. A. Wax, C.H. Yang, V. Backman, K. Badizadegan, C.W. Boone, R.R. Dasari, M.S. Feld, Cellular organization and substructure measured using angle-resolved low-coherence interferometry. Biophys. J. **82**, 2256–2264 (2002)
30. V. Backman, M.B. Wallace, L.T. Perelman, J.T. Arendt, R. Gurjar, M.G. Muller, Q. Zhang, G. Zonios, E. Kline, T. McGillican, S. Shapshay, T. Valdez, K. Badizadegan, J.M. Crawford, M. Fitzmaurice, S. Kabani, H.S. Levin, M. Seiler, R.R. Dasari, I. Itzkan, J. Van Dam, M.S. Feld, Detection of preinvasive cancer cells. Nature **406**, 35–36 (2000)
31. F. Charriere, N. Pavillon, T. Colomb, C. Depeursinge, T.J. Heger, E.A.D. Mitchell, P. Marquet, B. Rappaz, Living specimen tomography by digital holographic microscopy: morphometry of testate amoeba. Opt. Expr. **14**, 7005–7013 (2006)
32. V.V. Tuchin, *Tissue Optics* (Belingham, WA:SPIE–The International Society for Optical Engineering, 2000)
33. J.W. Pyhtila, K.J. Chalut, J.D. Boyer, J. Keener, T. D'Amico, M. Gottfried, F. Gress, A. Wax, In situ detection of nuclear atypia in Barrett's esophagus by using angle-resolved low-coherence interferometry. Gastrointest. Endosc. **65**, 487–491 (2007)
34. L.Q.H. Fang, E. Vitkin, M.M. Zaman, C. Andersson, S. Salahuddin, L.M. Kimerer, P.B. Cipolloni, M.D. Modell, B.S. Turner, S.E. Keates, I. Bigio, I. Itzkan, S.D. Freedman, R. Bansil, E.B. Hanlon, L.T. Perelman, Confocal light absorption and scattering spectroscopic microscopy. Appl. Opt. **46**, 1760–1769 (2007)
35. H.F. Ding, F. Nguyen, S.A. Boppart, G. Popescu, Optical properties of tissues quantified by Fourier-transform light scattering. Opt. Lett. **34**, 1372–1374 (2009)
36. G.A. Dunn, D. Zicha (eds.), *Using DRIMAPS System of Transmission Interference Microscopy to Study Cell Behavior* (Academic, London, 1997)
37. G.A. Dunn, D. Zicha, Dynamics Of fibroblast spreading. J. Cell Sci. **108**, 1239–1249 (1995)
38. T.E. Gureyev, A. Roberts, K.A. Nugent, Phase retrieval with the transport-of-intensity equation – matrix solution with use of Zernike polynomials. J. Opt. Soc. Am. A-Opt. Image Sci. Vis. **12**, 1932–1941 (1995)
39. T.E. Gureyev, A. Roberts, K.A. Nugent, Partially coherent fields, the transport-of-intensity equation, and phase uniqueness. J. Opt. Soc. Am. A-Opt. Image Sci. Vis. **12**, 1942–1946 (1995)
40. J.W. Goodman, R.W. Lawrence, Digital image formation from electronically detected holograms. Appl. Phys. Lett. **11**, 77 (1967)
41. D. Gabor, A new microscopic principle. Nature **161**, 777 (1948)
42. I. Yamaguchi, T. Zhang, Phase-shifting digital holography. Opt. Lett. **22**, 1268–1270 (1997)
43. C.J. Mann, L.F. Yu, C.M. Lo, M.K. Kim, High-resolution quantitative phase-contrast microscopy by digital holography. Opt. Expr. **13**, 8693–8698 (2005)
44. D. Carl, B. Kemper, G. Wernicke, G. von Bally, Parameter-optimized digital holographic microscope for high-resolution living-cell analysis. Appl. Opt. **43**, 6536–6544 (2004)
45. N. Lue, W. Choi, G. Popescu, R.R. Dasari, K. Badizadegan, M.S. Feld, Quantitative phase imaging of live cells using fast Fourier phase microscopy. Appl. Opt. **46**, 1836 (2007)
46. Y.K. Park, G. Popescu, K. Badizadegan, R.R. Dasari, M.S. Feld, Diffraction phase and fluorescence microscopy. Opt. Exp. **14**, 8263 (2006)
47. K. Creath, Phase-measurement interferometry techniques. Prog. Opt. **26**, 349–393 (1988)
48. G. Popescu, K. Badizadegan, R.R. Dasari, M.S. Feld, Observation of dynamic subdomains in red blood cells. J. Biomed. Opt. Lett. **11**, 040503 (2006)
49. M. Hammer, D. Schweitzer, B. Michel, E. Thamm, A. Kolb, Single scattering by red blood cells. Appl. Opt. **37**, 7410–7418 (1998)
50. F. Brochard, J.F. Lennon, Frequency spectrum of the flicker phenomenon in erythrocytes. J. Phys **36**, 1035–1047 (1975)
51. A. Zilker, H. Engelhardt, E. Sackmann, Dynamic reflection interference contrast (Ric-) microscopy - a new method to study surface excitations of cells and to measure membrane bending elastic-moduli. J. Phys **48**, 2139–2151 (1987)

52. G. Popescu, T. Ikeda, K. Goda, C.A. Best-Popescu, M. Laposata, S. Manley, R.R. Dasari, K. Badizadegan, M.S. Feld, Optical measurement of cell membrane tension. Phys. Rev. Lett. **97**, 218101 (2006)
53. R. Nowakowski, P. Luckham, P. Winlove, Imaging erythrocytes under physiological conditions by atomic force microscopy. Biochim. Biophys. Acta **1514**, 170–176 (2001)
54. P. Matarrese, E. Straface, D. Pietraforte, L. Gambardella, R. Vona, A. Maccaglia, M. Minetti, W. Malorni, Peroxynitrite induces senescence and apoptosis of red blood cells through the activation of aspartyl and cysteinyl proteases. FASEB J. **19**, 416–418 (2005)
55. C.A. Best, Fatty acid ethyl esters and erythrocytes: metabolism and membrane effects. Ph.D. Thesis, Northeastern University, Boston, 2005
56. R. Baber, Interference microscopy and mass determination. Nature **169**, 366–367 (1952)
57. H.G. Davies, M.H.F. Wilkins, Interference microscopy and mass determination. Nature **161**, 541 (1952)
58. E. Cuche, F. Bevilacqua, C. Depeursinge, Digital holography for quantitative phase-contrast imaging. Opt. Lett. **24**, 291–293 (1999)
59. G. Popescu, Y.K. Park, N. Lue, C.A. Best-Popescu, L. Deflores, R.R. Dasari, M.S. Feld, K. Badizadegan, Optical imaging of cell mass and growth dynamics. Am. J. Physiol.-Cell Physiol. **295**, C538 (2008)
60. I.J. Conlon, G.A. Dunn, A.W. Mudge, M.C. Raff, Extracellular control of cell size. Nat. Cell Biol. **3**, 918–921 (2001)
61. D.E. Ingber, Fibronectin controls capillary endothelial cell growth by modulating cell shape. Proc. Natl. Acad. Sci. USA **87**, 3579–3583 (1990)
62. D. Boal, *Mechanics of the Cell* (Cambridge University Press, Cambridge, 2002)
63. R.M. Hochmuth, R.E. Waugh, Erythrocyte membrane elasticity and viscocity. Ann. Rev. Physiol. **49**, 209–219 (1987)
64. R. Lipowsky, The conformation of membranes. Nature **349**, 475–481 (1991)
65. E. Sackmann, Supported membranes: scientific and practical applications. Science **271**, 43–48 (1996)
66. N. Gov, A.G. Zilman, S. Safran, Cytoskeleton confinement and tension of red blood cell membranes. Phys. Rev. Lett. **90**, 228101 (2003)
67. N. Gov, Membrane undulations driven by force fluctuations of active proteins. Phys. Rev. Lett. **93**, 268104 (2004)
68. N.S. Gov, S.A. Safran, Red blood cell membrane fluctuations and shape controlled by ATP-induced cytoskeletal defects. Biophys. J. **88**, 1859–1874 (2005)
69. R. Lipowski, M. Girardet, Shape fluctuations of polymerized or solidlike membranes. Phys. Rev. Lett. **65**, 2893–2896 (1990)
70. K. Zeman, E.H., E. Sackman, Eur. Biophys. J. **18**, 203 (1990)
71. A. Zilker, M. Ziegler, E. Sackmann, Spectral-analysis of erythrocyte flickering in the 0.3-4-Mu-M-1 regime by microinterferometry combined with fast image-processing. Phys. Rev. A **46**, 7998–8002 (1992)
72. H. Strey, M. Peterson, E. Sackmann, Measurement of erythrocyte membrane elasticity by flicker eigenmode decomposition. Biophys. J. **69**, 478–488 (1995)
73. D.E. Discher, N. Mohandas, E.A. Evans, Molecular maps of red cell deformation: hidden elasticity and in situ connectivity. Science **266**, 1032–1035 (1994)
74. R.M. Hochmuth, P.R. Worthy, E.A. Evans, Red cell extensional recovery and the determination of membrane viscosity. Biophys. J. **26**, 101–114 (1979)
75. H. Engelhardt, E. Sackmann, On the measurement of shear elastic moduli and viscosities of erythrocyte plasma membranes by transient deformation in high frequency electric fields. Biophys. J. **54**, 495–508 (1988)
76. H. Engelhardt, H. Gaub, E. Sackmann, Viscoelastic properties of erythrocyte membranes in high-frequency electric fields. Nature **307**, 378–380 (1984)
77. S. Suresh, J. Spatz, J.P. Mills, A. Micoulet, M. Dao, C.T. Lim, M. Beil, T. Seufferlein, Connections between single-cell biomechanics and human disease states: gastrointestinal cancer and malaria. Acta Biomater. **1**, 15–30 (2005)

78. H.W.G. Lim, M. Wortis, R. Mukhopadhyay, Stomatocyte-discocyte-echinocyte sequence of the human red blood cell: evidence for the bilayer- couple hypothesis from membrane mechanics. Proc. Natl. Acad. Sci. USA **99**, 16766–16769 (2002)
79. Y. Kaizuka, J.T. Groves, Hydrodynamic damping of membrane thermal fluctuations near surfaces imaged by fluorescence interference microscopy. Phys. Rev. Lett. **96**, 118101 (2006)
80. A. Zidovska, E. Sackmann, Brownian motion of nucleated cell envelopes impedes adhesion. Phys. Rev. Lett. **96**, 048103 (2006)
81. C.A. Best, J.E. Cluette-Brown, M. Teruya, A. Teruya, M. Laposata, Red blood cell fatty acid ethyl esters: a significant component of fatty acid ethyl esters in the blood. J. Lipid Res. **44**, 612–620 (2003)
82. N. Gov, A. Zilman, S. Safran, Cytoskeleton confinement of red blood cell membrane fluctuations. Biophys. J. **84**, 486A–486A (2003)
83. T.G. Mason, K. Ganesan, J.H. vanZanten, D. Wirtz, S.C. Kuo, Particle tracking microrheology of complex fluids. Phys. Rev. Lett. **79**, 3282–3285 (1997)
84. J.C. Crocker, M.T. Valentine, E.R. Weeks, T. Gisler, P.D. Kaplan, A.G. Yodh, D.A. Weitz, Two-point microrheology of inhomogeneous soft materials. Phys. Rev. Lett. **85**, 888–891 (2000)
85. N. Lue, J. Bewersdorf, M.D. Lessard, K. Badizadegan, K. Dasari, M.S. Feld, G. Popescu, Tissue refractometry using Hilbert phase microscopy. Opt. Lett. **32**, 3522 (2007)
86. B. Rappaz, P. Marquet, E. Cuche, Y. Emery, C. Depeursinge, P.J. Magistretti, Measurement of the integral refractive index and dynamic cell morphometry of living cells with digital holographic microscopy. Opt. Exp. **13**, 9361–9373 (2005)
87. W. Choi, C.C. Yu, C. Fang-Yen, K. Badizadegan, R.R. Dasari, M.S. Feld, Feild-based angle-resolved light-scatteiring study of single live COS. Opt. Lett. **33**, 1596–1598 (2008)
88. H.F. Ding, Z. Wang, F. Nguyen, S.A. Boppart, G. Popescu, Fourier transform light scattering of inhomogeneous and dynamic structures. Phys. Rev. Lett. **101**, 238102 (2008)
89. G. Popescu, in *Methods in Cell Biology*, vol. 87, ed. by P.J. Bhanu. (Elsevier, Heidelberg, 2008)
90. L.O. Reynolds, N.J. Mccormick, Approximate 2-parameter phase function for light-scattering. J. Opt. Soc. Am. **70**, 1206–1212 (1980)
91. J.M. Schmitt, G. Kumar, Optical scattering properties of soft tissue: a discrete particle model. Appl. Opt. **37**, 2788–2797 (1998)
92. M. Hunter, V. Backman, G. Popescu, M. Kalashnikov, C.W. Boone, A. Wax, G. Venkatesh, K. Badizadegan, G.D. Stoner, M.S. Feld, Tissue self-affinity and light scattering in the born approximation: a new model for precancer diagnosis. Phys. Rev. Lett. **97**, 138102 (2006)
93. J.D. Wilson, T.H. Foster, Characterization of lysosomal contribution to whole-cell light scattering by organelle ablation. J. Biomed. Opt. **12**(3), 030503 (2007)
94. J.R. Mourant, J.P. Freyer, A.H. Hielscher, A.A. Eick, D. Shen, T.M. Johnson, Mechanisms of light scattering from biological cells relevant to noninvasive optical-tissue diagnostics. Appl. Opt. **37**, 3586–3593 (1998)
95. G. Popescu, A. Dogariu, Scattering of low coherence radiation and applications. Invited review paper, E. Phys. J. **32**, 73–93 (2005)
96. S.A. Alexandrov, T.R. Hillman, D.D. Sampson, Spatially resolved Fourier holographic light scattering angular spectroscopy. Opt. Lett. **30**, 3305–3307 (2005)
97. H. Ding, Z. Wang, F. Nguyen, S.A. Boppart, L.J. Millet, M.U. Gillette, J. Liu, M. Boppart G. Popescu, Fourier Transform Light Scattering (FTLS) of Cells and Tissues, J. Comput. Theor. Nanosci. **7**(12), 2501–2511 (2010)
98. I.T. Jolliffe, *Principal Component Analysis*, 2nd edn. (Springer, Heidelberg, October 1, 2002
99. T.W. Loong, Understanding sensitivity and specificity with the right side of the brain. Br. Med. J. **327**, 716–719 (2003)
100. B.J. Berne, R. Pecora, *Dynamic Light Scattering with Applications to Chemistry, Biology and Physics* (Wiley, New York, NY, 1976)
101. T. Kuriabova, A.J. Levine, Nanorheology of viscoelastic shells: applications to viral capsids. Phys. Rev. E **77**, 031921 (2008)

102. D. Mizuno, C. Tardin, C.F. Schmidt, F.C. MacKintosh, Nonequilibrium mechanics of active cytoskeletal networks. Science **315**, 370–373 (2007)
103. J. Liu, M.L. Gardel, K. Kroy, E. Frey, B.D. Hoffman, J.C. Crocker, A.R. Bausch, D.A. Weitz, Microrheology probes length scale dependent rheology. Phys. Rev. Lett. **96**, 118104 (2006)
104. X. Trepat, L.H. Deng, S.S. An, D. Navajas, D.J. Tschumperlin, W.T. Gerthoffer, J.P. Butler, J.J. Fredberg, Universal physical responses to stretch in the living cell. Nature **447**, 592 (2007)
105. B.D. Hoffman, G. Massiera, K.M. Van Citters, J.C. Crocker, The consensus mechanics of cultured mammalian cells. Proc. Natl. Acad. Sci. USA. **103**, 10259–10264 (2006)
106. M.L. Gardel, J.H. Shin, F.C. MacKintosh, L. Mahadevan, P. Matsudaira, D.A. Weitz, Elastic Behavior of cross-linked and bundled actin networks. Science **304**, 1301–1305 (2004)
107. T.D. Pollard, G.G. Borisy, Cellular motility driven by assembly and disassembly of actin filaments. Cell **112**, 453–465 (2003)
108. J.A. Cooper, D.A. Schafer, Control of actin assembly and disassembly at filament ends. Curr. Opin. Cell Biol. **12**, 97–103 (2000)
109. T.J. Mitchison, L.P. Cramer, Actin-based cell motility and cell locomotion. Cell **84**, 371–379 (1996)
110. J.R. Kuhn, T.D. Pollard, Real-time measurements of actin filament polymerization by total internal reflection fluorescence microscopy. Biophys. J. **88**, 1387–1402 (2005)
111. T.D. Pollard, The cytoskeleton, cellular motility and the reductionist agenda. Nature **422**, 741–745 (2003)
112. K.J. Amann, T.D. Pollard, Direct real-time observation of actin filament branching mediated by Arp2/3 complex using total internal reflection fluorescence microscopy. Proc. Natl. Acad. Sci. USA. **98**, 15009–15013 (2001)
113. J.A. Theriot, T.J. Mitchison, Actin microfilament dynamics in locomoting cells. Nature **352**, 126–131 (1991)
114. J.A. Theriot, T.J. Mitchison, L.G. Tilney, D.A. Portnoy, The rate of actin-based motility of intracellular listeria-monocytogenes equals the rate of actin polymerization. Nature **357**, 257–260 (1992)
115. D. Uttenweiler, C. Veigel, R. Steubing, C. Gotz, S. Mann, H. Haussecker, B. Jahne, R.H.A. Fink, Motion determination in actin filament fluorescence images with a spatio-temporal orientation analysis method. Biophys. J. **78**, 2709–2715 (2000)
116. C.C. Wang, J.Y. Lin, H.C. Chen, C.H. Lee, Dynamics of cell membranes and the underlying cytoskeletons observed by noninterferometric widefield optical profilometry and fluorescence microscopy. Opt. Lett. **31**, 2873–2875 (2006)
117. H. Ding, L.J. Millet, M.U. Gillette, G. Popescu, Actin-driven cell dynamics probed by Fourier Transform light scattering. Biomed. Opt. Express **1**(1), 260–267 (2010)
118. A.J. Levine, T.C. Lubensky, One- and two-particle microrheology. Phys. Rev. Lett. **85**, 1774–1777 (2000)
119. P. Dieterich, R. Klages, R. Preuss, A. Schwab, Anomalous dynamics of cell migration. Proc. Natl. Acad. Sci. USA. **105**, 459–463 (2008)
120. N. Watanabe, T.J. Mitchison, Single-molecule speckle analysis of Aactin filament turnover in lamellipodia. Science **295**, 1083–1086 (2002)
121. L.C.L. Lin, N. Gov, F.L.H. Brown, Nonequilibrium membrane fluctuations driven by active proteins. J. Chem. Phys. **124**, 074903 (2006)
122. G. Popescu, Y.K. Park, R.R. Dasari, K. Badizadegan, M.S. Feld, Coherence properties of red blood cell membrane motions. Phys. Rev. E. **76**, 031902 (2007)
123. M. Mir, Z. Wang, K. Tangella, G. Popescu, Diffraction phase cytometry: blood on a CD-ROM. Opt. Expr. **17**, 2579 (2009)
124. M. Mir, M. Ding, Z. Wang, K. Tangella, G. Popescu, Blood screening using diffraction phase cytometry. J. Biomed. Opt. **15**(2), 027016 (2010)

125. Y.K. Park, M. Diez-Silva, G. Popescu, G. Lykorafitis, W. Choi, M.S. Feld, S. Suresh, Nanoscale cell membrane fluctuation is and indicator of disease state. PNAS (under review)
126. G. Popescu, Y.K. Park, W. Choi, R.R. Dasari, M.S. Feld, K. Badizadegan, Imaging red blood cell dynamics by quantitative phase microscopy. Blood Cell. Mol. Dis. **41**, 10–16 (2008)

Part III
Image Formation and Super Resolution

Chapter 10
Coherent Microscopy for 3-D Movement Monitoring and Super-Resolved Imaging

Yevgeny Beiderman, Avigail Amsel, Yaniv Tzadka, Dror Fixler, Mina Teicher, Vicente Micó, Javier García, Bahram Javidi, Mehdi DaneshPanah, Inkyu Moon, and Zeev Zalevsky

Abstract In this chapter we present three types of microscopy-related configurations while the first one is used for 3-D movement monitoring of the inspected samples, the second one is used for super-resolved 3-D imaging, and the last one presents an overview digital holographic microscopy applications. The first configuration is based on temporal tracking of secondary reflected speckles when imaged by properly defocused optics. We validate the proposed scheme by using it to monitor 3-D spontaneous contraction of rat's cardiac muscle cells while allowing nanometric tracking accuracy without interferometric recording. The second configuration includes projection of temporally varying speckle patterns on top of the sample and by proper decoding exceeding the diffraction as well as the geometrical-related lateral resolution limitation. In the final part of the chapter, we overview applications of digital holographic microscopy (DHM) for real-time non-invasive 3-D sensing, tracking, and recognition of living microorganisms such as single- or multiple-cell organisms and bacteria.

10.1 Introduction

Microscopy is a vital part of almost any application in biology, chemistry, neuroscience, bioengineering as well as in nanoscale-related research fields. Dynamic microscopy or movement monitoring is an important part of modern research trends aiming toward comprehensive understanding of live cells interaction, like the pumping of heart muscle cells, neuron activity, and other biology-related processes. For instance, heart cells are dynamic and moving objects which consist of a broad variety of cell types such as muscle cells and connective tissue. Its functioning depends on the cooperation between its components [1]. It is possible to isolate thesetypes

Z. Zalevsky (✉)
School of Engineering, Bar Ilan University, Ramat Gan, 52900, Israel
e-mail: zalevsz@eng.biu.ac.il

P. Ferraro et al. (eds.), *Coherent Light Microscopy*, Springer Series in Surface Sciences 46, DOI 10.1007/978-3-642-15813-1_10,

of heart cells by using proteolytic enzymes such as trypsin and collagenase [2]. "Myocyte" is one of the types of cells that can be extracted from an embryonic heart using a proteolytic enzyme. Cardiovascular research is frequently based on models involving myocyte cells [3], as these cells have the ability to contract independently, while the contraction of each individual cell involves the activity of ion channels. Moreover, these cells contract in a synchronized way, as gap junctions are formed between adjacent cells [4].

Movement tracking of cells in microscopy involves several methods. Some of them are obtained using various interferometric and holographic approaches applied mainly in transmission [5, 6]. Digital interferometry is the most common approach used in either on- or off-axis phase shift configuration [7, 8]. Improvement and adjustment of those approaches for real-time measurement were presented in [9–11]. However, all those approaches are working in transmission, and thus they allow detection of only transversal movement.

The interference state of the resulting image in microscopy depends on a change in the refraction index of the cells. However, neither axial movement nor tilts do not involve change of the refraction index, and therefore these movements cannot be mapped using those approaches applied in a transmissive configuration.

One way to track movement may involve usage of speckles. The speckles are self-interference random patterns [12] and have the remarkable quality that each individual speckle serves as a reference point from which one may track the changes in the phase of the light that is being scattered from the surface [12]. Because of that, speckle techniques such as electronic speckle pattern interferometry (ESPI) have been widely used for displacement measuring and vibration analysis (amplitudes, slopes, and modes of vibration) as well as characterization of deformations [13–19]. ESPI is widely used in case of objects' deformation measurement. If one subtracts the speckle pattern before the deformation has occurred (due to change in loading, change in temperature, etc.) from the pattern after loading has occurred, it produces correlation fringes that correspond to the object's local surface displacements between the two exposures. From the fringe pattern both the magnitude and the direction of the object's local surface displacement are determined [14, 15]. Usage of speckles was also applied for improving the resolving capabilities of imaging sensors [20] as well as ranging and 3-D estimation [21].

In Sect. 10.2 of this chapter we propose a novel configuration which includes projection of laser beam and observation of the movement of the secondary speckle patterns that are created on top of the target. The new configuration and concept allowing high-accuracy detection of axial and tilted movement from remote were developed [22, 23]. This approach uses a property of defocusing of the imaging lens of the camera. Usage of speckles for movement tracking was demonstrated already before [24] without applying defocusing. However, this method (without applying defocusing) does not allow tilt extraction or on-axis movement estimation when the relative tilt angle is small. Tracking of tilting or on-axis movement becomes a simple task, when the defocusing of the lens is being used. It allows translation of the tilt movement into lateral movement of the random speckle pattern rather than the variation of this random pattern as obtained without performing the proper defocusing. Tracking this lateral movement is a simple numerical task that can be fulfilled using basic correlation-based algorithmic.

Note that in Sect. 10.2 of this chapter we discuss the implementation of this approach not only in tracking of 2-D movement of the speckle patterns to allow detection of tilting (if exists) and transversal movement, but in addition, we detect both the position and the value of the correlation peak between the sequential frames. The value of the correlation peak designates the relative change in the randomness of the speckle pattern (and not the shift of the pattern). This change in the pattern itself corresponds to the axial movement. It is obtained by observing the value of the correlation peak and comparing it to a calibration reference look-up table.

Therefore, Sect. 10.2 includes the development of 3-D movement monitoring module providing nanometric accuracy which can be applied for transmissive microscope configuration. Then, the proposed tracking technique is used for 3-D movement tracking of *several spatial segments* within cardiac muscle cells.

In Sect. 10.3 of this chapter we present a super-resolved approach projecting temporally varying speckle pattern to improve the obtainable lateral resolution.

With respect to using the speckles in order to obtain lateral super-resolved imaging it is well known that the resolution of an optical imaging system can be defined as the smallest spatial separation distance between two features that still can be resolved in the imaging system. This minimal separation distance is proportional to the optical wavelength and the F number of the imaging lens [25].

The various approaches for overcoming those limitations are called super-resolving techniques. In order to exceed the resolving limitations of an imaging system one needs to convert the spatial degrees of freedom into other domains (for encoding), to transmit them through the optical system, and then to reconstruct the image (to decode) [26–28]. The domains that may be used in order to multiplex the spatial information are time [29–31], polarization [32, 33], wavelength [34, 35], field of view or spatial dimensions [36–39], and code [40–42].

The time-multiplexing approach can be extended to projection of random speckle patterns [43–45]. This random pattern encoded the high spatial frequencies existing in the object. After time multiplexing and by knowing the encoding pattern one may decode the information and construct the high-resolving image of the object.

The final part of this chapter (Sect. 10.4) focuses on over viewing applications of digital holographic microscopy (DHM) for real-time non-invasive 3-D sensing, tracking, and recognition of living microorganisms such as single- or multiple-cell organisms and bacteria [46–53]. Digital holography allows for measurement of both phase and amplitude of an unknown wavefront and post-processing for wavefront reconstruction [6–8]. In particular, digital holographic microscopy (DHM) has been successfully applied to non-invasive real-time 3-D sensing, recognition, and tracking of microorganisms such as cells without the need for staining or destroying the cells [46–53]. In one arrangement, an on-axis, single-exposure Mach–Zehnder interferometer can be used to record digital holograms of microorganisms [46–53]. Statistical methods are applied to digital reconstruction of holograms for recognition of biological microorganisms [46–53]. Marker-free detection of diseased cells (e.g., malaria), cell sorting, and long-term non-invasive monitoring are some other applications of this approach.

10.2 Speckle-Based Movement Monitoring

In Sect. 10.2.1 we describe the experimental setup of the speckle-based configuration. Its theoretical modeling is shown in Sect. 10.2.2. The experimental results for the 3-D tracking of the entire cell are presented in Sect. 10.2.3. In Sect. 10.2.4 we experimentally extract the 3-D map and flow distribution of spatial segments *inside* cardiac muscle cells.

10.2.1 Experimental Setup

The work presented in this section describes an experimental setup that was build for speckle-based movement monitoring. We have installed this setup on a microscope. It combines the proposed configuration suited to an inverted microscope with an oblique illumination of the inspected cells with green laser at wavelength of 532 nm. An image of the experimental setup can be seen in Fig. 10.1. The setup consists of two principal components. The first is a green laser, which is used to illuminate the inspected object in order to generate the secondary reflected speckles. The second is a camera that captures video frames which later on are analyzed by MATLAB software. The software tracks the movement of the reflected speckle patterns and extracts the 2-D projection of the 3-D movement of the cells (including tilting

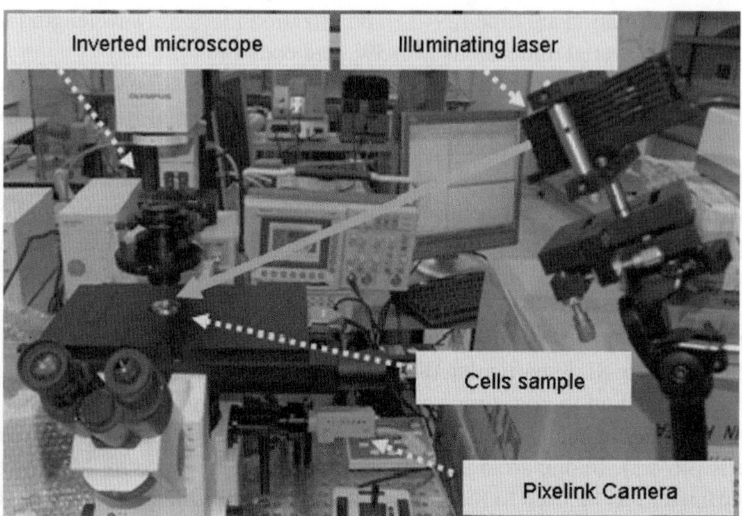

Fig. 10.1 Image of the experimental setup

movement if exists). In addition, the system can track the axial movement of the objects by observing the value (and not the position) of the correlation peak between sequential speckle images.

The proposed configuration has two significant advantages over existing approaches. First, although the proposed method is demonstrated in transmissive configuration it is similar to reflective illumination configuration. We illuminate obliquely the sample at an angle higher than the one defined by the NA of the lens. That is, the DC term of the illumination (non-diffracted illumination beam) is not transmitted through the microscope lens and only diffracted components are. Thus, it is similar to a reflective configuration where the image becomes without a background. Such illumination procedure (high oblique angle illumination in transmission or reflection configurations) maintains detection of cells' transversal movement with the same accuracy as in transmission interferometry, and additionally it allows nanometric accuracy in the detection of axial movements and tilts of the cells which cannot be detected at all in regular transmissive interferometry.

Second, the proposed setup is not an interferometric one, and thus it needs no reference arm. This is a very important difference in comparison with previous work since it significantly simplifies the experimental setup while retaining the nanometric accuracy of holographic approaches.

Rat heart cells have been put on test, due to its dynamic contradictions. Rat cardiac myocytes were isolated as previously described in [54]. Briefly, hearts from newborn rats were rinsed in phosphate-buffered saline (PBS), cut into small pieces, and incubated with a solution of proteolytic enzymes-RDB (Biological Institute, Ness Ziona, Israel). Separated cells were suspended in Dulbecco's Modified Eagle's Medium (DMEM) containing 10% inactivated horse serum (Biological Industries, Kibbutz Beit Haemek, Israel) and 2% chick embryo extract and centrifuged at $300 \times g$ for 5 min. Precipitant (cell pellet) was resuspended in growth medium and placed in culture dishes on collagen/gelatin-coated cover glasses. On day 4, cells were treated with 0.5–5 M of DOX for 18 h and then with drug-free growth medium for an additional 24 h. Cell samples were grown on microscope coverslips and placed under a microscope where they exhibited spontaneous contractions.

Cell culture and sample preparation for measurements were performed for the speckle-based configuration. The illuminated field was confined to approximately the cross section of a single cardiomyocyte. The microscope we used was an Olympus inverted epifluorescence IX81 microscope equipped with an external green laser module (see Fig. 10.1).

In order to avoid collecting the backscattered light, we installed the laser in such a manner that the illumination angle was higher than the angle defined by the NA of the objective lens. Cells were illuminated with green laser at wavelength of 532 nm from an oblique angle of about 30° and imaged through the objective lens of the microscope with proper defocusing. Thus, the illumination angle is higher than the angle defined by the NA of the microscope lens ($\text{NA} = 0.4 \Rightarrow \theta = 23.5°$, approximately). In the experiment we have used Olympus objectives of UPlanSApo 10×/0.40/0.17/FN26.5. Image was captured with Pixelink PL-A741-E camera having pixel size of $6.7 \times 6.7\ \mu\text{m}$ which was connected to the microscope. The output of the Pixelink camera was connected to a computer which was capturing video frames using the manufacturer software. The captured video was analyzed by MATLAB software.

10.2.2 Theoretical Modeling

We distinguish between three kinds of movements: longitudinal (axial), transversal (lateral), and tilting. As previously explained our 3-D movement estimation includes computation of the correlation peak for the subsequent captured speckle patterns. From the relative position of the correlation peak we extract the information regarding the transversal movement and the tilting of the object [22, 23], and from its value we obtain the information regarding the longitudinal movement.

The following mathematical model proves that when slightly defocusing, instead of changing, the speckle pattern is moving. Figure 10.2 explains our considerations. We will denote by (x, y) the coordinates of the transversal plane while the axial axis will be denoted by Z. Laser spot with diameter of D is illuminating normally a diffusive object using optical wavelength of λ . A speckle pattern was obtained onto a detector from the light reflected from the object. This random amplitude and phase pattern are generated due to the random phase of the surface of the diffusive object. In the regular case the imaging system is imaging the plane close to the object determined by distance Z_1, and in this case the amplitude distribution of the speckles equals to the Fresnel integral performed over the random phase ϕ that is created by the surface roughness.

$$\begin{aligned} T_{\mathrm{m}}(x_o,\ y_o) &= \iint \exp[i\phi(x,\ y)] \exp\left[\frac{\pi i}{\lambda Z_1}\left((x-x_o)^2+(y-y_o)^2\right)\right] \mathrm{d}x\mathrm{d}y \\ &= A_{\mathrm{m}}(x_o,\ y_o)\exp[i\psi(x_o,\ y_o)] \end{aligned} \tag{10.1}$$

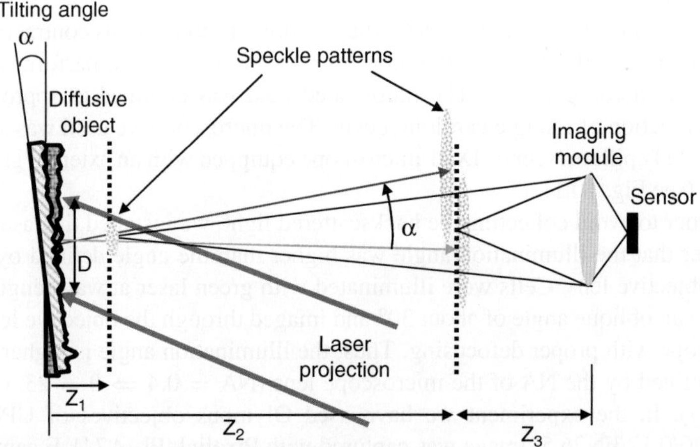

Fig. 10.2 Schematic description of the system

where a uniform reflectivity over the object's illuminated area as well as paraxial approximation has been assumed. The intensity distribution of the obtained image, which is viewed by the imaging device, equals to

$$I(x_{\mathrm{s}}, y_{\mathrm{s}}) = \left| \iint T_{\mathrm{m}}(x_{\mathrm{o}}, y_{\mathrm{o}}) h(x_{\mathrm{o}} - Mx_{\mathrm{s}}, y_{\mathrm{o}} - My_{\mathrm{s}}) \right|^2 \tag{10.2}$$

where h is the spatial impulse response of the system and M is the inverse of the magnification of the imaging system. The parameter h takes into account the blurring due to the optics as well as due to the size of the pixels in the sensor, and it is computed in the sensor plane (x_{s},y_{s}).

$$M = \frac{(Z_2 + Z_3 - Z_1) - F}{F} \tag{10.3}$$

F is the focal length of the imaging lens. Note that in the case of remote inspection, typically the distance between the object and the lens is much longer than any other distance involved in the process.

As mentioned before, assuming a rigid body movement, the movement of the object can be classified into three types of movements which cannot be separated and they occur simultaneously: transverse, axial, and tilt. Transversal movement is commonly observed in microscopy. However, under transversal movement the amplitude distribution of the speckle pattern T_{m} will simply shift in x and y by the same distance as the movement of the object, as can be checked in (10.1). Under normal imaging conditions and small vibrations, this movement will be demagnified by the imaging systems resulting in barely detectable shifts on the image plane. The second type of movement is axial movement in which the speckle pattern will remain basically the same since the variations in Z_1 (which will only scale the resulted distribution) are significantly smaller in comparison to the magnification of the camera:

$$M = \frac{Z_2 + Z_3}{F} \approx \frac{Z_2 - Z_1 + Z_3}{F} \tag{10.4}$$

The third type of movement is the tilt of the object which may be expressed as follows:

$$\begin{aligned} A_{\mathrm{m}}(x_{\mathrm{o}}, y_{\mathrm{o}}) &= \left| \iint \exp\left[\mathrm{i}\phi(x, y)\right] \exp\left[\mathrm{i}(\beta_x x + \beta_y y)\right] \right. \\ &\quad \left. \times \exp\left[\frac{\pi i}{\lambda Z_1}\left((x - x_{\mathrm{o}})^2 + (y - y_{\mathrm{o}})^2\right)\right] \mathrm{d}x\mathrm{d}y \right| \\ \beta_x &= \frac{4\pi \tan \alpha_x}{\lambda} \\ \beta_y &= \frac{4\pi \tan \alpha_y}{\lambda} \end{aligned} \tag{10.5}$$

The angles α_X and α_Y are the tilt in the X- and Y-axes, respectively, and the factor of 2 accounts for the back and forth axial distance change. In this case it is well seen from the last equation that the resulting speckle pattern will change completely. The magnification factor of the system as described in (10.2) can be very large (M can be a few hundred) due to blurring, thus causing the small speckles of the imaging system to have a large size.

The speckle pattern actually varies randomly, since the three types of movements basically cannot be separated. It should be noted that for small Z_1 values the size of the speckle pattern at the Z_1 plane will be very small and will not be visible in the sensor after imaging with large demagnification. The speckles associated with the aperture of the lens are dominant, due to the blurring width of $\lambda F_{\#}$ which is properly magnified when transferred to Z_1 plane. Under small Z_1 values the speckles are under-sampled on the detector plane.

Assuming now that we strongly defocus the image captured by the camera. Defocusing brings the plane of the imaging from position at distance of Z_1 into a plane positioned at distance of Z_2. In this case several changes occur: first, the magnification factor M is relatively small (it is reduced at least by one order of magnitude); second, the plane at which the speckle pattern that is imaged by the camera is formed by the far-field approximation regime (the relevant speckle plane is far from the object). Therefore, the equivalent of (10.1) and (10.2) is

$$\begin{aligned} T_{\mathrm{m}}(x_{\mathrm{o}}, y_{\mathrm{o}}) &= \iint \exp\left[\mathrm{i}\phi(x, y)\right] \exp\left[\frac{-2\pi i}{\lambda Z_2}(x x_{\mathrm{o}} + y y_{\mathrm{o}})\right] \mathrm{d}x\mathrm{d}y \\ &= A_{\mathrm{m}}(x_{\mathrm{o}}, y_{\mathrm{o}}) \exp\left[\mathrm{i}\psi(x_{\mathrm{o}}, y_{\mathrm{o}})\right] \end{aligned} \tag{10.6}$$

and

$$I(x_{\mathrm{s}}, y_{\mathrm{s}}) = \left| \iint T_{\mathrm{m}}(x_{\mathrm{o}}, y_{\mathrm{o}}) h(x_{\mathrm{o}} - M x_{\mathrm{s}}, y_{\mathrm{o}} - M y_{\mathrm{s}}) \right|^2 \tag{10.7}$$

Note that now also

$$M = \frac{Z_3 - F}{F} \approx \frac{Z_3}{F} \tag{10.8}$$

Therefore, in the case of transversal movement the speckle pattern is almost unchanged since shift does not affect the amplitude of the Fourier transform and because the magnification of the blurred function h is much smaller. Axial movement does not affect the distribution at all as well since Z_2 is much larger than the shifts of the movement (only a constant phase is added in (10.6)).

The shifting property of Fourier transform states that a linear phase factor in the original function will be converted to a constant shift of the Fourier of the object in the Fourier domain [55]. Using this property one can easily see how the tilt movement of speckles (projected onto object) translates into the spatial shift. Thus, tilting is simply expressed as shifts of the speckle pattern. It can be seen in Fig. 10.2 as well as understood from the following equation:

$$
\begin{aligned}
A_{\mathrm{m}}(x_{\mathrm{o}}, y_{\mathrm{o}}) &= \left| \iint \exp\left[\mathrm{i}\phi(x, y)\right] \exp\left[\mathrm{i}(\beta_x x + \beta_y y)\right] \right. \\
&\quad \left. \times \exp\left[\frac{-2\pi i}{\lambda Z_2}(x x_{\mathrm{o}} + y y_{\mathrm{o}})\right] \mathrm{d}x\mathrm{d}y \right| \\
\beta_x &= \frac{4\pi \tan\alpha_x}{\lambda} \\
\beta_y &= \frac{4\pi \tan\alpha_y}{\lambda}
\end{aligned}
\tag{10.9}
$$

The three types of movement are not separated, but since now two of them produce negligible variations in the imaged speckle pattern, the overall effect of three of them is only the pure shift which may easily be detected by spatial pattern correlation.

The resolution or the size of the speckle patterns that is obtained at Z_2 plane and imaged to the sensor plane equals to

$$\delta x = \frac{\lambda Z_2}{D} \cdot \frac{1}{M} = \frac{\lambda F}{D} \cdot \frac{Z_2}{Z_3} \tag{10.10}$$

This is of course assuming that this size of δx is larger than (and therefore is not limited by) the optical as well as the geometrical resolution of the imaging system.

The conversion of angle to the displacement of the pattern on the camera is as follows:

$$d = \frac{Z_2 \alpha}{M} = \frac{Z_2 F}{Z_3} \tag{10.11}$$

Note that assuming that Δx is the size of the pixel in the detector then the requirement for the focal length is (we assume that every speckle in this plane will be seen at least by K pixels)

$$F = \frac{K \Delta x \, Z_3 D}{Z_2 \lambda} \tag{10.12}$$

Note that Z_2 fulfills the far-field approximation:

$$Z_2 > \frac{D^2}{4\lambda} \tag{10.13}$$

The number of speckles in every dimension of the spot equals to

$$N = \frac{\phi}{M \delta x} = \frac{\phi D}{\lambda Z_2} = \frac{F \cdot D}{F_{\#} \lambda Z_2} \tag{10.14}$$

where ϕ is the diameter of the aperture of the lens. $F_{\#}$ is the F number of the lens. The term of $M\delta x$ is the speckle size obtained at plane of Z_2. This relation is obtained since the spot of the lens is covered by the light coming from the reflecting surface of the object.

Equations (10.12) and (10.14) determine the requirements of the focal length of the imaging system. There are contradictory requirements on both equations. On one hand, from (10.12) it is better to have small F; since then the speckles are large in the Z_2 plane (especially for large Z_2), and it is preferred to increase the demagnification factor such that it will be easier to see the speckles with the pixels of the detector. In (10.14) we prefer larger F in order to have more speckles per spot. Therefore, a point of optimum may be found.

Another interesting problem is a longitudinal and transversal movement recovery based on observing changes in the correlation function's amplitude. Let us assume a circular uniform intensity at the object plane, being the diffusive source that generates the secondary speckles. Then, the dependence of the value of the correlation function on the transversal and the longitudinal positions may be modeled as follows: the correlation function of the intensity between two points separated in the transverse plane by a radial distance s and which are located at distance z from the diffusive object may be estimated as follows [56]:

$$\Gamma_{\text{Transversal}}(s) = \bar{I}^2 \left(1 + 2\left|\frac{J_1\left(\pi \Phi s/\lambda z\right)}{\pi \Phi s/\lambda z}\right|\right) \tag{10.15}$$

where $\bar{I}$ is the mean of the intensity in the output plane, Φ is the diameter of the illumination beam, J_1 is the first kind Bessel function, λ is the optical wavelength, and z is the axial distance.

With respect to the longitudinal extent of the speckles, the problem reduces to the calculation of the axial correlation function of the intensity between two points separated axially by Δz. In this case, with the same assumptions as in the transverse case plus the requirement that Δz will be small compared with z, the correlation of intensity results with [56]

$$\Gamma_{\text{Longitudinal}}(\Delta z) = \bar{I}^2 \left(1 + 2\left|\sin c\left(\Delta z \cdot \Phi^2/8\lambda z^2\right)\right|\right) \tag{10.16}$$

10.2.3 Experimental Results

Rat heart muscle cells as presented in Sect. 10.2.1 were dynamically mapped under their repeatable contraction. The cells were observed using inverted microscope (as mentioned in Sect. 10.2.1). In Fig. 10.3 we present image sequence extracted from video recording using regular white light illumination (not laser). In Fig. 10.3a one may see the image captured just before contraction. In Fig. 10.3b we present the image of differences between "before contraction" and "at maximal contraction"

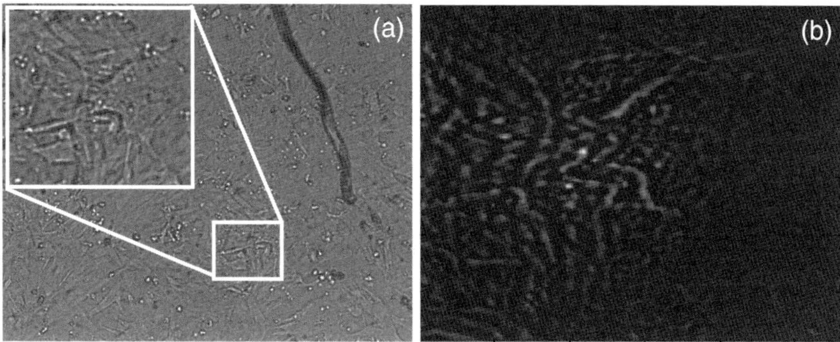

Fig. 10.3 Image sequence extracted from video using regular white light illumination (not laser). (**a**) Before contraction. (**b**) Image of differences between "before contraction" and "at maximal contraction" states for the zoomed region appearing in the *white rectangle* of (a)

states for the zoomed region appearing in the white rectangle of Fig. 10.3a in order to demonstrate that indeed a contraction has occurred.

The contraction presented in Fig. 10.3 was tracked using the speckle-based approach that was previously described in [22, 23]. The obtained results can be seen in Fig. 10.4. The movement tracking presented in Fig. 10.4 is the 2-D projection of the 3-D movement of the cell (X- and Y-axes projection). In the figure one may see how these cells repeatedly expand and shrink in fixed intervals. The vertical units in Fig. 10.4 are distance of movement in micrometers as calculated from the pixels of the camera (using the given magnification of the lens), and the horizontal axis is the temporal frames that are captured at rate of 30 fps. The analyzed region of interest was of 80×80 pixels. In the lower part of the image one may see a single diffused speckle pattern that was used to perform the movement tracking.

If we observe Fig. 10.4 again one may see that movement of even 1/20 of a pixel size (in the vertical axis) was detected. Since an objective of $10\times$ was used, this means that the equivalent cell movement was of $0.67\,\mu\text{m}/20 = 33.5\,\text{nm}$. As mentioned before, this accuracy capability of the proposed optical configuration in tracking cell movement includes tracking not only this movement in the transversal plane but also movement due to their tilt and axial-related shifts.

In Fig. 10.5 we present the experimental calibration which allows the measurement of the object's axial movement. The measurement was performed using the same configuration as in the previous experiment. In the chart of Fig. 10.5 we have positioned a reflective object connected to a controlled piezoelectric stage capable of shifting the object in the axial dimension with nanometric resolution. For each movement we have measured the obtained correlation peak while trying to show the relation existing between the value of this peak and the axial displacement.

To demonstrate repeatability we have repeated the experiment with different movement steps of 50, 100, and 200 nm. One may see that for all steps almost full coincidence occurs. If the signal-to-noise ratio allows detection of 1% change in the height of the correlation peaks, it means that the detection accuracy for the axial movement is about 120 nm.

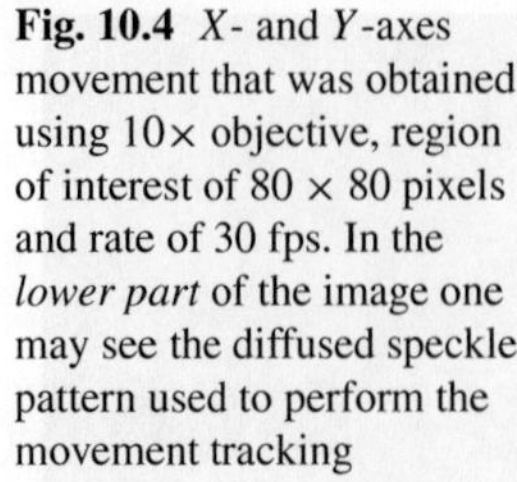

Fig. 10.4 X- and Y-axes movement that was obtained using 10× objective, region of interest of 80 × 80 pixels and rate of 30 fps. In the *lower part* of the image one may see the diffused speckle pattern used to perform the movement tracking

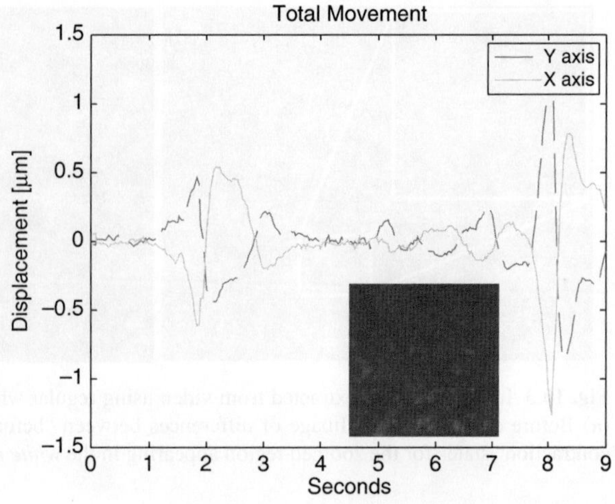

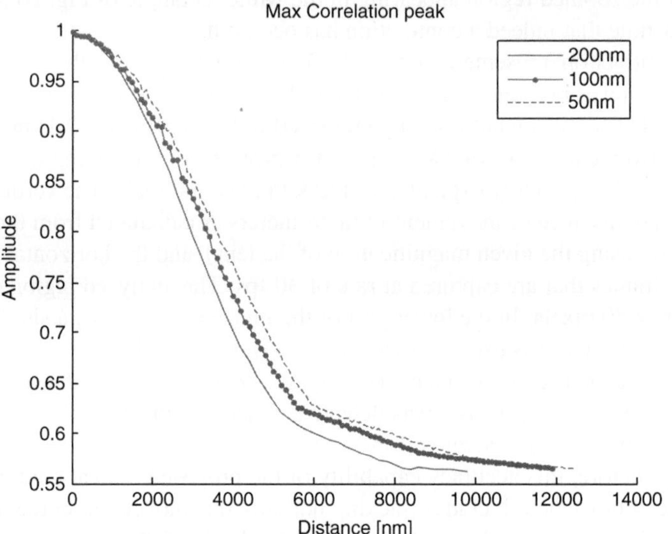

Fig. 10.5 Experimental measurement of the axial movement that is obtained by mapping the amplitude of the correlation peak versus controlled axial shifts

In Fig. 10.6 we present how the Z-movement calibration chart of Fig. 10.5 was applied to track the Z-movement of rat's cardiac muscle cells. The figure shows the normalized correlation peak of first frame with other frames. By comparing the change in the amplitude of the correlation peak one may see that for instance the 0.4 change in the value (from 1 to 0.6) of the correlation peak corresponds to axial movement of approximately 6 μm (following the calibration presented in Fig. 10.5). Note that when we say movement (e.g., of 6 μm) we refer to the change in the optical path, i.e., the product between the change in the refraction index (obtained due to the contraction of the cell) and the actual displacement.

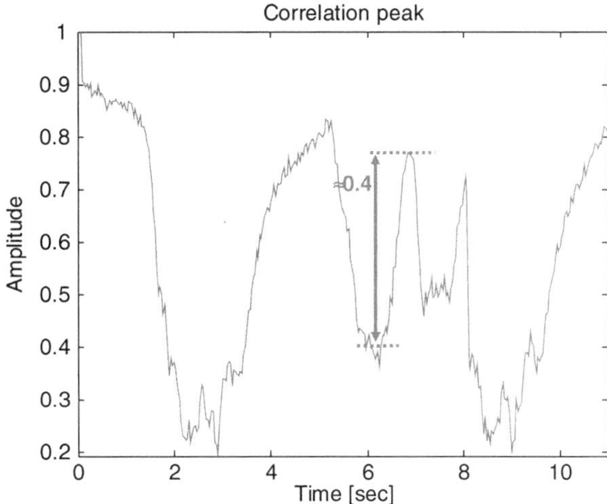

Fig. 10.6 Normalized correlation peak of first frame with other frames that is used for tracking of axial movement of rat's cardiac muscle cells

10.2.4 Measuring 3-D Map and Flow Distribution

In Figs. 10.4 and 10.6 we measured the movement of the entire cardiac muscle cell (or of its center of mass). Now we used the constructed module to map the movement of spatial segments within the cell in order to provide a tool to study movement interaction between adjustment cell parts. The extracted results were obtained by spatially dividing each image in the sequence into 4×4 resolution regions. In each region correlation was computed between subsequent frames in the image sequence. From the position of the correlation peak we estimated the in-plane 2-D movement of the cells, and from the value of the correlation peak we extracted the axial movement of the relevant segment. The conversion of the correlation peak value to axial displacement was done following the calibration presented in Fig. 10.5. Since the proposed processing was applied in each one of the 4×4 resolution regions, a 3-D displacement map of the cell was extracted.

Figure 10.7a presents the white illumination muscular cell image, with marked area of inside part of the cell that had been chosen for fragmented processing. Figure 10.7b is a zoom over the inner marked area in Fig. 10.7a.

The obtained experimental results for the 3-D mapping can be seen in Figs. 10.8 and 10.9. In Fig. 10.8 we present the overall time-variant 3-D movement of fragmented cell areas. The red lines represent the X–Y movement, starting from the origin (center of each area). The yellow bars represent the axial movement versus time (starting from left to right). In the left bottom corner of the figure we marked the scale for the lines and bars as 300 nm and 5 μm, respectively. Note once again that when we say movement we refer to the change in the optical path.

In Fig. 10.9 we present the temporal evolving of the 3-D movement mapping of the cells. A dot represents a geometrical center of the processed region of interest (ROI). Background is a picture of the call taken in white illumination. The red lines represent the relative X–Y movement in a time frame between two subsequent

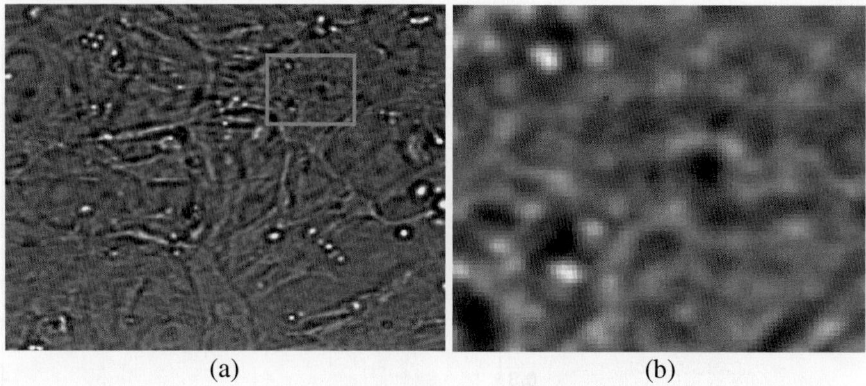

Fig. 10.7 (**a**) White illumination muscular cell image, with *marked area* of the internal part of the cell that had been chosen for fragmented processing. (**b**) Zoom over the *inner marked area* in (a)

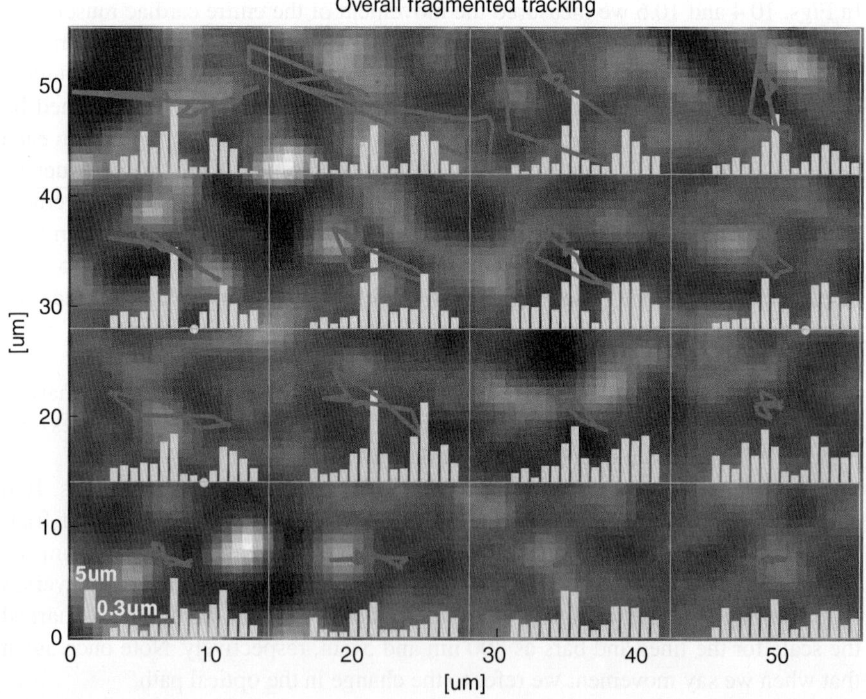

Fig. 10.8 Overall time-variant 3-D movement of fragmented cell areas. The *red lines* represent the X–Y movement, starting from origin (center of each area). The *yellow bars* represent the axial movement versus time (*starting from left to right*). In the *left bottom corner* we marked the scale for the lines and bars as 300 nm and 5 μm, respectively

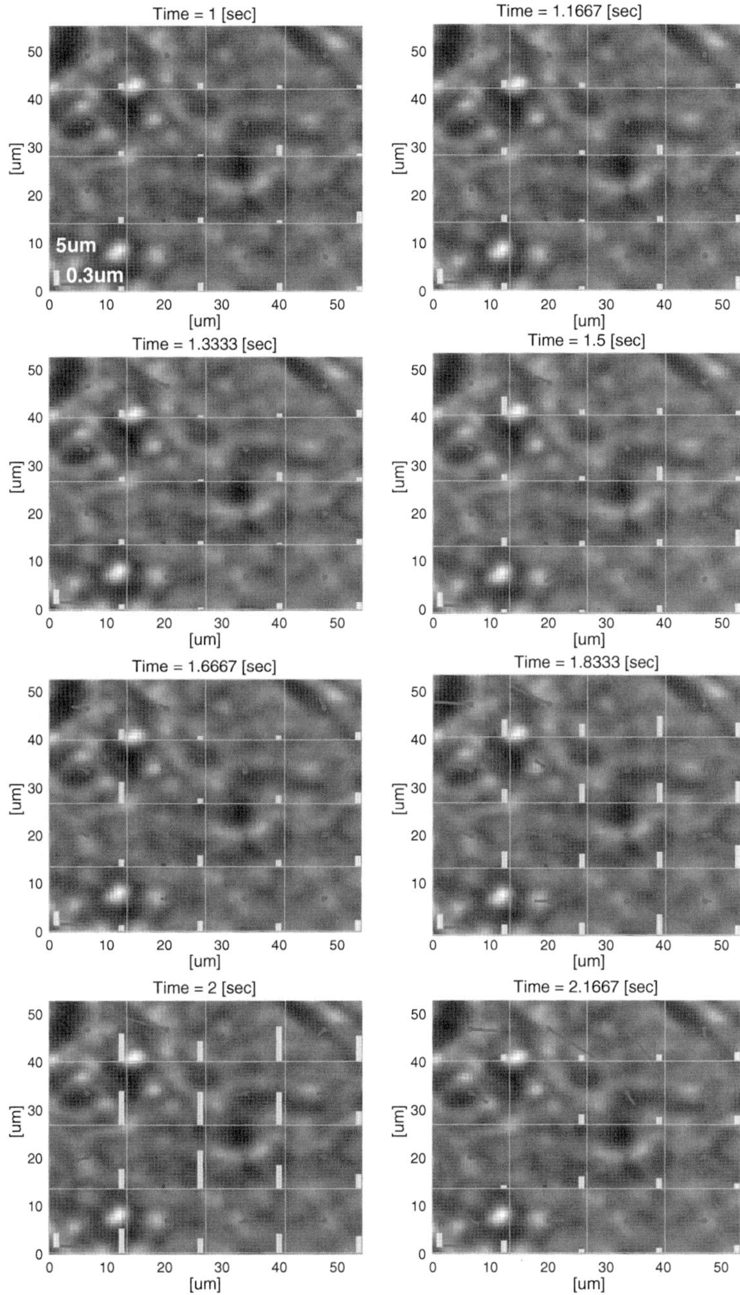

Fig. 10.9 Temporal evolving of the 3-D movement mapping of the cells. The *dot* represents a geometrical center of the processed ROI. Background is a picture of the cell taken in white illumination. The *red lines* represent the X–Y movement (starting from the center of each area). The *yellow bars* represent the axial movement. In the *left bottom corner* we marked the scale for the lines and bars as 300 nm and 5 μm, respectively (those numbers appear in the first image obtained for time = 1 s)

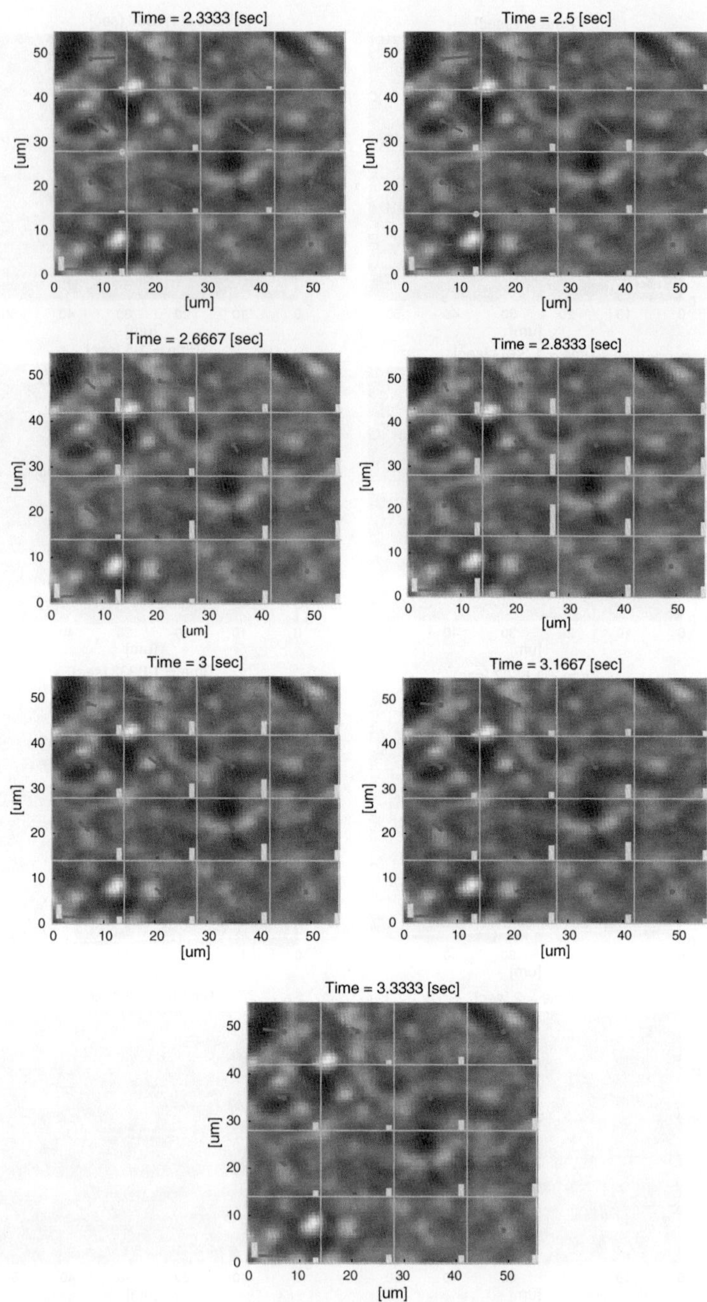

Fig. 10.9 (continued)

figures. This movement is presented as a relative vector change from last processed time point (the origin of each vector is in the center of each area). The yellow bars represent the axial movement. In the left bottom corner we marked the scale for the lines and bars as 300 nm and 5 μm, respectively (those numbers appear in the first image obtained for time = 1 s). The sampling time of each movement map is designated on top of the image. The time difference between images is of about 0.1667 s.

10.3 Super-Resolution by Means of Projected Speckle Patterns

In previous section we have described a novel technique for high-resolution mapping of tilting and axial movement of objects by usage of properties of secondary speckle pattern and a proper optical adjustments. In this section we present a super-resolution technique improving transversal resolution using high-resolution time-varying random speckle pattern.

Projection of high-resolution time-varying random speckle pattern can be used to improve the imaging resolution of a given lens. The technique basically includes capturing a set of low-resolution images (obtained after projecting the random high-resolution speckle pattern), multiplying each by the decoding pattern that equals to the one that was projected and summing all the images. After this summation which is actually a time-averaging procedure one may obtain high-resolution reconstruction.

The presented analysis is made in one dimension for simplicity, the extension to two dimensions being straightforward. We denote by $g(x)$ the object and by $h(x)$ the PSF of the low-resolution imager (due to diffraction as well as due to the geometry of the pixels). $f_m(x)$ is the projected high-resolution random pattern while m denotes the index of the projected pattern in the random sequence (the time-multiplexing sequence). We denote by $o(x)$ the captured field distribution. The intensity at the detector equals to

$$|o(x)|^2 = \iint h(x_1')h^*(x_2')f_m(x-x_1')f_m^*(x-x_2')g(x-x_1')g^*(x-x_2')\mathrm{d}x_1'\mathrm{d}x_2' \tag{10.17}$$

We do decoding (the decoding pattern is reconstructed from the low-resolution image due to the low duty cycle ratio). The obtained expression is

$$|o(x)|^2 = \iint h(x_1\prime)h^*(x_2')g(x-x_1')g^*(x-x_2')f_m(x-x_1')f_m^*(x-x_2')|f_m(x)|^2\mathrm{d}x_1'\mathrm{d}x_2' \tag{10.18}$$

After time averaging (i.e., summation) one obtains

$$\sum_m f_m(x-x_1')f_m^*(x-x_2')f_m(x)f_m^*(x) = \delta(x_1')\cdot\delta(x_2') + k \tag{10.19}$$

where k is a constant. The final result of this derivation becomes

$$|o(x)|^2 = \left|\int h(x_1)\mathrm{d}x_1\right|^2 \cdot |g(x)|^2 + k \cdot \left|\int g(x_1)h(x - x_1)\mathrm{d}x_1\right|^2 \qquad (10.20)$$

The last equation shows that the reconstructed intensity equals to the intensity of the object (in high resolution) multiplied by a constant. On top of the reconstructed image we have the low-resolution image (the image that is seen through the low-resolution imager without applying the super-resolving technique). To obtain the proper reconstruction of the high-resolution image one needs of course to subtract this low-resolution term.

Below we show an experimental demonstration of the discussed idea. In Fig. 10.10a we show the original image that when imaged through the sensor becomes the low-resolution image of Fig. 10.10b. After applying the super-resolving approach with 400 projections we obtain the result of Fig. 10.10c that clearly shows the resolution enhancement effect. In the experiment we used a UI1220 CMOS Micron camera with 752×480 pixels of 6 μm. The lens had focal length of 2.8 mm, and the camera was positioned 28 cm away from the target.

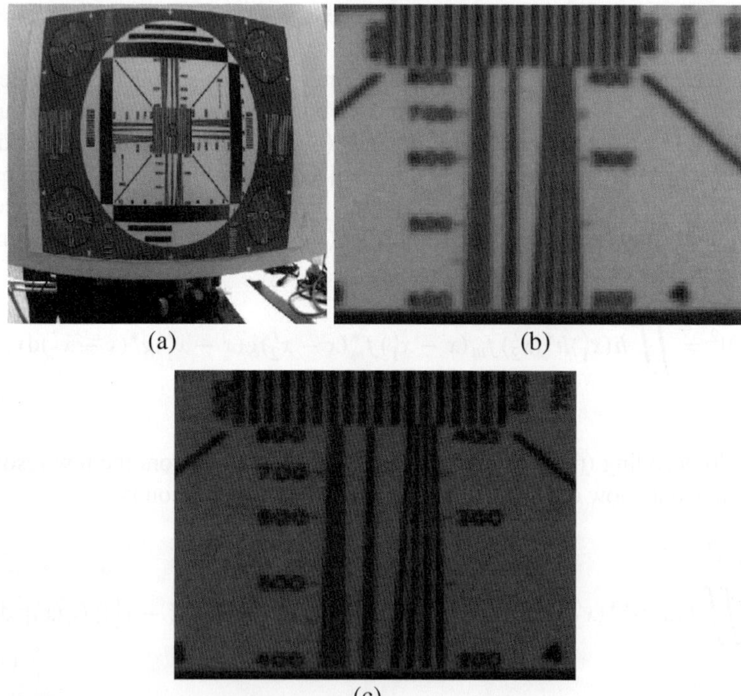

Fig. 10.10 Experimental demonstration. (**a**) The original test target. (**b**) The low-resolution image. (**c**) The super-resolved reconstruction

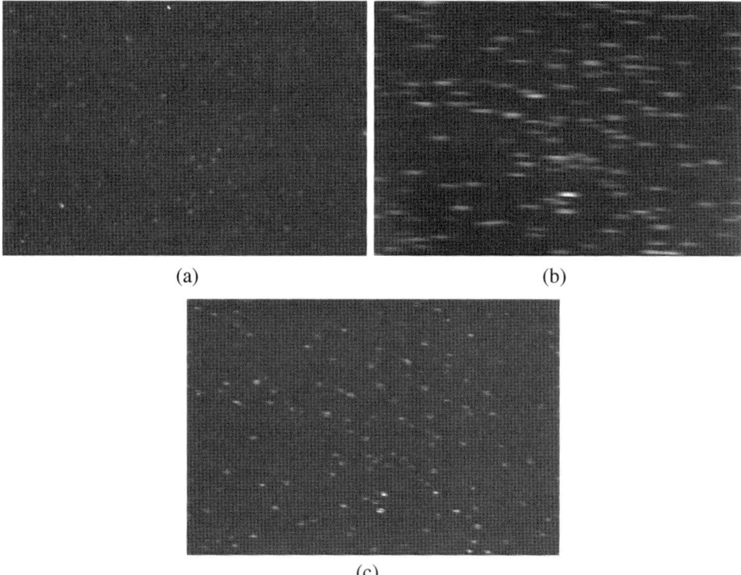

Fig. 10.11 Two-dimensional resolution enhancement with cells. (**a**) High-resolution reference image. (**b**) Low-resolution image. (**c**) Reconstruction after image processing

Obviously the technique fits well to microscopy-related applications as well in which the projections as well as the decoding are performed from a closer range. Thus, in Fig. 10.11 we repeated similar 2-D super-resolving experiment with cell samples when the laser projecting the speckles was installed on top of an inverted microscope. In Fig. 10.11a one may see the high-resolution reference image when high NA objective lens is used (a 20× microscope objective with NA of 0.4). The low-resolution images captured when a low NA lens is placed are shown in Fig. 10.11b. Figure 10.11c depicts the reconstructed image after some image processing manipulations. One may clearly see the high resemblance between the reference image of Fig. 10.11a and the reconstructed image of Fig. 10.11c.

10.4 Real-Time Non-invasive Identification of Cells and Microorganism Using Digital Holographic Microscopy

In this section, we present an overview of applications of digital holographic microscopy (DHM) for real-time non-invasive 3-D sensing and recognition of living microorganisms such as single- or multiple-cell organisms and bacteria [46–53]. This approach is attractive for non-invasive real-time identification of microorganisms without the need for staining or destroying the cells. Both coherent illumination and partially coherent illumination can be used. Use of partially coherent light for 3-D cell identification with DHM is demonstrated in [53]. A filtered white light source (e.g., xenon lamp), LED, or laser diode can provide illumination with line width ranging from 1 to 20 nm. White light or a broadband sources allow us to use various spectral bands which may be useful in identification of the microorganisms. The short coherence length of the source is not as critical in Gabor geometry in which the object and reference beams propagate on the same path. The illuminating beam is partially diffracted by the specimen, and meanwhile a fraction of the beam passes through without diffraction (semitransparent specimen) which forms the reference wavefront. Figure 10.12 shows an on-axis DHM setup with filtered white light source that is more compact and stable. In one experiment, digital holograms of plant stem cells are recorded by partially coherent DHM method [53].

The specimen is sandwiched between two transparent coverslips, and the exiting wavefront is magnified using a microscope objective with NA $= 0.80$ and $100\times$ magnification. Figure 10.13 shows two planes of corn stem cells reconstructed at 25 and 29 μm from the hologram plane. In our implementation, a xenon lamp is filtered with 10-nm filter centered at 520 nm [53].

A number of methods can be used for computational reconstruction including Fresnel transformation, convolution, and angular spectrum approaches. Since the

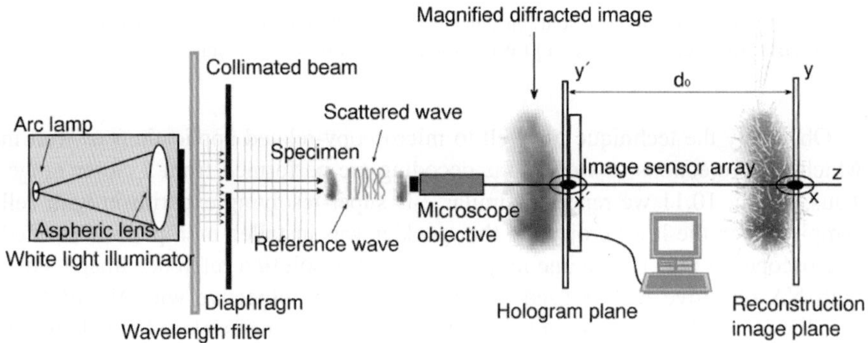

Fig. 10.12 Partially coherent DHM in the Gabor configuration provides a compact and stable tool

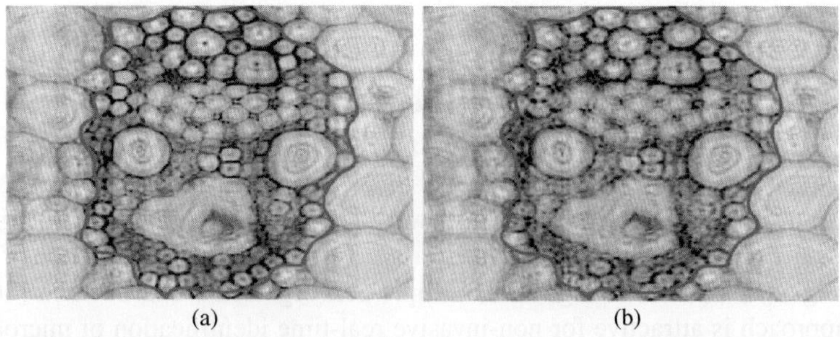

Fig. 10.13 Amplitude contrast image of corn stem cell reconstructed from partially coherent DHM. The central cells are in focus in reconstruction at distance (**a**) 25 μm, while they are defocused at (**b**) 29 μm from the hologram plane [53]

reconstruction of the wavefront is computational, most of the unwanted effects and aberrations can be compensated for digitally.

In [52, 53] it has been shown that statistical hypothesis testing can be applied to distinguish between different classes of microorganisms. The Gabor transformed complex amplitude reconstruction information is fed into the hypothesis testing block for recognition of different species (see Fig. 10.14).

The test statistics for null hypothesis of reference and input data are given, respectively, by [52]

$$D_{\mathrm{ref}} = E\left[F_S^{\mathrm{ref}}(u) - \hat{F}_S^{\mathrm{ref}}(u)\right]^2 \quad \text{and} \quad D_{\mathrm{inp}} = E\left[F_S^{\mathrm{ref}}(u) - F_S^{\mathrm{inp}}(u)\right]^2 \tag{10.21}$$

where $F_S^x(u)$ is the empirical cumulative distribution function of the object x reconstruction data. Monte Carlo technique is used to compute the associated p value for each hypothesis. To evaluate the performance of the identification system,

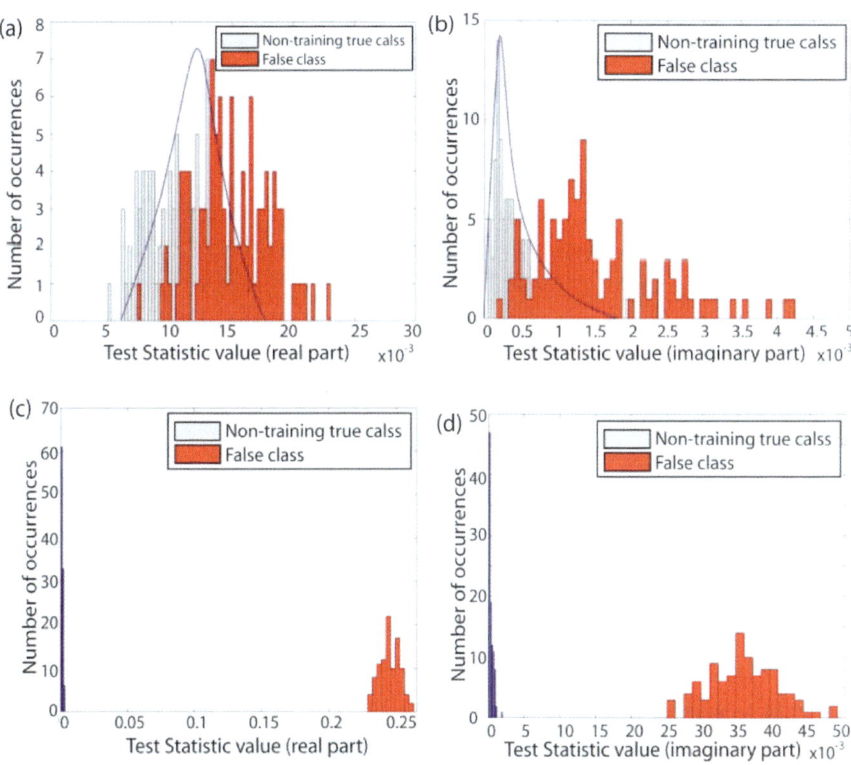

Fig. 10.14 Effectiveness of Gabor transformation. (**a**) Real and (**b**) imaginary part of the test statistic D_{inp} for corn and sunflower cells without Gabor transform. (**c**) Real and (**d**) imaginary part of the same statistic obtained from Gabor transformed holograms. Ratio R for the non-training true and false classes using the (**e**) real part and (**f**) imaginary part of the reconstructions from their Gabor filtered digital hologram

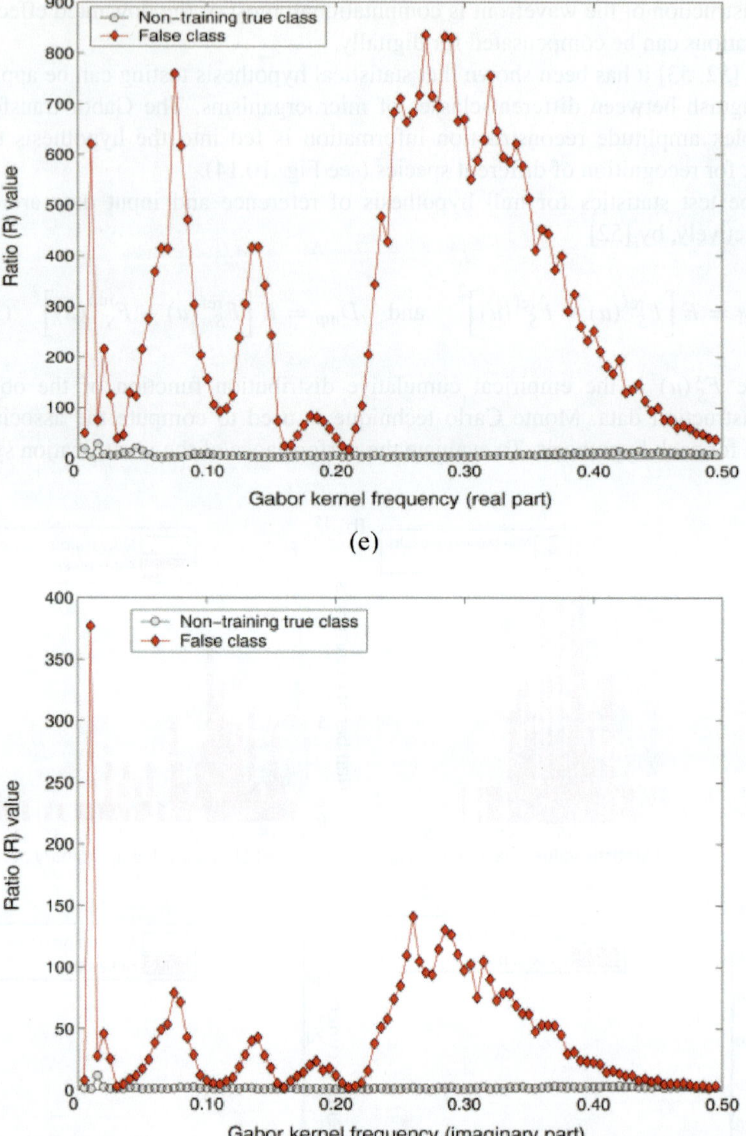

Fig. 10.14 (continued)

the following ratio is defined between the null hypothesis (true class) and the unknown input:

$$R = \frac{\text{Averaged test statistic's value for the unknown input data}}{\text{Averaged test statistic's value for the null hypothesis}} \tag{10.22}$$

It should also be noted that the decision about classification of a sample could be based on the whole range of different Gabor kernel frequencies. Thus, the aggregated information in a frequency range (and not a single frequency) would easily discriminate between the two different species.

In summary, digital holographic microscopy provides a rich content for applying different statistical methods for detection and recognition of biological microorganisms. The applications range from detection of diseased cells, marker-less cell sorting, pathogen identification, etc. Statistical pattern recognition techniques are used to track and recognize biological species from one another.

References

1. B.S. Burlew, K.T. Weber, Connective tissue and the heart. Functional significance and regulatory mechanisms. Cardiol. Clin. **18**, 435–442 (2000)
2. J.O. Bustamante, T. Watanabe, T.F. McDonald, Nonspecific proteases: a new approach to the isolation of adult cardiocytes. Can. J. Physiol. Pharmacol. **60**, 997–1002 (1982)
3. J.S. Mitcheson, J.C. Hancox, A.J. Levi, Cultured adult cardiac myocytes: future applications, culture methods, morphological and electrophysiological properties. Cardiovasc. Res. **39**, 280–300 (1998)
4. S.L. Jacobson, H.M. Piper, Cell cultures of adult cardiomyocytes as models of the myocardium. J. Mol. Cell. Cardiol. **18**, 661–678 (1986)
5. T. Ikeda, G. Popescu, R.R. Dasari, M.S. Feld, Hilbert phase microscopy for investigating fast dynamics in transparent systems. Opt. Lett. **30**, 1165–1167 (2005)
6. P. Marquet, B. Rappaz, P.J. Magistretti, E. Cuche, Y. Emery, T. Colomb, C. Depeursinge, Digital holographic microscopy: a noninvasive contrast imaging technique allowing quantitative visualization of living cells with subwavelength axial accuracy. Opt. Lett. **30**, 468–470 (2005)
7. I. Yamaguchi, T. Zhang, Phase-shifting digital holography. Opt. Lett. **22**, 1268–1270 (1997)
8. U. Schnars, W. Jüptner, Direct recording of holograms by a CCD target and numerical reconstruction. Appl. Opt. **33**, 179–181 (1994)
9. N.T. Shaked, M.T. Rinehart, A. Wax, Dual-interference-channel quantitative-phase microscopy of live cell dynamics. Opt. Lett. **34**, 767–769 (2009)
10. N.T. Shaked, M.T. Rinehart, A. Wax, Dynamic quantitative phase microscopy of biological cells, in *Conference on Lasers and Electro-Optics/International Quantum Electronics Conference*, OSA Technical Digest (CD) (Optical Society of America, 2009), paper CFA4, Baltimore, Maryland, United States. 2–4 June, 2009
11. N.T. Shaked, Y. Zhu, M.T. Rinehart, A. Wax, Two-step-only phase-shifting interferometry with optimized detector bandwidth for microscopy of live cells. Opt. Express **17**, 15585–15591 (2009)
12. J.C. Dainty, *Laser Speckle and Related Phenomena*, 2nd edn. (Springer, Berlin, 1989)
13. H.M. Pedersen, Intensity correlation metrology: a comparative study. Opt. Acta **29**, 105–118 (1982)
14. J.A. Leedertz, Interferometric displacement measurements on scattering surfaces utilizing speckle effects. J. Phy. E Sci. Instrum. **3**, 214–218 (1970)
15. P.K. Rastogi, P. Jacquot, Measurement on difference deformation using speckle interferometry. Opt. Lett. **12**, 596–598 (1987)
16. T.C. Chu, W.F. Ranson, M.A. Sutton, Applications of digital-image-correlation techniques to experimental mechanics. Exp. Mech. **25**, 232–244 (1985)
17. W.H. Peters, W.F. Ranson, Digital imaging techniques in experimental stress analysis. Opt. Eng. **21**, 427–431 (1982)

18. N. Takai, T. Iwai, T. Ushizaka, T. Asakura, Zero crossing study on dynamic properties of speckles. J. Opt. (Paris) **11**, 93–101 (1980)
19. K. Uno, J. Uozumi, T. Asakura, Correlation properties of speckles produced by diffractal-illuminated diffusers. Opt. Commun. **124**, 16–22 (1996)
20. J. García, Z. Zalevsky, P. García-Martínez, C. Ferreira, M. Teicher, Y. Beiderman, A. Shpunt, 3D Mapping and range measurement by means of projected speckle patterns. Appl. Opt. **47**, 3032–3040 (2008)
21. J. Garcia, Z. Zalevsky, D. Fixler, Synthetic aperture superresolution by speckle pattern projection. Opt. Exp. **13**, 6073–6078 (2005)
22. Z. Zalevsky, J. Garcia, Motion detection system and method. Israeli Patent Application No. 184868 (July 2007); WO/2009/013738 International Application No PCT/IL2008/001008 (July 2008)
23. Z. Zalevsky, Y. Beiderman, I. Margalit, S. Gingold, M. Teicher, V. Mico, J. Garcia, Simultaneous remote extraction of multiple speech sources and heart beats from secondary speckles pattern. Opt. Expr. **17**, 21566–21580 (2009)
24. J.A. Leedertz, Interferometric displacement measurements on scattering surfaces utilizing speckle effects. J. Phy. E Sci. Instrum. **3**, 214–218 (1970)
25. E. Abbe, Beitrage zur theorie des mikroskops und der mikroskopischen wahrnehmung, Arch. Mikrosk. Anat. **9**, 413–468 (1873)
26. Z. Zalevsky, D. Mendlovic, *Optical Super Resolution* (Springer, Heidelberg, 2002)
27. Z. Zalevsky, D. Mendlovic, A.W. Lohmann, *Progress in Optics*, vol. XL, Chapter 4 (Elsevier, 1999)
28. D. Mendlovic, A.W. Lohmann, Z. Zalevsky, SW – Adaptation and its application for super resolution – Examples. JOSA **14**, 563–567 (1997)
29. W. Lukosz, Optical systems with resolving powers exceeding the classical limits. J. Opt. Soc. Am. **56**, 1463–1472 (1967)
30. M. Francon, Amelioration de resolution d'optique, Nuovo Cimento, Suppl. **9**, 283–290 (1952)
31. D. Mendlovic, I. Kiryuschev, Z. Zalevsky, A.W. Lohmann, D. Farkas, Two dimensional super resolution optical system for temporally restricted objects. Appl. Opt. **36**, 6687–6691 (1997)
32. W. Gartner, A.W. Lohmann, An experiment going beyond Abbe's limit of diffraction. Z. Physik. **174**, 18 (1963)
33. A. Zlotnik, Z. Zalevsky, E. Marom, Superresolution with nonorthogonal polarization coding. Appl. Opt. **44**, 3705–3715 (2005)
34. A.I. Kartashev, Optical systems with enhanced resolving power. Opt. Spectry. **9**, 204–206 (1960)
35. D. Mendlovic, J. Garcia, Z. Zalevsky, E. Marom, D. Mas, C. Ferreira, A.W. Lohmann, Wavelength multiplexing system for a single mode image transmission. Appl. Opt. **36**, 8474–8480 (1997)
36. M.A. Grimm, A.W. Lohmann, Super resolution image for 1-D objects. JOSA **A56**, 1151–1156 (1966)
37. H. Bartelt, A.W. Lohmann, Optical processing of 1-D signals. Opt. Commun. **42**, 87–91 (1982)
38. W. Lukosz, Optical systems with resolving powers exceeding the classical limits. II J. Opt. Soc. Am. **57**, 932–941 (1967)
39. Z. Zalevsky, D. Mendlovic, A.W. Lohmann, Super resolution optical systems using fixed gratings. Opt. Commun. **163**, 79–85 (1999)
40. Z. Zalevsky, E. Leith, K. Mills, Optical implementation of code division multiplexing for super resolution. Part I. Spectroscopic method. Opt. Commun. **195**, 93–100 (2001)
41. Z. Zalevsky, E. Leith, K. Mills, Optical implementation of code division multiplexing for super resolution. Part II. Temporal method. Opt. Commun. **195**, 101–106 (2001)
42. J. Solomon, Z. Zalevsky, D. Mendlovic, Super resolution using code division multiplexing. Appl. Opt. **42**, 1451–1462 (2003)
43. J. García, Z. Zalevsky, D. Fixler, Synthetic aperture superresolution by speckle pattern projection. Opt. Expr. **13**, 6073–6078 (2005)

44. D. Fixler, J. Garcia, Z. Zalevsky, A. Weiss, M. Deutsch, Speckle random coding for 2-D super resolving fluorescent microscopic imaging. Micron **38**, 121–128 (2007)
45. E. Ben-Eliezer, E. Marom, Aberration-free superresolution imaging via binary speckle pattern encoding and processing. Appl. Opt. **24**, 1003–1010 (2007)
46. I. Moon, M. Daneshpanah, B. Javidi, A. Stern, Automated three dimensional identification and tracking of micro/nano biological organisms by computational holographic microscopy. Proc. IEEE **97**(6), 990–1010 (2009)
47. B. Javidi, I. Moon, S. Yeom, E. Carapezza, Three-dimensional imaging and recognition of microorganism using single-exposure on-line (SEOL) digital holography. Opt. Expr. **13**, 4492–4506 (2005)
48. I. Moon, B. Javidi, Shape tolerant three-dimensional recognition of biological microorganisms using digital holography. Opt. Expr. **13**, 9612–9622 (2005)
49. B. Javidi, I. Moon, S. Yeom, Real-Time 3D Sensing and identification of microorganisms. Opt. Photon. News **17**, 16–21 (2006)
50. B. Javidi, S. Yeom, I. Moon, Real-time 3D sensing, visualization and recognition of biological microorganisms. Proc. IEEE **94**(3), 550–567 (2006)
51. B. Javidi, S. Yeom, I. Moon, M. Daneshpanah, Real-time automated 3D sensing, detection, and recognition of dynamic biological micro-organic events. Opt. Expr. **14**(9), 3806–3829 (2006)
52. I. Moon, B. Javidi, 3D identification of stem cells by computational holographic imaging. J R Soc Interface, R Soc Lond. **4**(13), 305–313 (2007)
53. I. Moon, B. Javidi, 3-D visualization and identification of biological microorganisms using partially temporal incoherent light in-line computational holographic imaging. IEEE Trans. Med. Imaging **27**(12), 1782–1790 (2008)
54. D. Fixler, R. Tirosh, T. Zinman, A. Shainberg, M. Deutsch, Differential aspects in ratio measurements of [Ca2+]i relaxation in cardiomyocyte contraction following various drug treatments. Cell Calcium **31**, 279–287 (2002)
55. R.N. Bracewell, *The Fourier Transform and Its Applications*, 3rd edn. (McGraw-Hill, Boston, MA, 2000)
56. L. Leushacke, M. Kirchner, Three dimensional correlation coefficient of speckle intensity for rectangular and circular apertures. J. Opt. Soc. Am. A **7**, 827–833 (1990)

Chapter 11
Image Formation and Analysis of Coherent Microscopy and Beyond – Toward Better Imaging and Phase Recovery

Shan Shan Kou, Shalin B. Mehta, Shakil Rehman, and Colin J.R. Sheppard

Abstract Applications of phase microscopy based on either coherent or partially coherent sources are widely distributed in today's biological and biomedical research laboratories. But the quantitative phase information derivable from these techniques is often not fully understood, because in general, no universal theoretical model can be set up, and each of the techniques has to be treated specifically. This chapter is dedicated to the fundamental understanding of the methodologies that derive optical phase information using imaging techniques and microscopic instrumentation. Several of the latest and most significant techniques are thoroughly studied through the theoretical formalism of the optical transfer function. In particular, we classify these systems into two main categories: those based on coherent illumination, such as digital holographic microscopy (DHM) and its extension into tomography, and those based on partially coherent illumination, such as differential interference contrast (DIC) and differential phase contrast (DPC). Our intention is that the models described in this chapter give an insight into the behaviour of these phase imaging techniques, so that better instrumentation can be designed and improved phase retrieval algorithms can be devised.

11.1 Introduction

Imaging is an inverse problem, which involves estimation of the specimen properties from recorded intensity images. For design of successful approaches for phase measurement, an accurate forward analysis is imperative. With that goal in mind, in this chapter, we present image formation analysis of some important phase

C.J.R. Sheppard (✉)
Division of Bioengineering, Optical Bioimaging Laboratory, National University of Singapore, 7 Engineering Drive 1, Singapore 117576, Singapore; Graduate School for Integrative Sciences and Engineering (NGS), National University of Singapore, 10 Medical Drive, Singapore 117597, Singapore; Department of Biological Sciences, National University of Singapore, 14 Science Drive 4, Singapore 117543, Singapore
e-mail: colin@nus.edu.sg

The authors Shan Shan Kou and Shalin B. Mehta have contributed equally to the work.

P. Ferraro et al. (eds.), *Coherent Light Microscopy*, Springer Series in Surface Sciences 46, DOI 10.1007/978-3-642-15813-1_11,

imaging methods. We describe how this analysis improves our understanding of key properties of the imaging systems as well as paving the way for better phase retrieval methods. We consider two major classes of methods – those based on coherent illumination and those based on partially coherent illumination.

Spatially coherent methods such as holography (discussed in Sect. 11.2) produce an amplitude in which the phase is linearly dependent on the optical path length (OPL) of the specimen, allowing measurement of phase information – from which refractive index or thickness can be recovered. Quick and dynamic as it is, direct single-shot holography has poor 3D imaging performance. As shown in Sect. 11.2.1, the spatial frequency coverage of single-shot holography along the axial direction is inadequate to provide true 3D imaging. One way to enhance the spatial frequency coverage is to use a polychromatic source, but this approach is beyond the scope of this chapter. Another way is to introduce tomography (Sect. 11.2.3), through either object rotation or illumination scanning. We present the coherent optical transfer function for holographic tomography for different geometries, both paraxial and high-aperture treatments are presented for an insightful view of such diffractive tomographic systems.

On the other hand, performance of phase imaging systems can be improved by employing simultaneous illumination from a large range of directions. Such an illumination is engineered by using an incoherent source in conjunction with high-NA illumination optics, leading to *partially* coherent field at the specimen plane. Partially coherent methods, in contrast to coherent methods, produce an image which depends bi-linearly on the specimen's transmission (as discussed in Sect. 11.3). These methods can be designed to be sensitive to the gradient or curvature of the specimen phase, allowing retrieval of the phase via integration.

In Sect. 11.3 image formation and phase retrieval in partially coherent systems are discussed based on our recently developed phase-space model. In Sect. 11.3.2–11.3.4, we focus our attention on three partially coherent methods that have shown promise for providing quantitative phase information – differential interference contrast (DIC), differential phase contrast (DPC), and use of the transport of intensity equation (TIE).

11.2 Imaging and Phase Reconstruction in Digital Holography and Holographic Tomography

A key component of today's coherent imaging regime, digital holography (DH) has its advantage in the instantaneous and quantitative acquisition of both amplitude and the phase by wavefront reconstruction. Originally invented by Dennis Gabor in 1948 as a method for recording and reconstructing of a wave field using photographic plates for recording and reconstruction [1], its numerical counterpart has come a long way through the development of computer technology. Digital holographic method was first proposed by Goodman and Laurence [2]. Another pioneering work was conducted by Schnars and Juptner [3], when direct recording of Fresnel holograms with a charged coupled devices (CCDs) was implemented. This method enables full digital recording and processing of holograms, as is widely used today. In particular, microscopy based on digital holography (DHM) has proved to be a convenient method in phase contrast imaging for samples in biological and biomedical studies. Imaging of phase distributions with high spatial resolution can be used to determine refractive index variations as well as the thickness of the specimen. The ability to detect very minute phase variations also allows quantitative phase imaging to reveal structural characteristics of cells and tissues which, in turn, may have potential implications in biomedical diagnosis. Some of the applications of digital holography are microscopy of biological material [4–7], shape measurement [8], quantitative phase estimation [9–11], and aberration compensation [12, 13].

Such rapid advancements and prevalent applications of DHM demand a fundamental understanding of its imaging behavior and how it is different from other alternative coherent phase imaging methods such as interference or confocal microscopy. In particular, tomography has been applied to holography for enhanced 3D imaging performance. Analysis of image formation for such systems should account for diffraction effects which are often neglected in direct tomographic projection algorithms [14]. Making use of the concept of the 3D coherent transfer function (CTF) [15], we present a theoretical formalism with full analytical expressions for holographic and holographic tomography systems, through which an insight into the system behavior can be appreciated, and better phase restoration schemes may be envisioned.

11.2.1 Image Formation in Digital Holographic Microscopy (DHM)

The following analysis is based on a basic configuration of the DHM using a Mach–Zehnder interferometer (Fig. 11.1 [9]). The collimated laser beam is divided by a beam splitter (BS1). The microscope objective (MO) collects the transmitted

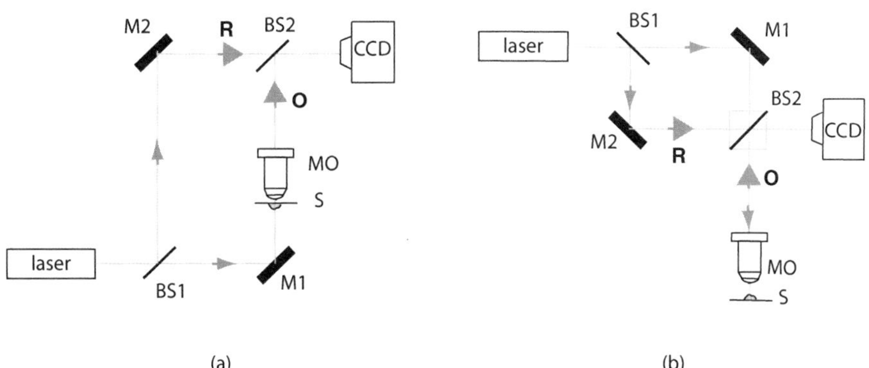

Fig. 11.1 Basic configuration of the DHM in (**a**) transmission or (**b**) reflection mode. BS1, BS2, beam splitters; M1, M2, mirrors; MO, microscope objective; S, sample

or reflected light and forms the object wave (**O**), which then interferes in an on-axis configuration (derivation for off-axis configuration is similar) with the reference wave (**R**) to produce a hologram intensity that is recorded by a CCD camera. This is very distinct from a conventional interference microscope, where a condenser is usually used [16]. Because of the use of collimated beams and the lack of condenser, the reference beam wavefront plays a trivial role in the overall imaging of DHM after reconstruction.

The complex field amplitude recorded on the CCD is given by

$$\begin{aligned} t(\mathbf{r}) &= |O(\mathbf{r}) + R(\mathbf{r})|^2 \\ &= |O(\mathbf{r})|^2 + |R(\mathbf{r})|^2 + O(\mathbf{r})R^*(\mathbf{r}) + O^*(\mathbf{r})R(\mathbf{r}) \end{aligned} \tag{11.1}$$

where $O(\mathbf{r})$ is the complex field amplitude generated by the object beam, $R(\mathbf{r})$ is the complex field amplitude of the reference beam, and the asterisk ($*$) denotes complex conjugation. The information content in the wavefront from the sample is contained in the interference terms. In the following analysis, we will focus on the image content inside these interference terms.

11.2.1.1 3D CTF for DHM

There are several approaches to derive 3D CTFs [17]. Here we apply a direct 3D Fourier transform (FT) to the 3D amplitude point spread function (APSF). Our attention is focused on a scalar theory but including the spherical converging wavefront instead of the paraxial (paraboloidal) approximation. In addition, an apodization effect for systems satisfying the sine condition for practical lens design is also considered. We assume an aberration-free microscope objective (MO) in first-order optics which has magnification M, numerical aperture NA, and a maximum subtended half-angle α. The image space 3D APSF for an optical system with a high-aperture circular lens can be derived as [18]:

$$h(v,u) = \int_0^{\alpha_0} P(\theta) J_0\left(\frac{v\sin\theta}{\sin\alpha_0}\right) \exp\left[-\frac{\mathrm{i}u\cos\theta}{4\sin^2(\alpha_0/2)}\right] \sin\theta \,\mathrm{d}\theta \tag{11.2}$$

where $P(\theta)$ is the pupil function or apodization function related to the angle of convergence of the light ray, J_0 is the zeroth order Bessel function of the first kind, $\mathrm{i} = \sqrt{-1}$, and v and u are defined as optical coordinates of transverse and axial directions:

$$\begin{aligned} v &= kr\sin\alpha_0 \\ u &= 4kz\sin^2(\alpha_0/2) \end{aligned} \tag{11.3}$$

where r and z are radial and axial coordinates in the image space, $k = 2\pi/\lambda$, and the NA is equal to $\sin\alpha_0$. We define normalized spatial frequencies m, n, s in the x-, y-, z-directions, and cylindrical transverse spatial frequency as

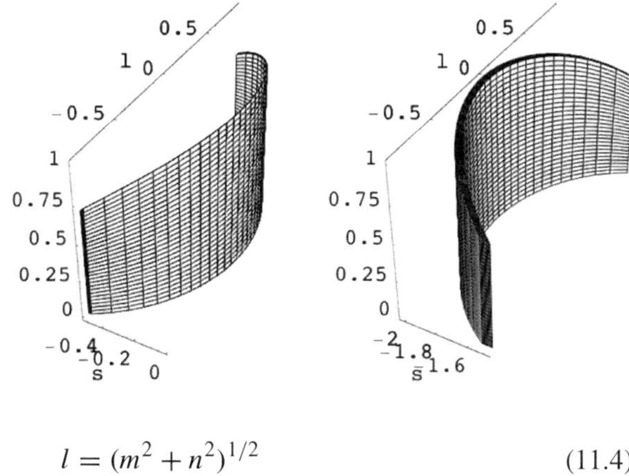

Fig. 11.2 Quasi-monochromatic CTF for holographic microscope at $\alpha_0 = \pi/3$ for (**a**) transmission and (**b**) reflection mode

$$l = (m^2 + n^2)^{1/2} \tag{11.4}$$

where such conventions of notation will be used throughout this chapter.

Applying a direct inverse 3D FT, the 3D CTF for DHM can then be derived, with consideration of axial shift, and after normalization [19],

$$c(l, s) = \delta\left(s + k \mp \sqrt{k^2 - l^2}\right) P(\theta) \tag{11.5}$$

where $\mp$ accounts for transmission and reflection systems, respectively. In the above, the support of the CTF (where the δ function has a value) can be represented by the cap of a sphere, which is consistent with Wolf's theory [15]. The final CTF with apodization function $P(\theta) = \sqrt{\cos\theta}$ can be visualized as in Fig. 11.2.

It should be noted that the above 3D CTF shows an incremental band of wavelength due to a quasi-monochromatic light source. This results in a thickness of the shells in reciprocal wavelength space. In broadband DHM, the integral effect of coherence gating increases the support in imaged spatial frequencies and, hence, more spatial frequencies will be included [19]. Different spectral densities can be assumed with real experimental conditions, but when the coherence length of a broadband source is limited, it may create difficulties in using an off-axis setup. An alternative means to increase spatial frequency coverage is to conduct tomography, and this will be presented in Sect. 11.2.3.

11.2.1.2 Comparison with Interference Microscopes

Let us now consider the comparison of 3D imaging capacity between holography and a normal interference microscope. We can do this by observing the support of 3D CTFs for the interference terms [20]. An incident plane wave denoted as **k1** vector is diffracted by an object, which can be considered as a grating, producing a scattered wave in the direction of vector **k2**. Vector **k** represents the characteristic wave vectors of the grating, and the two end points of **k** lie on a cap of a sphere due to the conservation of momentum. The range of angles detected by the holographic medium truncate the spherical shell. Putting this to the framework of a holographic

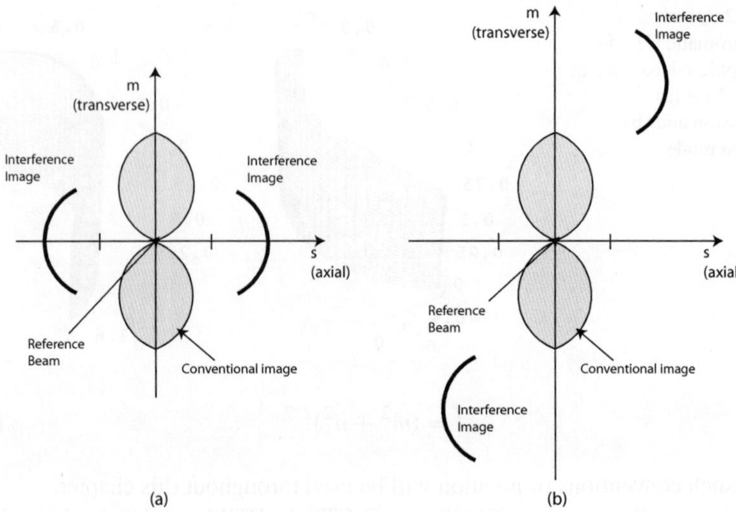

Fig. 11.3 Spatial frequency cutoffs in holographic microscope for an on-axis configuration (**a**) and an off-axis configuration (**b**)

imaging system based on interference, there will be four terms after 3D FT of (11.1), where the support of the interference image is represented by two thin shells for each of the conjugate pair (Fig. 11.3). The reference beam corresponds to a δ-function at the origin; the ordinary image transforms to a region around the origin.

In comparison, the 3D FT of the image of a conventional interference microscope, i.e., when a condenser is used (for example, the coherence probe microscope (CPM) [21]), also consists of four parts where the reference and ordinary image parts are similar to those of holography. But the interference terms transform to two regions each of which contain the object amplitude information that has far greater support than for single-shot holography (Fig. 11.4). Although quantitative phase information derived from DHM can provide knowledge on optical thickness or surface height, it has very limited spatial frequency coverage, especially in the axial direction. Compared with other types of interference microscopy, DHM does not result in true 3D imaging, but rather its advantage lies in its quick acquisition and reconstruction for dynamic processes.

11.2.2 Example of Quantitative Phase Imaging Using Lensless Quasi-Fourier Holography

Here a digital holographic microscopy technique is described using image correction. The underlying principle of this method is that if a blurred image is corrected in the hologram plane, a magnified image of the object can be obtained. A phase profile can be mapped out using a numerical reconstruction method. Phase information of an object is needed for perfect recovery of the blurred image and the holographic method is advantageous in this regard.

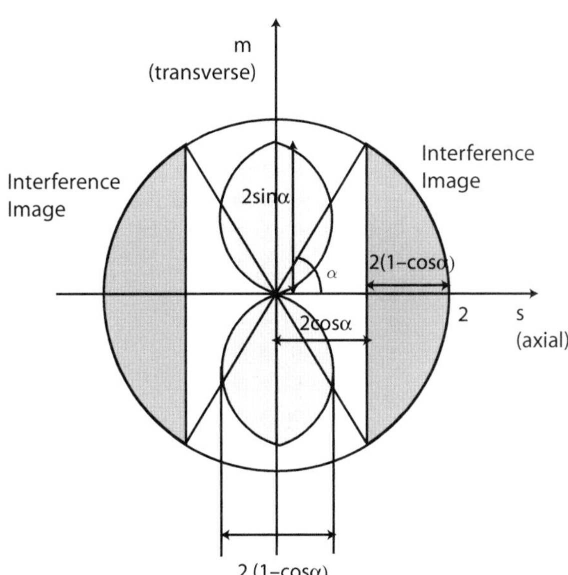

Fig. 11.4 Spatial frequency cutoffs in conventional interference microscope with geometry labels. α is the maximum subtended half-angle for numerical aperture. The maximum N.A in air required for the interference image to be separated from the object image is 0.943 (which corresponds to $\cos\alpha = 1/3$)

The basic principle in the lensless quasi-Fourier transform holography is to match the spherical phase factors that are due to the propagation of the object wave and the reference wave at the hologram plane. Point source illumination ensures the utilization of full spatial bandwidth of the sampling medium like a CCD. A quasi-Fourier hologram has been described as a hologram recorded with a point source acting as a spherical reference located in the same plane of the object and using a plane wave for the reconstruction [22]. In a digital hologram the reconstruction beam is simulated numerically for the digital reconstruction of a quasi-Fourier hologram, for example, of large objects [8]. Lensless Fourier holography has been used for reconstruction of digitally recorded holograms for shape and the deformation measurements [23] and microscopy of biological samples [24].

A lensless Fourier transform digital holographic microscope [25], in which a phase object is placed inside an expanding beam and hologram of the magnified and blurred image of the object is sampled by a CCD sensor, is also termed as quasi-Fourier transform holographic microscope. The field distribution at the hologram plane can be estimated by Fresnel approximation to the scalar diffraction theory [26]. Reconstruction of the hologram is carried out numerically to obtain amplitude or phase contrast image of the object. This technique eliminates the requirement to use a microscope objective lens for magnification giving the advantage that the working distance is increased and the imaging is aberration free [27].

11.2.2.1 Principle

Figure 11.5a shows the optical scheme that can be used to build a lensless digital holographic microscope. A phase object $o(x, y)$ is placed in an expanding beam.

The magnified and blurred image of the object is given by the convolution of the object beam and the transfer function of free-space propagation, $o(x_1/m, y_1/m) * h(x_1, y_1)$ where $*$ indicates convolution and m is the magnification of the object at the CCD plane. The phase of the object is recorded using a reference beam $r(x_1, y_1)$ in the holographic recording setup. The intensity of the hologram to be recorded is given by

$$I(x_1, y_1) = \left| r(x_1, y_1) + o\left(\frac{x_1}{m}, \frac{y_1}{m}\right) * h(x_1, y_1) \right|^2 \tag{11.6}$$

where

$$h(x_1, y_1) = -\frac{1}{\lambda z_1 z_2 m^2} \exp\left[jk(z_1 + z_2)\right] \exp\left[\frac{jk(x_1^2 + y_1^2)}{2z_2}\right] \tag{11.7}$$

λ is the wavelength of the light used, k is the wave number and the distance between a point source located at $(0, 0)$ and the CCD plane is z_1, and between the object and the CCD is z_2. The reconstruction algorithm for the hologram is given by the following equation where a Fourier transform is carried out after multiplying the recorded intensity pattern at the CCD by a reference signal r and then dividing by H yielding

$$\frac{\mathcal{F}\{r \cdot I\}}{H} = \mathcal{F}\left\{|r|^2 + |o|^2 \cdot r\right\} + \mathcal{F}\left\{r^2 o_0 * (mx, my)\right\} + O_0 \tag{11.8}$$

where $H = \mathcal{F}\{h\}$ and $O_0 = \mathcal{F}\{o_0(mx, my)\}$. $\mathcal{F}$ denotes the Fourier transform and $r(x_1, y_1)$ is the amplitude of the spherical reference beam at the hologram plane. Equation (11.8) suggests that selecting only the fourth term by filtering in the spatial frequency domain and carrying out the Fourier inverse transform retrieves the blurred phase object. Thus, a magnified image is obtained since it is expressed by $\mathcal{F}^{-1}\{O_0\} = o_0(mx, my)$. In the spatial frequency domain, selecting output of the center pixel in the zero-order light beam and expressing it by a delta function $\delta(x, y)$ and selecting a pass band of the output O_0 given by the fourth term in (11.8) and carrying out Fourier inverse transform can be expressed by

$$\left| \mathcal{F}^{-1}\{\delta + O_0\} \right|^2 = |a + o_0(mx, my)|^2 = A + B \cos \Phi \tag{11.9}$$

where $\mathcal{F}^{-1}[\delta(x, y)] = a$, $A = |a|^2 + |o_0(mx, my)|^2$, $B = 2|a||o_0(mx, my)|$ and Φ is the phase of the magnified object $o_0(mx, my)$. A quantitative and magnified image of the phase object can now be obtained by mapping the phase profile extracted from the interferogram given by (11.9). The lateral resolution in digital holography depends on the wavelength, CCD specifications, and reconstruction distance, however, due to the speckle noise associated with any coherent imaging system it is not possible to reach this theoretical limit. The lateral resolution ÏĄ can be written as

$$\rho = \frac{\lambda z}{D} \tag{11.10}$$

where λ is the wavelength of the source, z is the object to sensor distance, and D is the size of the CCD sensor. The formula give by (11.10) depends on the numerical aperture of the CCD and is based on the diffraction-limited resolution in optical microscopy. For a coherent imaging system the diffraction-limited resolution is given by $1.22\lambda/\mathrm{NA}$, where NA is the numerical aperture of the imaging optics. This factor determines the resolution of a coherent optical microscope. The optical resolution in a lensless holographic recording setup is given by (11.10) is also governed by this factor.

When a hologram is recorded by a point source close to the object and reconstructed by performing a Fourier transform of the recorded intensity one gets a quasi-Fourier hologram. In our case, we illuminate the object by an expanding spherical wave to achieve magnification instead of using lenses that cause aberrations. The hologram is recorded with the help of a reference wave that is a point source located at the reference point source plane but given a small tilt for recording an off-axis hologram. The resulted hologram is, therefore, termed as a quasi-Fourier transform hologram. The complex amplitude of this hologram is then sampled by a CCD device. The spatial frequency spectrum of the hologram, that is determined by the angle between the object and reference beams, must be smaller than the modulation transfer function of the CCD for correct Nyquist sampling.

Figure 11.5b shows the optical setup for a lensless quasi-Fourier transform microscope. A single hologram is recorded in the CCD and later processed in a computer. The digitized hologram is numerically processed by taking its Fourier transform and dividing by optical transfer function H which is calculated by measuring the distance from the object plane to the CCD. After a filtering process, in which the zero-order and first-order diffracted beams are selected and inverse Fourier transformed, complete amplitude and phase information of the object is obtained in the resulting interferogram. The magnified image of the object can be obtained by mapping out the phase profile (the phase contrast image) extracted from this interferogram, using the Fourier transform method.

11.2.2.2 Experimental Results

The optical scheme shown in Fig. 11.5a that can be realized in a Mach–Zehnder interferometric configuration as shown in Fig. 11.5b. A reflection type phase object was used that is a 48 nm step height standard (VLSI Standards Inc. model SHS440QC). The lateral width of the step size is 100 μm. Figure 11.6a shows a lensless Fourier transform hologram in which the magnified and blurred image of the phase object is recorded. Figure 11.6b shows the hologram obtained after the recovery of the blurred image. Figure 11.6c shows the phase profile of the step height standard and Fig. 11.6d shows the phase contrast image of a red blood cell.

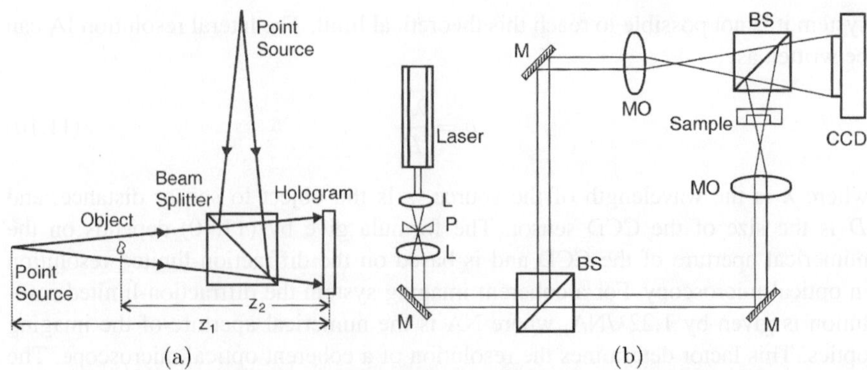

Fig. 11.5 (**a**) Principle of a lensless holographic microscope. (**b**) Optical setup for a lensless holographic microscope with; L: lens, P: pin-hole, M: mirror, BS: beam splitter, MO: microscope objective (only for point source illumination), and CCD: charge-coupled device camera

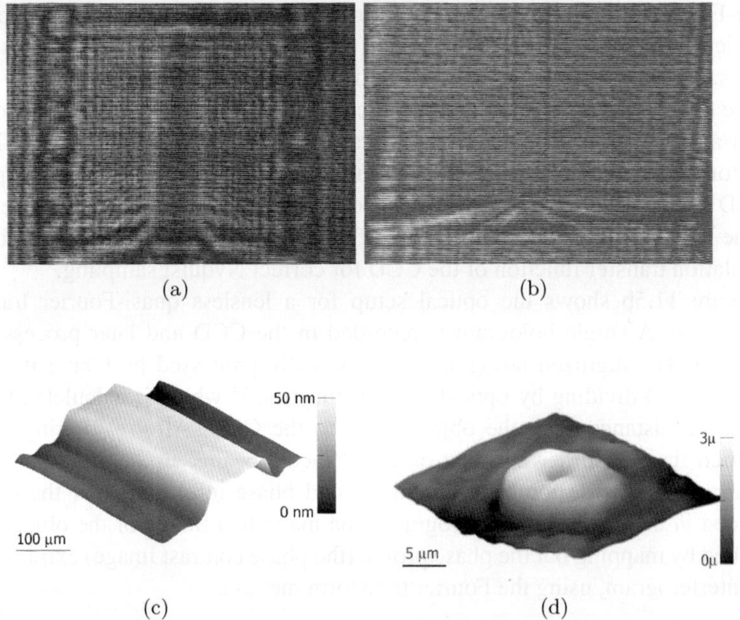

Fig. 11.6 (**a**) Lensless Fourier transform hologram of a 48 nm step height standard (VLSI Standards Inc. model SHS440QC), (**b**) filtered hologram obtained by carrying out the image correction technique, (**c**) phase map of a part of the hologram in (**b**), and (**d**) phase profile of an erythrocyte (red blood cell)

A magnified phase contrast image of an object can be obtained by a digital holographic microscope using an error correction method based on perfect recovery of the blurred image. In such a microscope, an objective lens is not required for magnification. Thus, an optical microscope with a long working distance and free of focusing elements can be obtained that can make magnified images of phase objects without aberrations caused by optical components. Quantitative phase profiles with nanometer order resolution in the axial direction can be obtained and this technique may be applied in imaging systems where magnification by a lens is difficult, as in the case of X-ray microscopes.

11.2.3 Image Formation in Holographic Tomography

It can be seen from previous section that imaging performance in single-shot holography is rather restricted in the axial direction and cannot provide true 3D imaging. Coherence gating or tomography can be applied to improve the spatial frequency coverage [19], while the latter in a transmission configuration has become an increasingly promising approach for observing live and unstained samples [28–30]. Such a system must be distinguished from the commonly known computed tomography (CT) technology, as optical diffraction (and refraction) effects should be accounted for [31]. Reconstruction of optical path length (OPL) from holographic tomography is also known as an inverse problem in optical diffraction tomography (ODT) [14, 32, 33].

The following analysis differs from previous published work in ODT in that the diffractive tomography system is analyzed using optical transfer functions, without considering any reconstruction of the object data as yet. In a way, the "filter response" of the system is presented through the location of spatial frequency cutoffs in Fourier space. Through visualization of 3D CTFs we can understand the behavior of different setups, but one needs to take further approximations such as Born or Rytov models for actual data reconstruction in the inverse problem [34]. The analysis presented below focuses on the transmission case but the reflective case can be derived in similar manners [35].

While fixing the detector position, one can either rotate the illumination or the object to acquire 3D object information tomographically [19]. While scanning the illumination, one can further distinguish the systems through the degrees of freedom in the scanning beam: 1D for scanning about one axis, 2D for scanning about two perpendicular axes, etc. Similarly, object rotation can also include 1D, 2D, and higher dimensional cases. The 1D case in both object rotation and illumination rotation are the most common, as a result of its efficiency and reduction of redundancy. Hence we will focus our analysis on the 1D case with particular attention to illumination rotation setup. The case of 1D object rotation gives a CTF which has been termed as the *apple core* [36], and the case of 1D illumination rotation gives a CTF called the *peanut* [37], both after the geometrical shapes of their 3D CTFs, respectively.

11.2.3.1 Paraxial Treatment

The 3D coherent transfer function can be calculated using 1D FT of the defocused transfer function [38]. For the case of the *peanut*, the defocused CTF can be calculated as a convolution of the imaging and illuminating pupils,

$$c(m,n;u) = \frac{1}{2}\int\int \exp\left[\frac{iu}{2}\left(m'^2 + n'^2\right)\right] \mathrm{circ}(m',n')\delta(n-n')\mathrm{rect}(m-m') \times \exp\left[-\frac{iu(m-m')^2}{2}\right] dm'\,dn' \tag{11.11}$$

Here the line of illumination is taken in the x-direction. m, n, α and u follows our conventions as previously defined in Sect. 11.2.1.1. u is the defocus distance. The functions $\mathrm{circ}(x,y) = 1, \sqrt{x^2+y^2} < 1$ and zero otherwise, and $\mathrm{rect}(x) = 1, |x| < 1$ and zero otherwise. Expanding the squares and performing the integral in n', we then have

$$c(m,n;u) = \frac{1}{2}\exp\left[\frac{iu}{2}(n^2-m^2)\right]\int_{A,B,C} \exp(iumm')\,dm' \tag{11.12}$$

The limits of integration are considered for three cases as $A \in (m-1, \sqrt{1-n^2})$, $B \in (-\sqrt{1-n^2}, m+1)$, $C \in (-\sqrt{1-n^2}, \sqrt{1-n^2})$. Cases A and B can be combined, and we can then obtain the defocused transfer functions after some mathematical manipulation [37],

$$c_{A,B}(m,n;u) = \frac{1}{2iu|m|}\exp\left[iu\left(\frac{n^2-m^2}{2} + |m|\sqrt{1-n^2}\right)\right] - \frac{1}{2iu|m|}\exp\left[iu\left(\frac{m^2+n^2}{2} - |m|\right)\right] \tag{11.13}$$

$$c_C(m,n;u) = \frac{1}{2iu|m|}\exp\left[iu\left(\frac{n^2-m^2}{2} + |m|\sqrt{1-n^2}\right)\right] - \frac{1}{2iu|m|}\exp\left[iu\left(\frac{n^2-m^2}{2} - |m|\sqrt{1-n^2}\right)\right] \tag{11.14}$$

It can be seen from (11.13) and (11.14) that the defocused CTF decays as $1/iu|m|$, so that the CTF is weighted by $1/|m|$ after applying the 1D FT. The cutoff of the 3D CTF is

$$A \text{ and } B: \frac{m^2-n^2}{2} - |m|\sqrt{1-n^2} < s < |m| - \frac{m^2+n^2}{2}, \sqrt{1-n^2} > 1-|m| \tag{11.15}$$

$$C: \frac{m^2-n^2}{2} - |m|\sqrt{1-n^2} < s < \frac{m^2-n^2}{2} + |m|\sqrt{1-n^2}, \sqrt{1-n^2} < 1-|m| \tag{11.16}$$

and its shape shown in Fig. 11.7. There is a line singularity in the plane $m = 0$, along the curved line $s = -n^2/2$, so there is a missing (curved) wedge of spatial frequencies. The spatial frequency coverage is non-isotropic. This contrasts with the case

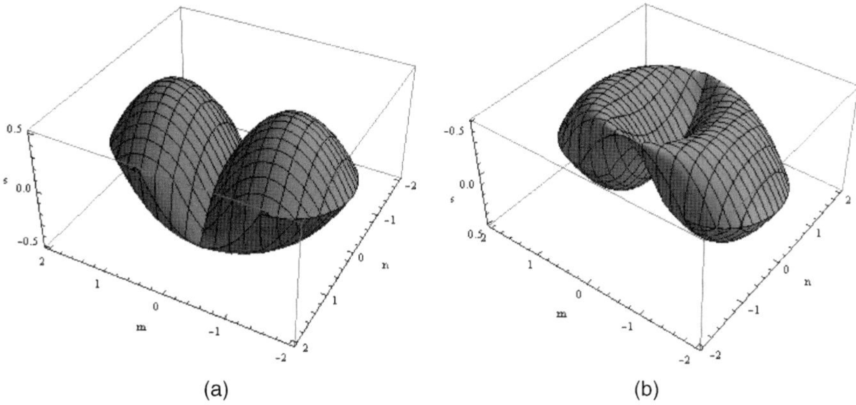

Fig. 11.7 3D transfer function cutoff for 1D illumination rotation *peanut* case with (**a**) *top* and (**b**) *flipped bottom* views

of 1D object rotation [36], where there is a point singularity at $m = n = s = 0$, and the region of missing frequencies is bounded by the figure of rotation of a parabola (for the paraxial case) about a tangent through its vertex (an apple core shape) [39].

11.2.3.2 High-Aperture Treatment

In the previous section, the full aperture condition of a sphere is reduced to a paraboloid in paraxial approximation. The more general case in high-aperture imaging implies incorporation of spherical 3D pupils [40]. The direct Fourier integration method, as used in last section under spherical convolution, proves to be difficult for deriving an analytical solution under spherical convolution. Instead, 3D analytical geometry is deployed as spherical caps are translated and rotated, and their motional trajectories being calculated.

The simpler case of the object being uniformly imaged over a complete sphere of directions is first considered. Then, by symmetry, the situation is the same whether the object is rotated through a complete circle, or the illumination is rotated over a complete circle. Geometrically the CTF can be visualized as being formed by rotating a sphere about an axis through a point on its surface (Fig. 11.8a). The 3D pupil rotates about the axis of rotation (Fig. 11.8b). The final result is a spindle torus [41]. If the axis of rotation is taken as the n, then the final CTF is rotationally symmetric about the n axis. Consider that the sphere intersects the n, s plane in a circle, with the equation

$$n^2 + s^2 - 2s\cos\theta = 0 \tag{11.17}$$

filling in the interior of the torus. We then have the 3D CTF for imaging the complete sphere after integrating angle θ from $-\pi/2$ to $\pi/2$ of the above circle [35] (Fig. 11.9),

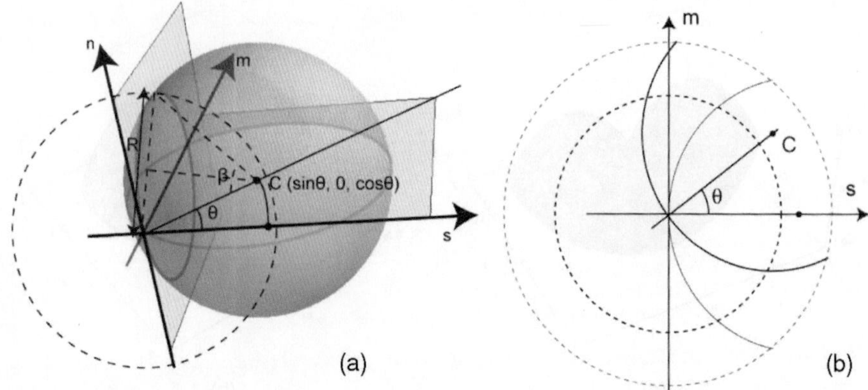

Fig. 11.8 Analytical geometry for deriving high-aperture full-circle rotation case (**a**) 3D view (**b**) projective 2D view

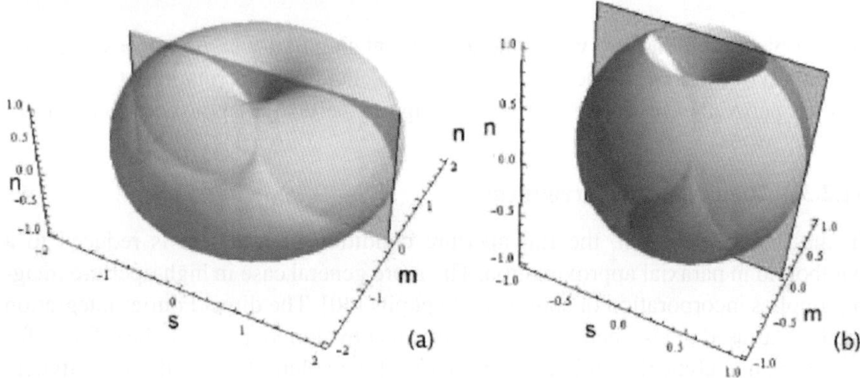

Fig. 11.9 The 3D transfer function cutoff for object rotation in uniform sphere. (**a**) Spindle torus for 4π configuration. (**b**) Torus for object rotation with aperture semi-angle equals to $\pi/3$

$$c(l,n) = \frac{1}{|\sin\theta|} = \frac{2}{\left[4l^2 - \left(l^2+n^2\right)^2\right]^{1/2}}, \quad 2l > (l^2+n^2) \qquad (11.18)$$

Turning now to the case of the high-aperture *peanut* for illumination rotation in 1D, the axis of the lens pupil now remains fixed relative to the s axis, so that the 3D pupil translates rather than rotates [19], its centre traveling through an arc of angle $\theta \leq \alpha$, where $\sin\alpha$ is numerical aperture of the objective. Let us set up a coordinate system in 3D Fourier space where the rotation direction is about an axis in the transverse plane, i.e., n for this case, so that s is the axial direction, and m is the other direction in the transverse plane (Fig. 11.10). θ is the maximum physical rotation angle one would implement for the incident wave, which is limited by the numerical aperture of the system. In the m, s plane we can consider contributions to the spatial frequency cut-off for $m > 0$ only for simplicity, since the final 3D

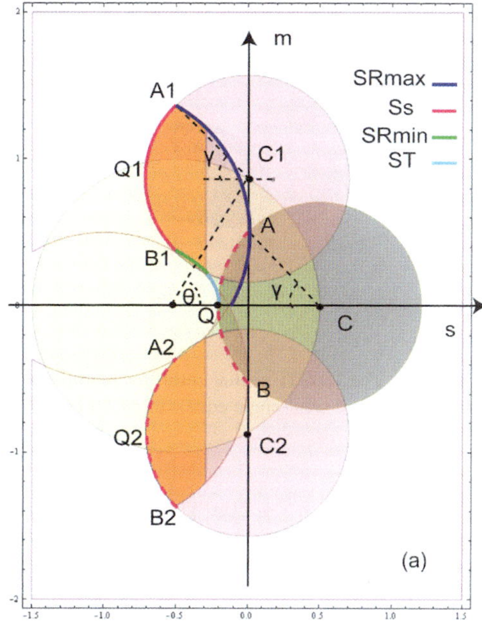

Fig. 11.10 Detailed analytical geometry for deriving high-aperture *peanut*. The figure depicts a specific plane at normalized n value of 0.6

CTF will be symmetrical about m. Translation of the cap of the sphere can now be visualized as the translation of an arc AQB (dashed magenta colored) in a plane with center C moving to $C1$ (or $C2$ for m negative). The cutoff at the maximum value of s is given by a translation of A to its maximum possible extent to A1 or A2, i.e., a translation of the rim of the cap of the sphere in 3D and denoted $S_{R\,\max}$ (for $m > 0$, dark blue colored) in Fig. 11.10. The cutoff for a minimum value of s is more complicated and turns out to be constituted of three parts. Its major part is determined by the surface of the sphere after translation, i.e., the arc AQB at its maximum possible position A1Q1B1 or A2Q2B2 after translation (magenta colored, denoted S_S for $m > 0$). There is also a part of the minimum value cutoff determined by the rim of the cap of the sphere, i.e., A moving downward to A2 or B moving upward to B1 (green colored, denoted for $m > 0$). However, the final part of the cutoff for minimum s is extended by the translation of points on the surface of the sphere, i.e., points on AQ moving downwards or points on BQ moving upwards (cyan colored, denoted S_T) when θ is smaller than α. This gives a cutoff the same as the toroidal apple core shape for rotation of the sample. The different regions of behavior are shown in Fig. 11.10, and each of their analytical expressions is given [35]. Combining all the regions of boundary, we obtain three analytical inequalities that completely define the overall shape of the high-aperture *peanut*,

$$\left[\cos^2\alpha - m^2 + n^2 + 2|m|\left(\sin^2\alpha - n^2\right)^{1/2}\right]^{1/2} - \cos\alpha > s$$

$$> \cos\alpha - \left[\cos^2\alpha - m^2 - n^2 + 2|m|\sin\alpha\right]^{1/2}$$

$$\sin\alpha + \left(\sin^2\alpha - n^2\right)^{1/2} > |m| > \sin\alpha - \left(\sin^2\alpha - n^2\right)^{1/2} \qquad (11.19)$$

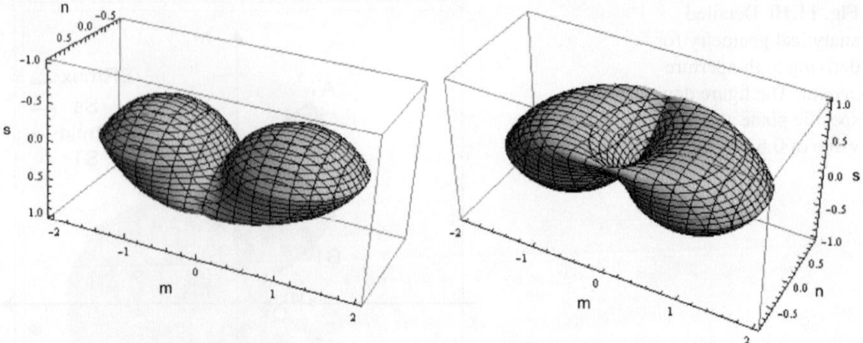

Fig. 11.11 The 3D CTF spatial cutoff for illumination rotation in high-aperture transmission imaging with aperture semi-angle equals to $\pi/3$. *Left*: top view *Right*: flipped bottom view

$$\left[\cos^2\alpha - m^2 + n^2 + 2|m|\left(\sin^2\alpha - n^2\right)^{1/2}\right]^{1/2} - \cos\alpha > s$$
$$> \left[\cos^2\alpha - m^2 + n^2 - 2|m|\left(\sin^2\alpha - n^2\right)^{1/2}\right]^{1/2} - \cos\alpha$$
$$\sin\alpha - \left(\sin^2\alpha - n^2\right)^{1/2} > |m| > \left(\sin^2\alpha - n^2\right)^{1/2}\frac{\left[1-(1-n^2)^{1/2}\right]}{(1-n^2)^{1/2}} \tag{11.20}$$

$$\left[\cos^2\alpha - m^2 + n^2 + 2|m|\left(\sin^2\alpha - n^2\right)^{1/2}\right]^{1/2} - \cos\alpha > s$$
$$> \left\{\left[1-\left(1-n^2\right)^{1/2}\right]^2 - m^2\right\}^{1/2}$$
$$\left(\sin^2\alpha - n^2\right)^{1/2}\frac{\left[1-(1-n^2)^{1/2}\right]}{(1-n^2)^{1/2}} > |m| > 0 \tag{11.21}$$

From these three inequalities one can plot the geometrical representation of the 3D cutoff. The final CTF is illustrated in Fig. 11.11, and the shape is interestingly similar to our previous paraxial *peanut* (Sect. 11.2.3), but exhibiting some significant differences. In particular, there is no longer a line singularity in the plane of n, s when $m = 0$. The 3D imaging performance for objects extended in the s direction is therefore better than that predicted by the paraxial treatment.

11.3 Partially Coherent Phase Microscopy

Imaging of transparent specimens requires special optical methods to transform the phase information into intensity information – which is detectable by detectors such as human eye and camera. Until recently, quantitative measurement of phase information has been carried out with spatially coherent or nearly spatially coherent illumination – due to ease of inverting the linear image formation process applicable to coherent imaging. However, coherent illumination suffers from some serious drawbacks, especially when imaging microscopic specimens, viz., lack of resolving power, artifacts due to impurities in the light path, speckle and lack of optical sectioning. To limit the undesirable effects of speckle and mottle (noise due to impurities in the optical train), sources of slightly reduced coherence have been employed for holography [42, 43]. However, imaging systems that use partially coherent illumination using a large aperture (i.e., illuminate the specimen simultaneously from a finite range of angles) can overcome all of the above disadvantages. Therefore, partially coherent methods have been popular in microscopic imaging of biological specimens – examples being phase contrast and differential interference contrast (DIC). However, when partially coherent illumination is employed, image formation is no longer linear in the amplitude transmittance of the specimen, but is rather bi-linear [44, 45]. The complication of bi-linear imaging, together with lack of extensive research into bi-linear image formation of phase imaging methods, has impeded development of appropriate phase-reconstruction methods that can account for the spatial frequency filtering effects of the microscope.

The transport of intensity equation (TIE) [46–49] and differential interference contrast (DIC) [50, 51] are two partially coherent methods that have been extensively investigated for quantitative phase measurement. In scanning electron and optical microscopy, the partially coherent method of differential phase contrast (DPC) [52, 53] has been employed. We have recently developed [54] a full-field version of DPC which we termed, asymmetric illumination-based DPC (AIDPC). In the following sections, the working principles and image formation in these partially coherent methods are described. Alongside the discussion on image formation, we also survey progress in phase retrieval using these methods. Note that while coherent methods are linearly dependent on phase, partially coherent methods often detect derivatives of the phase information from which phase is retrieved by integration.

11.3.1 Phase-Space Imager Model for Partially Coherent Systems

A schematic of a general partially coherent method (which can be based on scanning or full-field geometry) is shown in Fig. 11.12.

We model the specimen as being a thin transparency,

$$t(x, y) = a(x, y)e^{i\theta(x,y)} \tag{11.22}$$

where $a(x, y)$ is the absorption and $\theta(x, y) = -(2\pi/\text{lambda})O(x, y)$ the phase delay introduced by the specimen due to its optical thickness $O(x, y)$.

Let us write a space domain model for the system shown in Fig. 11.12. The image intensity is the sum of the intensities due to individually incoherent condenser points, the image formation by each point of the condenser being spatially coherent. This leads us to *the space domain model* of partially coherent systems as follows:

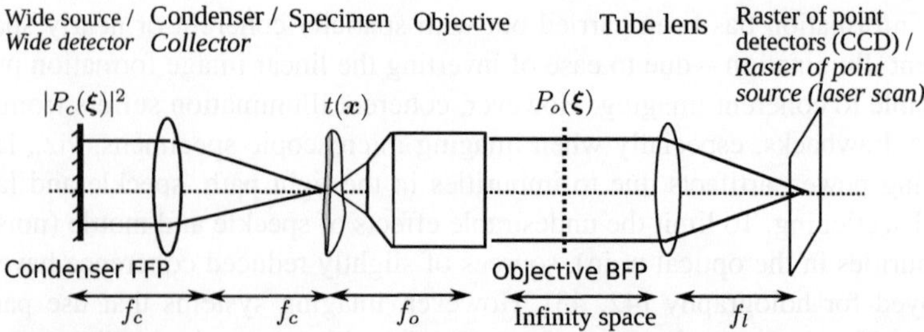

Fig. 11.12 Schematic representation of a general partially coherent system. A scanning-based (light propagation from right to left) or wide-field (light propagation from left to right) partially coherent imaging system is described by intensity distribution in the condenser pupil and amplitude distribution in the objective pupil. Labels in *italics* indicate terms used for the scanning microscope. f_c, f_o, and f_l are focal lengths of the condenser/*collector*, objective, and tube lens, respectively

$$I(\mathbf{x}) = \int |P_c(\boldsymbol{\xi})|^2 \left| e^{2\pi i \boldsymbol{\xi}.\mathbf{x}} t(\mathbf{x}) \otimes h(\mathbf{x}) \right|^2 d\boldsymbol{\xi}$$

$$= \iiint |P_c(\boldsymbol{\xi})|^2 t(\mathbf{x_1}) t^*(\mathbf{x_2}) h(\mathbf{x} - \mathbf{x_1}) h^*(\mathbf{x} - \mathbf{x_2}) e^{2\pi i \boldsymbol{\xi}.(\mathbf{x_1} - \mathbf{x_2})} dx_1 dx_2 d\xi \tag{11.23}$$

where $|P_c(\boldsymbol{\xi})|^2$ is the intensity distribution of the condenser aperture and $h(\mathbf{x})$ is the amplitude point spread function (PSF) of the imaging path. $e^{2\pi i \boldsymbol{\xi}.\mathbf{x}}$ is the oblique plane wave illumination produced by the illumination pupil coordinate $\boldsymbol{\xi}$. All integrals range from $-\infty$ to ∞ unless otherwise noted. Different partially coherent systems essentially differ in the structure of the condenser and objective pupils. Aberrations (including defocus employed in TIE) are accounted for by employing a complex effective objective pupil. The phase distribution at the condenser aperture is unimportant due to its incoherence. We express all spatial domain coordinates in normalized unit of $\lambda/\mathrm{NA_o}$. All pupil and spatial-frequency coordinates are expressed in unit of $\mathrm{NA_o}/\lambda$. Above equation shows that partially coherent imaging is inherently nonlinear (precisely, bi-linear) due to presence of large illumination aperture. The bi-linearity of the image formation is manifested in the dependence of the image intensity at a given point on *pairs of specimen points* ($\mathbf{x_1}$ *and* $\mathbf{x_2}$,) rather than individual specimen points.

The space-domain model of (11.23) makes it difficult to segregate effects of the specimen and the system. Hopkins developed a spatial-frequency model based on concept of the transmission cross-coefficient (TCC) [44] that segregates the contributions from the specimen and the system. Sheppard and Choudhury showed that the same model is applicable to a scanning microscope which does not employ a pin-hole [45]. However, the scanning microscope that employs a pin-hole (confocal microscope) has very different imaging properties as discussed in [45], although it, and other partially coherent imaging systems, can still be regarded as a generalization of the TCC model. In the TCC model, bi-linearity is manifested by the dependence of the spatial frequencies of the image on pairs of spatial frequencies of the specimen, as seen from the following equation [34, 44, 45].

$$I(\mathbf{x}) = \iint T(\mathbf{m_1})T^*(\mathbf{m_2})C(\mathbf{m_1}, \mathbf{m_2})\mathrm{e}^{2\pi \mathrm{i}(\mathbf{m_1}-\mathbf{m_2}).\mathbf{x}}\mathbf{dm_1 dm_2} \tag{11.24}$$

In the above, $C(\mathbf{m_1}, \mathbf{m_2}) = \frac{1}{C_N}\int |P_c(\boldsymbol{\xi})|^2 P_o(\mathbf{m_1} + \boldsymbol{\xi})P_o^*(\mathbf{m_2} + \boldsymbol{\xi})\mathrm{d}\boldsymbol{\xi}$ is the normalized transmission cross-coefficient (TCC) of the imaging system. The TCC is also called the partially coherent transfer function (PCTF) or bi-linear transfer function. Interpreting geometrically, the TCC for a conventional (i.e., not confocal) system is the area of overlap of three pupils – the condenser pupil (purely real) assumed to be situated at the center, the objective pupil shifted by $\mathbf{m_1}$, and the conjugate objective pupil shifted by $\mathbf{m_2}$.

Interpretation of image formation in terms of pairs of points or pairs of frequencies of the specimen is somewhat involved and non-intuitive. It is well-known that certain phase-space descriptions (exemplified by the Wigner distribution) are also bi-linear [55]. These phase-space distributions are physically more insightful and offer computational advantages. We have recently discovered a phase-space equivalent of the TCC model, termed the phase-space imager (PSI).

As per the PSI model [56, 57] the image intensity is given by

$$I(\mathbf{x}) = \int \Psi(\mathbf{m}, \mathbf{x}; c)\mathbf{dm} \tag{11.25}$$

where,

$$\Psi(\mathbf{m}, \mathbf{x}; C) = \int S_m(\mathbf{m}, \mathbf{m}')C(\mathbf{m}, \mathbf{m}')\mathrm{e}^{2\pi \mathrm{i}\mathbf{m}'.\mathbf{x}}\mathbf{dm}' \tag{11.26}$$

is the *windowed Wigner distribution of the specimen transmission* [55]. This distribution is the phase-space imager.

The quantity

$$S_m(\mathbf{m}, \mathbf{m}') = T\left(\mathbf{m} + \frac{vbfm'}{2}\right) T^*\left(\mathbf{m} + \frac{vbfm'}{2}\right) \tag{11.27}$$

is termed the mutual spectrum of the specimen. The imaging system can be said to filter the mutual spectrum with the window $C(\mathbf{m}, \mathbf{m}')$, which we call the *phase-space imager window (PSI window)*. The PSI-window is given by

$$C(\mathbf{m}, \mathbf{m}') = \frac{1}{C_N}\int |P_c(\boldsymbol{\xi})|^2 P_o\left(\mathbf{m} + \frac{\mathbf{m}'}{2} + \boldsymbol{\xi}\right) P_o^*\left(\mathbf{m} - \frac{\mathbf{m}'}{2} + \boldsymbol{\xi}\right)\mathrm{d}\boldsymbol{\xi} \tag{11.28}$$

The PSI-window can be computed as an area of overlap of three pupils, the condenser pupil $|P_c(\boldsymbol{\xi})|^2|$ situated at the center, the objective pupils $P_o(\boldsymbol{\xi})$ and $P_o^*(\boldsymbol{\xi})$

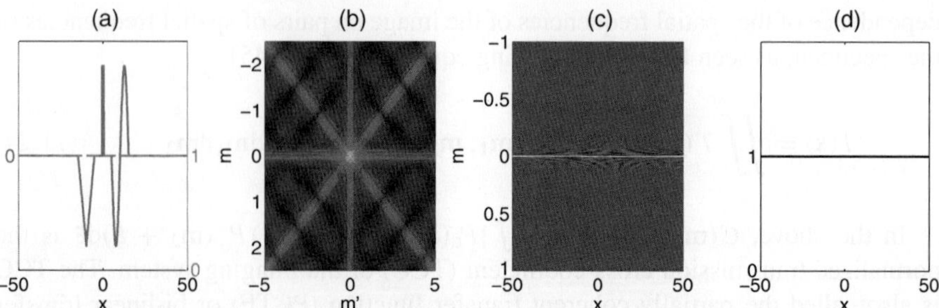

Fig. 11.13 A hypothetical 1D phase specimen used to illustrate the image formation in DIC, DPC, and TIE: **(a)** magnitude (*dashed line*) and phase (*solid line*) of the transmission, **(b)** the log-magnitude of the mutual spectrum of the specimen, **(c)** Wigner distribution of the specimen, and **(d)** the 'ideal' image obtained as a spatial marginal of the Wigner distribution

shifted from the center by vector **m** and from that point, $P_o(\boldsymbol{\xi})$ shifted by $\frac{vbfm'}{2}$ and $P_o^*(\boldsymbol{\xi})$ shifted by $-\frac{vbfm'}{2}$. We adapted our previously published algorithm for computing the TCC [58] to compute the PSI-window as discussed in [57].

In Sect. 11.3.2, 11.3.3 and 11.3.4, we discuss image formation in DIC, DPC, and TIE in terms of the above model using a simple hypothetical 1D phase-specimen depicted in Fig. 11.13. The interference structure in the Wigner distribution contains the information about phase-gradient of the specimen as a function of space. It is interesting to note that when computing the "ideal" image as a spatial marginal of the Wigner distribution, this phase information is canceled perfectly, as it is known to do. As seen in Figs. 11.15, 11.16, 11.17 and 11.18 the imaging system can be thought of as modifying the specimen's Wigner distribution, to provide a distribution whose spatial marginal does have phase information. In the following text, we use the terms point source, spatially coherent source, and coherent illumination interchangeably to imply a condenser aperture closed down to a point on the optical axis.

11.3.2 Differential Interference Contrast

DIC is a polarization-based wavefront shearing interferometer [50, 51]. It employs polarizing optics to produce two orthogonally polarized beams that traverse the specimen, a sub-resolution distance apart from each other. These two beams "perceive" slightly different optical path lengths of the specimen. Both beams are combined after traversing the specimen and upon interference, produce an intensity distribution that depends on the difference of phase between the two beams. Since the beams "see" the same phase profile but shifted by a sub-resolution distance, the intensity distribution produced by their interference represents the specimen's phase-gradient to the first approximation. Image formation in DIC is rather involved [16, 58, 59] and the image depends on the absorbance, phase, and birefringence of the specimen.

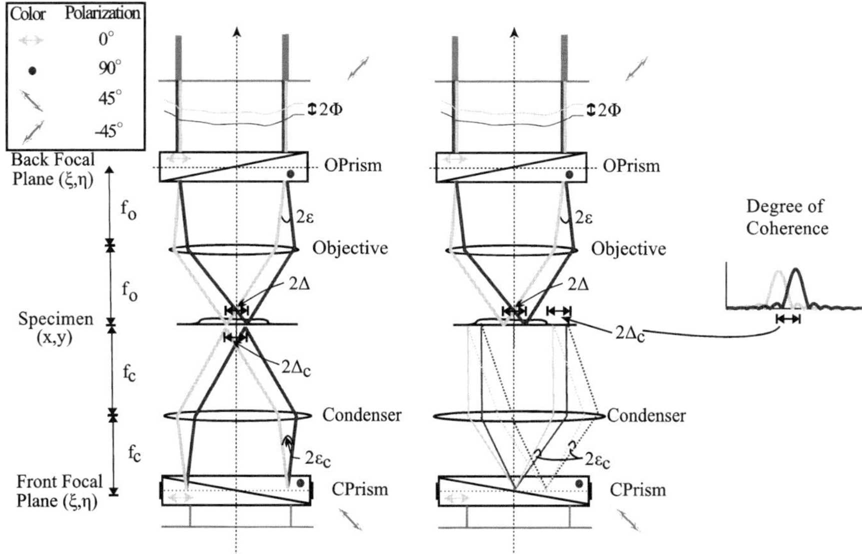

Fig. 11.14 Schematic representation of image formation in DIC. The color is used to indicate polarization of light: (**a**) Popular view of image formation in DIC and (**b**) Coherence view of image formation in DIC

The popular description of image formation in DIC is as follows (depicted in Fig. 11.14a): The polarizer and the Nomarski prism on the condenser-side (CPrism) create two orthogonally polarized beams (indicated by yellow and green lines). These two beams are focused at the specimen plane at sub-resolution distance apart. The beams traverse the specimen, are collected by the objective, and are combined by the objective-side prism (OPrism) and the analyzer. However, it is known that DIC employs Köhler illumination – thus the beams illuminate the whole specimen, rather than being focused at two points. Therefore, a more accurate description of image formation is as follows [60] (depicted in Fig. 11.14b): the plane wave originating from each point of the condenser aperture is split into two orthogonally polarized plane waves. Two such plane waves originating from the on-axis and off-axis points of the condenser aperture are shown in the figure. The beams comprising of plane waves of given polarization are individually partially coherent. However, they are *mutually coherent* since they are derived from the same scalar field produced by the polarizer. The objective side light path is *entirely* responsible for the shearing interferometry process [51, Chapter 7]. The role of illumination-side polarization optics is to create a field that is mutually coherent at distance equal to the shear [58]. We term this view of the image formation a *coherence view*.

11.3.2.1 Partially Coherent Model

Although DIC is preferably used with partially coherent illumination, initial models have assumed coherent illumination to study the effects of shear and bias on properties of the imaging path [61, 62]. Recently, partially coherent models have been developed that account for properties of the illumination as well [16, 58, 59, 63]. Sheppard & colleagues [16, 59] assumed matched illumination ($S = 1$) and accounted for the effects of both the condenser-side and the objective-side prisms. Preza [63] accounted for arbitrary illumination aperture, but did not account for the effects of the condenser-side prism. We elucidated the effects of both the condenser geometry and the condenser-side prism on coherence of illumination to develop a more complete model of imaging in various configurations of DIC [58]. We termed the standard DIC setup, Nomarski-DIC, and the DIC setup lacking the condenser-side prism (as simulated by Preza), as Köhler-DIC [58, figure 1]. We have shown with simulations of TCC and experimental images that the role of condenser-side prism is to illuminate the points that are separated by a distance equal to shear in the spatially coherent manner. The objective side light path, per se, performs the shearing interferometry operation. As an additional benefit of segregation of the specimen and system, we developed an accurate method of calibrating shear and bias of a DIC setup without having to use a sample [60].

In the following analysis, we assume that the shear is along the x-direction and that a hypothetical specimen being imaged has phase-profile variation along x as shown in Fig. 11.13. The specimen is assumed to be constant along the y-direction. If the shear and bias employed by the DIC microscope are 2Δ and 2ϕ, let us define a kernel

$$\Delta_{\mathrm{DIC}}(\mathbf{x}) = \delta(x + \Delta, y)\mathrm{e}^{-\mathrm{i}\phi} - \delta(x - \Delta, y)\mathrm{e}^{\mathrm{i}\phi} \tag{11.29}$$

which accounts for the spatial-shift and the phase-shift introduced by the DIC components. Note the use of scalar space variables (x, y) and the scalar spatial frequency m along the x-direction. The effect of shear is only along x-direction, and the imaging in y-direction remains the same as in bright-field setup.

In Nomarski-DIC configuration, both prisms employ the same shear and effectively image the following transmission function of the specimen [58, equation 7].

$$t_N(\mathbf{x}) = t(\mathbf{x}) \otimes \Delta_{\mathrm{DIC}}(\mathbf{x}) \tag{11.30}$$

In the frequency domain, the convolution above leads to a modification of the specimen spectrum by factor $2\mathrm{i}\sin(2\pi m\Delta - \phi)$. The modification of the specimen spectrum can be accounted for by modifying the bright-field TCC [58, equation 11], and equivalently by modifying the bright-field PSI-window. Therefore, the PSI-window for Nomarski-DIC is

$$C_{\mathrm{ndic}}(\mathbf{m}, \mathbf{m}') = C(\mathbf{m}, \mathbf{m}') \sin\left[2\pi\left(m + \frac{m'}{2}\right)\Delta - \phi\right] \sin\left[2\pi\left(m - \frac{m'}{2}\right)\Delta - \phi\right] \tag{11.31}$$

where m and m' are center and difference spatial frequencies along the x-direction. Figure 11.15 illustrates the process of computation of the image of the hypothetical

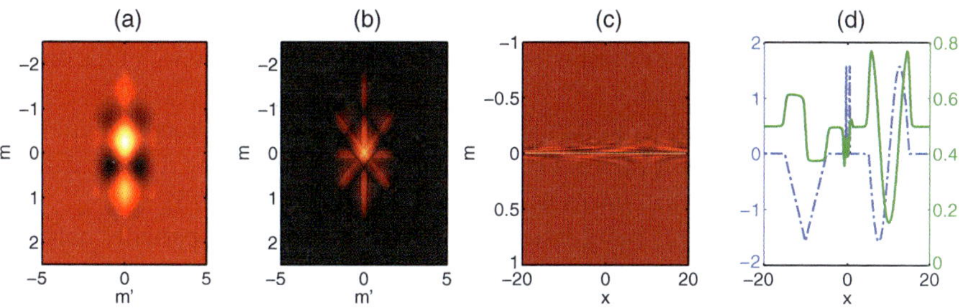

Fig. 11.15 Computation of the partially coherent image of the specimen shown in Fig. 11.13 under Nomarski-DIC: DIC components are assumed to provide shear of $2\Delta = 0.4\lambda/\mathrm{NA_o}$ and bias $2\phi = \pi/2$. **(a)** PSI-window, **(b)** log-magnitude of the filtered mutual spectrum, **(c)** the PSI, and **(d)** partially coherent image obtained as a spatial marginal of the PSI

specimen under a Nomarski-DIC system. Note that the PSI-window shown in Fig. 11.15a was computed by first computing the bright-field PSI-window [56, figure 2], followed by modulation as per (11.31).

11.3.2.2 Phase Retrieval with DIC

Due to difficulty in inverting the full bi-linear model, various approximations have been employed for estimating phase from DIC images.

The most widely used assumption is to ignore diffraction and use a geometric model to describe image formation in DIC. With this assumption, the image formation in DIC becomes similar to interferometry with two beams of the form, $t(x+\Delta, y)\mathrm{e}^{-\mathrm{i}\phi}$ and $t(x-\Delta, y)\mathrm{e}^{\mathrm{i}\phi}$, where $t(x, y)$ is the transmission of the specimen (Eq. 11.22). The problem of estimating the phase difference between two beams $\left(\frac{\mathrm{d}\theta(x,y)}{\mathrm{d}x}\right)$ becomes similar to phase-shifting interferometry. By measuring the DIC images at different biases, the phase-shifting DIC method [64–67] obtains an image that linearly represents the gradient of the image along one direction. By measuring the phase-gradient along two orthogonal directions with phase-shifting DIC [67–71], one can obtain the vector gradient field of the phase of the specimen. The vector gradient field can be integrated with non-iterative Fourier integration [67, 72] or iterative methods originating from the field of 2D phase unwrapping [73].

A semi-quantitative approach is developed by Shribak & colleagues [74] to produce high-resolution orientation-independent maps of the phase of the specimen. In this approach, assuming a geometric model as well as a small shear allows estimation of the linear gradient along the shear-direction from two DIC images. The estimated gradients are assumed to arise by a convolution of a phase map with a point spread function (having positive and negative peaks separated by shear distance) and oriented along X and Y. By applying iterative constrained optimization developed by Autoquant, their approach produces an image that represents deconvolved phase variation of the specimen. Note that the goal of this approach is to produce an image that represents the dry mass of biological specimens and to eliminate directional effect present in DIC due to shear – not to provide quantitative retrieval of the optical path length of the specimen.

By assuming coherent illumination, Preza et al. have proposed iterative approaches for phase retrieval from non-absorbing specimens using rotational diversity [71] and absorbing specimens using two-level optimization [75].

Ishiwata et al. [76, 77] have assumed a weak specimen, but partially coherent source, to develop a phase-retrieval approaches based on iterative optimization. The assumption of a weak specimen allows isolation of phase information (affected by the transfer function of the microscope) by measuring two images at equal and opposite bias, from which one can retrieve phase information by deconvolution with help of the transfer function [76].

11.3.3 Differential Phase Contrast

As is evident from the discussion in the previous section, DIC has a rather complex image formation mechanism that depends on the shear, the bias, and the system's pupils. Because of this complexity, the DIC image does not linearly represent the specimen's phase-gradient. Therefore, quantitative imaging with DIC necessitates use of approaches such as phase-shifting, which require acquisition of multiple images at different bias values. As a simpler and more quantitative alternative, DPC was developed using a split-detector in scanning optical microscopy [52, 53]. A schematic diagram of DPC system is shown in Fig. 11.16. Scanning DPC is sensitive to small changes in the specimen's phase-gradients [78] and is inherently more quantitative than DIC. We developed a reciprocal equivalent of the DPC system that uses a "split-source" – which is positive in one half and negative in the other. The negative half of the split-source is synthesized by acquiring two images by blocking opposite-halves of the condenser aperture and then taking their difference. Since this method uses asymmetric illumination (from halves of the condenser aperture),

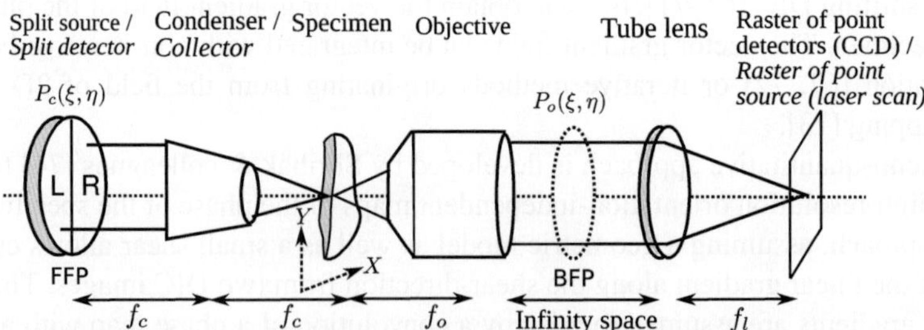

Fig. 11.16 Equivalence of a full-field system using split-source (AIDPC) and a scanning system with split-detector (scanning DPC). The light propagates from left to right in the full-field system and from right to left in the scanning system. The direction of phase differentiation is assumed to be the X-direction

we call it asymmetric illumination-based differential phase contrast (AIDPC) [54]. Since both methods are reciprocal to one another, their imaging properties are identical. AIDPC differs from the Schlieren-type imaging methods, e.g., Hoffman modulation contrast [79] and asymmetric illumination contrast [80], in terms of the synthesized negative condenser aperture that allows quantitative imaging of the phase-gradient. In presence of birefringence, phase-shifting DIC fails to provide an image proportional to phase-gradient [54, 81]. DPC on the other hand is insensitive to birefringence and provides an image linearly dependent on the phase-gradient of the specimen along the direction of split. By measuring the phase gradient in two orthogonal directions with either DPC or DIC, one obtains the vector gradient field of phase information. By appropriately integrating the gradient field one can recover the specimen's phase. By controlling the sensitivity of the detector in scanning system or intensity profile of the source in full-field system it is possible to improve the linearity of the image formation [82–84].

11.3.3.1 Partially Coherent Model

The partially coherent model of DPC is simpler in comparison to DIC and is illustrated in [57, section 5].

Figure 11.17 shows computed image of the specimen shown in Fig. 11.13. From Fig. 11.17(a) we notice that the PSI-window of AIDPC has a simple structure which is odd along the instantaneous spectrum (**m**). This odd symmetry results in sensitivity of AIDPC to the phase-gradient information as is evident from the PSI shown in Fig. 11.17c and simulated images shown in Fig. 11.17d. Thus, AIDPC provides nearly linear measurement of the directional gradient of the specimen and is simple to implement (no modification on the objective-side light path is required). Since AIDPC does not modify the objective side light path, it provides high light throughput and is easy to integrate with other imaging modalities such as fluorescence microscopy. AIDPC offers an attractive alternative to DIC in this respect.

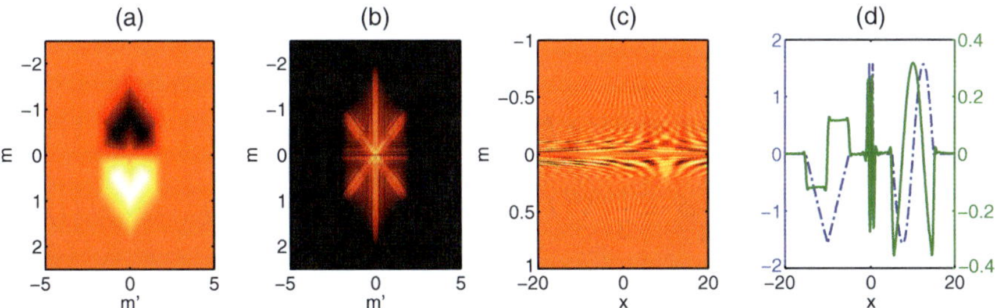

Fig. 11.17 Computation of the partially coherent image of the specimen shown in Fig. 11.13 under DPC: The arrangement of this figure is the same as Fig. 11.15. AIDPC image is seen to be linearly dependent on the gradient of the specimen, albeit effects of filtering especially at high spatial frequencies are visible

11.3.3.2 Phase Retrieval with DPC/AIDPC

In the geometric approximation, DPC/AIDPC provides a nearly linear measurement of the specimen's phase gradient. Therefore, one does not require phase-shifting DIC. The single gradient image can be integrated to provide phase image [85]. Also, by measuring the gradient in two orthogonal directions, one can use the Fourier integration algorithm (used for DIC) to retrieve the phase. To the best of our knowledge, phase-retrieval approaches that take diffraction into account within DPC setup (even approximately) have not yet been published. A possible reason is that in X-ray and electron regimes, where this method is currently used more frequently than in optics, the geometric assumption suffices.

11.3.4 Transport of Intensity

TIE relies on the fact that the phase of the wavefront introduces variation in the intensity distribution as it propagates. By measuring intensity distribution along the optical axis at multiple (usually three) positions (i.e., axial derivative of the intensity), one can infer the wavefront shape (transverse Laplacian of the phase). Thus, TIE allows measurement of second derivative of the phase. Note, however, that this is not the phase of the specimen, but rather phase of the wavefront in the *image plane*. The commonly known *transport of intensity* (TIE) equation was proposed by Teague [46] as part of the scheme in optical phase-retrieval problems. The aim is to deduce optical phase from only irradiance measurements using non-interferometric techniques. Compared to many other phase retrieval algorithms [86], TIE is non-iterative, purely computational and no auxiliary device needs to be introduced.

The original set of TIE equations proposed by Teague consists of two equations that are developed from parabolic wave equation. Let wave amplitude be expressed in terms of the irradiance I and the phase ϕ, which are real-valued quantities

$$U_z(\mathbf{r}) = \left[I_z(\mathbf{r})\right]^{1/2} \exp\left[\mathrm{i}\phi_z(\mathbf{r})\right] \tag{11.32}$$

The two TIE equations are repeated here

$$\frac{2\pi}{\lambda}\frac{\partial}{\partial z}I = -\nabla \bullet I\nabla\phi \tag{11.33}$$

$$\frac{4\pi}{\lambda}I^2\frac{\partial}{\partial z}\phi = \frac{1}{2}I\nabla^2 I - \frac{1}{4}(\nabla I)^2 - I^2(\nabla\phi)^2 + kI^2 \tag{11.34}$$

with the first one being the *transport of intensity* and the second one being the *transport of phase*. All the ∇ operators are in 2D transverse direction, and the vector coordinates are $(\mathbf{r}_\perp, z)$. The first one can be simplified into a 2D Poisson equation and hence is the choice for a solution of ϕ while the latter equation does not enjoy the same popularity and is rarely used mainly due to its higher orders.

11.3.4.1 Partially Coherent Model

Most of the current approaches of reconstructing phase from TIE equation ignore the diffraction effects. Sheppard [87] discussed the defocused transfer functions assuming a weak specimen and proposed an inversion approach. Recently, Nugent, Arhatari, and colleagues presented a similar analysis by assuming a weak specimen [88, 89]. In the following, we present our preliminary results related to bi-linear image formation in TIE using the PSI model.

Effects of defocus can be accounted for by modulating the objective pupil with the aberration term of $e^{-iu\rho^2}$, where u is the normalized defocus co-ordinate. Thus, equal and opposite defocus give rise to conjugate objective pupils. Consequently, it follows from (11.24) and (11.28) that the TCC and PSI-windows will also be complex conjugates for equal and opposite values of u.

We have assumed a non-absorbing specimen. Therefore, the in-focus image is uniformly bright and the difference of images taken across with equal and opposite bias equal the transverse Laplacian of the phase. Since the computation of the Laplacian involves only linear operation when the specimen is non-absorbing, we can associate a PSI-window with it. Since the PSI-windows for equal and opposite bias are complex conjugate of one another, the effective PSI-window (their difference) for TIE measurement is simply the imaginary part. Figure 11.18 shows the PSI-window for TIE measurement and an image computed using that PSI-window. It is seen that image does represent the second derivative of the phase; however, the filtering effects of the finite apertures are also visible.

11.3.4.2 Phase Retrieval With TIE

Experimental implementations of TIE started with Ichikawa et al. [90] who demonstrated phase retrieval when an amplitude grating is inserted into a plane $z = 0$ right after the pupil plane. At about the same time, another group of people were working independently on developing methods for wavefront sensing through its curvature values [91], and it turns out TIE and wavefront curvature sensing are unified problems [92]. Recently, a group in Australia revived the interest of TIE by integrating it

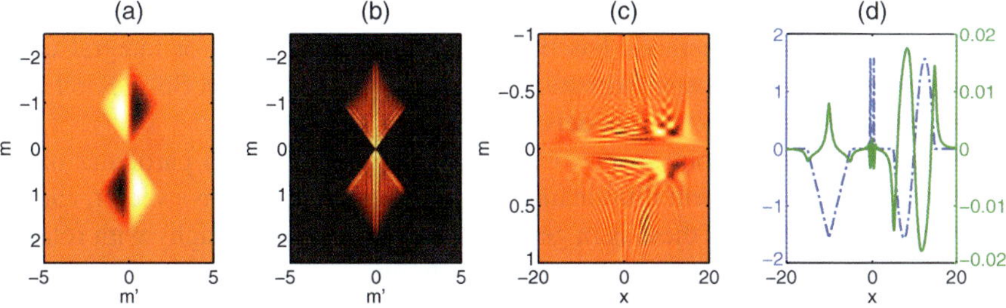

Fig. 11.18 Computation of the partially coherent image of the specimen shown in Fig. 11.13 under TIE with defocus of $u = 4$. The arrangement of this figure is the same as Fig. 11.15, except that (**a**) depicts the imaginary PSI-window

directly to X-ray microscopy and bright-field optical microscopy, coining the term quantitative phase microscopy (QPM) [93]. The fact that TIE is not restricted to coherent regime and works also with a partially coherent source [47, 48] makes it accessible to a wide range of experiments in the optics [94], and X-ray regimes [95], as well as electron-beam microscopy [96].

11.4 Discussion and Conclusion

We have presented a transfer function analysis for some key approaches for today's coherent imaging systems, namely the basic digital holography technique and also holographic tomography based on rotation either of the object or of the illumination. Through visualizations of the 3D spatial content of the transfer functions, we can comprehend the image formation behavior of these systems better. Ideally one would like to cover with these 3D CTF entities as large a range of spatial frequencies as possible. We can observe the strength and weakness of each imaging modality, and perhaps devise better inverse filtering algorithms to fill in the gaps. Proper usage of these 3D CTFs will lead to enhanced imaging performance and more accurate phase retrieval [35]. A specific experimental result was obtained with lensless holography to demonstrate quantitative phase retrieval.

We have also described an emerging class of quantitative methods based on partially coherent illumination. These methods provide useful advantages of higher resolution and improved signal to noise ratio over coherent methods based on simple holography when imaging microscopic specimens. We have discussed image formation and transfer properties of three important partially coherent methods (DIC, DPC, and TIE) using the phase-space imager model. This analysis should also lead to deeper insight into their quantitative behavior and useful inversion approaches. The phase-space imager model provides a unified and promising framework for design and analysis of the quantitative partially coherent systems.

The coherent transfer function approach for imaging in holography with a single illumination direction is based on the first Born approximation for scattering by the object. The geometries of the CTFs are unchanged for other theories for the scattering of the illuminating wave by the sample. However, the relative strength of the CTF depends on the scattering model. The different models range from rigorous or approximate methods based on modal, coupled wave, or integral equation approaches, to other approximations including those based on Born, Kirchhoff, Rytov, Raman-Nath, or Rytov models [97–99].

For tomography, a point in Fourier space corresponding to a particular object grating vector can represent illumination by waves of different wave vector, and the strength of scattering for these is in general different. For tomography with rotation about a single axis two possible directions of illumination correspond to a particular point in Fourier space. For tomography with rotation about two axes, there are in general a range of illumination directions corresponding to different Ewald spheres through the point representing the grating vector. Reconstruction based on the Rytov approximation can be performed by processing holograms for different illumination directions independently. For partially coherent imaging, illumination from different directions is performed simultaneously, so that information from the different illumination directions is lost in the image recording process. This may constitute a disadvantage of the partially coherent approach for image reconstruction. Similarly, confocal interference imaging, in which the object is illuminated coherently simultaneously from different directions, also loses the information about different illumination directions.

References

1. D. Gabor, A new microscopic principle. Nature. **161**(4098), 777–778 (1948)
2. J.W. Goodman, R.W. Lawrence, Digital image formation from electronically detected holograms. Appl. Phys. Lett. **11**, 77–79 (1967)
3. U. Schnars, W. Juptner, Direct recording of holograms by a ccd target and numerical reconstruction. Appl. Opt. **33**(2), 179–181 (1994)
4. E. Cuche, F. Bevilacqua, C. Depeursinge, Digital holography for quantitative phase-contrast imaging. Opt. Lett. **24**, 291–293 (1999)
5. B. Kemper, D. Carl, J. Schnekenburger, I. Bredebusch, M. Schafer, W. Domschke, G. von Bally, Investigation of living pan- creas tumor cells by digital holographic microscopy. J. Biomed. Opt. **11**, 034005 (2006)
6. P. Marquet, B. Rappaz, P. Magistretti, E. Cuche, Y. Emery, T. Colomb, C. Depeursinge, Digital holographic microscopy: a noninvasive contrast imaging technique allowing quantitative visualization of living cells with subwavelength axial accuracy. Opt. Lett. **30**, 468–470 (2005)
7. A. Stern, B. Javidi, Theoretical analysis of three-dimensional imaging and recognition of micro-organisms with a single-exposure on-line holo-graphic microscope. J. Opt. Soc. Am. A Opt. Image Sci. Vis. **24**, 163–168 (2007)
8. G. Pedrini, P. Fröning, H.J. Tiziani, F.M. Santoyo, Shape measurement of microscopic structures using digital holograms. Opt. Comm. **164**(4–6), 257–268 (1999)
9. E. Cuche, P. Marquet, C. Depeursinge, Simultaneous amplitude and quantitative phase-contrast microscopy by numerical reconstruction of Fresnel off-axis holograms. Appl. Opt. **38**, 6994–7001 (1999)
10. G. Indebetouw, Y. Tada, J. Leacock, Quantitative phase imaging with scanning holographic microscopy: an experimental assessment. BioMed. Eng. OnLine. 5(1), 63 (2006)
11. G. Popescu, Quantitative phase imaging of nanoscale cell structure and dynamics. Methods. Cell Biol. **90**, 87–115 (2008)
12. T. Colomb, F. Montfort, J. Kühn, N. Aspert, E. Cuche, A. Marian, F. Charrière, S. Bourquin, P. Marquet, C. Depeursinge, Numerical parametric lens for shifting, magnification, and complete aberration compensation in digital holographic microscopy. J. Opt. Soc. Am. A. **23** (12), 3177–3190 (2006)
13. A. Stadelmaier, J.H. Massig, Compensation of lens aberrations in digital holography. Opt. Lett. **25** (22), 1630–1632 (2000) Opt. Lett.
14. A.C. Kak, M. Slaney, *Principles of Computerized Tomographic Imaging* (Society of Industrial Applied Mathematics, 2001), http://www.slaney.org/pct/pct-toc.html
15. E. Wolf, Three-dimensional structure determination of semi-transparent objects from holographic data. Opt. Commun. **1** (4), 153–156 (1969)
16. C.J.R. Sheppard, T. Wilson, Fourier imaging of phase information in scanning and conventional optical microscopes. Phil. Trans. R. Soc. Lond. Ser. A **295** (1415), 513–536 (1980)
17. C.J.R. Sheppard, M. Gu, The significance of 3-d transfer functions in confocal scanning microscopy. J. Microsc. **165** 377–390 (1991)
18. C.J.R. Sheppard, M. Gu, Imaging by high-aperture optical system. J. Modern Opt. **40**, 1631–1651 (1993)

19. S.S. Kou, C.J.R. Sheppard, Imaging in digital holographic microscopy. Opt. Expr. **15**, 13640–13648 (2007)
20. S.S. Kou, C.J.R. Sheppard, Comparison of three dimensional transfer function analysis of alternative phase imaging methods. in *Three-Dimensional and Multidimensional Microscopy: Image Acquisition and Processing XIV*, **6443**, 64430Q, SPIE (2007)
21. M. Davidson, K. Kaufman, I. Mazor, The coherence probe microscope. Solid State Tech. **30** (9), 57–59 (1987)
22. W.T. Cathey, *Optical Information Processing and Holography, Chapter 9* (Wiley, New York, NY, 1974)
23. C. Wagner, S. Seebacher, W. Osten, W. Juptner, Digital recording and numerical reconstruction of lensless Fourier holograms in optical metrology. Appl. Opt. **38** (22), 4812–4820 (1999)
24. D. Dirksen, H. Droste, B. Kemper, H. Delere, M. Deiwick, H.H. Scheld, G. von Bally, Lensless fourier holography for digital holographic interferometry on biological samples. Opt. Laser Eng. **36**, 241–249 (2001)
25. S. Rehman, K. Matsuda, T. Nakatani, M. Yamauchi, K. Homma, Lensless quasi fourier transform digital holographic microscope. in *Proceedings of Asian and Pacific Rim Symposium on Biophotonics*, Cairns, Australia, 2007)
26. J.W. Goodman, *Introduction to Fourier Optics*. (McGraw-Hill, New York, NY, 1996)
27. S. Rehman, K. Matsuda, H. Ohashi, C.J.R. Sheppard, Quantitative phase imaging with lensless quasi fourier transform hologram method. in *Proceedings of Optics Within Life Sciences (OWLS)*, Singapore, 2008
28. V. Lauer, New approach to optical diffraction tomography yielding a vector equation of diffraction tomography and a novel tomographic microscope. J. Microsc. (Oxf) **205**, 165–176 (2002)
29. F. Charrière, A. Marian, F. Montfort, J. Kühn, T. Colomb, E. Cuche, P. Marquet, C. Depeursinge, Cell refractive index tomography by digital holographic microscopy. Opt. Lett. **31** (2), 178–180 (2006)
30. W. Choi, C. Fang-Yen, K. Badizadegan, S. Oh, N. Lue, R. Dasari, M. Feld, Tomographic phase microscopy. Nat Methods. **4** (9), 717–719 (2007)
31. A. Devaney, A filtered backpropagation algorithm for diffraction tomography. Ultrason Imaging. **4** (4), 336–50 (1982)
32. R. Mueller, M. Kaveh, G. Wade, Reconstructive tomography and applications to ultrasonics. *Proc. IEEE.* **67**, 567–587 (April 1979)
33. A.J. Devaney, Optics in four dimensions. in *AIP Conference Proceedings*, eds. by M.A. Machado, L.M. Narducci vol. 65, (American Institute of Physics, New York, 1980)
34. M. Born, E. Wolf, *Principles of Optics*, 7 ed. (Cambridge University Press, Cambridge 2005)
35. S.S. Kou, C.J.R. Sheppard, Image formation in holographic tomography:high-aperture image conditions. Appl. Opt. **48**, H168–H175 (2009)
36. S. Vertu, J.-J. Delaunay, I. Yamada, O. Haeberlé, Diffraction microtomography with sample rotation: influence of a missing apple core in the recorded frequency space. Cent. Eur. J. Phys. **7** (1), 22–31 (2009)
37. S.S. Kou, C.J.R. Sheppard, Image formation in holographic tomography. Opt. Lett. **33** (20), 2362–2364 (2008)
38. M. Gu, *Advanced Optical Imaging Theory* (Optical Sciences, Springer, 2000)
39. S. Vertu, I. Yamada, J.J. Delaunay, O. Haeberle, Tomographic observation of transparent objects under coherent illumination and reconstruction by filtered backprojection and fourier diffraction theorem - spie. (San Jose, CA 2008,), **6861**, 86103–86103 SPIE. Proc. (2008)
40. C.W. McCutchen, Generalized aperture and the three-dimensional diffraction image. J. Opt. Soc. Am. **54** (2), 240–242 (1964)
41. A. Gray, E. Abbena, S. Salamon, *Modern Differential Geometry of curves and Surfaces with Mathematica*, 3rd ed. (Chapman and Hall CRC, London 2006)
42. F. Dubois, L.Joannes, J.C. Legros, Improved three-dimensional imaging with a digital holography microscope with a source of partial spatial coherence. Appl. Opt. **38**, 7085–7094 (1999)
43. F. Dubois, M.N. Requena, C. Minetti, O. Monnom, E. Istasse, Partial spatial coherence effects in digital holographic microscopy with a laser source. Appl. Opt. **43** 1131–1139 (February 2004)

44. H.H. Hopkins, On the diffraction theory of optical images. Proc. R. Soc. Lond. A. **217** 408–432 (May 1953)
45. C.J.R. Sheppard, A. Choudhury, Image formation in the scanning microscope. J. Mod. Opt. **24** (10), 1051–1073 (1977)
46. M.R. Teague, Deterministic phase retrieval: a green's function solution. J. Opt. Soc. Am. **73** 1434–1441 (November 1983)
47. N. Streibl, Phase imaging by the transport of equation of intensity. Opt. Commun. **49** 6 (1984)
48. D. Paganin, K.A. Nugent, Noninterferometric phase imaging with partially coherent light, Phys. Rev. Lett. **80**, 2586 (March 1998)
49. K.A. Nugent, X-ray noninterferometric phase imaging: a unified picture. J. Opt. Soc. Am A. **24**, 536–547 (February 2007)
50. G. Nomarski, Interference polarizing device for study of phase objects, U.S. Patent 2924142, Feb 1960
51. M. Pluta, *Advanced Light Microscopy*, Specialized Methods. vol. 2 (PWN-Polish Scientific Publishers, Warszawa 1989)
52. N.H. Dekkers, H. de Lang, Differential phase contrast in a STEM. *Optik*. **41** (4), 452–456 (1974)
53. D. Hamilton, C. Sheppard, Differential phase contrast in scanning optical microscopy. J. Microsc. **133** (1), 27–39 (1984)
54. S.B. Mehta, C.J.R. Sheppard, Quantitative phase retrieval in the partially coherent differential interference contrast (DIC) microscope, (April 2009). (Focus on Microscopy, Krakow, Poland) URL: http://www.focusonmicroscopy.org/2009/PDF/209_Mehta.pdf
55. L. Cohen, *Time-Frequency Analysis: Theory and Applications*, (Prentice Hall, Englewood Cliffs, NJ, 1995)
56. S.B. Mehta, C.J.R. Sheppard, Phase-space representation of partially coherent imaging systems using the Cohen class distribution. Opt. Lett. **35**, 348–350 (February 2010)
57. S.B. Mehta, C.J.R. Sheppard, Using the phase-space imager to analyze partially coherent imaging systems: brightfield, phase-contrast, differential interference contrast, differential phase contrast, and spiral phase contrast. J. Modern Opt. **57**, 718–739 (2010)
58. S.B. Mehta, C.J.R. Sheppard, Partially coherent image formation in differential interference contrast (DIC) microscope. Opt. Express. **16** (24), 19462–19479 (2008)
59. C. Cogswell, C. Sheppard, Confocal differential interference contrast(DIC) microscopy: including a theoretical analysis of conventional and confocal DIC imaging. J. Microsc. **165** 81–101 (1992)
60. S.B. Mehta, C.J.R. Sheppard, Sample-less calibration of the differential interference contrast microscope. Appl. Opt. **49**(15), 2954–2968 (2010)
61. W. Galbraith, The image of a point of light in differential interference contrast microscopy: computer simulation. Microsc. Acta. **85** (3), 233–254 (1982)
62. T.J. Holmes, W.J. Levy, Signal-processing characteristics of differential-interference-contrast microscopy. Appl. Opt. **26** (18), 3929 (1987)
63. C. Preza, D.L. Snyder, J. Conchello, Theoretical development and experimental evaluation of imaging models for differential-interference-contrast microscopy. J. Opt. Soc. Am. A. **16** (9), 2185–2199 (1999)
64. P. Hariharan, M. Roy, Achromatic phase-shifting for two-wavelength phase-stepping interferometry. Opt. Commun. **126** (4–6), 220–222 (1996)
65. C.J. Cogswell, N.I. Smith, K.G. Larkin, P. Hariharan, Quantitative DIC microscopy using a geometric phase shifter. Proc. SPIE **2984**, 72 (1997)
66. Y. Xu, Y.X. Xu, M. Hui, X. Cai, Quantitative surface topography determination by differential interference contrast microscopy. Opt. Precis. Eng. **9** (3), 226–229 (2001)
67. M.R. Arnison, K.G. Larkin, C.J.R. Sheppard, N.I. Smith, C.J. Cogswell, Linear phase imaging using differential interference contrast microscopy. J. Microsc. **214**, 7–12 (April 2004)
68. J.S. Hartman, R.L. Gordon, D.L. Lessor, Quantitative surface topography determination by Nomarski reflection microscopy. 2: microscope modification, calibration, and planar sample experiments. Appl. Opt. **19** (17), 2998–300 (1980)

69. W. Shimada, T. Sato, T. Yatagai, in *Optical Testing and Metrology III: Recent Advances in Industrial Optical Inspection* ed. by C.P. Grover. Optical surface microtopography using phase-shifting nomarski microscope., vol 1332 (SPIE, San Diego, CA, 1991), 525–529, pp. (1991)
70. C. Preza, Phase estimation using rotational diversity for differential interference contrast microscopy. PhD thesis, Washington University, 1998
71. C. Preza, Rotational-diversity phase estimation from differential-interference-contrast microscopy images. J. Opt. Soc. Am. A. **17** (3), 415–424 (2000)
72. S.V. King, A. Libertun, R. Piestun, C.J. Cogswell, C. Preza, Quantitative phase microscopy through differential interference imaging. J. Biomed. Opt. **13** (2), 024020 (2008)
73. D.C. Ghiglia, M.D. Pritt, *Two-Dimensional Phase Unwrapping: Theory, Algorithms, and Software*, (Wiley, New York NY, 1998)
74. M. Shribak, J. LaFountain, D. Biggs, S. Inoue, Orientation-independent differential interference contrast microscopy and its combination with an orientation-independent polarization system. J. Biomed. Opt. **13** (1), 014011 (2008)
75. J.A. O'Sullivan, C. Preza, Alternating minimization algorithm for quantitative differential-interference contrast (DIC) microscopy. Proc. SPIE **6814**, 68140Y (2008)
76. H. Ishiwata, M. Itoh, T. Yatagai, A new method of three-dimensional measurement by differential interference contrast microscope. Opt. Commun. **260** 117–126 (April 2006)
77. H. Ishiwata, M. Itoh, T. Yatagai, A new analysis for extending the measurement range of the retardation-modulated differential interference contrast (RM-DIC) microscope. *Opt. Commun.* **281** (6) 1412–1423 (2008)
78. C.J.R. Sheppard, D.K. Hamilton, H.J. Matthews, Scanning optical microscopy of low-contrast samples. *Nature*. **334** (6183) 572 (1988)
79. R. Hoffman, L. Gross, Modulation contrast microscope. *Appl. Opt.* **14** 1169–1176 (May 1975)
80. B. Kachar, Asymmetric illumination contrast: a method of image formation for video light microscopy. *Science*. **227** 766–768 (February 1985)
81. X. Cui, M. Lew, C. Yang, Quantitative differential interference contrast microscopy based on structured-aperture interference. *Appl. Phys. Lett.* **93** (9) 091113 (2008)
82. D. Hamilton, C. Sheppard, T. Wilson, Improved imaging of phase gradients in scanning optical microscopy. *J. microsc.* **135** (3) 275–286 (1984)
83. J. Chapman, I. McFadyen, S. McVitie, Modified differential phase contrast Lorentz microscopy for improved imaging of magnetic structures. Magn. IEEE Trans. **26** (5) 1506–1511 (1990)
84. S.B. Mehta, C.J.R. Sheppard, Linear phase-gradient imaging with asymmetric illumination-based differential phase contrast (AIDPC). in *OSA Technical Digest*, (Novel Techniques in Microscopy, Vancouver, Canada), p. NTuA5, (April 2009)
85. F. Pfeiffer, T. Weitkamp, O. Bunk, C. David. Phase retrieval and differential phase-contrast imaging with low-brilliance x-ray sources. *Nat Phys.* **2** 258–261 (April 2006)
86. J.R. Fienup, Phase retrieval algorithms: a comparison. *Appl. Opt.* **21** 2758–2769 (1982)
87. C.J.R. Sheppard, Defocused transfer function for a partially coherent microscope and application to phase retrieval. *J. Opt. Soc. Am. A.* **21** (5) 828–831 (2004)
88. K.A. Nugent, B.D. Arhatari, A.G. Peele, A coherence approach to phase-contrast microscopy: Theory, *Ultramicrosc.* **108** 937–945 (August 2008)
89. B. Arhatari, A. Peele, K. Hannah, P. Kappen, K. Nugent, G. Williams, G. Yin, Y. Chen, J. Chen, Y. Song, A coherence approach to phase-contrast microscopy II: experiment. *Ultramicrosc.* **109** 280–286 (February 2009)
90. K. Ichikawa, A.W. Lohmann, M. Takeda, Phase retrieval based on the irrandiance transport equation and the fourier transform method: experiments. *Appl. Opt.* **27** 3433–3436 (1988)
91. F. Roddier, Curvature sensing and compensation: a new concept in adaptive optics. *Appl. Opt.* **27** 1223–1225 (1988)
92. F. Roddier Wavefront sensing and the irradiance transport equation. *Appl. Opt.*, **29** 1402–1403 (1990)

93. K. Nugent, T. Gureyev, D. Cookson, D. Paganin, Z. Barnea, Quantitative phase imaging using hard x rays. *Phys Rev Lett.* **77** (14) 2961–2964 (1996)
94. A. Barty, K.A. Nugent, D. Paganin, A. Roberts, Quantitative optical phase microscopy. *Opt. Lett.* **23** (11) 817–819 (1998)
95. L.J. Allen, M.P. Oxley, Phase retrieval from series of images obtained by defocus variation. Opt. Commun. **199** (1–4), 65–75 (2001)
96. M. Beleggia, M.A. Schofield, V.V. Volkov, Y. Zhu, On the transport of intensity technique for phase retrieval. Ultramicroscopy, **102** (1), 37–49 (2004)
97. C.J.R. Sheppard, T.J. Connolly, M. Gu, Scattering by a one-dimensional rough surface and surface reconstruction by confocal imaging. Phys. Rev. Lett. **70**, 1409–1412 (1993)
98. C.J.R. Sheppard, T.J. Connolly, M. Gu, Imaging and reconstruction for rough surfaces scattering in the Kirchhoff approximations by confocal microscopy. J. Mod. Opt. **40**, 2407 (1993)
99. C.J.R. Sheppard, T.J. Connolly, M. Gu, The scattering potential for imaging in the reflection geometry. Opt. Commun. **117**, 16–19 (1995)

Chapter 12
Improving Numerical Aperture in DH Microscopy by 2D Diffraction Grating

Melania Paturzo, Francesco Merola, Simonetta Grilli, and Pietro Ferraro

Abstract In this chapter an approach using a 2D phase grating to enhance the resolution in digital holographic microscopy exploiting the electro-optic effect is proposed. We show that, by means of a tunable lithium niobate phase grating, it is possible to increase the numerical aperture of the imaging system, thus improving the spatial resolution of the images in two dimensions. The enhancement of the numerical aperture of the optical system is obtained by recording spatially multiplexed digital holograms. Furthermore, thanks to the flexibility of the numerical reconstruction process, it is possible to selectively use the diffraction orders carrying useful information for optimizing the diffraction efficiency and to increase the final spatial resolution.

12.1 Super-Resolution Methods in Digital Holography

Recently, important results have been achieved for increasing the optical resolution in DH imaging. Essentially, the resolution of the optical systems is limited by the numerical aperture (NA). In fact, because of the finite aperture of the imaging system, only the low-frequency parts of the object spectrum are transmitted and then recorded by the sensor. Therefore, the corresponding reconstructed images are band limited in the frequency domain.

Several strategies have been defined and different approaches have been tested to increase the NA of the optical system in order to get super-resolution. Most of the methods are essentially aimed at increasing synthetically the NA of the light sensor. They can be divided into three groups according to the kind of strategy adopted.

In the first group's techniques, multiple subsequent acquisitions are performed while the CCD camera is moved with respect to the object, in order to collect light scattered at wider angles. For example, Massig et al. increased the NA by recording nine holograms with a camera (CCD array) translated to different positions and by recombining them in a single synthetic digital hologram [1]. Martínez et al. trans-

M. Paturzo (✉)
Istituto Nazionale di Ottica del CNR (INO-CNR), Via Campi Flegrei 34, 80078 Pozzuoli (NA), Italy
e-mail: melania.paturzo@ino.it

P. Ferraro et al. (eds.), *Coherent Light Microscopy*, Springer Series in Surface Sciences 46, DOI 10.1007/978-3-642-15813-1_12,

lated the camera by a few microns in order to increase both the spatial resolution and the sampling in the recoding process [2].

In the second group we find all the methods in which the object is illuminated by different directions and the light is collected from the same imaging system.

For example, Alexandrov et al. were able to break the diffraction limit by rotating the sample and recording a digital hologram for each position in order to capture the diffraction field along different directions. Then, the multiple recorded holograms, each registering a different, limited region of the sample object's Fourier spectrum, are 'stitched together' to generate the synthetic aperture [3, 4]. A different approach was proposed by Kuznetsova et al. who rotated the sample with respect to the optical axis in order to re-direct the rays scattered at wider angles into the aperture of the optical system, thus going beyond its diffraction limit [5].

Mico et al. proposed and demonstrated two methods for enhancing the resolution of the aperture limited imaging systems based on the use of tilted illumination [6, 7]. In [6], they use vertical-cavity surface-emitting laser (VCSEL) arrays as a set of coherent sources, mutually incoherent. The technique accomplishes the transmission of several spatial frequency bands of the object's spectrum in parallel by use of spatial multiplexing that occurs because of the tilted illumination of the source array (see Fig. 12.1). In [7] a time multiplexing super-resolved approach to overcome the Abbe's diffraction limit, based on the object's spectrum shift produced by tilted illumination, is presented (see Fig. 12.2).

Price et al. [8] described a technique to computationally recombine multiple reconstructed object waves, acquired under a variety of illumination angles, that increase spatial resolution from a physical NA of 0.59 to an effective NA greater than 0.78.

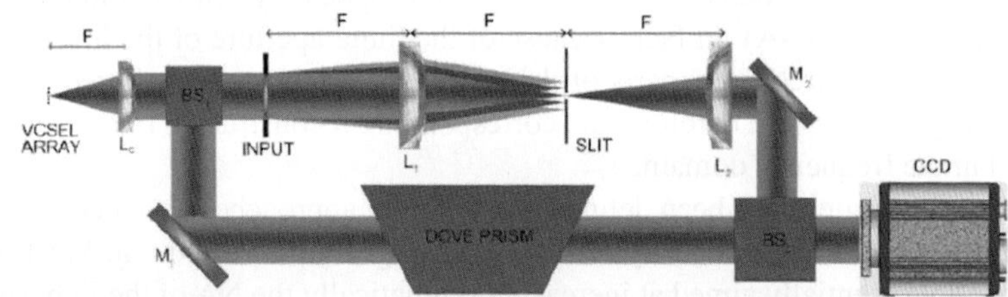

Fig. 12.1 Experimental setup used in [6]: five lit VCSELs are used simultaneously as light sources

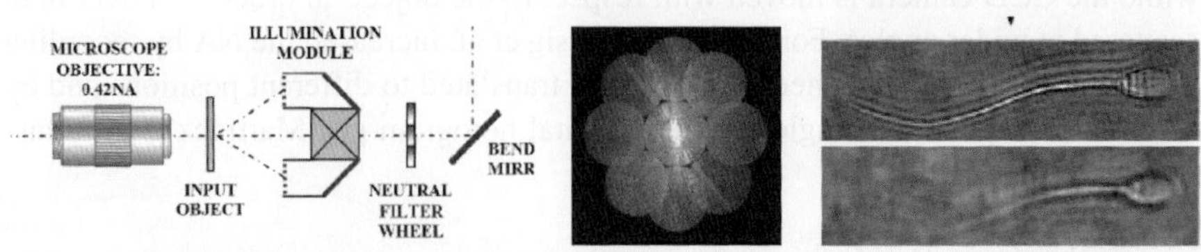

Fig. 12.2 Sketch of the experimental arrangement; synthetic object spectrum and comparison between synthetic aperture and conventional images [7]

In all the aforementioned techniques, the Fourier spectra of the different object waves have to be joined together in order to get the super-resolved image, and, often, to this aim, complex numerical superimposing operations are needed.

The third group includes all techniques that exploit a diffraction grating, inserted in the optical path, for increasing the NA.

In this case, no handling of Fourier spectra is necessary. In fact, the super-resolved images can be obtained simply by using the diffraction effect of an appropriate grating [9–11]. Essentially, this technique allows one to collect parts of the spectrum diffracted by the object, which otherwise would fall outside the CCD array. This was achieved by inserting a diffraction grating in the recording DH setup. The basic principle is simple but effective. In fact, the diffraction grating allows one to re-direct toward the CCD array the information that otherwise would be lost. Basically, three digital holograms are recorded and spatially multiplexed onto the same CCD array. Super-resolved images can be obtained by the numerical reconstruction of those multiplexed digital holograms, by increasing three times the NA.

In [9], a 1D diffraction grating was used, thus enabling to increase the NA of the recording system only along one direction, while in [10] a special diffraction grating is used. It has three important characteristics that allow one to improve the optical resolution behind the limit imposed by the recording system. First, it is a 2D hexagonal grating that allows one to obtain super-resolution in two dimensions. Second, it is a phase grating, instead of the amplitude one. The main drawback of the amplitude gratings is the smaller overall diffraction efficiency. In fact, only the light passing through the openings is used for imaging formation, that is, only a part of the total amount of light illuminating the grating. This characteristic can turn out useful when the light gathering is critical. Third, it is made of an electro-optic substrate (lithium niobate) and therefore it has a tuneable diffraction efficiency, allowing to adjust the relative intensities of the multiplexed holograms in order to improve the quality of the super-resolved images.

12.2 Synthetic Aperture Digital Holography by Phase and Amplitude Gratings

The holographic setup adopted in our experiments is shown in Fig. 12.3. The recording process was carried out by using a Fourier configuration in off-axis mode. The laser source is a He–Ne laser emitting at 632 nm. The specimen is illuminated with a collimated plane laser beam, and a spherical laser beam from a pinhole is used as the reference beam. The distance between the pinhole and the CCD is the same as that between the object and the CCD, according to the Fourier holography configuration. The CCD array has (1024×1024) pixels with $7.6\,\mu\text{m}$ pixel size.

The diffraction grating G is inserted in the optical path between the object and the CCD. We used two different kinds of diffraction gratings: a phase grating and an amplitude one. Both of them are 2D array to allow us to increase the resolution in two dimensions. The first one consists of a 2D array of hexagonally periodically

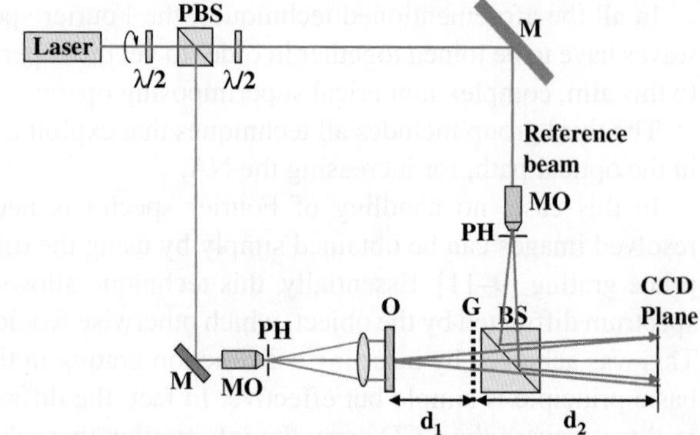

Fig. 12.3 Scheme of the DH recording setup: Fourier configuration in off-axis mode [10]

poled domains in lithium niobate (LN) crystal (with a pitch of 35 μm), while the second one is a standard amplitude grating, commercially available, with a pitch of 25 μm.

As to the numerical reconstruction of the multiplexed digital hologram, it is divided into two steps. First, the wavefield in the plane just behind the grating is obtained through the formula

$$b(x_1, y_1) = \frac{1}{i\lambda d_2} e^{\frac{i\pi}{\lambda d_2}(x_1^2+y_1^2)} \iint r(x_2, y_2) h(x_2, y_2) \mathrm{e}^{\frac{i\pi}{d_2\lambda}[x_2^2+y_2^2]} \mathrm{e}^{-\frac{2i\pi}{\lambda d_2}[x_2 x_1 + y_2 y_1]} \mathrm{d}x_2 \mathrm{d}y_2 \tag{12.1}$$

where $r(x_2, y_2)$ is the reference wave while $h(x_2, y_2)$ is the intensity of the digital hologram acquired by the CCD.

The complex amplitude distribution of the plane immediately before the grating can be obtained by multiplying $b(x_1, y_1)$ with the transmission function of the grating, $T(x_1, y_1)$, and the reconstructed image in the object plane x_0, y_0 can be obtained by computing the Fresnel integral of $b(x_1, y_1)T(x_1, y_1)$ according to

$$b(x_0, y_0) = \frac{1}{i\lambda d_1} e^{\frac{i\pi}{\lambda d_1}(x_0^2+y_0^2)} \iint b(x_1, y_1) T(x_1, y_1) e^{\frac{i\pi}{d_1\lambda}[x_1^2+y_1^2]} \mathrm{e}^{\frac{2i\pi}{\lambda d_1}[x_1 x_0 + y_1 y_0]} \mathrm{d}x_1 \mathrm{d}y_1 \tag{12.2}$$

However, in Sect. 12.2.2, we demonstrate that, when the numerical aperture improvement is exactly equal to 3, it is not necessary to introduce the grating transmission function in the reconstruction process.

In fact, in this case it results that the reconstructed images corresponding to the different diffraction orders are automatically and precisely superimposed.

The double-step reconstruction algorithm is adopted to make the reconstruction pixel (PR) in the image plane independent of the distance between the object and the CCD, different from what occurs in a typical single-step Fresnel reconstruction process, where PR $= \lambda d/(N P_{CCD})$. In fact, in our case, the PR only depends on the ratio d_1/d_2, according to the formula PR $= P_{CCD}\, d_1/d_2$ [12]. We fix d_1 equal to d_2 so that PR $= P_{CCD} = 7.6\,\mu$m. In this way, we assure that the PR is the minimum achievable without decreasing the field of view, corresponding to the pixel size of the CCD.

12.2.1 Super-Resolution by 2D Dynamic Phase Grating

The hexagonal LN phase grating was prepared by standard electric field poling at room temperature following the fabrication process explained in [13, 14]. After poling, transparent ITO (indium tin oxide) electrodes were deposited on both z faces of the sample in order to apply an external field across the crystal preserving the optical transmission along the z-axis.

The phase step between opposite ferroelectric domains can be varied by changing the applied voltage across the z-axis of the crystal. When no voltage is applied to the crystal, no diffraction occurs since the diffraction grating is inactive (*switched-off*). When voltage is applied, the grating becomes active (*switched-on*).

It is able to generate different diffraction orders. Essentially, each diffraction order produces a corresponding digital hologram and all of the holograms are spatially multiplexed and recorded simultaneously by the CCD.

The schematic view of the object waves is shown in Fig. 12.4a, b where, for sake of simplicity, only one point object P is discussed. In case of conventional holographic configurations, all of the rays scattered by the object freely propagate forward to the CCD plane, but only the central ray fan reaches the area of the hologram and can be digitally recorded, as shown in Fig. 12.4a. Therefore, because of the limited aperture of the CCD array, the recorded object wave beams are only a portion of the total light scattered by the object. However, when the grating is placed between the object and the CCD array six further fan beam waves can reach the CCD. The sketch of this configuration is depicted in Fig. 12.4b. Each of the six waves is produced by the first diffraction orders of the grating. The resulting digital hologram is essentially formed by seven digital holograms that are spatially multiplexed and coherently superimposed. The digital hologram is numerically reconstructed to obtain the in-focus real image of the tested target.

The holographic system in Fig. 12.4b clearly exhibits higher NA compared to that in Fig. 12.4a. In fact, the CCD aperture augments up to three times along each of the three directions at 120°, thanks to the hexagonal geometry. Consequently, the reconstructed image of the point P has a resolution enhanced up to three times compared to the usual DH system without the diffraction grating.

12.2.1.1 Experimental Results Demonstrating the Resolution Enhancement

Figure 12.5a shows the amplitude reconstruction of the digital hologram of the object when no voltage is applied to the electro-optic grating. The object is a

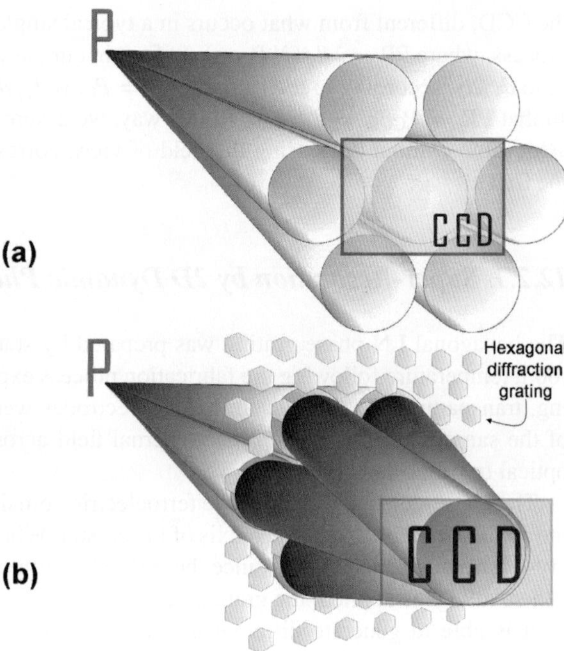

Fig. 12.4 Ray diagrams of the object waves: (**a**) without the grating in the setup and (**b**) with the grating in the setup [10]

microscopy target with different spatial frequencies ranging from 12.59 to 100 lines/mm. The amplitude reconstruction clearly shows that the resolution is limited up to the maximum value of 31.6 lines/mm, as evidenced by the magnified view in Fig. 12.5b, while pitches with 25.1, 20.0, and 15.8 μm are clearly below the resolution limit of the system. However, as explained in the previous section, the DH system is expected to reconstruct correctly up to the extreme limit of

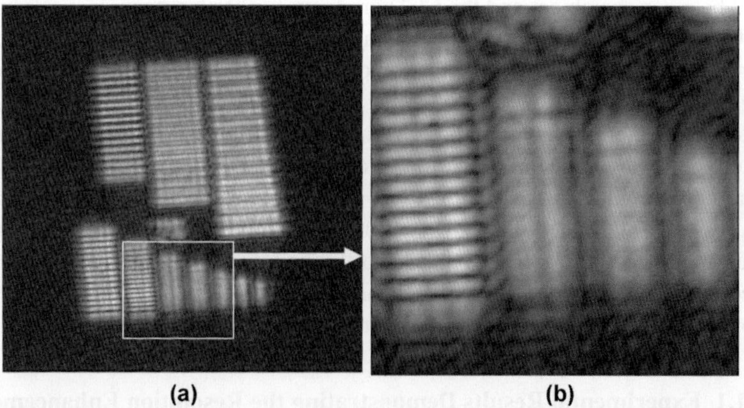

Fig. 12.5 (**a**) Amplitude reconstruction of the digital hologram when no voltage is applied to the electro-optic grating; (**b**) magnified view showing the reticules with the shortest pitches (31.6, 25.1, 20.0, 15.8 μm) [10]

$2 \times 7.6\,\mu\text{m} = 15.2\,\mu\text{m}$, according to the Nyquist criteria (at least two pixels per period), and taking into account that PR $= 7.6\,\mu\text{m}$. Nevertheless, the reticules with pitches below $31.6\,\mu\text{m}$ are clearly unresolved due to the limited NA of the system.

When the phase grating is switched-on, seven spatially multiplexed digital holograms are recorded by the CCD array simultaneously. Figure 12.6 shows the amplitude reconstruction of the multiplexed digital hologram when a voltage is applied (2.5 kV in this case).

In fact, as explained above, the diffraction grating used in this experiment is an electro-optically tuneable binary phase grating. Therefore, by changing the applied voltage across the z-axis of the LN crystal, it is possible to tune the phase step. Consequently, the efficiency of the diffraction orders is also adjustable since it is proportional to $\sin^2(\Delta\varphi/2)$, where $\Delta\varphi$ is the phase step between inverted domains that changes proportionally to the applied voltage V. The maximum value of the efficiency is obtained for $\Delta\varphi = \pi(V = 2.5\,\text{kV})$ and therefore the images coming from the diffracted orders, that contain the information about the high spatial frequencies of the object spectrum, have a high weight in the superimposition process.

The numerical reconstruction is performed by using (12.2) without introducing the transmission function $T(x, y)$ of the phase diffraction grating (i.e., $T(x, y) = 1$).

As explained before, each diffraction order produces one of the multiplexed holograms. Consequently the reconstruction process shows up seven corresponding images, one for each of the multiplexed holograms: one for the 0th order and six for the first orders of diffraction, that are along the three typical directions of the hexagonal grating (see Fig. 12.6).

The images along the three different directions, encircled, respectively, by the yellow, blue, and red ellipses in Fig. 12.7a, carry different information about the object spatial frequencies spectrum. In fact, the corresponding holograms originate

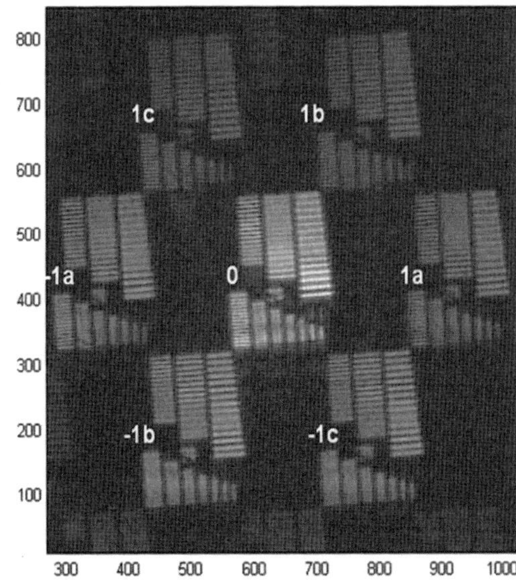

Fig. 12.6 Amplitude reconstruction of the multiplexed digital hologram when the phase grating is switched-on (applied voltage of 2.5 kV). This numerical reconstruction has been obtained without introducing the transmission function of phase diffraction grating in the reconstruction algorithm (i.e., $T(x, y) = 1$). The label of the reconstructed images indicates the corresponding diffraction orders [10]

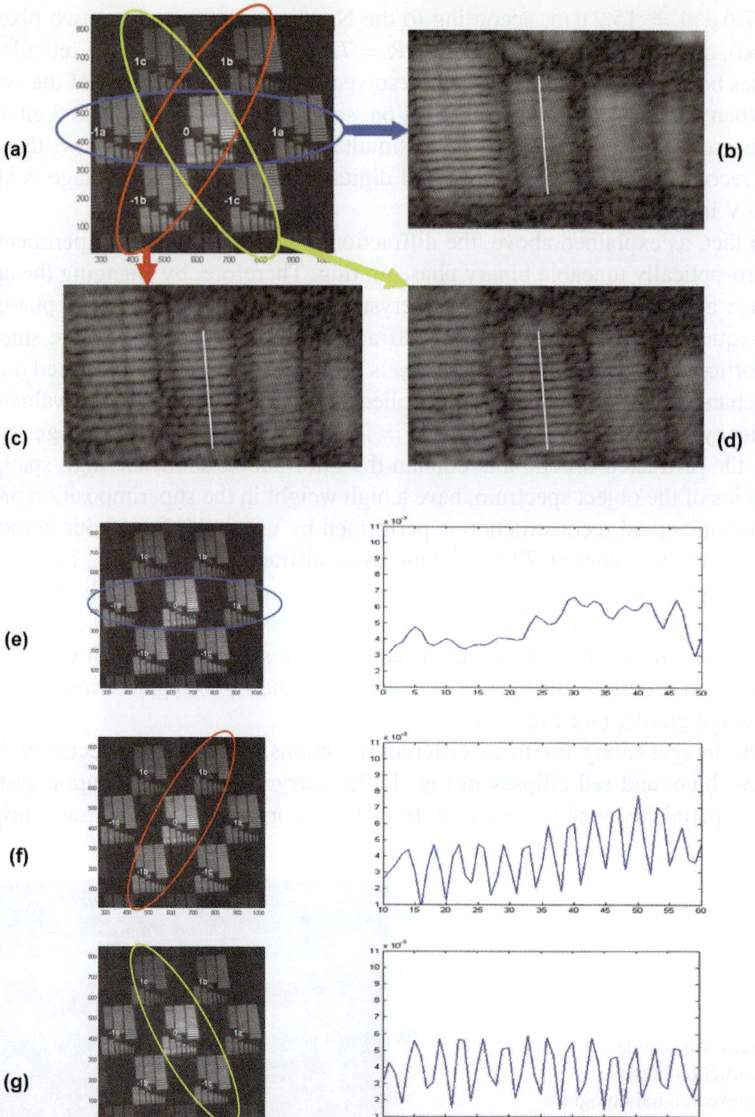

Fig. 12.7 (**a**) The colored ellipses encircle the reconstructed images along the three typical directions of the hexagonal grating. Magnified view of the image obtained by superimposing only the −1a, 0th, +1a diffraction orders (*blue ellipse* in (**a**)); (**c**) the −1b, 0th, +1b orders (red ellipse); (**d**) the −1c, 0th, +1c orders (*yellow ellipse*). The reticule with a pitch of 25.1 µm, completely blurred in (**b**), is resolved in (**c**) and (**d**) thanks to an improvement of the optical resolution. Plot of intensity profile along the *white lines* in the images (**b**), (**c**), and (**d**) is shown in (**e**), (**f**), and (**g**), respectively. Axes of the plots have atomic units for intensities on ordinates while pixel number on abscissa [10]

from rays scattered by the object along different directions and re-directed onto the CCD area by means of the grating.

The resolution enhancement can be obtained only by selectively superimposing the different reconstructed images obtained by the digital holograms. In fact, in this way, we can increase the NA of the optical system and therefore the optical resolution of the resulting image in 2D. Even though the target used here for the imaging experiments has 1D geometry, the 2D capability of the technique is demonstrated by setting intentionally the 1D target along a direction different from the three diffraction directions of the hexagonal grating. This configuration reveals that the best image resolution can be obtained by using at least two diffraction directions, as shown by the following results. The superimposition of the reconstructed multiplexed holograms is obtained by using (12.2). In this first experiment the distances are such that the increment in the numerical aperture should be 1.6.

Therefore, the transmission function of the grating has to be introduced in the second step of the reconstruction process, that is, (12.2).

In this case, we assume that the grating has a transmission function that can be written as

$$\begin{aligned} T(x_1,\ y_1) = 1 + a\cos(2\pi x_1/p) + b\cos\left(\left(x_1 + \sqrt{3}y_1\right)\pi/p\right) \\ + c\cos\left(\left(x_1 - \sqrt{3}y_1\right)\pi/p\right) \end{aligned} \tag{12.3}$$

Equation (12.3) is made of four terms. The first term is a constant offset. The second takes into account the diffraction along the horizontal direction, while the third and fourth terms consider the two other directions. In (12.3) $\boldsymbol{p}$ is the period of the grating, while **a**, **b**, and **c** are the diffraction efficiencies along the three different directions typical of the hexagonal pattern of our diffraction grating, respectively.

12.2.1.2 Flexibility of the Numerical Reconstruction

According to the particular geometry of the object, it is also possible to superimpose only some of the reconstructed images that effectively possess and carry the useful information with the aim at resolving the details of the object under examination. In fact, the numerical reconstruction algorithm can be considered for one, two, or all of the three directions, simply assigning appropriate values to the diffraction efficiency coefficients in (12.3) (i.e., a, b, c, respectively).

For example, Fig. 12.7b shows the magnified view of the reconstructed image obtained by superimposing the diffraction orders −1a, 0th, and +1a (blue ellipse) only, which means we are considering just the horizontal direction. Differently, Fig. 12.7c, d shows the reconstructions obtained by taking into account the −1b, 0th, +1b and −1c, 0th, +1c orders, respectively, corresponding to the inclined directions (red and yellow ellipses). Only the last two reconstructions lead to an improvement of the optical resolution allowing one to resolve the reticule with 25.1 μm pitch, otherwise completely blurred in Fig. 12.7b. This is clearly evidenced by the profiles in Fig. 12.7e–g, respectively. This result demonstrates that the collection of the rays diffracted along the horizontal direction is not useful for increasing the resolution of such target. In fact, since the rulings have only lines parallel to the horizontal direction, the rays scattered from finer rulings are directed mainly at higher angles along the vertical direction. Therefore the object frequencies have components only along the directions of the diffraction orders **±1b** and **±1c**. Consequently, in order to obtain the best signal/noise ratio in the super-resolved image, the superimposition of the orders **0th**, **±1b**, and **±1c** without the **±1a** is better.

Figure 12.8a shows clearly the reconstruction obtained by superimposing all of the first diffraction orders, **±1a, ±1b**, and **±1c**, on the zero order 0th, while

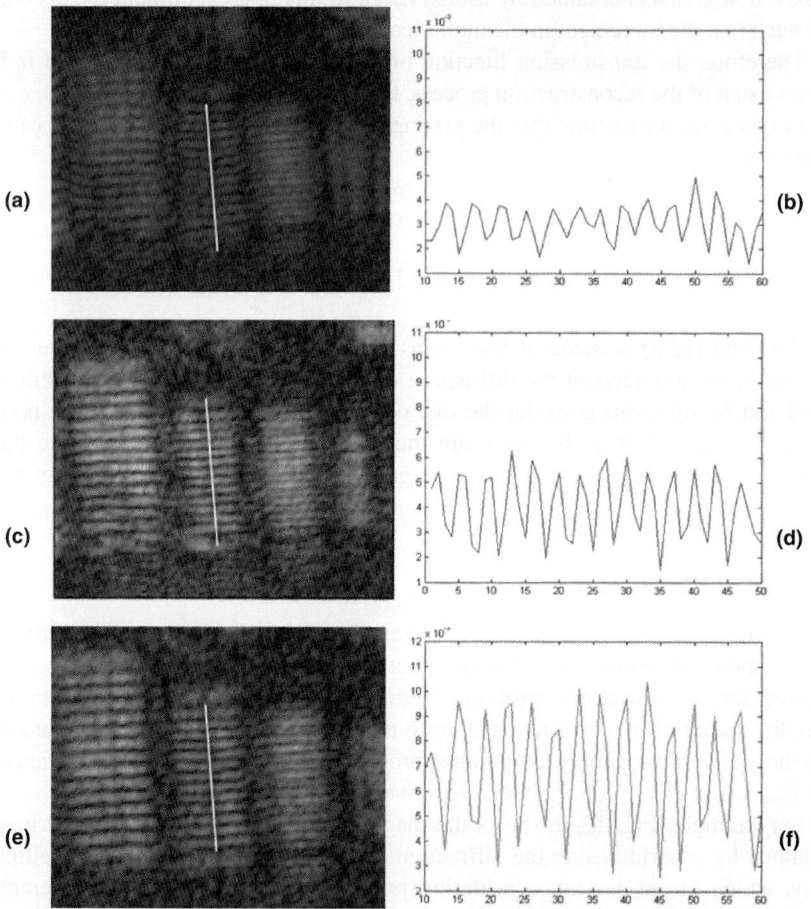

Fig. 12.8 (**a**) Amplitude reconstruction of the target obtained by superimposing all the first diffraction orders, ±1a, ±1b, and ±1c, on the zero order 0th; (**c**), (**e**) amplitude reconstruction obtained by ignoring the ±1a orders and by using different or same weights for the orders ±1b, ±1c, respectively; (**b**), (**d**), (**f**) corresponding profiles calculated along the ruling with 25.1 μm pitch [10]

Fig. 12.8c, e shows the reconstructions where the $\pm$**1a** orders (corresponding to the horizontal direction) are not considered. By comparing the reconstruction in Fig. 12.8a with those in Fig. 12.8c, e, it is possible to notice that, involving a useless order ($a \neq 0$) in the reconstruction, only noise is added without any benefit for the resolving power. Figure 12.8b, d, f shows the profiles calculated along the ruling with a 25.1 μm pitch for each of the corresponding reconstructed images in Fig. 12.8a, c, e. The super-resolved images shown in Fig. 12.8c, e differ from each other because they are obtained using two different sets of parameters in (12.3), namely ($a = 0$, $b = 2$, $c = 4$) for Fig. 12.8c and ($a = 0$, $b = 4$, $c = 4$) for Fig. 12.8e.

The results show clearly that for the used object, the signal/noise ratio in the super-resolved image is increased when the two orders $\pm$**1b** and $\pm$**1c** have the same weight in the superimposition. This depends on the particular geometry of the object which has spatial frequencies all along the vertical direction, that is, exactly along the bisector of the angle between the directions of the diffraction orders b and c. Therefore, the spatial frequency components along the directions of the diffraction orders $\pm$**1b** and $\pm$**1c** are the same. However, for some particular experimental conditions, the possibility to modulate the weight of each diffraction order could be useful with the aim at recovering the best super-resolved image. This selective superimposition, both in terms of selected directions and in terms of weights to be assigned to each of the considered diffraction orders, is uniquely allowed by the flexibility of such numerical reconstruction process.

12.2.2 Super-Resolution by 2D Amplitude Grating

The technique described above allows to increase the numerical aperture up to three times along each diffraction direction. One of the main advantage of this methods in respect to the others is that by introducing the correct grating transmission function into the reconstruction diffraction integral, the spatial frequencies of the objects are assembled together automatically without cumbersome and tedious numerical superimposing operations [1–8]. Nevertheless, the introduction of the transmission function of the grating produces some deleterious numerical noise that can affect the final super-resolved image, as will be shown below.

However, when the numerical aperture improvement is exactly equal to 3, it is not necessary to introduce the grating transmission function in the numerical reconstruction formulas. In fact, in this case it results that the reconstructed images corresponding to the different diffraction orders are automatically and precisely superimposed.

We show that it was possible exploiting the typical wrapping effect in the reconstructed image plane when the bandwidth of the image does not fit entirely into the reconstructed window. This novel approach is demonstrated by performing two different experiments with a target and a biological sample, respectively. To experimentally validate the proposed approach a 1D resolution improvement is performed, employing a amplitude grating, commercially available, with a pitch of 25 μm.

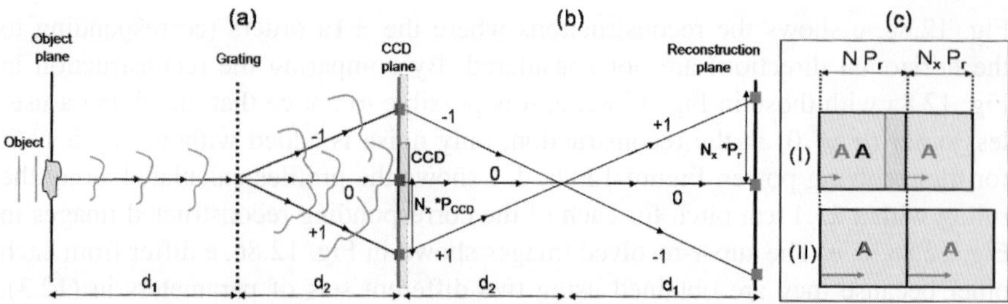

Fig. 12.9 DH recording setup with a grating (**a**); geometric scheme of the reconstruction without a grating in self-assembling mode (**b**); geometric scheme to show the automatic superimposition in the self-assembling mode (**c**) [11]

Then, we show a 2D super-resolved image of a biological sample, using two 1D amplitude gratings with the same pitch, placed in close contact but one at 90° with respect to the other. This method has the significant advantage to avoid the use of the transmission function by reducing the noise considerably.

The sketch in Fig. 12.9 describes the object wave optical path during the recording (Fig. 12.9a) and the reconstruction processes (Fig. 12.9b). The CCD records three different holograms simultaneously (i.e., spatially multiplexed holograms) corresponding to the three diffraction orders, respectively. Each hologram carries different information about the object. In fact, because of the limited numerical aperture of the CCD, the hologram corresponding to the zero order contains low frequencies of the wave scattered by the object, while the holograms corresponding to the first diffraction orders collect the marginal rays in the wavefront scattered by the object at wider angles and therefore carry information about higher frequencies.

Three distinct images corresponding to each hologram are numerically reconstructed, as shown in Fig. 12.9b. Our aim is to get together these three images in an appropriate way in order to get super-resolution. In Fig. 12.9 the blue squares in the CCD plane indicate the center of the three holograms, while the red squares in the reconstruction plane are the center of the reconstructed images. The distance between two contiguous holograms is $Nx \times P_{\text{CCD}}$, where P_{CCD} is the CCD pixel size and Nx is a real number, while the distance between two reconstructed images is $Nx \times \text{Pr}$, where Pr is the pixel of reconstruction.

When Nx is equal to N, that is, the number of CCD pixels, it means that the relative distance between the three holograms created by the grating is exactly equal to the lateral dimension of the CCD array, as shown in Fig. 12.9. Therefore, only when $Nx = N$, we obtain a resolution enhancement equal to 3. By a simple geometrical computation, it results that the distance between the zero and the first diffracted orders in the CCD plane is $\Delta x = \lambda d_2/p$ where p is the grating pitch and d_2 is the distance between the grating and the CCD. Since to obtain a numerical aperture improvement equal to 3 it is necessary to fulfill the condition $\Delta x = NP_{\text{CCD}}$, the distance d_2 has to be

$$d_2 = \frac{pNP_{\text{CCD}}}{\lambda} \tag{12.4}$$

In the general case, when d_2 does not match exactly (12.4), we have $\Delta x = N_x P_{\mathrm{CCD}}$ and the improvement of the numerical aperture is given by $1 + 2\,(N_x/N)$ with $N_x \leq N$. Only if we get the resolution improvement of 3, it is possible to obtain at the same time an automatic self-assembling of the various spatial frequency of the objects in the reconstruction image plane. In Fig. 12.9c the geometric diagram, concerning the reconstructed images, to explain how it happens is shown. It results that the lateral distance between the reconstruction images, corresponding to the three diffracted orders, is equal to

$$\Delta\xi = N_x \,\mathrm{Pr} \tag{12.5}$$

where $\mathrm{Pr} = \frac{d_1}{d_2} P_{\mathrm{CCD}}$ is the reconstruction pixel for the double-step reconstruction process we use [11, 12]. In fact, all the reconstructions are obtained by the two steps process computing two Fresnel integrals: first we reconstruct the hologram in the grating plane and then we propagate this complex field up to the object plane [11].

In the double-step reconstruction, Pr depends only on the ratio d_1/d_2 while it does not depend on the distance $D = d_1 + d_2$ between the object and the CCD, different from what occurs in a typical single-step Fresnel reconstruction process, where $\mathrm{Pr} = \lambda D/(N P_{\mathrm{CCD}})$.

It is worth to note that, in DH, when the reconstruction window is not large enough to contain the entire spatial frequency band of the object, the object signal is wrapped within the reconstruction window.

In our case, the object lateral dimension is $(N + 2N_x)\mathrm{Pr}$ while the reconstruction window is only N Pr. In the sketch of Fig. 12.9c(I), for simplicity, only the central and right reconstruction image windows, corresponding to $(N + N_x)\mathrm{Pr}$, are shown. Left reconstructed image has been omitted for clarity. Letter A represents the central point of the field of view of each reconstructed hologram.

In Fig. 12.9c(I), the portion of the right reconstructed image that includes the red letter A falls outside the reconstruction window and, therefore, is wrapped and re-enters in the reconstruction window from the left side. In this case an observer will see, in the reconstruction plane, multiple reconstructed images due to each of the three multiplexed holograms incorrectly superimposed. Only when $N_x = N$ (as in Fig. 12.9c(II), when the red letter is perfectly superimposed on the black one), the reconstructed images, corresponding to the different diffraction orders, are automatically and perfectly superimposed. The fulfillment of the condition $N_x = N$ assures the self-assembling of the spatial frequencies of the object in the reconstructed plane and allows the achievement of an enhancement of super-resolution with a factor of 3, but without the need to introduce the grating transmission function in the diffraction propagation algorithm.

With the aim to demonstrate that, we performed different experiments. In Fig. 12.10, different numerical reconstructions obtained for different values of the distance d_1 and d_2 are shown. The grating we used has a pitch of 25 μm and therefore to fulfill the condition of (12.1), d_2 has to be equal to 21 cm.

Looking at Fig. 12.10, it is clear that, only when $d2$ has to be equal to 21 cm, the three reconstructions are superimposed. Moreover, we can assert that the value of

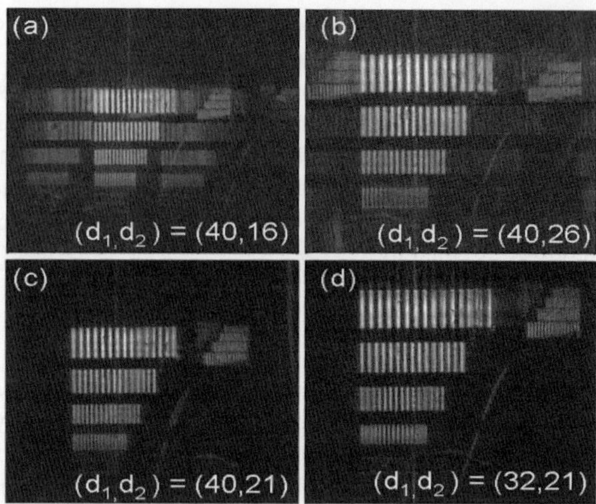

Fig. 12.10 Reconstructions adopting different values (d_1,d_2) (in cm). In (**a**) and (**b**) the spatial frequencies are wrongly assembled. To fulfill the condition of (12.1), d_2 has to be equal to 21 cm as shown in (**c**) and (**d**) [11]

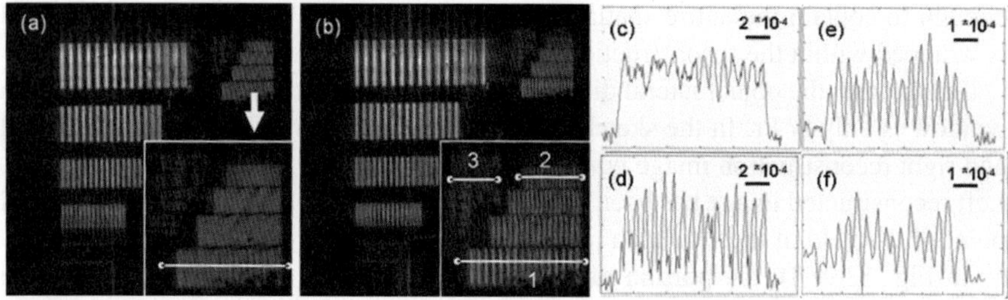

Fig. 12.11 How the reconstructions appear (**a**) without a grating (no super-resolution) and (**b**) with the grating (super-resolved image). In (**c**) profile of the 100 μm along the line in the inset of (**a**) showing that the grating is not well resolved. In (**d**), (**e**), and (**f**) profiles of line 1, 2, and 3 indicated in (**b**) showing the super-resolving capability [11]

distance d_1 does not affect the self-assembling properties, but only the value of the pixel of reconstruction and, therefore, the field of view width. Fixing the values (d_1, d_2) = (32, 21) (unit are given in cm), we have measured the effective resolution improvement. In Fig. 12.11a, b, the numerical reconstructions without and with the grating in the setup are shown.

The profiles corresponding to the reticule with a pitch of 100 μm (line 1) are shown in Fig. 12.11c, d, respectively. Moreover, in Fig. 12.11d, e the profiles of the reticules with a pitch of 50.14 (line 2) and 31.63 (line 3) μm, respectively, obtained by the reconstruction of the hologram recorded when the grating is inserted in the setup are shown. It results that the grating with a pitch of 31.63 μm has more or less the same contrast of the grating with a pitch of 100 μm obtained by the

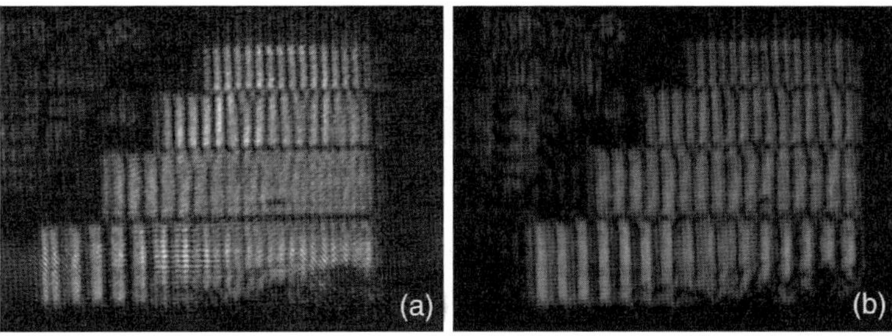

Fig. 12.12 (**a**) Super-resolved image obtained by the method reported in Sect. 12.2.1 using grating transmission function in the reconstruction. (**b**) Super-resolved image obtained in self-assembling approach [11]

reconstruction of the hologram acquired without the grating in the setup. Therefore, we can assert that the resolution enhancement is effectively equal to 3.

Moreover, with the aim to show how the introduction of the grating transmission function in the reconstruction process can affect the image quality, a magnified view of a portion of the reconstructed target by using the standard method (see Sect. 12.2.1) and the self-assembling approach is shown in Fig. 12.12a, b, respectively. Figure 12.12a shows the holographic reconstruction that adopts the grating transmission function into numerical diffraction integral, while the self-assembled reconstruction method proposed in this section is shown in Fig. 12.12b. It is clear that the noise introduced by the transmission function hinders some relevant information in the super-resolved image.

Finally, we made an additional experiment recording an hologram of a biological sample, formed of a slice of a fly head. The image of the object such as it appears to an optical microscope is shown in Fig. 12.13a. In this case we introduced in the optical setup a 2D grating with the same pitch (25 μm). Therefore, the resolution increases three times along both the diffraction directions.

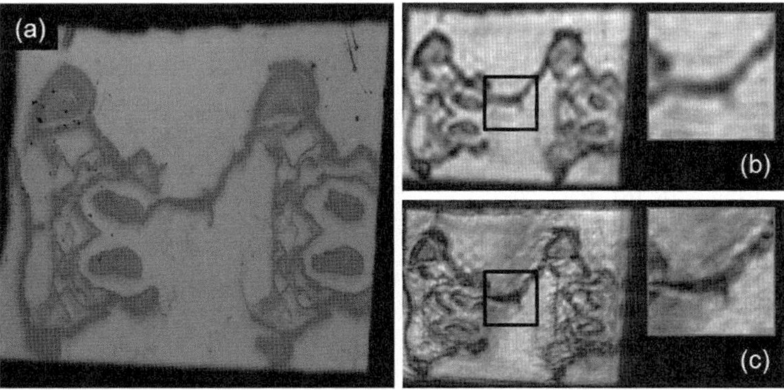

Fig. 12.13 (**a**) Slice of fly head eye at optical microscope; (**b**) DH reconstructed image without super-resolution; (**c**) DH super-resolved image by using a 2D square grating shows spatial frequencies self-assembled correctly in two dimensions [11]

In Fig. 12.13b, c, the amplitude reconstructions corresponding to the holograms acquired with and without the grating are shown. The insets in Fig. 12.13b, c clearly show how fine details of the object are clearly unresolved in the optical configuration without a grating while they are clearly resolved in Fig. 12.13b when a 2D grating is in the setup and self-assembling of spatial frequencies in 2D works well.

12.3 Conclusion

Super-resolution can be obtained by adopting a diffraction grating that allows one to augment the NA of the optical system. We demonstrate that the improvement is possible in two dimensions, adopting two different kinds of diffraction gratings: a phase hexagonal grating and an amplitude one. In the first case we are able to increase the NA along three different directions (i.e., the three directions typical of the hexagonal geometry). The phase levels and therefore the diffraction efficiency can be tuned by electro-optic effect, allowing one to optimize the recording process of the digital holograms. Moreover, by an appropriate handling of the transmission function of the numerical grating in the reconstruction algorithm, it is possible to further improve the signal/noise ratio in the final super-resolved image.

When the numerical aperture improvement is exactly equal to 3, it is not necessary to introduce the grating transmission function in the numerical reconstruction formulas.

The spatial frequencies of the object are naturally self-assembled in the image plane by exploiting the reconstruction of multiplexed holograms with different angular carriers. Thanks to this novel approach we were able to obtain a resolution improvement factor of 3 and to reduce significantly the noise in the final image that otherwise can hinder the super-resolved signal. Results have been proofed in lensless configuration for a 1D optical target and also with a true 2D biological sample and a 2D grating to get super-resolution in 2D.

References

1. J.H. Massig, Digital off-axis holography with a synthetic aperture. Opt. Lett. **27**, 2179–2181 (2002)
2. L. Martínez-León, B. Javidi, Synthetic aperture single-exposure on-axis digital holography. Opt. Express **16**, 161–169 (2008)
3. S.A. Alexandrov, T.R. Hillman, T. Gutzler, D.D. Sampson, Synthetic aperture Fourier holographic optical microscopy. Phys. Rev. Lett. **97**, 168102 (2006)
4. T.R. Hillman, T. Gutzler, S.A. Alexandrov, D.D. Sampson, High-resolution, wide-field object reconstruction with synthetic aperture Fourier holographic optical microscopy. Opt. Express **17**, 7873–7892 (2009)
5. Y. Kuznetsova, A. Neumann, S.R. Brueck, Imaging interferometric microscopy–approaching the linear systems limits of optical resolution. Opt. Express **15**, 6651–6663 (2007)
6. V. Mico, Z. Zalevsky, P. Garcia-Martinez, J. Garcia, Single-step superresolution by interferometric imaging. Opt. Express **12**, 2589–2596 (2004)

7. V. Mico, Z. Zalevsky, C. Ferreira, J. García, Superresolution digital holographic microscopy for three-dimensional samples. Opt. Express **16**, 19260–19270 (2008)
8. J.R. Price, P.R. Bingham, C.E. Thomas, Jr., Improving resolution in microscopic holography by computationally fusing multiple, obliquely illuminated object waves in the Fourier domain. Appl. Opt. **46**, 827–833 (2007)
9. C. Liu, Z. Liu, F. Bo, Y. Wang, J. Zhu, Super-resolution digital hològraphic imaging method. Appl. Phys. Lett. **81**, 3143 (2002)
10. M. Paturzo, F. Merola, S. Grilli, S. De Nicola, A. Finizio, P. Ferraro, Super-resolution in digital holography by a two-dimensional dynamic phase grating. Opt. Express **16**, 17107–17118 (2008)
11. M. Paturzo, P. Ferraro, Correct self-assembling of spatial frequencies in super-resolution synthetic aperture digital holography, Opt. Lett. **34**, 3650–3652 (2009)
12. F. Zhang, I.Yamaguchi, L.P. Yaroslavsky, Algorithm for reconstruction of digital holograms with adjustable magnification. Opt. Lett. **29**, 1668–1670 (2004)
13. S. Grilli, M. Paturzo, L. Miccio, P. Ferraro, In situ investigation of periodic poling in congruent LiNbO3 by quantitative interference microscopy. Meas. Sci. Technol. **19**, 074008 (2008)
14. M. Paturzo, P. De Natale, S. De Nicola, P. Ferraro, S. Mailis, R.W. Eason, G. Coppola, M. Iodice, M. Gioffré, Tunable two-dimensional hexagonal phase array in domain-engineered Z-cut lithium niobate crystal. Opt. Lett. **31**, 3164–3166 (2006)

Chapter 13
Three-Dimensional Mapping and Ranging of Objects Using Speckle Pattern Analysis

Vicente Micó, Zeev Zalevsky, Javier García, Mina Teicher, Yevgeny Beiderman, Estela Valero, Pascuala García-Martínez, and Carlos Ferreira

Abstract In this chapter, we present two novel approaches for 3-D object shape measurement and range estimation based on digital image processing of speckle patterns. In the first one, 3-D mapping and range measurement are retrieved by projecting, through a ground glass diffuser, random speckle patterns on the object or on the camera for a transmissive and reflective configuration, respectively. Thus, the camera sensor records in time sequence different speckle patterns at different distances, and by using correlation operation between them, it is possible to achieve 3-D mapping and range finding. In the second one, the 3-D mapping and ranging are performed by sensing the visibility associated with the coherence function of a laser source used to illuminate the object. In this case, the object depth is encoded into the amplitude of the interference pattern when assembling a typical electronic speckle pattern interferometric (ESPI) layout. Thus, the 3-D object shape is reconstructed by means of a range image from the visibility of the image set of interferograms without the need for depth scanning. In both cases, we present experimental implementation validating the proposed methods.

13.1 Introduction

In industry, there are numerous applications demanding 3-D measurement of objects [1]. Some examples are image encryption [2, 3], object recognition [4, 5] and 3-D mapping [6–8]. All of those possibilities have been allowed by the development of digital sensors and, in particular, by digital holography [9]. Digital holography offers highly attractive advantages such as high resolution, non-contact measurement and versatility of implementation depending on the type of object. This way, digital holography has been successfully applied to surface contouring of macro-, small- and micro-objects [10, 11].

But not only digital holography deals with this task. Triangulation provides 3-D object shape measurement using both white light structured illumination [6] and

J. García (✉)
Departamento de Óptica, Universitat de Valencia, C/Doctor Moliner 50, 46100 Burjassot, Spain
e-mail: javier.garcia.monreal@uv.es

P. Ferraro et al. (eds.), *Coherent Light Microscopy*, Springer Series in Surface Sciences 46, DOI 10.1007/978-3-642-15813-1_13,

coherent illumination [12]. The general concept of triangulation is that the light projection and the observation directions are different. As the projected pattern is distorted by the object shape, the lateral positions of the measured pattern details include depth information of the object. But triangulation suffers from several problems such as shadows incoming from the projection angle between illumination and detection directions, and discontinuities and occlusions due to specific geometry of the object which generates additional shadowing. In any case, shadowing implies object information losses. Moreover, the axial resolution of the system is limited by triangulation accuracy, and the method needs a wide measuring head for accurate measurements at long distances.

One way to minimize the shadowing problems is to n-plicate the measurement system [13] or rotate the object [14]. The former method implies both a high computational cost and expensive solution while the latter means a time-consuming process to obtain the final result.

Shadowing problems (not occlusions) can be completely removed by projecting the illumination coaxial with the observation direction. However, the spatially dependant structured illumination will not help and an alternative coding must be used. The most common one is the time coding either in light pulse shape or in illumination spectral content. Examples are the time gating methods [15], time of flight [16] or the use of two wavelength illuminations [17]. Some of these methods do not provide directly a 3-D mapping but an object contouring.

Although coaxial illumination and observation is complex to implement when considering conventional white light structured illumination, it is easy to achieve with coherent illumination because there is no need to project a structured illumination pattern onto the object surface: one can split the coherent beam before the object, illuminate the object coaxially with the observation direction and reinsert a reference beam at the digital recording device. This is the underlying principle of ESPI [18]. In ESPI, phase shifting and Fourier transformation are the two main methods to recover the object phase distribution and correspond with on-axis and off-axis reference beam reinsertion geometry, respectively [19, 20]. However, both procedures allow access to the wrapped phase distribution of the object (as is the case in contouring methods), that is, phase values lying in the range of $-\pi$ to π, as a consequence of the depth-to-phase coding. The phase map needs to be unwrapped to recover the 3-D object information. Once again, phase-unwrapping algorithms will not be useful if the image contains unknown relative heights and/or the object contains sudden jumps on its profile, that is, discontinuities, occlusions and/or shadows.

To avoid phase ambiguity problem, several approaches had been proposed. Saldner and Huntley proposed a method named as temporal phase unwrapping where the basic idea is to vary the pitch of the projected fringes over time [21]. They evaluated independently the object phase at each pixel in order to provide an absolute measurement of the surface height. Thus, by recording a sequence of phase maps, a 3-D phase distribution is obtained by unwrapping the phase at each pixel along the time axis. This procedure yields on an absolute measurement of the object surface height. On the other hand, Sjödahl and Synnergren solved the discontinuity problem of the phase map by means of the projection of a random pattern instead of periodic fringes [22, 23]. Since the intensity of a random pattern varies without a regular period, every small region in the projected pattern becomes unique. Using an initial calibration process of the projected random pattern along the working volume, it is possible to trace the movement of each small region at the detector by simple correlation algorithm when the object under study is illuminated by such random pattern. Thus, assuming that the motion of every small area can be detected, the 3-D object shape can be retrieved. Such techniques are named as digital speckle photography (DSP) and are based on optical triangulation.

Other interesting methods that solve the discontinuity problem of the phase map are based on the coherence of the light source. Coherence coding has been applied to spatial information transmission and superresolution [24–27], optical coherence tomography [28, 29] and 3-D topography in digital holographic microscopy [30, 31]. Yuan et al. introduced a new approach to achieve 3-D surface contouring of reflecting micro-objects [32]. They proposed a digital lensless Fourier holographic system with a short coherence light source that can be used to record different layers of a micro-object through changing the object position. Thus, by bringing into the coherence spatial interval different sections of the object, it is possible to digitally reconstruct the whole 3-D surface contouring of the object. This approach takes the advantage provided by a limited source coherence length which allows optical sectioning of the 3-D object profile.

Recently, Zalevsky et al. presented a new method for 3-D imaging based on partial coherence in digital holography [33]. In their paper, Zalevsky et al. reported on two different approaches, both of which aimed to remove the phase ambiguity problem in digital holography. In one of the proposed concepts, they achieve 3-D profile of a reflective object using an approach based on illuminating the object with a light source having a relatively long coherence length. As the contrast of the fringes within this length will vary, this fact can be used for 3-D estimation after performing a calibration of the contrast of the interference fringes in the whole coherence length axial interval. Thus, by comparing the contrast of the fringes provided by the object with that one obtained in the calibration process, they were able to extract 3-D information. They proposed a lensless digital holographic optical setup in which no lenses are needed to image the input object because the approach is focused on mapping the 3-D information of reflective objects. This fact disables the reported approach when diffuse objects are considered. Moreover, because of the lensless geometry, there will be restrictions between the observation distance and the object size in order to optimize the space-bandwidth product of the system.

The projection of random (speckle) patterns onto the tested object can also be interpreted in a different sense. The combination of phase-retrieval algorithm with the recording of different intensity planes inside the diffracted volume speckle field provided by an object which becomes illuminated with coherent light provides a tool for the reconstruction of the complex wavefront diffracted by the object [34–36]. And the recovery of phase information distribution allows 3-D object information [37, 38] and capabilities for object deformation analysis [38, 39].

In this chapter we propose two additional approaches for 3-D object shaping and range measurement based on projecting a random pattern instead of a periodic one.

Although the projection of random patterns was previously described by Sjödahl and Synnergren [22, 23] they proposed the use of a white light optical projector and optical triangulation architecture. As consequence, the measured 3-D object shape suffers from previously commented problems arising from the use of triangulation. On the other hand, speckle techniques have been used for 3-D shape measurement [40, 41]. Named as speckle pattern sampling (or Laser Radar 3-D imaging), 3-D object information is achieved by using the principle that the optical field in the detection plane corresponds to a 2-D slice of the object's 3-D. A speckle pattern is measured using a CCD array for each different laser wavelength, and the individual frames are added up to generate a 3-D data pattern [41].

In our first proposed method, we considered the projection of the random speckle pattern through a ground glass diffuser [42]. Speckle patterns happen when coherent light is reflected or transmitted from a rough surface and inherent changes in propagation of the speckle pattern will uniquely characterize each specific location in the volume speckle field. Due to this, it is possible to determine the 3-D object shape by analyzing the speckle pattern itself (not that one produced by the object). Unlike other methods that use cross-correlation function for the speckle pattern before and after the deformation of an object and try to reduce the speckle decorrelation since it is the single most important limiting parameter in measuring systems based on laser speckles [43], the proposed method studies the differences between uncorrelated speckle patterns that were projected on a certain object as source of 3-D object information. As result, this procedure allows the representation of the 3-D object shape in a range image composed of gray levels in which each level is representative of a given object depth. This method defines a resolution which is independent of the object observation distance or the angle between the illumination and the observation direction because it is based on local correlations of speckle patterns. Moreover, no shadowing problem is produced because it can be performed using parallel architecture between illumination and detection. Also, the proposed method was validated for transparent samples and reflective diffuse objects.

The second reported method in this chapter uses the change in visibility of the interference speckle pattern in the volume speckle field when using a relatively long but limited source coherence length as source of 3-D object information [44]. When a classical temporal ESPI configuration is considered, the object areas falling inside the coherence volume range of the illumination source will produce speckle interference with the reference beam. A phase modulation of the reference will result in blinking intensity only for those points inside the coherence range. Thus, the proposed approach performs coherence-to-depth coding of the 3-D object shape in such a way that the 3-D object information is related with the amplitude of the speckle interference. And the way to decode such information coding is by computing the visibility of the speckle interference along the whole volume range. After some digital post-processing to smooth speckle noise, it is possible to get a range image where each gray level contains information of a different object depth. The proposed approach has three major advantages. First and as in the case of triangulation methods, the resolution of the method does not depend on the observation distance because the coherence volume range can be easily tuned from near to far distance by simple adding optical path in the reference beam. Second, no shadowing problem and minimization of the occlusions are produced because the illumination and observation directions are collinear. And third, as the method implies the obtaining of a range image, there is no problem with sudden jumps or occlusions in the object shape and as in those methods implying wrapped phase distribution of the object. In addition, unlike other coherence-based methods like optical coherence tomography [45, 46], the proposed method does not require axial distance [45] or wavelength [46] scanning.

The chapter is organized as follows. Section 13.2 presents theory (Sect. 13.2.1) and experimental validation (Sect. 13.2.2) on how changes in the speckle pattern provide means for measurement and characterization of 3-D objects. In particular, we present how applying those speckle pattern changes to 3-D mapping and range estimation by observing the differences and similarities in the speckled patterns when an object is illuminated using light coming from a diffuser. In Sect. 13.3 we present a novel method for 3-D diffuse object mapping and ranging based on the degree of temporal coherence of the interference speckle pattern reflected by the object. Sect. 13.3.1 includes a system description and its theoretical foundation while Sect. 13.3.2 provides experimental demonstration. Essentially, since each small area in the coherence volume of the illumination source produces an interference pattern with different amplitude, it is possible to estimate the 3-D object shape by computing the visibility of recorded interference pattern. The chapter finishes with the most relevant bibliography concerning the two proposed approaches.

13.2 3-D Mapping by Means of Projected Speckle Patterns

13.2.1 Theory

The speckle statistics depends upon the ground glass that is used to project the patterns (primary speckle) and on the surface characteristics on which they are impinged. In this section we show the evolution of the speckle patterns when the object is axially translated. A ground glass object $g(x, y)$ (positioned in a plane that we will call plane G) normal to the optical axis (see Fig. 13.1) is transilluminated by a monochromatic parallel beam of laser light of wavelength λ. Speckle patterns are recorded at plane C by which we denote the plane of the CCD camera and which is parallel to the object plane. In the plane C, at a distance z from T (the plane at which the transparent object is being positioned), the amplitude distribution obtained by Fresnel diffraction equation is proportional to

$$U(\xi, \eta) = \iint g(x, y) \exp\left(j\frac{\pi}{\lambda z}\left[(\xi - x)^2 + (\eta - y)^2\right]\right) \mathrm{d}x\, \mathrm{d}y \tag{13.1}$$

and the irradiance recorded by the CCD is given by

$$I(\xi, \eta) = \left|\iint g(x, y) \exp\left(j\frac{\pi}{\lambda z}\left[x^2 + y^2\right]\right) \exp\left(-j\frac{2\pi}{\lambda z}\left[\xi x + \eta y\right]\right) \mathrm{d}x\, \mathrm{d}y\right|^2 \tag{13.2}$$

To check how the irradiance $I(\xi,\eta)$ changes with the distance from the ground glass, let us move the glass a small distance of z_0. Thus, the irradiance at C is obtained by replacing z by z-z_0 as

$$I(\xi,\eta) = \left| \iint g(x,y) \exp\left(j\frac{\pi}{\lambda z}\left[x^2+y^2\right]\right) \right. \\ \times \exp\left(j\frac{\pi z_0}{\lambda z^2}\left[x^2+y^2\right]\right) \\ \left. \times \exp\left(-j\frac{2\pi}{\lambda z}\left[\xi\left(1+\frac{z_0}{z}\right)x+\eta\left(1+\frac{z_0}{z}\right)y\right]\right) \mathrm{d}x\,\mathrm{d}y \right|^2 \tag{13.3}$$

The axial translation of g has two main effects. One is a radial shift of the speckle linked to the magnification due to the change of angular size of the object in the recording plane. The other effect is a decorrelation of the corresponding speckle due to the term $\exp\left(j\frac{\pi z_0}{\lambda z^2}\left[x^2+y^2\right]\right)$.

Taking into account the decorrelation between different speckle distributions, the idea of this chapter for range estimation and 3-D mapping is the following. In transmissive systems as well as in reflective systems, the movement of the object changes the speckle distribution. If we denote this distribution as $g_Z(x,y)$ where z is the longitudinal position of the sample, then

$$\frac{1}{S}\iint_{\Sigma} g_{Z_1}(x_0-x,y_0-y)\,g_{Z_2}(x_0,y_0)\,\mathrm{d}x_0\mathrm{d}y_0 = \begin{cases} k\delta(x,y) & \text{for } Z_1 = Z_2 \\ 0 & \text{for } Z_1 \neq Z_2 \end{cases} \tag{13.4}$$

where the integral is normalized according to the surface (S) of the region of integration (Σ). This equation states that the complex amplitude is completely decorrelated in two different planes, provided that the axial distance between the two planes is large compared with the axial speckle size. If the intensity distribution is considered, instead of the amplitude one, the decorrelation is not complete, owing to the non-zero average value. Thus in this case small cross-correlation value is expected. This property is used for both 3-D mapping and range finding.

13.2.2 Experimental Validation

Let us consider now the problem of 3-D mapping of a transparent object (T), placed at the distance z of the ground glass, as shown in Fig. 13.1. After placing the object, illuminated by a spot of light through the ground glass, the plane imaged by the camera changes by an amount that depends on the index of refraction and the thickness from one region of the object to the other. Note that if the object is imaged using an optical system, the speckle lying in the image plane suffers a decorrelation when

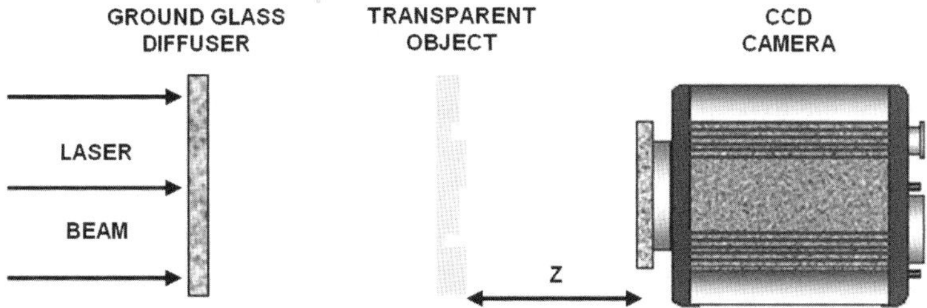

Fig. 13.1 Transmissive object illumination for 3-D mapping

the object is axially translated through a distance greater than λ/NA^2M^2 (which is the average length of speckle in z direction) where NA is the numerical aperture of the imaging lens and M is the magnification between the object and the image plane. Because of that the optical parameters have been adjusted correctly in order to register the decorrelation between speckle patterns.

In Fig. 13.2 we show some of the images used for the calibration of the system. In Fig. 13.2a we present the captured speckle pattern when no transmissive object is placed. In Fig. 13.2b we placed instead of the object glass with width of 0.5 mm.

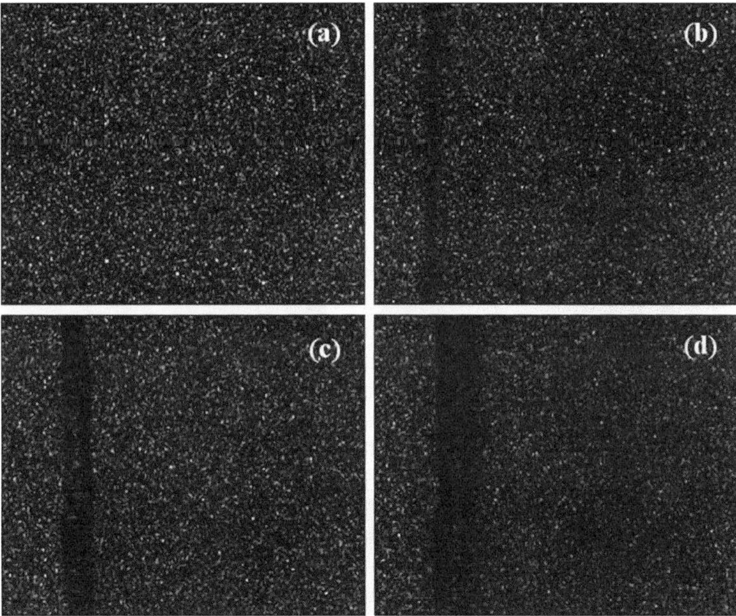

Fig. 13.2 Calibration reference readout for transmissive microscopic configuration for 3-D mapping using speckle random coding. The calibration was done with glasses with varying widths constructed out of slides of 0.5 mm in width. (**a**) Speckle pattern without any glass in the optical path. (**b**) Speckle pattern when one glass slide is inserted into the optical path. (**c**) Speckle pattern when two glass slides are inserted into the optical path. (**d**) Speckle pattern when three glass slides are inserted into the optical path

In Fig. 13.2c, d we placed glass with width of 1 and 1.5 mm, respectively. The refraction index of the glass that we used as the inspected object was 1.48. The laser we used for the experiment was He–Ne laser with wavelength of 632 nm. As one may see from the obtained images, the patterns are neatly different.

Presented in Fig. 13.3 is the constructed object containing spatial regions with no glass, glass of 0.5 mm and glass of 1 mm width. It is denoted in Fig. 13.3a as region 0, region 1 and region 2, respectively. Figure 13.3b presents the image captured in the CCD camera.

Figure 13.4 is the reconstruction results while the reconstruction is done by subtraction between the captured patterns and the reference patterns (obtained in the calibration process). The dark zones (marked with a solid wide white line) indicate detection. In Fig. 13.4a we present subtracting the reference pattern of zero width glass in the optical path. In Fig. 13.4b, c we show subtraction of 0.5 mm and 1 mm width glass slide. As one may see the dark region appeared where they were supposed to following the spatial construction of the object as indicated in Fig. 13.3a. Note that the images have a vertical black band. We have used this band as a reference to add at that position the different glass panels and it plays no role in the speckle pattern analysis.

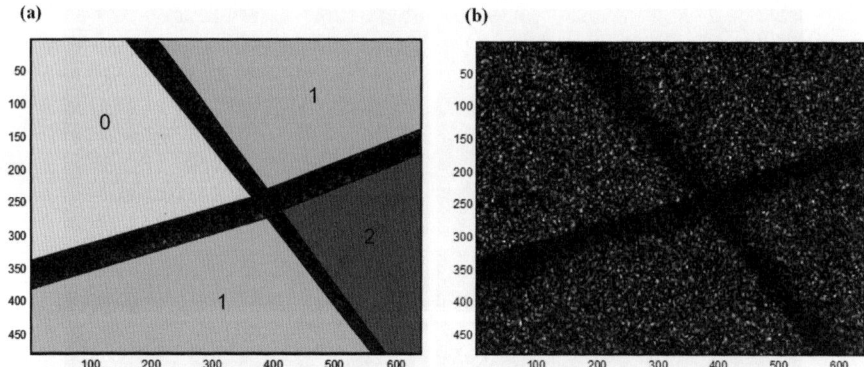

Fig. 13.3 The captured image containing (**a**) the original input pattern that contains a structure of transmissive glasses having 0, 1 and 2 slides (each slide is 0.5 mm wide) and (**b**) the image captured by the CCD camera when projected with speckle pattern

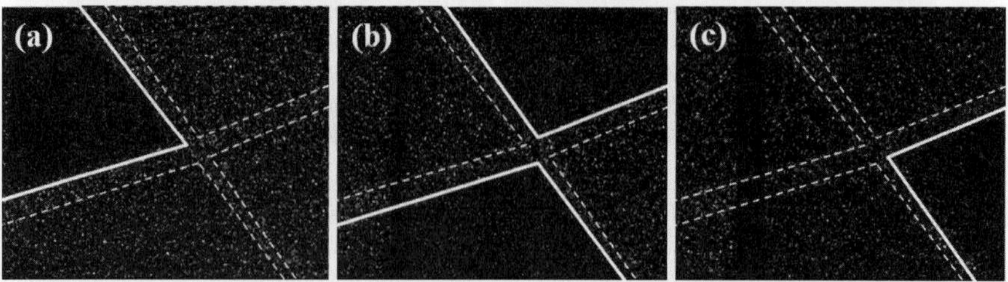

Fig. 13.4 Segmentation of different regions by subtraction between the captured patterns and the reference patterns: (**a**) segmentation of region 0, (**b**) segmentation of region 1 and (**c**) segmentation of region 2

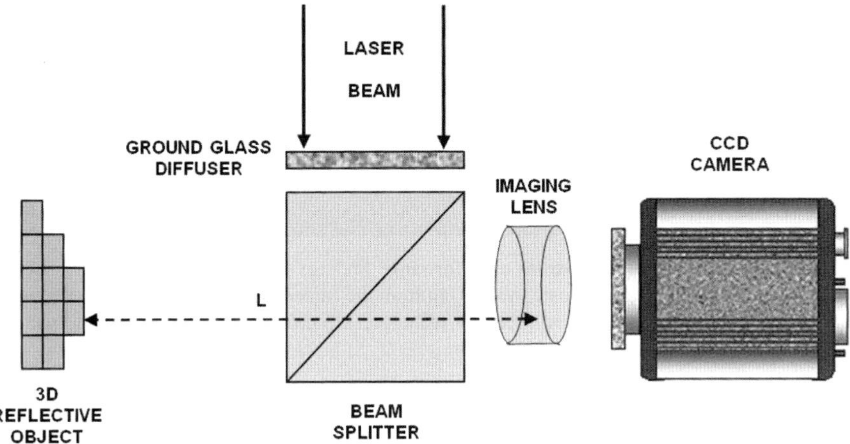

Fig. 13.5 Optical setup for range estimation

The experiment of range estimation was performed in reflective configuration in which the ground glass (diffuser) was used to project the speckle pattern on top of a reflective object and the reflection was imaged on a CCD camera. The system is shown in Fig. 13.5. Light from a laser source passes through the diffuser. The speckle patterns coming from that diffuser are projected on the reflective object and then registered onto a CCD camera. For the case of reflected configuration the optical parameters are such that the speckle statistics coming from the diffuser (creating the primary projected speckle pattern) is the dominant rather than the influence of the statistics of the object texture itself (creating the secondary speckle pattern). Such a situation may be obtained, for instance, by choosing proper diameter of the spot that illuminates the diffuser (since this diameter determines the size of the projected speckle) such that it generates primary speckle that are larger than the secondary speckle generated due to the transmission from the object. That way if the diameter of the imaging lens fits the diameter of the spot on top of the diffuser, the primary speckle are fully seen by the diffraction resolution limit of the camera (the F number) while the secondary speckle which are smaller will be partially filtered by the camera since its F number will be too large to contain their full spatial structure.

For range finding the object was positioned 80 cm away from the CCD camera. In order to capture the decorrelated speckle patterns a calibration process was done. It consists of mapping the object by capturing images reflected from 11 planes separated 5 mm apart. In Fig. 13.6a we present the reflected reference speckle pattern that corresponds to plane positioned 80 cm away from the camera. In Fig. 13.6b we depict the auto-correlation between the reflected speckle pattern from a certain plane and its reference distribution obtained in the calibration process. In Fig. 13.6c we present the cross-correlation of speckle patterns reflected from two planes separated by 5 mm apart. As we have mentioned previously, there is a diagonal black band in the optical experiments. We have used it as a reference.

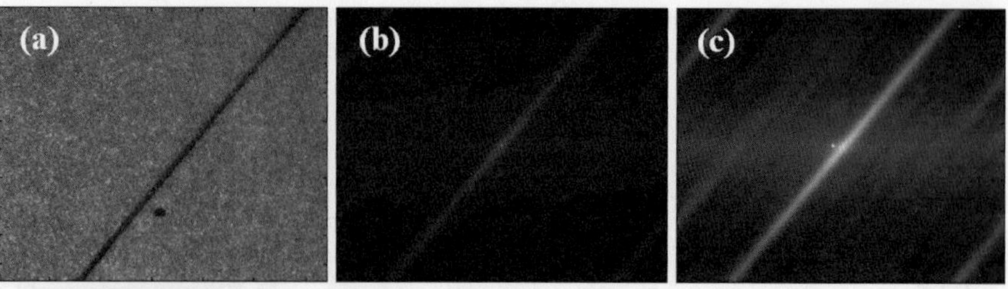

Fig. 13.6 Reflective speckle patterns used for range finding: (**a**) the reflected reference speckle pattern corresponds to a plane positioned 80 cm from the camera, (**b**) an auto-correlation between the reflected speckle pattern from a certain plane and (**c**) cross-correlation of speckle patterns reflected from two planes separated by 5 mm

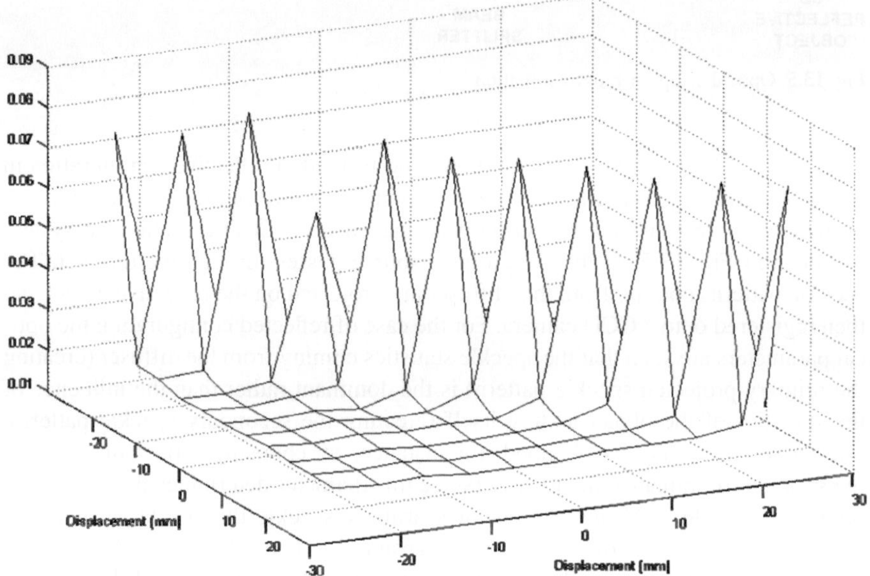

Fig. 13.7 The range finding results. All the cross-correlation combinations of the reflected patterns from all possible 11 planes positioned 5 mm apart

In Fig. 13.7 we display the range finding results. All the cross-correlation combinations of the reflected patterns from all possible 11 planes positioned 5 mm apart. As one may see the detection peaks appeared only at the diagonal (auto-correlation) while the distribution outside of the diagonal is small (cross-correlation). For further clarification Fig. 13.7 is actually a cross-correlation matrix of 11 by 11 while the index (1–11) represents the plane number. The diagonal of the matrix is the auto-correlation of each plane with itself. The non-diagonal terms are related to the cross-correlation between the 11 planes.

In Fig. 13.8 we show the results that were obtained for a 3-D object made up with toy building blocks. The object occupied a volume of 40 × 25 × 30 mm. We use a calibration procedure similar to the previous one, but in this case the correlation is

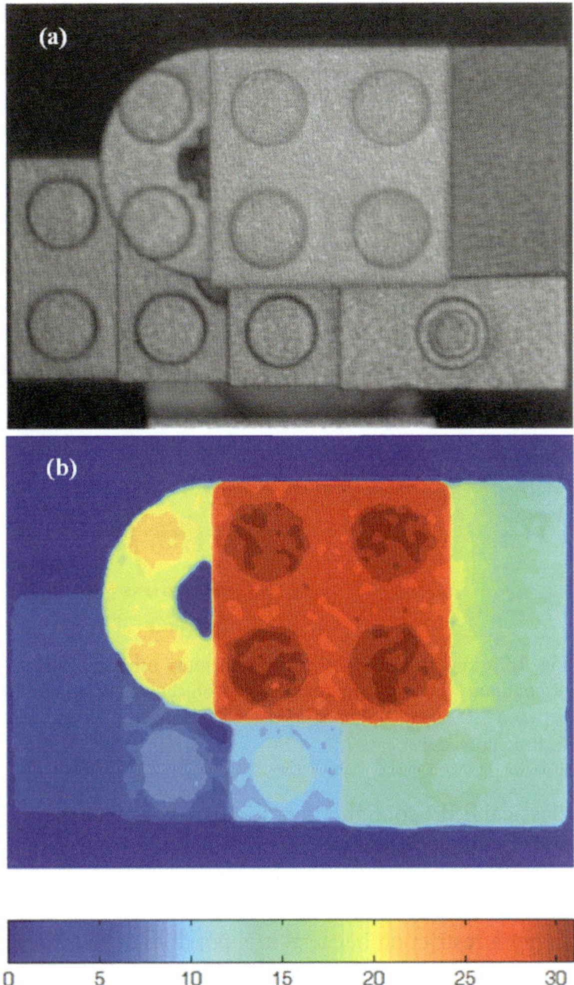

Fig. 13.8 (**a**) The object illuminated with the projected speckle pattern and (**b**) the 3-D reconstruction. Colour bar represents object depth in mm

not performed for the full image as in the range finding case, but we rather perform a correlation of local windows of the object with the same region of each of the calibration patterns to give the final 3-D mapping. The proposed procedure has virtually no shadowing as the illumination and image recording are performed from the same location.

13.3 3-D Shaping by Means of Coherence Mapping

13.3.1 Theory

To test the capabilities of the proposed method from an experimental point of view the basic setup depicted in Fig. 13.9 has been assembled at the laboratory. It is a

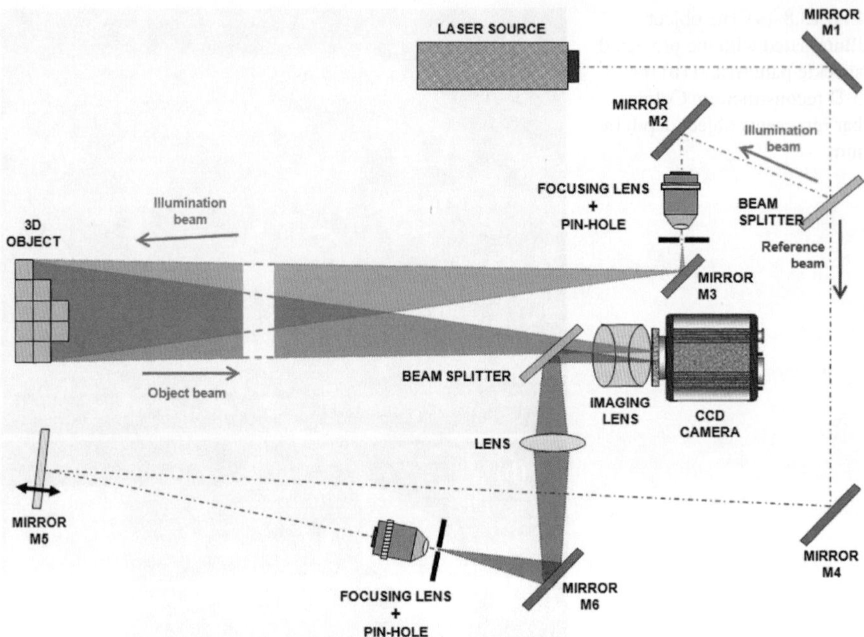

Fig. 13.9 Temporal ESPI experimental setup used for validation of the proposed approach. The discontinuity in both illumination and object beams is representative of the arbitrary distance between 3-D object and imaging device. The mirror M5 is a piezo-electrical one to provide phase-shifting procedure

classical ESPI configuration. Light coming from a laser source is split into reference and illumination beams. The illumination beam reaches the 3-D object after being spatially filtered and projected onto it. Notice that, in this configuration, the illumination direction is not strictly collinear with the observation one. This fact does not imply a restriction because it is possible to illuminate using the beam splitter just before the imaging system. However, to gain in simplicity and to better matching the optical paths, we have selected the depicted configuration.

Thus, a diffuse reflective object is illuminated by a light beam having a given coherence length. And at the same time, a reference beam is inserted in on-axis mode onto the detection plane with a divergence that equals the one provided by the 3-D object. So, an on-axis hologram having no carrier frequency fringes but dominated by speckle is recorded by the CCD. Taking into account that the coherence length of the illumination light is relatively long, it is possible to well-match the optical path length of the reference beam in such a way that the whole object depth falls inside half the coherence length interval. In this case, different speckles coming from different depths of the 3-D object will interfere with different amplitudes once they are imaged by the CCD camera. We can say the proposed approach is performing a coding of the 3-D object depth into the coherence of the illumination light since the 3-D object information is related with the amplitude of the speckle interference.

In order to evaluate such variable amplitude with the distance when considering different regions of the 3-D interference object field of view, we introduce a slow perturbation in time at the reference beam, that is, a temporal phase shifting, by using a piezo-electric mirror. Thus, computing the visibility of the recorded set of interferograms, it is possible to finally obtain a range image where each small area of the 3-D object is represented by a given gray level incoming from the coherence-to-depth encoding that has been performed. And this range image becomes a 3-D mapping of the object under test.

Additionally, by mapping the whole coherence volume with a reference object (a plane calibration plate, for instance) in order to obtain the visibility of the interference versus axial position, we can obtain a gray-level chart corresponding with different depths in the working volume. This record will make the depth calibration of the system. Thus, once the range image is obtained, it is possible to directly assign each gray level to a given depth, and the 3-D information of the object under test becomes recovered.

Considering now the proposed approach from a theoretical point of view, we can say that the resulting field distribution impinging at the CCD when considering two interference beams in on-axis mode is

$$A(x, y; t) = A_1 \exp\left(i\phi_1(t)\right) + A_2 \exp\left(i\phi_2(t)\right) \tag{13.5}$$

where A_i, ϕ_i are the amplitude and phase distributions of each beam, respectively. Then, the CCD performs intensity recording, that is

$$I(x, y; t) = |A(x, y; t)|^2 = A_1^2 + A_2^2 + 2A_1A_2 \langle\cos\left(\phi_1(t) - \phi_2(t)\right)\rangle \tag{13.6}$$

where $< \ldots >$ designates time averaging operation in a time span comparable to the inverse of the optical frequencies. Just as examples, if both beams are fully coherent: $\langle\cos\left(\phi_1(t) - \phi_2(t)\right)\rangle = \cos\left(\phi_1(t) - \phi_2(t)\right)$, while if both beams are incoherent: $\langle\cos\left(\phi_1(t) - \phi_2(t)\right)\rangle = 0$. Therefore it is clear that the degree of coherence determines the contrast of the interference. Defining γ as the degree of temporal coherence, it is clear that two completely coherent or incoherent beams can be represented by $\gamma = 1$ and $\gamma = 0$, respectively. In our case, the degree of temporal coherence is a function of the axial distance since we are making interference between different reflected beams incoming from different axial positions. So, considering $\gamma = \gamma(z)$, (13.6) can be rewritten as

$$I(x, y, z; t) = I_1 + I_2 + 2\sqrt{I_1 I_2}\gamma(z)\cos[\phi(t)] = I_T\{1 + V_0(t)\gamma(z)\cos[\phi(t)]\} \tag{13.7}$$

where $\phi(t) = \phi_1(t) - \phi_2(t)$ is the phase difference between both interferometric beams at a given instant, $I_i = A_i^2$ is the intensity of each beam and $I_T = I_1 + I_2$ is the total intensity in case of incoherent addition of the beams. Notice that $V_0 = 2\sqrt{I_1 I_2}/(I_1 + I_2)$ is the visibility of the recorded interference pattern in case of perfect coherence.

Equation (13.7) shows that the degree of temporal coherence modulates the total intensity recorded at the CCD detector in such a way that different object depths will produce a different intensity in the recorded interferogram as we are mapping the coherence volume of the illumination light. Thus, it is possible to obtain a range image of the 3-D object by determining the interference visibility and from this value the degree of temporal coherence $\gamma(z)$.

To do this, we measure in time sequence the intensity of the object beam (I_1) by occluding the reference one, the intensity of the reference beam (I_2) by occluding the object one and the visibility of the interference pattern by capturing a sequence of phase-shifted interferograms and computing the recorded intensities. With this procedure, we obtain an image which is proportional to the degree of temporal coherence, which is related to the object depth. Finally, the profile of the object can be estimated by comparing the result with the gray-level chart obtained by calibration.

The visibility extraction can be performed without any particular phase shifting on the reference arm. Note that, eventually, only the minimum and maximum intensities for each point are needed, regardless of the phase, which achieve those values. If a random phase is applied over the reference then it is enough to grab enough number of interferograms so that maximum and minimum intensities are obtained for each camera pixel. Obviously this is hard to warrant in case of random phase shift, but simple for a linear phase, which is easily achieved by means of a piezo-mirror, as shown in Fig. 13.9. In case of a linear phase the intensity for each pixel can be fit to a sinusoidal function (13.7) giving as a result higher accuracy in the visibility calculation.

It is important to stress that, despite the need for several frames (at least three) the method does not relies on scanning over the depth or the wavelength. The multiple frames are used just for determining the contrast of the interferences. Thus, the expected range for depth is given by the coherence length of the light used, while the accuracy will be governed by the stability in the coherence shape of the light source and the noise introduced by the system. In contrast, in an OCT-based system the accuracy is determined by the coherence length and the range by the scanning range, either in wavelength or in axial distance. In our case, a major noise source is the speckle, because the system is based on a coherent imaging system.

13.3.2 Experimental Validation

13.3.2.1 Validation and Calibration

Figure 13.10 images a picture of the experimental setup assembled at the laboratory for the short measurement distance case and for calibration. One can easily identify the different components according to the layout provided in Fig. 13.9. A He–Ne laser source (632 nm wavelength) is used as illumination light in the experimental setup. Object is placed at a distance of 90 cm approximately in front of a imaging system composed of an imaging lens (50 mm focal length video lens) and a CCD camera (Kappa DC2, 12 bits dynamic range, 1352 × 1014 pixels with 6.7 μm pixel size).

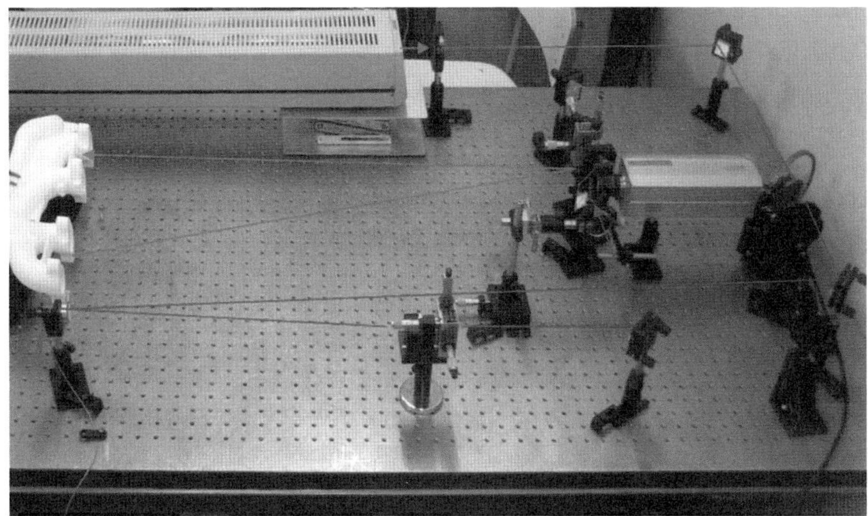

Fig. 13.10 The experimental setup assembled at the laboratory for measurements at short distance where the main optical paths of both beams are depicted for clarity. The different components depicted in Fig. 13.9 can be easily identified in the image where additional neutral filters are included for intensity beam ratio adjustment

In a first step we perform a testing of the method by using a planar diffuse calibration plate. The plate normal is aligned with the line of sight and is translated along the said line of sight during calibration. The set of axial location of the plate includes the location at null optical path difference between reference and imaging arms (which will provide maximum visibility interference). This location distance is considered as origin for axial distances. As the calibration plate is progressively far from the origin the visibility is expected to diminish following the coherence function of the light source. As explained in the previous section, a set of images with variations in the reference phase by means of a piezo-mirror are taken and the visibility is extracted in each image point.

Figure 13.11 shows the visibility extracted as a function of the depth around the equal path length point. Owing to the inherent coherent noise from the speckle imaging, the visibility for a single point would suffer excessive noise. Instead a spatial averaging (smoothing) over a small size window is needed. We use a smoothing with a Gaussian window with diameter (HWMH) of 5 pixels. This still renders high lateral resolution and reasonable axial resolution. As can be seen from the calibration curve, an axial range of approximately 120 mm can be mapped at each side of the origin without ambiguity. For each axial location in the calibration the standard deviation is also plotted. Note that increasing the averaging window size would give an enhanced axial resolution, but at the expense of a diminished lateral resolution.

13.3.2.2 Short Range Measurement

The same setup as for calibration is used for short range mapping. Figure 13.12 images the two objects used to test the proposed method in both cases, short and

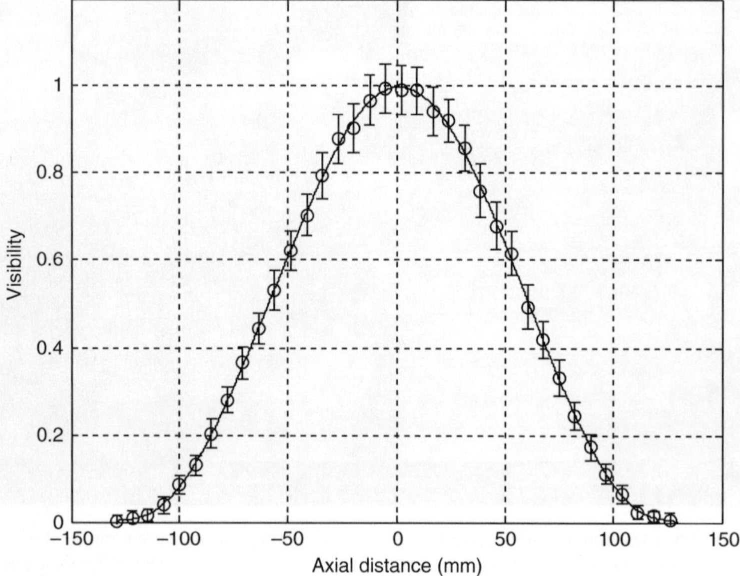

Fig. 13.11 Calibration curve: visibility averaged over a weighted window with diameter of 5 pixels for different axial distances

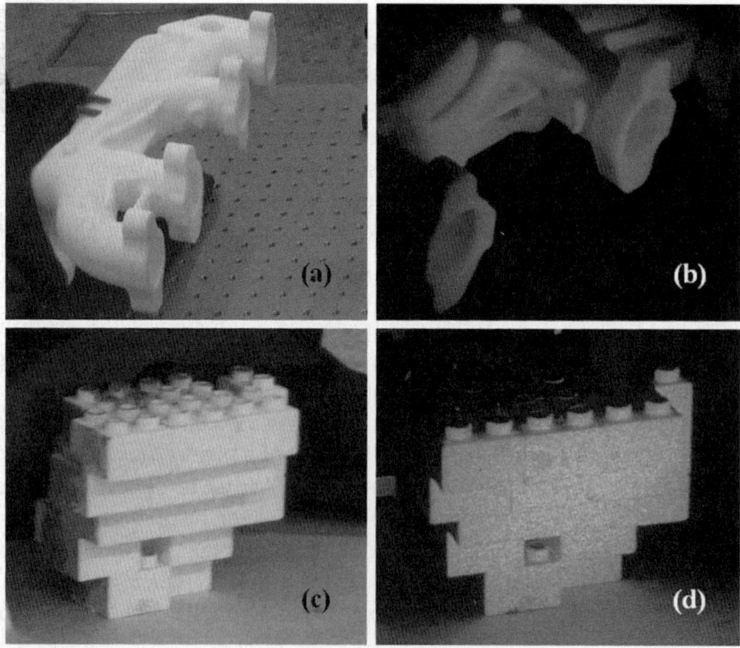

Fig. 13.12 Objects used in the experimental validation of the proposed approach: (**a–b**) and (**c-d**) are a 3-D object composed of building blocks (blocks) and a gas collector prototype of an engine (collector), respectively

long measurement distance. One is a 3-D object built from single blocks of a toy object and the other one is a gas collector prototype of an engine. The blocks in the first case permit the construction of an object with steps separated by 16 mm in depth. Note in this object, the hole in the lower part, that is closed in the back side, giving a rectangular hole with also 16 mm depth with respect to the borders.

Figure 13.13 shows the reconstructions obtained for the proposed objects: a and b are the range images of the collector and blocks, respectively, and c depicts a 3-D plot of the relief for the blocks object. It is especially noticeable the ability to map the inside of the holes facing the camera. This situation happens for both objects and demonstrates the absence of shadow problem with this method.

13.3.2.3 Long-Range Measurement

For long-range measurements the objects were located at 5 m approximately from the camera (limited by the laboratory size). The system is rearranged to add this distance to the reference arm by means of a delay line, providing a neat interference

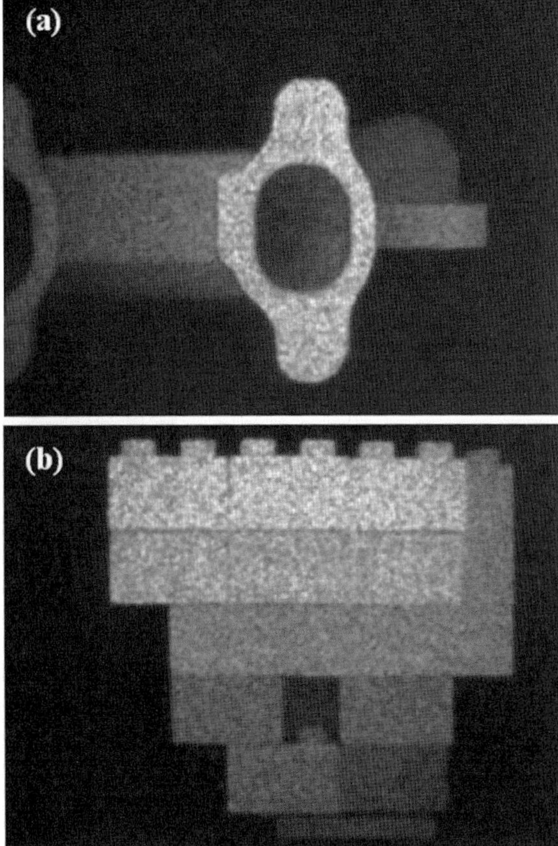

Fig. 13.13 Mapping obtained for the objects in Fig. 13.12 when positioned at short distance (0.9 m). (**a**) Blocks, (**b**) collector and (**c**) 3-D plot of the mapping obtained for the blocks

Fig. 13.13 (continued)

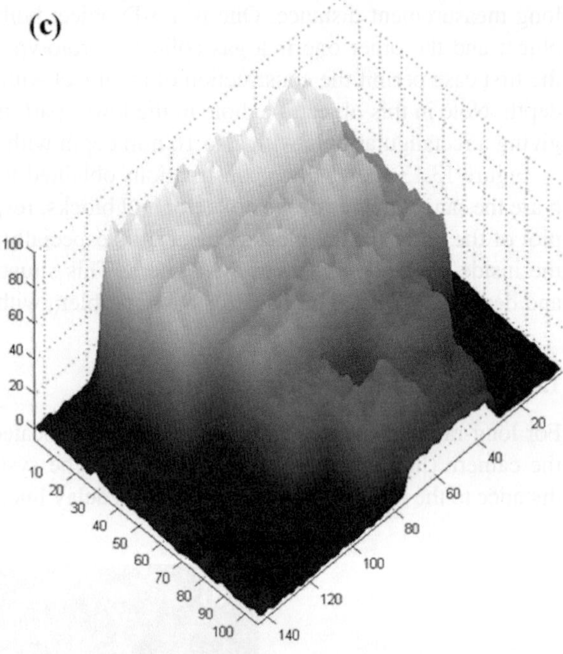

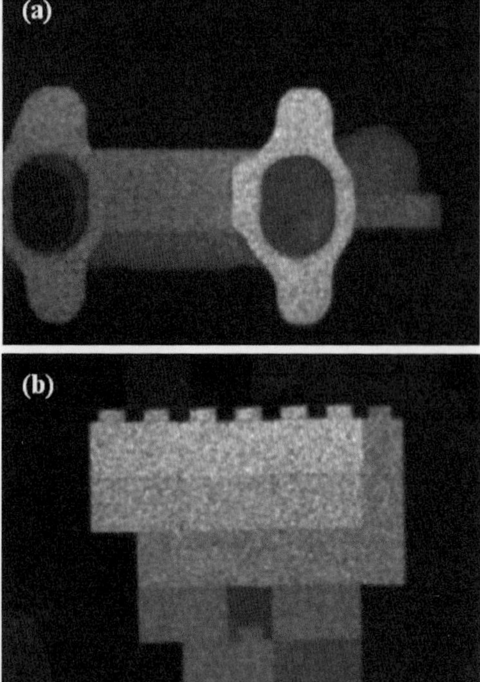

Fig. 13.14 Mapping obtained for the objects in Fig. 13.12 when positioned at long distance (5 m): (**a**) Blocks and (**b**) Collector

speckle pattern from the object. Additionally a longer focal length imaging lens is used, in order to provide a similar field of view as in the short range case. The resulting images are depicted in Fig. 13.14. As can be seen, aside of the scaling factor due to different imaging magnification, the results are analogous to those obtained at short distances. Again the lack of shadowing is patent as well. It is worth noticing that the additional distance between the system and the object does not introduce any change in the performance of the method.

Acknowledgments Part of this work was supported by the Spanish Ministerio de Educación y Ciencia under the project FIS2007-60626 and by the Instituto de la Mediana y Pequeña Industria Valenciana under the project IMIDIC/2008/52.

References

1. F. Chen, G.M. Brown, M. Song, Overview of three-dimensional shape measurement using optical methods. Opt. Eng. **39**, 10–22 (2000)
2. O. Matoba, B. Javidi, Encrypted optical memory system using three-dimensional keys in the Fresnel domain. Opt. Lett. **24**, 762–764 (1999)
3. E. Tajahuerce, B. Javidi, Encrypting three-dimensional information with digital holography. Appl. Opt. **39**, 6595–6601 (2000)
4. T. Poon, T. Kim, Optical image recognition of three-dimensional objects. Appl. Opt. **38**, 370–381 (1999)
5. J.J. Esteve-Taboada, J. García, C. Ferreira, Rotation-invariant optical recognition of three-dimensional objects. Appl. Opt. **39**, 5998–6005 (2000)
6. M. Takeda, K. Mutoh, Fourier transform profilometry for the automatic measurement of 3 D object shapes. Appl. Opt. **22**, 3977–3882 (1983)
7. V. Srinivasan, H.C. Liu, M. Halioua, Automated phase-measuring profilometry of 3-D diffuse objects. Appl. Opt. **23**, 3105–3108 (1984)
8. W.H. Su, Color-encoded fringe projection for 3D shape measurements. Opt. Express **15**, 13167–13181 (2007)
9. U. Schnars, W. Jüptner, Direct recording of holograms by a CCD target and numerical reconstruction. Appl. Opt. **33**, 179–181 (1994)
10. G. Pedrini, P. Fröning, H.J. Tiziani, F.M. Santoyo, Shape measurement of microscopic structures using digital holograms. Opt. Commun. **164**, 257–268 (1999)
11. C. Quan, X.Y. He, C.F. Wang, C.J. Tay, H.M. Shang, Shape measurement of small objects using LCD fringe projection with phase-shifting. Opt. Commun. **189**, 21–29 (2001)
12. G. Indebetouw, Profile measurement using projection of running fringes. Appl. Opt. **17**, 2930–2933 (1978)
13. W. Su, Ch. Kuo, Ch. Wang, Ch. Tu, Projected fringe profilometry with multiple measurements to form an entire shape. Opt. Express **16**, 4069–4077 (2008)
14. M. Halioua, R.S. Krishnamurthy, H. Liu, F. Chiang, Automated 360° profilometry of 3-D diffuse objects. Appl. Opt. **24**, 2193–2196 (1985)
15. G. J. Iddan, G. Yahav, 3D imaging in the studio. Three-dimensional image capture and applications IV. Proc. Soc. Photo-Opt. Instrum. Eng. **4298**, 48–55 (2001)
16. N. Abramson, Time reconstructions in light-in-flight recording by holography. Appl. Opt. **30**, 1242–1252 (1991)
17. B.P. Hildebrand, K.A. Haines, Multiple wavelength and multiple source holography applied to contour generation. J. Opt. Soc. Am. **57**, 155–159 (1967)
18. R.S. Sirohi, *Speckle Metrology* (Marcel Dekker, New York, NY, 1993)
19. I. Yamaguchi, T. Zhang, Phase-shifting digital holography. Opt. Lett. **22**, 1268–1270 (1997)

20. J.M. Huntley, in *Digital Speckle Pattern Interferometry and Related Techniques*, Chapter 2, ed. by P.K. Rastogi. Automated analysis of speckle interferograms. (Wiley, New York, NY, 2001), pp. 59–140
21. H.O. Saldner, J.M. Huntley, Temporal phase unwrapping: Application to surface profiling of discontinuous objects. Appl. Opt. **36**, 2770–2775 (1997)
22. M. Sjödahl, Electronic speckle photography: Increased accuracy by non-integral pixel shifting. Appl. Opt. **33**, 6667–6673 (1994)
23. M. Sjödal, P. Synnergren, Measurement of shape by using projected random patterns and temporal digital speckle photography. Appl. Opt. **38**, 1990–1997 (1999)
24. B.J. Guo, S.L. Zhuang, Image superresolution by using a source-encoding technique. Appl. Opt. **30**, 5159–5162 (1991)
25. Z. Zalevsky, J. García, P. García-Martínez, C. Ferreira, Spatial information transmission using orthogonal mutual coherence coding. Opt. Lett. **30**, 2837–2839 (2005)
26. V. Micó, J. García, C. Ferreira, D. Sylman, Z. Zalevsky, Spatial information transmission using axial temporal coherence coding. Opt. Lett. **32**, 736–738 (2007)
27. D. Sylman, Z. Zalevsky, V. Micó, C. Ferreira, J. García, Two-dimensional temporal coherence coding for super resolved imaging. Opt. Commun. **282**, 4057–4062 (2009)
28. D. Huang, E.A. Swanson, C.P. Lin, J.S. Schuman,W.G. Stinson, W. Chang, M.R. Hee, T. Flotte, K. Gregory, C.A. Puliafito, J.G. Fujimoto, Optical coherence tomography. Science **254**, 1178 (1991)
29. A.F. Fercher, C.K. Hitzenberger, Optical coherence tomography. Prog. Opt. **44**, 215–302 (2002)
30. P. Massatsch, F. Charrière, E. Cuche, P. Marquet, C.D. Depeursinge, Time-domain optical coherence tomography with digital holographic microscopy. Appl. Opt. **44**, 1806–1812 (2005)
31. L. Martínez-León, G. Pedrini, W. Osten, Applications of short-coherence digital holography in microscopy. Appl. Opt. **44**, 3977–3984 (2005)
32. C. Yuan, H. Zhai, X. Wang, L. Wu, Lensless digital holography with short-coherence light source for three-dimensional surface contouring of reflecting micro-objects. Opt. Commun. **270**, 176–179 (2007)
33. Z. Zalevsky, O. Margalit, E. Vexberg, R. Pearl, J. García, Suppression of phase ambiguity in digital holography by using partial coherence or specimen rotation. Appl. Opt. **47**, D154–D163 (2008)
34. G. Pedrini, W. Osten, Y. Zhang, Wave-front reconstruction from a sequence of interferograms recorded at different planes. Opt. Lett. **30**, 833–835 (2005)
35. P. Almoro, G. Pedrini, W. Osten, Complete wavefront reconstruction using sequential intensity measurements of a volume speckle field. Appl. Opt. **45**, 8596–8605 (2006)
36. P.F. Almoro, S.G. Hanson, Object wave reconstruction by speckle illumination and phase retrieval. J. Eur. Opt. Soc. – Rap. Public **4**, 09002 (2009)
37. P. Bao, F. Zhang, G. Pedrini, W. Osten, Phase retrieval using multiple illumination wavelengths. Opt. Lett. **33**, 309–311 (2008)
38. A. Anand, V.K Chhaniwal, P. Almoro, G. Pedrini, W. Osten, Shape and deformation measurements of 3D objects using volume speckle field and phase retrieval. Opt. Lett. **34**, 1522–1524 (2009)
39. P.F. Almoro, G. Pedrini, A. Anand, W. Osten, S.G. Hanson, Angular displacement and deformation analyses using a speckle-based wavefront sensor. Appl. Opt. **48**, 932–940 (2009)
40. T. Dressel, G. Hausler, H. Venzhe, Three dimensional sensing of rough surfaces by coherence radar. Appl. Opt. **31**, 919–925 (1992)
41. G.R. Hallerman, L.G. Shirley, A comparison of surface contour measurements based on speckle pattern sampling and coordinate measurement machines. Proc. SPIE **2909**, 89–97 (1996)
42. J. García, Z. Zalevsky, P. García-Martínez, C. Ferreira, M. Teicher, Y. Beiderman, Three-dimensional mapping and range measurement by means of projected speckle patterns. Appl. Opt. **47**, 3032–3040 (2008)

43. J.W. Goodman, *Speckle Phenomena in Optics* (Roberts and Company Publishers, Greenwood Village, USA, 2006)
44. E. Valero, V. Micó, Z. Zalevsky, J. García, Depth sensing using coherence mapping. Opt. Comm. **283**, 3122–3128 (2010)
45. D. Huang, E.A. Swanson, C.P. Lin, J.S. Schuman, W.G. Stinson, W. Chang, M.R. Hee, T. Flotte, K. Gregory, C.A. Puliafito, J.G. Fujimoto, Optical coherence tomography. Science **254**, 1178–1181 (1991)
46. T.C. Chen, B. Cense, M.C. Pierce, N. Nassif, B.H. Park, S.H. Yun, B.R. White, B.E. Bouma, G.J. Tearney, J.F. de Boer, Spectral domain optical coherence tomography – Ultra-high speed, ultra-high resolution ophthalmic imaging. Arch. Ophthalmol. **123**, 1715–1720 (2005)

Index

P. Ferraro et al. (eds.), *Coherent Light Microscopy*, Springer Series in Surface Sciences 46, DOI 10.1007/978-3-642-15813-1,

E

F

G

H

I

K

L

M

N

O

P

Q

R

S